AF749939

Hartley Library

UNIVERSITY OF
[illegible]mpton

Date due

University of Southampton
WITHDRAWN

The Scientific Legacy of
Poincaré

HISTORY OF MATHEMATICS
VOLUME 36

The Scientific Legacy of
Poincaré

Éric Charpentier
Étienne Ghys
Annick Lesne
Editors

Translated by
Joshua Bowman

AMS
American Mathematical Society
Providence, Rhode Island

The London Mathematical Society
London, England

Editorial Board

American Mathematical Society
Joseph W. Dauben
Peter Duren
Robin Hartshorne
Karen Parshall, Chair

London Mathematical Society
Jeremy J. Gray
June Barrow-Green
Tony Mann, Chair
Edmund Robertson

Original edition:
L'héritage scientifique de Poincaré
© Editions Belin - Paris, 2006

Ouvrage publié avec le concours du Ministère français chargé de la culture - Centre national du livre. [Published with financial support of the French Ministry of Culture - National Book Center.]

2000 *Mathematics Subject Classification.* Primary 30F35, 34C07, 37N05, 53C22, 32A27, 57M40, 16S30, 83A05, 00A30, 35J05, 01A60.

For additional information and updates on this book, visit
www.ams.org/bookpages/hmath-36

Library of Congress Cataloging-in-Publication Data

Héritage scientifique de Poincaré. English.
The scientific legacy of Poincaré / Éric Charpentier, Étienne Ghys, Annick Lesne, editors ; translated by Joshua Bowman.
p. cm. — (History of mathematics ; v. 36)
Includes bibliographical references.
ISBN 978-0-8218-4718-3 (alk. paper)
1. Poincaré, Henri, 1854–1912. 2. Chaotic behavior in systems. 3. Three-body problem. 4. Mathematical physics. I. Charpentier, Éric. II. Ghys, É. (Étienne) III. Lesne, Annick. IV. Title.
Q143.P7H4713 2010
510—dc22

2009027575

Copying and reprinting. Individual readers of this publication, and nonprofit libraries acting for them, are permitted to make fair use of the material, such as to copy a chapter for use in teaching or research. Permission is granted to quote brief passages from this publication in reviews, provided the customary acknowledgment of the source is given.

Republication, systematic copying, or multiple reproduction of any material in this publication is permitted only under license from the American Mathematical Society. Requests for such permission should be addressed to the Acquisitions Department, American Mathematical Society, 201 Charles Street, Providence, Rhode Island 02904-2294 USA. Requests can also be made by e-mail to `reprint-permission@ams.org`.

© 2010 by the American Mathematical Society. All rights reserved.
Printed in the United States of America.

The American Mathematical Society retains all rights
except those granted to the United States Government.
∞ The paper used in this book is acid-free and falls within the guidelines established to ensure permanence and durability.
The London Mathematical Society is incorporated under Royal Charter and is registered with the Charity Commissioners.
Visit the AMS home page at `http://www.ams.org/`

UNIVERSITY OF SOUTHAMPTON
LIBRARY

10 9 8 7 6 5 4 3 2 1 15 14 13 12 11 10

Contents

The Authors xi

Translator's Note xiii

Introduction 1

Chapter 1. Poincaré and His Disk
Étienne Ghys 17
1.1. An imaginary world 17
1.2. Some formulas 18
1.3. The Poincaré disk is ubiquitous 21
1.4. Lots and lots of models 22
1.5. Attempts to visualize in space 25
1.6. A little bit of triangle geometry 29
1.7. The disk is a tree 32
1.8. Some examples of dendrologic intuition 35
1.9. The disk is a curve 38
1.10. Arrival at the boundary 40
1.11. A few regrets... 42
Appendix: Sketch of a proof of the characterization theorem 42
Bibliography 43

Chapter 2. Differential Equations with Algebraic Coefficients over Arithmetic Manifolds
Nicolas Bergeron 47
2.1. Introduction 47
2.2. Algebraic linear differential equations 53
2.3. The step of an omnibus at Coutances 59
2.4. Quadratic forms and arithmetic surfaces 63
2.5. The question of uniformization 69
Bibliography 71

Chapter 3. Poincaré and Analytic Number Theory
Emmanuel Kowalski 73
3.1. Introduction 73
3.2. Poincaré series and Kloosterman sums 74
3.3. Prime numbers, arithmetic progressions, and a message 82
Bibliography 85

Chapter 4. The Theory of Limit Cycles
Jean-Pierre Françoise 87

4.1. Introduction 87
4.2. Periodic orbits 87
4.3. Limit cycles. Poincaré–Bendixson theorem 90
4.4. The local Hilbert's 16th problem, Bautin's theory 91
4.5. The Poincaré center-focus problem 93
4.6. Abel's equations and Hilbert's 16th problem in dimension 1 95
4.7. Complex Poincaré map 95
Bibliography 96

Chapter 5. Singular Points of Differential Equations: On a Theorem of Poincaré
DOMINIQUE CERVEAU 99
Introduction 99
5.1. The Poincaré–Lyapunov center theorem 101
5.2. A few pictures with commentary 103
5.3. Towards a proof of the center theorem 106
5.4. Examples of centers and connections with "integrability" 109
5.5. The "hyperbolic" Poincaré–Lyapunov theorem 111
Bibliography 112

Chapter 6. Periodic Orbits of the Three-Body Problem: Early History, Contributions of Hill and Poincaré, and Some Recent Developments
MICHAEL NAUENBERG 113
Introduction 113
6.1. Newton's periodic lunar orbit 115
6.2. Hill's periodic lunar orbits and Poincaré's loops 121
6.3. Poincaré's treatment of periodic orbits of the three-body problem 126
6.4. Poincaré's geometrical proofs for the existence of periodic orbits 131
6.5. Some recent developments in periodic orbit theory 135
Bibliography 140

Chapter 7. On the Existence of Closed Geodesics
NALINI ANANTHARAMAN 143
7.1. Introduction 143
7.2. *Sur les lignes géodésiques des surfaces convexes* (Poincaré, 1905) 144
7.3. The work of George D. Birkhoff 151
7.4. Using Morse theory 155
7.5. Return to the case of the sphere S^2: The work of Franks, Bangert, Hingston 157
Bibliography 159

Chapter 8. Poincaré's Memoir for the Prize of King Oscar II: Celestial Harmony Entangled in Homoclinic Intersections
FRANÇOIS BÉGUIN 161
8.1. Is the solar system stable? 161
8.2. Poincaré's memoir for the Prize of King Oscar II 165
8.3. The discovery of homoclinic intersections 175
8.4. "A kind of lattice, tissue, or web of infinitely tight stitches" 178
8.5. Have we untangled the web of homoclinic intersections? 183

8.6. Conclusion 188
Bibliography 190

Chapter 9. Variations on Poincaré's Recurrence Theorem
Étienne Ghys 193
9.1. Introduction 193
9.2. Poincaré's recurrence theorem 194
9.3. The example of *Scientific American* and *Science & Vie* 195
9.4. Why is the example surprising? 196
9.5. Why the example is very special 198
9.6. A question 202
9.7. Mixing 203
Bibliography 206

Chapter 10. Low-Dimensional Chaos and Asymptotic Time Behavior in the Mechanics of Fluids
Guido Boffetta, Guglielmo Lacorata, and Angelo Vulpiani 207
10.1. Introduction 207
10.2. Lagrangian chaos 207
10.3. Eulerian properties and Lagrangian chaos 210
10.4. Asymptotic time behavior: Transport and diffusion 217
10.5. Conclusions 221
Bibliography 222

Chapter 11. The Concept of "Residue" after Poincaré: Cutting across All of Mathematics
Alain Yger 225
11.1. Introduction 225
11.2. The Poincaré residue from the cohomological angle 231
11.3. The Poincaré residue revisited in the light of the theory of currents 233
11.4. Logarithmic differential forms; connections with logarithmic poles 236
11.5. Poincaré duality and global residues 237
11.6. Residues and duality: Conclusion 239
Bibliography 240

Chapter 12. The Proof of the Poincaré Conjecture, according to Perelman
Laurent Bessières, Gérard Besson, and Michel Boileau 243
12.1. Introduction 243
12.2. The flow associated to Ricci curvature 245
12.3. The work of R. Hamilton 247
12.4. G. Perelman's flow with surgery 249
12.5. The Poincaré conjecture 253
Bibliography 254

Chapter 13. Henri Poincaré and the Partial Differential Equations of Mathematical Physics
Jean Mawhin 257

13.1. Introduction 257
13.2. The equations of mathematical physics 257
13.3. The Dirichlet problem and the sweeping method 259
13.4. Eigenvalues and eigenfunctions of the Fourier problem: A heuristic approach 262
13.5. The Poincaré inequality 265
13.6. Eigenvalues and eigenfunctions of the Dirichlet problem: A rigorous approach 268
13.7. The Neumann method for the Dirichlet problem on an arbitrary domain 273
13.8. The heat equation and the telegraph equation 275
Bibliography 276

Chapter 14. Poincaré's *Calculus of Probabilities*
PIERRE CARTIER 279
14.1. The book's composition 279
14.2. A physicist's work 279
14.3. Mathematical methods of probability 281
14.4. Refutation of the classical theses on chance 282
14.5. First characteristic of chance 282
14.6. Second characteristic of chance 283
14.7. The mechanism of chance 284
14.8. Definition of probabilities 284
14.9. The case of continuous probabilities 285
14.10. Poincaré's response 286
14.11. Description of a probabilistic model according to Poincaré 287
14.12. Gauss's law of error 288
14.13. A jewel of group theory 289
14.14. Conclusion 290
Bibliography 291

Chapter 15. Poincaré and Geometric Probability
MICHEL MENDÈS FRANCE 293
15.1. Introduction 293
15.2. Buffon and his needle 293
15.3. Poincaré's theorem in the plane 297
15.4. The Edelman–Kostlan theorem 298
15.5. The real zeroes of a random real polynomial 299
15.6. Three remarks 300
15.7. Curves and entropy 301
15.8. Confusion 302
Bibliography 304

Chapter 16. Poincaré, and Lie's Third Theorem
PIERRE-PAUL GRIVEL 307
16.1. Introduction 307
16.2. Lie's three fundamental theorems 310
16.3. The exponential formula at the end of the 19th century 313
16.4. The universal enveloping algebra according to Poincaré 317

16.5. Poincaré's proof of Lie's third theorem 320
16.6. Poincaré's inheritors 324
16.7. From Lie to Leibniz 325
Bibliography 327

Chapter 17. The Poincaré Group
MICHEL LE BELLAC 329
17.1. Introduction 329
17.2. Genesis of the Poincaré group (1895/1905) 330
17.3. The 1905 articles and the Lorentz group 332
17.4. A deduction from the Lorentz group 335
17.5. The Poincaré group in quantum physics: Wigner's analysis (1939) 339
17.6. The theorems of Coleman and Mandula (1967) and of Haag, Łopuszański, and Sohnius (1975) 343
17.7. Galilean electromagnetism 345
17.8. Conclusion 348
Bibliography 349

Chapter 18. Henri Poincaré as an Applied Mathematician
YVES POMEAU 351
18.1. Introduction: The physical problem and its history 351
18.2. Poincaré's calculation: The solution in the form of a series and its estimation in the short wave limit 354
18.3. Fock's analysis 361
18.4. Dispersion of solitons in perturbed KdV 364
18.5. Final remarks and conclusion 370
Bibliography 370

Chapter 19. Henri Poincaré and His Thoughts on the Philosophy of Science
GERHARD HEINZMANN 373
19.1. Philosophy and Science 373
19.2. Poincaré's pragmatic occasionalism 374
19.3. Arithmetic 376
19.4. Geometry 379
19.5. Physics 383
19.6. Logic and abstraction 386
Bibliography 389

The Authors

Nalini Anantharaman is a *maître de conférences* in the CNRS Unité de Mathématiques Pures et Appliquées at the École Normale Supérieure in Lyon and an associate professor at the École Polytechnique (Palaiseau).

François Béguin is a *maître de conférences* in the topology and dynamics group of the Université d'Orsay.

Nicolas Bergeron is a professor at the Institut de Mathématiques de Jussieu (Université Paris 6).

Laurent Bessières is a *maître de conférences* at the Institut Fourier (Université de Grenoble).

Gérard Besson is the CNRS research director at the Institut Fourier (Université de Grenoble).

Guido Boffetta is a professor of general physics at the University of Turin.

Michel Boileau is a professor in the Laboratoire Émile Picard at the Université Paul Sabatier (Toulouse).

Pierre Cartier is a CNRS research director emeritus at the Institut de Mathématiques de Jussieu (Paris) and a visiting professor at the Institut des Hautes Études Scientifiques (IHES) at Bures-sur-Yvette.

Dominique Cerveau is a professor at the Institut de Recherche Mathématique de Rennes (Université de Rennes I).

Étienne Ghys is the CNRS research director at the CNRS Unité de Mathématiques Pures et Appliquées at the École Normale Supérieure in Lyon and a member of the Académie des Sciences.

Pierre-Paul Grivel is a former *chargé de cours* in the Mathematics Section of the Université de Genève.

Gerhard Heinzmann is a professor in the Laboratoire de Philosophie et d'Histoire des Sciences (Archives Henri Poincaré) of the Université Nancy 2.

Emmanuel Kowalski is a professor at the Institut de Mathématiques de Bordeaux (Université Bordeaux 1).

Guglielmo Lacorata is a CNR researcher at the Istituto di Scienze dell'Atmosfera e del Clima (ISAC) in Lecce (Italy).

Michel Le Bellac is a professor emeritus of physics at the Institut Non Linéaire de Nice (Université de Nice Sophia Antipolis).

Jean Mawhin is a professor in the Mathematics Department of the Catholic University of Louvain.

Michel Mendès France is a professor emeritus at the Institut de Mathématiques de Bordeaux (Université Bordeaux 1).

Michael Nauenberg is a research professor and professor emeritus of physics at the University of California in Santa Cruz.

Yves Pomeau is the CNRS research director in the Laboratoire de Physique Statistique of the École Normale Supérieure de Paris and an associate member of the Académie des Sciences.

Angelo Vulpiani is a professor of theoretical physics at Sapienza University of Rome.

Alain Yger is a professor at the Institut de Mathématiques de Bordeaux (Université Bordeaux 1).

Translator's Note

It has been a privilege and a pleasure to translate this collection of essays on Poincaré's mathematical and scientific legacy. I thank the AMS for providing me with this opportunity.

Poincaré himself wrote "popular" works on mathematics, its practice, and its philosophy for the broader scientific community. Two of these are mentioned throughout the pages of these essays: *Science and Hypothesis* and, to a lesser extent, *Science and Method.* Both are recommended for reading in their entirety. Both exist in English translations: the first by J. Larmor and the second by F. Maitland. I have chosen, however, to provide my own translation of Poincaré's words when they are quoted here, although I cannot deny some influence by these earlier translations.

In the course of this project, it has at times been necessary to make "cultural" translations as well as linguistic ones. When this has occurred, I have made a note to indicate the content of the original text. I have not attempted to preserve the original numbering of footnotes (indeed, some of the original footnotes have become obsolete, for example, when they contained the English text for a quote translated into French). Those footnotes which I have added are marked accordingly.

My thanks to Ina Mette and Marcia Almeida of the AMS for their aid, as well as to the editors and authors of the original texts for their responsiveness to my questions, and to the translation committee for their helpful comments. Natalya Pluzhnikov of the AMS did a superb job of proofreading the manuscript.

I also thank Hannah for her support during all stages of this project.

Introduction

Henri Poincaré[1] (Nancy, 1854; Paris, 1912) dominated the mathematics and the theoretical physics of his time; he was without a doubt the most admired mathematician while he was alive, and he remains today one of the world's most emblematic scientific figures.

[1]This name comes, so it seems, from *"poing carré"* ["square fist"]: cf. Poincaré's *Œuvres*, vol. 2, p. ix, note (1).

His work, slowly assimilated over the course of the twentieth century, is of phenomenal depth, but also of extraordinary variety.[2] In mathematics, nearly all of whose existing parts he enriched, he created entirely new branches (automorphic functions, dynamical systems, algebraic topology) and he opened the way to the theory of functions of several complex variables and that of asymptotic expansions. He completely renovated celestial mechanics, discovered on this occasion deterministic chaos, found new forms of equilibrium of celestial bodies, and proposed a scenario for the formation of double stars. In physics, he is one of the fathers of the theory of special relativity, for which he envisioned consequences, even to the motion of the stars. He moreover exerted a constant influence on the spectacular development of the physics of his time: he participated in all of the great debates, provided the first correct explanation of several experiments, and even elicited new ones.[3]

His teaching has remained legendary. In his lectures on mathematical physics (then in celestial mechanics) at the Sorbonne, he sifted the principal existing theories, reformulated them, corrected them; he presented and discussed the most recent experiments and observations. These courses, almost all of which were written up[4] and then published and widely distributed, contributed much (either by themselves or by the ideas that he conceived on their occasion and which he distributed in parallel) to the definitive formulation and acceptance of new theories (of Maxwell, Lorentz, Boltzmann, ...).

In the philosophy of sciences, he was one of the major actors in the great epistemological debates of his time; in particular, his *pragmatic occasionalism* (as it is called by Gerhard Heinzmann) nourished reflections throughout the twentieth century. In addition, his popular works on science, written in an extraordinarily clear style, had an enormous success before the wide public: *Science and Hypothesis* was translated into at least 23 languages. In a few years, more than 16 000 copies were sold in France. People read it in the public gardens and the cafés.

A "universal specialist", it was he who investigated (as a mining engineer) the causes of the firedamp explosion in the shafts of Magny (1879); it was to him that not only mathematicians, but physicists turned, such as Hertz when he did not succeed in calculating the speed of the propagation of a wave along a twisted wire, or Becquerel when he found himself in disagreement with Alfred Potier over a question of rotational polarization; it was from him (along with Darboux and Appell) that the court requested an expert opinion on the scientific validity of Alphonse Bertillon's arguments in the Dreyfus affair;[5] it was he that the Académie

---

[2]One can find a list of Poincaré's publications as well as numerous bibliographic references concerning his life and work on the site of the Laboratoire d'Histoire des Sciences at de Philosophie–Archives Henri Poincaré (CNRS–Nancy-Université):

`http://www.univ-nancy2.fr/poincare/ENG/documents/bd1.html`

(from which we have also obtained, with gratitude, the photos on the previous page and on page 392).

[3]See volumes 9 and 10 of his *Œuvres*, and his correspondence with the physicists: this latter, collected by André Coret and Scott Walter, is in process of publication by the Archives Henri Poincaré: `http://www.univ-nancy2.fr/poincare/chp/prjregionh.html`.

[4]By students or by other audience members: Jules Blondin, Émile Borel, Jules Drach, René Baire, Albert Quiquet, and others.

[5]See the article *Autour de l'Affaire Dreyfus: Henri Poincaré et l'action politique*, by Laurent Rollet: `http://poincare.univ-nancy2.fr/Presentation/?contentId=2240`.

des sciences asked to supervise the new measurement of the meridian arc of Quito; he presided (three times) over the Bureau des Longitudes; etc.

His scientific *Œuvres*, in 10 volumes,[6] are divided as follows:

Volume 1:	Differential equations.
Volume 2:	Automorphic functions.
Volume 3:	Algebraic integration of differential equations; continuous groups; abelian integrals; residues of double integrals; integral equations.
Volume 4:	Analytic functions of one or several variables; abelian functions; trigonometric series.
Volume 5:	Number theory.
Volume 6:	Algebraic geometry, algebraic topology.
Volume 7:	Analytic mechanics; celestial mechanics; fluid masses in rotation; three-body problem; trigonometric series of celestial mechanics (work on the methods of Lindstedt and of Gyldén).
Volume 8:	Celestial mechanics and geodesy; perturbation function and periods of double integrals; shape of the Earth; theory of tides, of the Moon, and of planets; mechanical quadratures; cosmogonic hypotheses; reports on the geodesic operations at the Equator.
Volumes 9–10:	Mathematical physics; theoretical physics.

His treatises on mathematical and theoretical physics address nearly all the important themes of his era: potential theory and fluid mechanics, mathematical theory of light, electricity and optics, thermodynamics, theory of elasticity, theory of vorticity, electric oscillations, capillarity, analytic theory of heat, probability, theory of Newtonian potential (he also provided a course on the kinetic theory of gases, which was never written).

Let us add his *Lectures on Celestial Mechanics* (3 volumes), his *Lectures on the Cosmogonic Hypotheses*, and a *Course in General Astronomy* (autograph). His text *New Methods in Celestial Mechanics* is not the fruit of a course of instruction (it is a research work), but it is written up as a course.

Let us mention, finally, his works on the philosophy of science: *Science and Hypothesis*, *The Value of Science*, *Science and Method*, *Last Thoughts* (posthumous collection).

As one can see by comparing the preceding with the table of contents of the present work, we shall address here only *a part* of the themes to which Poincaré contributed (it would be impossible to be exhaustive in a single volume!), especially underlining the modernity of his ideas and some of their current repercussions. The texts, written by experts of international renown, are accessible to any first- or second-year graduate student (or even undergraduate) in mathematics or in physics. They contain several levels of reading, and researchers (in particular doctoral candidates) will also be able to find in them useful ideas for their daily work.

Here is an overview of the themes addressed in this book.

[6]Gauthier-Villars, 1916–1954. Republished by J. Gabay, 1995–2005. An 11th volume contains a few articles, an extract of his mathematical correspondence, and the *Livre du centenaire* of his birth.

Hyperbolic geometry, automorphic functions, applications to number theory (Chapters 1, 2, and 3)

Henri Poincaré made himself known on the mathematical scene in 1881–1882 with his theory of automorphic functions. It was a kind of apotheosis of 19th century mathematics, a meeting place for group theory, complex analysis, elliptic and modular functions, differential equations, Riemann surfaces, hyperbolic geometry, and quadratic forms; it would also be a privileged model for his reflections on the philosophy of science.

The essential property of automorphic functions on a domain D of $\mathbb{C}$ is a kind of generalized periodicity: $f(\frac{az+b}{cz+d}) = f(z)$ for a certain discontinuous group of projective transformations $z \mapsto \frac{az+b}{cz+d}$ mapping D to itself. We also require that f be meromorphic on D.

After the already known case of *elliptic functions*, which are automorphic functions on $\mathbb{C}$ for a group of translations (identified with a lattice $\mathbb{Z}\omega_1+\mathbb{Z}\omega_2$), the most useful case is that of automorphic functions on a disk:[7] this is what Poincaré called *"Fuchsian functions"*. Only very particular cases were known. By interpreting the projective transformations of the disk as motions of Lobachevsky's non-Euclidean plane (the hyperbolic plane), Poincaré managed to geometrically construct *all* the discontinuous groups of projective transformations of the disk ("*Fuchsian groups*"); he associated to them *"theta-Fuchsian"* functions (today called *automorphic forms*), which he obtained in the form of series (today known under the name of *Poincaré series*); and he thence derived *all* the Fuchsian functions as quotients of such series.

When D is no longer a disk (or a half-plane), Poincaré speaks of *"Kleinian"* functions, and the corresponding groups ("*Kleinian groups*") are obtained from hyperbolic geometry in dimension 3. Felix Klein, dissatisfied with the adjectives *"Fuchsian"* (he thought it gave too much honor to Fuchs) and *"Kleinian"* (which he considered as a consolation prize), replaced them with *"automorphic"* (a term borrowed from Cayley).[8]

The automorphic functions made it possible to obtain the famous *uniformization theorem*: every planar algebraic curve $P(x,y) = 0$, where P is an irreducible polynomial, can be parametrized in the form $x = f(z)$, $y = g(z)$, where f and g are rational functions (the case of genus 0 curves), or elliptic functions with the same lattice (curves of genus 1), or Fuchsian functions of the same group (curves of genus ≥ 2).[9] An equivalent geometric statement is that every compact, connected Riemann surface is analytically isomorphic to the Riemann sphere, or to the quotient of the complex plane by a lattice $\mathbb{Z}\omega_1 + \mathbb{Z}\omega_2$, or to the quotient of the disk by a Fuchsian group.

But Poincaré's true motivation was integrating linear differential equations with algebraic coefficients, that is,

$$\frac{d^n u}{dx^n} + a_{n-1}(x,y)\frac{d^{n-1}u}{dx^{n-1}} + \cdots + a_0(x,y)u = 0,$$

---

[7] Or on a half-plane: one passes between them by a projective transformation.

[8] Poincaré responded to Klein by these words of Faust, which could serve as an epigraph to his philosophy: *"Name ist Schall und Rauch,"* that is, "[the] *name is* [nothing but] *sound and smoke.*"

[9] This theorem, discovered without rigorous proof in 1881 by Poincaré and Klein, would be extended to analytic curves in 1907, by Poincaré and (independently) by Koebe (each of them giving a rigorous proof).

where the a_i are rational functions and where x and y are connected by an algebraic relation $P(x, y) = 0$. One can construct a Fuchsian (or rational, or elliptic, for the genera 0 and 1) parametrization $x = f(z), y = g(z)$ such that the solutions are meromorphic functions of z having particular transformation properties (*"zeta-Fuchsian"* functions): a projective transformation of the Fuchsian group of f and g translates into a change of basis in the space of solutions.

Automorphic functions and forms and their groups are tools used constantly today in various areas of mathematics and even in physics (they are involved in the calculation of collision amplitudes in string theory), and the current theory of automorphic representations (in full bloom, centered on the *Langlands program*) is a distant heir.

Chapter 1 presents the *Poincaré disk*, whose hyperbolic geometry is the key that led Poincaré to the construction of Fuchsian groups: it shows the extent to which this object is natural, rich, and ubiquitous in the mathematics of today.

Chapter 2 recounts the discovery of Fuchsian functions and groups, following and explaining Poincaré's own narrative about it (one will find there confirmation that mathematicians transform coffee into theorems, and a very interesting testimony on the role of the unconscious in mathematical discovery), all while evaluating its current aspects.

Chapter 3 presents the application of Poincaré series to analytic number theory, some examples of the resulting progress, and some new hopes (perhaps a connection with the question of *twin prime numbers!*).

Ordinary differential equations, dynamical systems (Chapters 4 and 5)

Most differential equations cannot be solved explicitly. Before Poincaré, one was therefore content to study solutions in a neighborhood of a point. Poincaré himself developed this approach in his doctoral thesis (1879), and he returned to it several times: his research (parallel to his study of trigonometric series of celestial mechanics) led him to set up the foundation, in 1885 and 1886, of the classical theory of asymptotic expansions.

But Poincaré also understood that there would always be the need to know the global behavior of integral curves. Yet for this, there were no tools. He would therefore create them (*Mémoire sur les courbes définies par une équation différentielle* [*Memoir on curves defined by a differential equation*], first part: 1881). He considered a differential equation $dx/X = dy/Y$, where X, Y are polynomials in x, y. To take into account the behavior at infinity, he compactified the (x, y)-plane into a sphere by central projection. Two orbits cannot meet except at a singular point (a point where X and Y vanish): Poincaré showed that generically there are only three types: the *nodes*, where infinitely many orbits meet; the *saddles*, where two orbits cross; and the *foci*, where the orbits wind around in spirals. Under certain conditions, there also exist *centers*, which are surrounded by periodic orbits; even though they are non-generic, centers are important in mechanics (they correspond to a form of stability): Poincaré therefore studied the conditions for their existence (center-focus problem, Poincaré–Lyapunov theorem).

Poincaré showed that for a generic field (X, Y) on the sphere, the numbers of nodes, saddles, and foci are connected by the relation $N - S + F = 2$, analogous to Euler's relation between the numbers of vertices, edges, and faces of a convex polyhedron. He noticed that the analogy persists on arbitrary algebraic surfaces (corresponding to differential equations of the form $P(x, y, y') = 0$, where P is a polynomial). Later, Heinz Hopf (1926) would show that it also persists on manifolds of all dimensions (the Poincaré–Hopf formula)—Poincaré having meanwhile generalized Euler's formula to polyhedra (convex or not) of arbitrary dimension and invented the necessary topological tools.

One of Poincaré's key ideas is that of a transverse section ("*arc without contact*"): this is an arc of a curve that is not tangent to the field (X, Y) at any point; any orbit that meets it necessarily crosses it. Poincaré reduced the study of an orbit to that of its successive points of intersection with such an arc ("*theory of consequents*"). This allowed him to prove that an orbit on the sphere can only have one of three fates: either it lands on a singular point, or it closes on itself (periodic orbit), or it winds asymptotically around an isolated periodic orbit (which Poincaré called a "limit cycle") or around a *"graph"* (a closed curve consisting of orbits joining singular points): this is the *Poincaré–Bendixson theorem* (because it was generalized later by Bendixson to continuously differentiable fields). This restricted the set of possible behaviors so much that Poincaré could determine the shape of integral curves for explicit examples of equations. In 1886, he partially generalized his methods to higher dimensions. The Poincaré–Bendixson theorem is no longer valid there[10] (it already was not on 2-dimensional surfaces of non-zero genus, such as the torus).

Chapter 4 offers an introductory survey of the theory of limit cycles—covering the Poincaré map, asymptotic phase, frequency locking, Poincaré–Bendixson, the problem of an upper bound for the number of limit cycles (Hilbert's 16th problem, Bautin's theory), the center-focus problem, and the connection with Abel's equations—while emphasizing recent advances.

Chapter 5 presents the Poincaré–Lyapunov center theorem, with the outline of a very beautiful geometric proof (Moussu, 1981); for this, one learns how to "see in $\mathbb{C}^2$" by tomography. In it, one can also find the connection with integrability in the sense of Liouville, degenerate centers, and the "hyperbolic" case.

Celestial mechanics and related problems (Chapters 6–10)

From the start of his research on curves defined by differential equations (1881), Poincaré mentioned possible applications to celestial mechanics, in particular the question of the stability of an idealized solar system over an indefinite time period. Up until then, it was thought that this problem would be solved as soon as someone had succeeded in expressing the solutions of the N-body problem by absolutely convergent trigonometric series, because it was believed that the sum of such a series was necessarily bounded. Poincaré made the observation that it is nothing of the sort (1882). In 1883, he proved, by a very simple geometric argument, the existence of infinite families of periodic solutions in the three-body problem with two sufficiently small masses. He judged it *"improbable"* that the trigonometric series of celestial mechanics converge for *neighboring* solutions to the periodic solutions: he had in mind the example of limit cycles that he had discovered in other dynamical systems (*"Indeed, I know completely analogous problems where convergence does not take place" Œuvres*, vol. 4, p. 590). But already he added at this point (without

[10] We now know that in the sphere S^3 (or in $\mathbb{R}^3$) the trajectories can wind around much more complicated objects than a periodic orbit, such as for example the famous Lorenz attractor, which is a kind of interlacing of infinitely many periodic orbits.

details, for the moment) that, even if divergent, these trigonometric series could provide excellent approximations in practice. In an 1884 article, he began studying the neighboring solutions of the periodic solutions that he had found in the three-body problem.

All this led him to the memoir *Sur le problème des trois corps et les équations de la dynamique* [*On the three-body problem and the equations of dynamics*], which was crowned with a prize of the king of Sweden and appeared in 1890.[11] Poincaré gave a new (analytic) proof of the existence of periodic solutions. Half of them are linearly unstable, and their study provided three important results:

(1) there exist *"asymptotic solutions"*, which wind around linearly unstable periodic solutions, in the future or the past: a behavior quite different from the almost periodic motions previously encountered in celestial mechanics;
(2) the expansions into the usual trigonometric series of celestial mechanics diverge in general (but he also showed what one could still extract from these, thanks to his theory of asymptotic series);
(3) there do not exist other holomorphic first integrals (quantities that are constant on *each* trajectory of the system) than those already known[12] (if there had been enough, and if they had satisfied certain conditions, a theorem of Liouville would have guaranteed that the system is integrable by quadratures after an appropriate change of coordinates).[13]

Poincaré then turned to the quantities conserved on *bundles* of trajectories, which he called *"integral invariants"*; in modern language, these are invariant measures on the phase space.[14] (He rediscovered, after Liouville and Boltzmann, the theorem on the invariance of volume in the phase space of a Hamiltonian system.) And he proved his *recurrence theorem*, in the following form: *If there exists a positive invariant measure, and if the phase space has finite measure, then the trajectories issuing from a region of non-zero measure in the phase space will pass infinitely often within this region, except for some initial positions whose probability (i.e., measure) is zero.* Poincaré showed that this theorem applies to the restricted three-body problem. (The trajectories that are asymptotic in the future, mentioned in point (1) above, are obviously among the exceptions.) With the integral invariants and his theory of consequents, Poincaré also proved the existence of infinitely

[11]The announcement of the competition, published in 1885, specified that the prize would be awarded on January 21, 1889 (the sixtieth birthday of the king of Sweden), that the memoirs must be submitted before June 1, 1888, and that they must not have been previously published. Poincaré therefore devoted himself to other work and did not take up the subject until the last moment. The publication was further delayed, after the prize had been given, because Poincaré, pressed for time, had not written the detailed proofs, and while doing so in order to answer the editors' questions he discovered and had to correct an important error (see Chapter 8 in this volume).

[12]Poincaré proved this, in his memoir, for the *restricted* three-body problem (planar motion, two bodies having circular orbits and the third a negligible mass). In the *Méthodes nouvelles de la mécanique céleste* (volume 1, 1892) he extended the result to the general three-body problem, on the condition that two among them have sufficiently small masses.

[13]These results by Poincaré do not, however, exclude the possibility of solving the N-body problem by convergent series for all values of time (which was moreover the question initially posed for the prize): such a solution was obtained (under generic conditions) by Karl Sundman in 1909 for $N = 3$, and by Qiudong Wang in 1991 for arbitrary N.

[14]Space of the states of the system (which Gibbs called *"phases"*). For a mechanical system with N degrees of freedom, it is the space of generalized coordinates and generalized velocities, or possibly a submanifold obtained by fixing the energy and the other invariants of motion.

many *doubly asymptotic "homoclinic"* solutions, that is, those which wind in both the past *and* the future around the *same* periodic solution: their study led him to the discovery of the entanglement of stable and unstable manifolds (formed of solutions that are asymptotic, respectively, in the future and in the past, for a given periodic solution) and of what is today called *deterministic chaos*. Poincaré again took up his methods and results and developed them considerably in the three volumes of his *Méthodes nouvelles de la mécanique céleste* (1892–1899), which remain today an important source of inspiration.

Poincaré started from the existence of periodic solutions, *"the sole breach from which we can try to penetrate into a place heretofore reputed to be inaccessible"* (*Méthodes nouvelles*, vol. 1, 1892). Thereafter, he continued to look for other families of such solutions. He tried (1896) a variational method, but it was better suited to the interactions in $1/r^{n+1}$ with $n > 1$ than with $n = 1$. It is only recently that we have been able to adapt such a method to the Newtonian problem (see below). In his last article on celestial mechanics (1912), he managed to prove, thanks to a theorem of geometry (the Poincaré–Birkhoff theorem), that the periodic solutions are extremely abundant, at least in the restricted three-body problem. Since a generalization of the Poincaré–Birkhoff theorem in higher dimensions was established by Charles Conley and Eduard Zehnder (1984), this abundance of periodic solutions has been found in much more general dynamical systems and is the object of active research today.

Chapter 6 presents a brief history of the work on periodic orbits of the three-body problem: the theory of the Moon by Newton (1687), who introduced the restricted three-body problem and obtained a solution that is periodic to first approximation; the theory of Hill (1877), who completed that of Newton and provided a family of rigorously periodic solutions; the work of Poincaré, which extended Hill's family of solutions, then the key ideas of his various proofs of the existence of periodic solutions of the three-body problem; finally recent examples of periodic orbits of the N-body problem, obtained by a variational approach.

Chapter 7 deals with an analogous problem to that of the existence of periodic orbits, namely the existence of closed geodesics on a deformed sphere. One will find there the essential ideas of Poincaré's two proofs (by a continuity method, then by a variational method) in dimension 2 and of George D. Birkhoff's proof (1917) valid in all dimensions. Then come the case of manifolds with more complicated topology (by Morse theory) and the question of the existence of infinitely many closed geodesics on a deformed sphere of dimension 2: Poincaré (1912), Birkhoff (1917), John Franks (1992), Nancy Hingston (1993), Victor Bangert (1993); the analogue for spheres of higher dimension remains to be done.

Chapter 8 presents some of the main results of the memoir crowned with the prize of the king of Sweden and explains Poincaré's error and his discovery, while correcting this error, of the "lattice of homoclinic intersections" (the entanglement of stable and unstable manifolds); it also explains to what extent we have been able, today, to "untangle" this lattice (work of Birkhoff, Steve Smale, Rufus Bowen). Finally, it presents two recent results that illustrate the current importance of Poincaré's discovery: on the one hand, the work of Jacques Laskar (around 1990) on chaos in the solar system, and on the other hand, a theorem of Sylvain Crovisier (2005) showing that the phenomenon of homoclinic intersections discovered by Poincaré is, in some sense, inherent in chaos.

Chapter 9 discusses Poincaré's recurrence theorem and its limitations, and notably explains why an illustration that was proposed for this theorem in popular articles does not *at all* illustrate Poincaré's true theorem.

The results obtained by Poincaré in celestial mechanics serve as guides in many other areas, where one can apply the concepts and methods that he invented. Chapter 10 gives a few current examples of this kind of transfer of ideas into fluid mechanics[15] to study different cases of chaotic advection and long-term diffusive behavior.

Among the numerous other contributions Poincaré made to celestial mechanics, not touched upon in this collection, let us mention his study of the forms of equilibrium of fluids in rotation (1885). To the forms already known (ellipsoids, annuli) Poincaré added new ones (pear-shaped[16]). He established his theorem on the "exchange of stability": under certain conditions, when two solution curves cross, the solutions of one of the branches lose stability whereas those of the other gain it (he therefore called the solution at the crossing point a *"bifurcation form"*). Poincaré deduced from this a possible scenario for the formation of certain double-star systems (which, notably, made precise a hypothesis of George Darwin, 1878): an ellipsoid, initially almost spherical, flattens as it cools, all while remaining axially symmetric; then it reaches the point where the stability passes to the ellipsoids of unequal axes; then it bifurcates toward a pear shape, widens further, and ends by separating into two distinct bodies.[17] Poincaré also applied his theorem on the exchange of stability to families of periodic orbits.

Functions of several complex variables (Chapter 11)

Poincaré's first incursion into the theory of several complex variables (1883) was his proof that a *meromorphic* function on $\mathbb{C}^2$ is the quotient of two holomorphic functions on $\mathbb{C}^2$. (Weierstrass had proved the analogous result for functions of a single variable but was unable to establish it for functions of two variables.)[18] Using the same method, he obtained in 1898 an entirely new proof of a theorem of Riemann and Weierstrass, according to which a meromorphic function of p variables, $2p$ times periodic, is a quotient of "theta functions".

In 1907, Poincaré became interested in the problem of mapping a real analytic (three-dimensional) hypersurface of $\mathbb{C}^2$ onto another by an analytic isomorphism, either locally (there is an isomorphism in a *neighborhood* of each point of the surface) or globally. He discovered that it is in general impossible, even locally (whereas for curves in $\mathbb{C}$ it is always possible locally), and that, under certain conditions, once the map is possible locally, then it is also possible globally. These results are at the source of the theory of CR-structures (for "Cauchy–Riemann"), whose involvement has subsequently been seen in a large number of problems in mathematics and physics (including the theory of general relativity).

But Poincaré's most famous contribution to the theory of functions of several complex variables is his theory of residues (1887). Since the time of Cauchy, it had been desirable to extend the residue theorem to functions of several variables,

[15]This is a just turn of events, since Poincaré had been inspired by fluid mechanics to study the "invariant integrals".

[16]TRANSLATOR'S NOTE: Poincaré used the term *piriform*.

[17]For a modern discussion of Poincaré's scenario, see for example `arXiv:astro-ph/9505008`.

[18]And it would have to wait for Pierre Cousin's thesis (1894) for the result to be extended to functions of n variables.

but after more than forty years of efforts almost no progress had been made. It was Poincaré who unblocked the situation. On the way, Poincaré obtained the conditions for a differential form of arbitrary degree to be closed (that is, for its integral over a boundary to always be zero). His reasoning leads to the generalized "Stokes formula": it is implicit in his 1887 memoir, and would be made explicit first by Volterra in 1889 (inspired by Poincaré's memoir), then (in an even simpler form) by Poincaré himself in 1899, in his *Méthodes nouvelles de la mécanique céleste* (apropos of the "invariant integrals"). Poincaré would also use a particular case of this formula in his work on topology (for calculating homology).

Chapter 11 presents the Poincaré residue, first from the cohomological viewpoint (due essentially to Poincaré and implemented by Jean Leray), then from the viewpoint of currents (more recent), and its application in contemporary mathematics. One can also see how it revived a problem already considered by Poincaré in 1885, namely a generalization of a famous theorem of Abel (*the "trace" of a rational form is rational*) and its converse.

Algebraic topology and the Poincaré conjecture (Chapter 12)

In almost all of his research, Poincaré encountered problems of topology (problems of *analysis situs*, as was said at the time). He wrote:

> *"A method that would make known to us the qualitative relations in space of more than three dimensions could, to a certain extent, render services analogous to those rendered by figures. This method can only be the* Analysis situs *of more than three dimensions.* [...] *I had need of the facts of this Science to pursue my studies on curves defined by differential equations and to extend them to differential equations of higher order and, in particular, to those of the three-body problem. I had need of them for the study of non-uniform functions of two variables. I had need of them for the study of periods of multiple integrals and the application of this study to the expansion of the perturbing function. Finally, I sensed within* Analysis situs *a means of approaching an important problem in the theory of groups, the search for discrete groups or finite groups contained in a given continuous group."*[19]

But this "Science" was in its infancy: one saw (or believed one could see) some properties, but almost nothing had been proved, for lack of reliable tools. Poincaré would create a powerful theory (algebraic topology) in two stages. In a first memoir (*Analysis situs*, 1895), he gave a definition of manifolds (in fact, analytic submanifolds of $\mathbb{R}^n$) by means of charts, fixed the precise definition of the integral of a differential form over a manifold, introduced the notion of homology, and studied its connection with the integrals of closed differential forms (again taking up the ideas of his 1887 memoir on residues of double integrals). He considered the dimensions of homology spaces, which he called the *"Betti numbers"* (because he believed they coincided with the numbers already introduced by Enrico Betti) and gave an important symmetry formula for the "Betti numbers" of *closed*[20] orientable manifolds (*Poincaré duality*). He studied decompositions of manifolds into pieces homeomorphic to polyhedra, letting himself be guided by the analogy with his

[19] Poincaré: *Œuvres*, vol. 6, p. 183.

[20] In the sense that one speaks of *closed surfaces*. More precisely, these are connected, compact manifolds without boundary.

theory (mentioned above[21]) of Fuchsian groups (where the hyperbolic plane was decomposed into polygons, images of each other by the group) and Kleinian groups (where 3-dimensional hyperbolic space was decomposed into polyhedra). This led him to define the *fundamental group* of a manifold. He observed that two homeomorphic manifolds have the same fundamental group (up to isomorphism). He noted that in the uniformization of a Riemann surface (see page 4) the Fuchsian group is nothing other than the fundamental group of the surface; from this he concluded that two *closed* manifolds of dimension 2 that have the same "Betti numbers" are homeomorphic. But he noted that this is no longer true for closed manifolds of dimension 3: he gave examples of such manifolds that have the same "Betti numbers" as the sphere S^3, but a different fundamental group. Finally, he generalized Euler's formula for polyhedra in arbitrary dimension, convex or not: the alternating sum of the numbers of faces of the different dimensions is constant, and the constant is the alternating sum of the "Betti numbers".

One of the examples Poincaré considered in his memoir shows that his "Betti numbers" are not always equal to those of Betti, and that the duality formula is not valid for the latter; but Poincaré did not realize it until Heegaard, using Betti's definition, found another counterexample to the duality formula (1898). Moreover, Heegaard pointed out a dubious assertion in the proof of the duality formula. Poincaré therefore decided to return to the question. In a first *Complément à l'Analysis situs* (1899), he redefined homology by means of triangulations (simplicial homology), thus founding algebraic topology on solid bases, and he gave a new proof of the duality formula (he also filled in the hole of his first proof). In a second *Complément* (1900), he added to his "Betti numbers" new topological invariants, the *torsion coefficients*. We recall that in his 1895 memoir, Poincaré had observed that two closed manifolds of dimension 3 can have the same "Betti numbers" without being homeomorphic. It was therefore natural to ask if two closed manifolds of dimension 3 having the same "Betti numbers" *and* the same torsion coefficients are homeomorphic. In his fifth *Complément* (1904), Poincaré showed that nothing of the sort is true: he constructed an example of a 3-dimensional manifold having the same "Betti numbers" and the same torsion coefficients as the sphere S^3 but a different fundamental group.[22] Poincaré concluded that there remained one question to address: does there exist a manifold of dimension 3 whose fundamental group is trivial (that is, all closed curves in this manifold can be continuously contracted to a point), and which is not, however, homeomorphic to S^3? The hypothesis that one does not exist became famous under the name of the *Poincaré conjecture*. Grigori Perelman (2003) proposed a proof of this conjecture, and, even more generally, of *Thurston's geometrization conjecture*, according to which eight homogeneous geometries suffice to construct all 3-dimensional manifolds (the only one of the eight whose fundamental group is trivial being the sphere).[23] Roughly, the strategy, elaborated by Richard Hamilton in 1982, consists in letting the metric evolve according to a differential equation connected to the curvature and analogous to the equation of heat propagation, which tends to homogenize the metric, and the idea is to show

[21]See the section on hyperbolic geometry and automorphic functions.

[22]This Poincaré *dodecahedral space* has recently (2003) been envisioned as a cosmological model.

[23]The verification of Perelman's proof is still in progress, but the ideas that he introduced on this occasion have already earned him the Fields Medal (which he moreover refused) in August 2006.

that one can asymptotically approach only the eight geometries of Thurston. But the curvature can blow up at certain places, and Perelman showed how one can repair the process by surgery and restart it, until the metric converges (in a certain sense).

Chapter 12 presents the outline of Hamilton's strategy and Perelman's proof.

Partial differential equations of mathematical physics (Chapter 13)

In 1886, Poincaré became the title-holder of the chair of *Mathematical physics and probability* at the Sorbonne. At the same time, he began research on the mathematical aspects of the theory of electric potential and the analytic theory of heat (1887).

Chapter 13 describes several contributions Poincaré made to the mathematical theory of the partial differential equations of physics, notably his solution to the *Dirichlet problem* (the existence of harmonic functions on a given domain with prescribed boundary conditions) by the *sweeping method* (1887, 1890); the Poincaré inequality (1890, 1894); the first rigorous proof of the existence of the entire sequence of eigenvalues of the Laplacian on an arbitrary bounded domain in $\mathbb{R}^3$ (with corresponding eigenfunctions zero on the boundary), which he obtained as zeroes of a certain analytic function (1894); and finally, the general solution of the telegraph equation (1893).

Probability (Chapters 14 and 15)

When, in his memoir *Sur le problème des trois corps et les équations de la dynamique*, Poincaré had proved his recurrence theorem (see page 7), he had specified that the initial conditions of non-recurrent trajectories had zero *probability*. *"This word has of itself no meaning: I will give a precise definition in my Memoir,"* he insisted in a report on his 1891 memoir. In fact, it seems that this was the question, and more precisely at the time of the revision of his memoir (see Chapter 3 of Anne Robadey's thesis[24]), which led Poincaré to reflect on the notion of probability and to forge a personal approach to the question. He observed, notably, that any positive and finite "integral invariant" (invariant measure) on the entire phase space provides a valid notion of probability.

It was in 1894 that Poincaré chose to teach probability. His course would be published in 1896; a modified and expanded edition would appear in 1912. It is from this course by Poincaré that we receive the expressions *"characteristic function"* (in the probabilistic sense) and *"normal law"*. Darboux, in his *Éloge historique d'Henri Poincaré*[25] (1913) wrote:

> *"This treatise by Poincaré does not appear to me to have been regarded for its full value. I am certain it will figure proudly next to the masterpieces of Laplace and Bertrand. I especially point out its very fine introduction on the laws and the definition of chance, its chapters on the probability of the continuous, where Poincaré sheds light on a famous paradox proposed by Bertrand; also those which he dedicated to the theory of errors and*

---

[24] *Différentes modalités de travail sur le général dans les recherches de Poincaré sur les systèmes dynamiques* [*Different ways of approaching the notion of genericity in Poincaré's research on dynamical systems*] (Paris, January 3, 2006), available at: `oai:hal.archives-ouvertes.fr:tel-00011380_v1`.

[25] *Œuvres* of Henri Poincaré, volume 2, p. xxxiv.

to the famous law of Gauss. Bertrand limited himself to criticism and demolition. Poincaré started to rebuild."

Chapter 14 presents these contributions and other original aspects of Poincaré's course, notably the relations that he introduced between probability theory and group theory.

Chapter 15 deals more particularly with *geometric probabilities.* Starting with the Buffon needle problem, it presents a theorem of Poincaré and gives two relatively recent applications of it: the first shows (following Edelman and Kostlan, 1995) how this theorem makes it possible to recover very simply an estimate (due to Mark Kac, 1943) of the average number of zeroes of a random real polynomial of large degree; the second, due to Michel Mendès France (1987), is an illustration of the subtle connection that exists between "disorder" and "confusion".

Continuous groups (Lie groups) and their algebras (Chapter 16)

In 1899, the date of Sophus Lie's death, Poincaré published an article on *"continuous groups"* (Lie groups) and their algebras: he embeds every finite-dimensional real Lie algebra in an algebra of symbolic polynomials, in which the Lie bracket takes the form of a simple commutator bracket ($[A, B] = AB - BA$), and which is nothing other than what is today called the *universal enveloping algebra* of the Lie algebra. Its existence constitutes the *Poincaré–Birkhoff–Witt theorem.* Poincaré derived from it a conceptually simple proof of Lie's third fundamental theorem (local version), according to which every finite-dimensional real Lie algebra is isomorphic to that of a (germ of a) Lie group.

Chapter 16 relates the historical context of this contribution of Poincaré to the theory of Lie groups and presents Poincaré's proof, certain developments that occurred during the 20th century, and finally a recent (2003) analogue of the Poincaré–Birkhoff–Witt theorem within the framework of Leibniz algebras, which are a generalization of Lie algebras.

The principle of relativity and the Poincaré group (Chapter 17)

The discovery of the relativity of time is like a surprising puzzle. In 1886, Poincaré became interested in the philosophical work of Auguste Calinon on the foundations of mechanics, and in particular on the quantitative notion of time; Poincaré also emphasized a qualitative problem, that of simultaneity. In 1893, he entered the Bureau des Longitudes. Measuring a longitude amounted to determining the Paris (or Greenwich) time to compare it with the local time: at the time, one used for this a telegraph by submarine cables.[26] The proximity between the philosophical problem and the practical problem led him, in 1898, to a philosophical article entitled *La mesure du temps* [*The measurement of time*], in which Poincaré explains that the notion of simultaneity for distant phenomena presupposes the choice of a procedure of synchronization by exchanging signals. In 1900, he perceived that the formal time variable that appears in the formulas of Lorentz transformations has a very simple physical interpretation: it is the time that one measures in the given reference frame if one synchronizes the watches in

[26]Einstein would also be confronted with practical problems of synchronization: having entered the Patent Office of Bern in 1902, he there had to assess the patent requests concerning electrical synchronization of city clocks. See Peter Galison's book *Einstein's Clocks, Poincaré's Maps: Empires of Time*, 2003.

it by an exchange of signals at the speed of light in a vacuum, ignoring the motion of the reference frame. Moreover, from 1895 Poincaré had been led (mainly by analysis of the Michelson–Morley experiment) to extend the principle of relativity to electromagnetism and even to all of physics. In 1904–1905, he obtained results that would be recorded in his article entitled *Sur la dynamique de l'électron*: the Lorentz transformations form a "continuous group" (a Lie group), whose Lie algebra he writes; he derives the formula for the addition of velocities, the transformations of the electromagnetic field and of the sources, the invariance of $\vec{E}^2 - \vec{B}^2$, $\vec{E} \cdot \vec{B}$, and $x^2 + y^2 + z^2 - c^2t^2$. He suggests the idea that by replacing time with an imaginary variable the Lorentz transformations can be interpreted as rotations in a four-dimensional space; the idea that the principle of relativity, being valid for all of physics, imposes a modification of the law of gravitation, while also limiting the choices; the application to the problem of the precession of Mercury's perihelion (but these corrections from special relativity only explain a part of this); the idea of gravitational waves, citing as a potentially observable consequence the acceleration of the orbital motions of celestial bodies.[27]

Chapter 17 presents some of these contributions by Poincaré. It gives a proof (due to Jean-Marc Lévy-Leblond) of the formulas for Lorentz transformations, under very simple and general hypotheses: space is homogeneous and isotropic, the transformations form a group (the Poincaré condition), and there exist causal chains. It then explains the role of the Poincaré group (group of space-time displacements) in quantum physics (Wigner's theorem) and mentions the fact that it is essentially the only possible symmetry group for the quantum theory of fields (theorems of Coleman–Mandula and Haag–Łopuszański–Sohnius). Finally, it presents an argument explaining why it is natural that Galilean electromagnetism was not discovered before that of special relativity.

Applied physics (Chapter 18)

Poincaré was interested in many questions of applied physics: telegraphy (he solved the telegraph equation in 1893, and from 1904 to 1910 he taught at the École Professionnelle des Postes et Télégraphes), electrotechnics, propagation of waves... To explain the phenomena observed in Gouy's experiment on polarization by diffraction, he made the first correct asymptotic calculation of diffraction by a screen with a sharp edge (1892, 1896).

In 1909, he calculated the diffraction of a Hertzian wave by the Earth: he showed that the amplitude decreases exponentially in the shadow zone, which agreed with the hypothesis that the waves transmitted by Marconi from one end of the Atlantic to the other were not transmitted by diffraction but by reflections on the ocean and the ionosphere. Poincaré's calculation inaugurated a new field of research, that of exponentially small effects in the solutions of differential equations.

Chapter 18 presents Poincaré's calculation, then Fock's subsequent method, confirming Poincaré's result; it then describes the method of Kruskal and Segur (1985)—which (in contrast with those of Poincaré and Fock) does not require first obtaining an explicit expression for the solutions—and it illustrates this method by

[27]It was actually by this means (acceleration of the orbital motions in a binary pulsar) that the existence of gravitational waves was verified, which earned the Nobel Prize in physics for Joseph H. Taylor and Russell A. Hulse in 1993. But the correct calculations result from Einstein's general theory of relativity.

an example (the Korteweg–de Vries wave equation, perturbed by a small term of order 5).

Philosophy of science (Chapter 19)

Poincaré seems to have always been interested in the philosophy of science. The unexpected involvement of hyperbolic geometry in his research on automorphic functions (1880) could only reinforce his interest in the foundations of geometry: questions of Euclidean geometry concerning systems of curvilinear triangles were interpreted simply in terms of hyperbolic geometry, and this interpretation allowed him to find the solution. In such cases, there is every interest in taking the point of view of hyperbolic geometry, not because it is more "true" than Euclidean geometry, but because it is more *convenient* (*"we can no more say that Euclidean geometry is true and the geometry of Lobachevsky false* [or the converse], *than we could say that Cartesian coordinates are true and polar coordinates false"*[28]). On the other hand, when one studies the motions of solid bodies (beginning with the motions of our own limbs), the most convenient description uses the language of Euclidean geometry, for *"experience teaches us that the various possible motions of these bodies are connected almost exactly by the same relations"* as Euclidean displacements;[29] *"we can state mechanical facts by relating them to a non-Euclidean space, which would be a less convenient coordinate system, but one just as legitimate as our ordinary space; the statement would thus become much more complicated, but it would remain possible."*[30] This is what would be done by beings who had been educated in a world where the most convenient geometry was non-Euclidean, and who had been transported into our world. (Chapters III and IV of *Science and Hypothesis* present a suggestive illustration of this.) Thus, the fundamental hypotheses of geometry are neither inevitable frameworks of thought nor experimental facts: they are convenient choices, taking observations into account.

The period was also a time of critically examining the foundations of mechanics: Mach's *Mechanik* appeared in 1883, Calinon's *Étude critique sur la mécanique* in 1885... Poincaré was closely interested. (Later, in 1897 and 1901, he would publish two articles on the foundations of mechanics, whose essential points he would repeat in *Science and Hypothesis* in 1902.) From mechanics, he passed to all of physics. In his teaching of physics, each time he would draw up a *"tableau d'ensemble"* of all the available theories; for *"it would be dangerous to limit oneself to one of them; one would thus risk feeling a blind, and consequently misleading, confidence in one's place. One must therefore study all of them, and it is their comparison that may above all be instructive."*[31] This obviously does not exclude pruning them; one should only preserve the *"fruitful"* hypotheses.

Poincaré also extended his reflections to mathematical analysis, arithmetic, and logic. Poincaré started publishing in the *Revue de métaphysique et de morale* in 1893 (the date of the journal's creation) and published one or two articles in it nearly every year until the end of his life. The essence of these articles (and

[28]Poincaré, *Sur les hypothèses fondamentales de la géométrie* [*On the Fundamental Hypotheses of Geometry*], 1887.

[29]Ibid.

[30]*Science and Hypothesis*, Chapter VI.

[31]Preface to the *Théorie mathématique de la lumière* [*Mathematical theory of light*], course from the year 1887–1888, published in 1889.

others, published elsewhere) would form the material of his famous collections on the philosophy of science (already mentioned above), published starting in 1902.

Chapter 19 proposes a survey of Poincaré's philosophy, which Gerhard Heinzmann justly calls *pragmatic occasionalism*, since according to this point of view, experience is for our minds only the occasion to exercise its faculties and to become aware of this, and because its criterion is pragmatic (one only chooses the fruitful hypotheses). It explains and illustrates this philosophy successively in arithmetic, in geometry, and in physics. It also presents Poincaré's view on logic.

Thanks

Several texts from this book originated in presentations made by the authors at the Colloque Henri Poincaré organized by Alain Comtet, Étienne Ghys, and Yves Pomeau at the Institut Henri Poincaré (Paris) in December 2004.

A great thanks to the authors, of course, for these texts of exceptional quality, as well as to Matthieu Gendulphe, Cyril Mauvillain, Laurent Mazliak, Michel Mendès France, and Pierre Mounoud, and to the Archives Henri Poincaré (Nancy) for their help.

Éric Charpentier, Étienne Ghys, Annick Lesne

CHAPTER 1

Poincaré and His Disk

Étienne Ghys

1.1. An imaginary world

Henri Poincaré did not invent non-Euclidean geometry. Even the famous *Poincaré disk* existed before him. However, his use of this geometry and its group of isometries[1] was so staggering that the name *Poincaré disk* is by no means undeserved. The goal of this chapter is not to describe the history of non-Euclidean geometry, because a number of excellent works are already devoted to that topic (see [**16, 24, 30, 48, 49, 52**], among my favorites). I would have liked to propose a guided tour of the disk, but it is much too vast and I myself have only explored a small part. Instead, I invite the reader on an excursion. If one takes a random walk in the Euclidean plane, the risk is that one will return to the point of origin over and over ceaselessly, but this risk does not exist in non-Euclidean geometry! We shall see that a random path in the disk does not take many side trips and almost surely leads somewhere. My main goal is to try to convey a geometric intuition for this object, which has progressively passed, in less than two centuries, from the status of a counterexample—whose very existence was doubtful—to that of a central concept that has invaded nearly all of mathematics.

Of course, this text is not directed to the experts. I have instead tried to make it accessible to a (motivated) undergraduate. For those who would like to know more, I provide a copious bibliography—in reality, a pretext to present some of my favorite books.

Let us first make the acquaintance of a non-Euclidean geometry by using the imagery that Poincaré created in *Science and Hypothesis* [**63**, Ch. 3]:

> *"THE NON-EUCLIDEAN WORLD*
>
> *Let us assume* [...] *a world enclosed in* [a large circle] *and subject to the following laws:*
>
> *The temperature in this world is not uniform; it is largest at the center, and it diminishes as one moves away from the center, so that it reduces to absolute zero when one reaches* [the circle] *where this world is enclosed.*
>
> *I will moreover specify the following law by which this temperature varies. Let R be the radius of the limit* [circle]*; let r be the distance from the point under consideration to the centre of* [this circle]. *The absolute temperature will be proportional to $R^2 - r^2$.*

[1] See Chapter 2.

> *I will additionally assume that, in this world, all bodies have the same coefficient of dilation, in such a way that the length of any ruler shall be proportional to its absolute temperature.*
>
> *Finally, I will assume that an object transported from one point to another, whose temperature is different, shall immediately reach thermal equilibrium with its new location.*
>
> *Nothing in these hypotheses is contradictory or unimaginable.*
>
> *A moving object will then become smaller and smaller as it approaches the limit* [circle].
>
> *Let us first observe that, if this world is finite from the point of view of our customary geometry, it will appear infinite to its inhabitants.*
>
> *In fact, when they wish to approach the limit* [circle], *they will get colder and become smaller and smaller. The steps they take are therefore also smaller and smaller, so that they can never reach the limit* [circle].
>
> *If, for us, geometry is merely the study of the laws by which non-deformable solids may move, for these imaginary beings, it will be the study of the laws that drive solids deformed by these differences in temperature about which I have just spoken.* [...] *For brevity, I shall, with the reader's permission, call such a movement a non-Euclidean displacement.*
>
> *Thus beings like us, whose education would take place in such a world, would not have the same geometry as us.* [If these imaginary beings] *founded a geometry,* [...] *it would be non-Euclidean geometry."*

It is in this world "neither contradictory nor unimaginable" that we shall take our excursion.

Here is a quote from Coxeter showing just how "real" this geometry is for mathematicians [**24**]:

> *"When Hamlet exclaims (in Act II, Scene II) "I could be bounded in a nutshell and count myself a king of infinite space" he is providing a poetic anticipation of Poincaré's inversive model of the infinite hyperbolic plane, using a circular "nutshell" for the Absolute."*

The reader is therefore warned that this world is vast...

1.2. Some formulas

Before beginning our excursion, we must introduce some definitions, notations, and formulas which will serve to paraphrase the imagery of Poincaré's description.

We denote by $\mathbb{D} = \{z \in \mathbb{C} \mid |z| < 1\}$ the open unit disk in the complex plane. If v is a tangent vector to $\mathbb{D}$ at a point z, with Euclidean norm $\|v\|_{\mathrm{Eucl}}$, then its *hyperbolic norm*[2] $\|v\|_{\mathrm{hyp}}$ is defined by

$$\|v\|_{\mathrm{hyp}} = \frac{1}{1-|z|^2}\,\|v\|_{\mathrm{Eucl}}\,.$$

[2]It appears that Klein is responsible for the term *"hyperbolic geometry"*. There are of course good reasons for this choice, but one can only regret the all-too-frequent use of the word "hyperbolic" in mathematics, with quite different meanings.

From this we define the *hyperbolic length* of a curve $\gamma : [0,1] \to \mathbb{D}$ by

$$\text{length}_{\text{hyp}}(\gamma) = \int_0^1 \left\| \frac{d\gamma}{dt} \right\|_{\text{hyp}} dt,$$

and the *hyperbolic distance* (or *Poincaré distance*) $\text{dist}_{\text{hyp}}(z_0, z_1)$ between two points z_0 and z_1 of the disk as the minimum of the hyperbolic lengths of curves joining z_0 and z_1. The *Poincaré disk* is the metric space thus obtained.

Why choose this factor $1/(1-|z|^2)$ *rather than another?* Quite simply because it is the "obvious" factor that guarantees that the object we have just defined is *homogeneous.* If α is a real number and a is an element of $\mathbb{D}$, the transformation of $\mathbb{C} \cup \{\infty\}$ defined by

$$f_{\alpha,a}(z) = \exp(i\alpha)\frac{z-a}{1-\bar{a}z}$$

preserves the unit disk (check this!). The set of these transformations forms a group that we will often encounter in this chapter. Let us observe for the moment that this group acts *transitively* on the disk: given any two points in $\mathbb{D}$, one of the group elements sends the first point to the second point (check this!). The derivative of $f_{\alpha,a}$ is

$$f'_{\alpha,a} = \exp(i\alpha)\frac{1-|a|^2}{(1-\bar{a}z)^2},$$

from which it follows that

$$\frac{|df_{\alpha,a}(z)|}{1-|f_{\alpha,a}(z)|^2} = \frac{|dz|}{1-|z|^2}$$

(again, check this...). In other words, the *Poincaré metric* is invariant under the group of the $f_{\alpha,a}$'s, and the disk is seen to be homogeneous: *its group of isometries acts transitively* and all of its points are equivalent.

We will see a bit later that this homogeneity is a crucial property that "almost" characterizes the Poincaré disk. For the moment, we will use the homogeneity in order to effortlessly obtain certain formulas that we will need.

It is easy to find the shortest curve (in the hyperbolic sense) joining the point 0 of $\mathbb{D}$ to the point r situated on the axis $[0,1) \subset \mathbb{D}$. In fact, it suffices to observe that if $\gamma : [0,1] \to \mathbb{D}$ joins 0 to r, the radial projection $|\gamma| : [0,1] \to [0,1) \subset \mathbb{D}$ also joins 0 to r, and that its hyperbolic length is less than that of γ (because the radial component of a vector is shorter than the vector itself). Thus, the unique curve minimizing the hyperbolic length between 0 and r is the radius $[0,r]$ with hyperbolic length

$$\int_0^r \frac{dt}{1-t^2} = \frac{1}{2}\log\left(\frac{1+r}{1-r}\right) = \tanh^{-1}(r).$$

This length tends to infinity as r tends to 1, as Poincaré's imagery describes.

To find the *geodesics* of $\mathbb{D}$, i.e., the curves with fixed endpoints that minimize the hyperbolic lengths, it suffices to apply the group of isometries. Given two points z_0 and z_1, we can find an isometry that sends the first to 0 and the second to a point of the positive real axis, as previously. The sought-after geodesic is therefore the image of a radial segment by an isometry; it is an arc of a circle orthogonal to the unit circle (or a radial segment). To be convinced of this, the

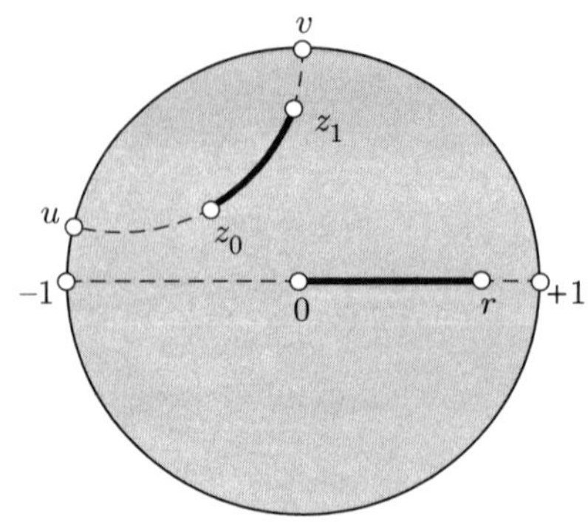

FIGURE 1.1. Geodesics in the disk.

reader should remember (or check for himself, or perhaps consult [**4, 24**]) that a projective transformation

$$z \in \mathbb{C} \cup \{\infty\} \mapsto \frac{Az+B}{Cz+D} \in \mathbb{C} \cup \{\infty\}$$

(with A, B, C, and D complex numbers such that $AD - BC \neq 0$) sends a circle to a circle (in the Riemann sphere $\mathbb{C} \cup \{\infty\}$) and preserves the angle of intersection between two circles. Thus, a diameter of the unit disk is sent by an isometry $f_{\alpha,a}$ onto an arc of a circle orthogonal to the unit circle (or onto another diameter). In the same way, it is easy to use homogeneity to establish the formula that gives the distance between two points z_0 and z_1. We already know that

$$\mathrm{dist}_{\mathrm{hyp}}(0, r) = \frac{1}{2}\log\left(\frac{1+r}{1-r}\right) = \tanh^{-1}(r)$$

which can also be written

$$\frac{1}{2}\left|\log\left(\frac{0-(-1)}{0-1}\cdot\frac{r-1}{r-(-1)}\right)\right|,$$

where we recognize $[-1:1:0:r]$, the *cross ratio* of four points in $\mathbb{C} \cup \{\infty\}$. Recall that the cross ratio of four distinct points x, y, z, t is defined by

$$[x:y:z:t] = \frac{z-x}{z-y}\cdot\frac{t-y}{t-x}$$

and that for any projective transformation f, we have

$$[f(x):f(y):f(z):f(t)] = [x:y:z:t].$$

Since the projective transformations $f_{\alpha,a}$ preserve the cross ratio, hyperbolic distances, and circles orthogonal to the unit circle, we easily obtain the hyperbolic distance between any two points z_0 and z_1 in $\mathbb{D}$. It suffices to consider the unique circle (or diameter) orthogonal to the unit disk that contains z_0 and z_1. This circle meets the unit circle in two points u, v (see FIG. 1.1), and we have

$$\mathrm{dist}_{\mathrm{hyp}}(z_0, z_1) = \frac{1}{2}\big|\log[u:v:z_0:z_1]\big|.$$

The reader who prefers a formula only involving z_0 and z_1 may express (as Poincaré did in his original article on the subject [**62**]) the distance as a function of the cross ratio of the (cocyclic) points z_0, $1/\overline{z_0}$, z_1, and $1/\overline{z_1}$.

Let us observe that the disk is in fact *2-homogeneous*: if $\mathrm{dist}_{\mathrm{hyp}}(z_0, z_1) = \mathrm{dist}_{\mathrm{hyp}}(z_0', z_1')$, there exists an isometry that sends z_0 to z_0' and z_1 to z_1'.

1.3. The Poincaré disk is ubiquitous

We know (at least since Euclid!) the geometry of the Euclidean plane. The geometry of the sphere is equally familiar because, after all, *geo–metry* is the science that *measures* the *earth.* It took two millennia for hyperbolic geometry to install herself among mathematicians (and a little among physicists). However, she deserves the same respect as her two older sisters.

It seems to me that the following "characterization theorem" fully justifies this respect. Its statement is simple, but its proof is not. It uses some of the most difficult theorems of the 20th century and goes well beyond the level of this elementary chapter. In the Appendix, I will nevertheless try to give a few indications of this proof.

THEOREM. *Let (X, d) be a metric space with the following properties:*

- *(X, d) is a* **surface**: *every point of X has a neighborhood homeomorphic to an open subset of $\mathbb{R}^2$;*
- *(X, d) is* **2-homogeneous**: *if $d(x, y) = d(x', y')$, there exists an isometry that sends x to x' and y to y';*
- *(X, d) is a* **geodesic space**: *for any pair of points (x, y), there exists a curve $\gamma : [0, \ell] \to X$ such that $\gamma(0) = x$, $\gamma(\ell) = y$, and which is an isometry onto its image (so that $d(x, y) = \ell$).*

Then (X, d) is isometric to one of the following three examples:

- *the* **Euclidean plane**;
- *the* **sphere** *of radius R in Euclidean space, or the quotient of this sphere in which one identifies pairs of diametrically opposed points (the Klein elliptic space);*
- *the* **Poincaré disk** *endowed with a constant multiple of hyperbolic distance.*

A few comments on the hypotheses of this theorem.

The property of having dimension 2 is fundamental. Later we will describe other geodesic 2-homogeneous metric spaces with larger dimension. The fact of being a surface, on the other hand, is not very important: one could assume, for example, that X is locally compact and has topological dimension[3] 2.

Homogeneity is of course essential, but 2-homogeneity is much less so. We will see that if we replace the second hypothesis by simple homogeneity, the theorem remains true as long as we add a few additional, less important examples (can the reader guess what they are?).

Nor is the third hypothesis very important. Let us remark that if (X, d) is a metric space, then $(X, \phi(d))$ is also a metric space, inducing the same topology, just as homogeneous as the first, provided that $\phi : \mathbb{R}^+ \to \mathbb{R}^+$ is subadditive, that is, such that $\phi(s + t) \leq \phi(s) + \phi(t)$. If we suppress the third hypothesis, then the theorem's conclusion is hardly weakened at all: the space (X, d) is then obtained by this construction from one of the preceding examples, Euclidean, elliptic, or hyperbolic.

---

[3]A topological space has topological dimension less than or equal to d if it admits arbitrarily fine coverings such that the intersection of $(d + 2)$ distinct open sets in a given covering is empty. For example, the product of two graphs has dimension 2 but is not, in general, a surface.

We shall come back to these "details" further on, but let us retain the fundamental fact that *the Euclidean, elliptic, and hyperbolic geometries are essentially the only homogeneous metric surfaces.*

In truth, these three geometries (X, d) are much more than homogeneous. If Y is a subset of X and if $f : Y \to X$ is an isometry onto its image, then f is the restriction of a global isometry of X. If Y contains only one point, this is the usual property of homogeneity; if Y contains two points, this is 2-homogeneity, etc.

1.4. Lots and lots of models

Homogeneous objects of dimension 2 arise naturally in many situations. The preceding theorem then allows us to identify such an object with one of the three geometries. Thus we find *models* of hyperbolic geometry abounding in the literature. Even though the goal of this chapter is not to discuss the epistemological aspect of non-Euclidean geometry, let us remark upon the common usage of the word "model" in this context: it is as though this geometry had an "ideal" intrinsic existence that mathematicians seek to understand by constructing several "illustrations" of it.

I will describe here a few of these models, but there are plenty of others (see for example [**2, 48, 71**]).

The first is nothing but a benign change of variables: the transformation

$$z \mapsto i\,\frac{1+z}{1-z}$$

sends the disk $\mathbb{D}$ onto the *Poincaré half-plane*[4] $\mathbb{H} = \{z \in \mathbb{C} \mid \operatorname{Im} z > 0\}$ because

$$\operatorname{Im}\left(i\,\frac{1+z}{1-z}\right) = \frac{1-|z|^2}{|1-z|^2}.$$

In this half-plane, the hyperbolic metric becomes $|dz|/\operatorname{Im} z$,[5] and the geodesics are half-circles (or rays) orthogonal to the boundary (see FIG. 1.2).

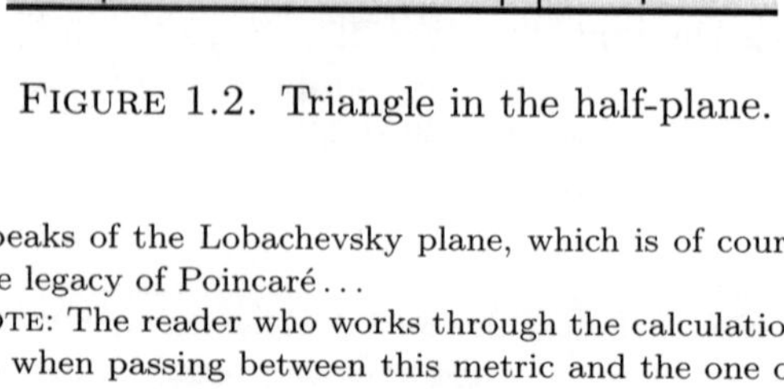

FIGURE 1.2. Triangle in the half-plane.

[4]One sometimes speaks of the Lobachevsky plane, which is of course entirely justified, but in a work devoted to the legacy of Poincaré...

[5]TRANSLATOR'S NOTE: The reader who works through the calculations will find that a factor of 2 (or of 1/2) appears when passing between this metric and the one on p. 18. This is slightly unfortunate, albeit interesting, but for purposes of this exposition can be ignored.

This model has the advantage of presenting the group of orientation-preserving isometries in the simplest way:

$$z \in \mathbb{H} \mapsto \frac{Az+B}{Cz+D} \in \mathbb{H}$$

with A, B, C, and D real numbers such that $AD - BC = 1$. We recognize the group $\mathrm{PSL}(2, \mathbb{R})$.

With the disk and half-plane models, one can already carry out many calculations. One may easily convince oneself that Euclid's fifth postulate is not true in this geometry: given a point x not on a geodesic γ, there are infinitely many geodesics that contain x and are "parallel" to γ, that is, they do not intersect γ (FIG. 1.3).

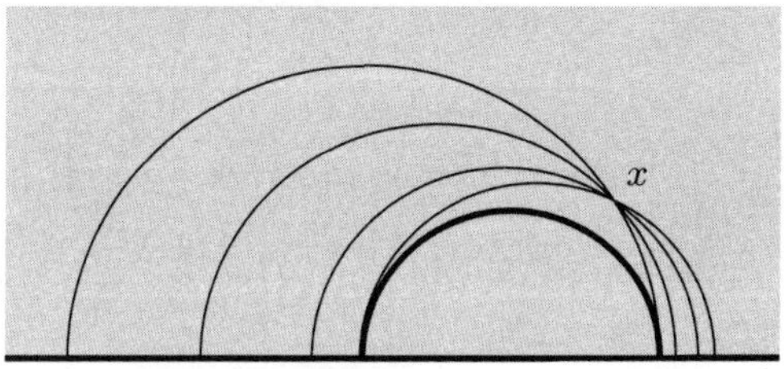

FIGURE 1.3. The fifth postulate.

Here is a less well-known model. We consider the set $\mathcal{E}$ of ellipses in the Euclidean plane, centered at the origin, and bounding a region of unit area. We may think of such an ellipse e as the unit sphere of a norm $\|\cdot\|_e$ on $\mathbb{R}^2$. If e_1 and e_2 are two elements of $\mathcal{E}$, we may compare them by setting

$$\mathrm{dist}(e_1, e_2) = \log \sup_{v \in \mathbb{R}^2 \setminus \{(0,0)\}} \frac{\|v\|_{e_1}}{\|v\|_{e_2}}.$$

With just a moment of reflection, the reader can verify that:

- dist is indeed a (geodesic?) distance;
- $(\mathcal{E}, \mathrm{dist})$ is a surface (because two parameters suffice to describe an element of $\mathcal{E}$);
- $(\mathcal{E}, \mathrm{dist})$ is homogeneous (there is only one ellipse, up to affine transformations).

Another moment of reflection will equally convince the reader that $(\mathcal{E}, \mathrm{dist})$ is neither Euclidean nor elliptic. Hence it must be a model of the Poincaré disk!

It remains to exhibit an explicit isometry between $(\mathcal{E}, \mathrm{dist})$ and $(\mathbb{D}, \mathrm{dist}_{\mathrm{hyp}})$. This, of course, poses no problem. To each point of the disk having the form $\rho \exp(i\phi)$, we associate the ellipse of ellipticity[6] $\rho < 1$ whose major axis makes an angle of $\phi/2$ with the real axis. The verification that this bijection is indeed an isometry is a routine exercise.

This model suggests a generalization: the space of symmetric convex sets having volume 1 in $\mathbb{R}^n$ is naturally a metric space.

[6]The ellipticity of an ellipse with axes $b \leq a$ is $(a-b)/(a+b)$. Do not confuse it with the eccentricity.

One other model: *the Hilbert metric* [**45**, App. 1]. Let C be a bounded open *convex* set in $\mathbb{R}^2$. If x and y are two points of C, the line (xy) that joins them intersects C on an open interval (u, v) (see FIG. 1.4). We set

$$\mathrm{dist}_{\mathrm{Hilbert}}(x, y) = |\log [u : v : x : y]|,$$

where this time we use the cross ratio of the four points u, v, x, y on the real line (before, we were using the cross ratio of four points on the complex line).

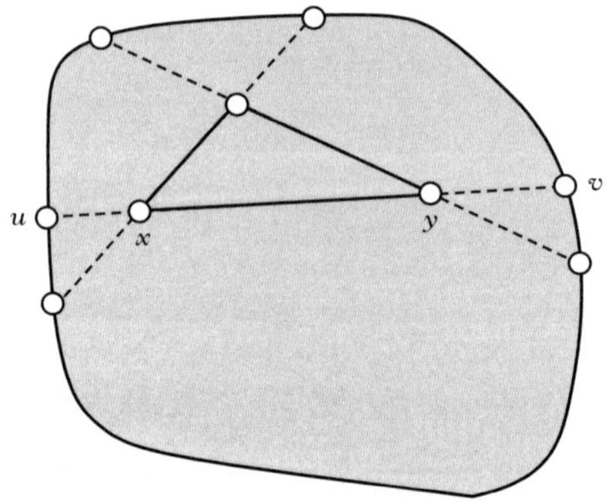

FIGURE 1.4. Hilbert distance.

Here again, it is not difficult to assure oneself that $\mathrm{dist}_{\mathrm{Hilbert}}$ is a geodesic distance (contemplate FIG. 1.4, not forgetting that the cross ratio is invariant under projection). Evidently, this distance is invariant under the group of projective transformations that preserve C. In the particular case where C is the interior of an ellipse, this group acts transitively on C (exercise),[7] and the metric space we obtain is homogeneous. We have thus found a version of the Poincaré disk in the interior of an ellipse, often called the "Klein model". It is not so easy to find an isometry between $(C, \mathrm{dist}_{\mathrm{Hilbert}})$ and $(\mathbb{D}, \mathrm{dist}_{\mathrm{hyp}})$, however. Note that the two models are very different: the geodesics in the Klein model are line segments, while those in the Poincaré disk are arcs of circles. It is not *a priori* clear that there is a homeomorphism of the disk that turns line segments into arcs of circles. Here is one method: given real numbers a and b, consider the second degree equation in an unknown z of the form $(1 - a)z^2 + 2bz + (1 + a) = 0$, with discriminant $4(a^2 + b^2 - 1)$. Thus to each point $(a + ib)$ in the unit disk, we can associate a second degree equation that has a unique solution z_{a+ib} in the upper half-plane $\mathbb{H}$. It is a pleasant exercise to show that the transformation $(a + ib) \in \mathbb{D} \mapsto z_{a+ib} \in \mathbb{H}$ realizes an isometry between the Klein metric and the Poincaré metric (as seen in the half-plane).

The Hilbert metric has many other interesting properties (see [**44**] for numerous examples and developments). Note that the ellipse is not the only projectively

[7] We cite here the lovely consequence Hilbert drew from this result: it is impossible to construct the center of a circle using only a ruler. In fact, if such a construction existed, its conjugate by a projective transformation preserving the circle could be used to construct any point of the disk: a contradiction!

homogeneous convex figure: so is the interior of the triangle.[8] What is the Hilbert metric in this case?

1.5. Attempts to visualize in space

Even though Riemann taught us that one can (should?) do geometry on an abstract Riemannian manifold, many of us like to visualize surfaces as embedded in Euclidean space.

A differentiable function $\phi : \mathbb{D} \to \mathbb{R}^N$ is called an *isometric immersion* if for any vector v tangent to the disk, we have $\|v\|_{\text{hyp}} = \|d\phi(v)\|_{\text{Eucl}}$. If, in addition, ϕ is injective, we call it an *isometric embedding.* This does not mean that ϕ is an isometry onto its image, but only that the hyperbolic length of a curve in $\mathbb{D}$ equals the Euclidean length of the image of the curve by ϕ. Moreover, there is no function $\phi : \mathbb{D} \to \mathbb{R}^N$ that is an isometry onto its image, because such a function would have to send a geodesic onto a line, and the image of $\mathbb{D}$ would therefore have to be contained in a Euclidean plane, which is of course impossible.

Early on, Beltrami sought to construct an isometric embedding of the hyperbolic disk into $\mathbb{R}^3$. By a very simple method, he succeeded in *locally* finding such an embedding. We consider a surface of revolution whose equation in cylindrical coordinates (r, z) has the form $r = F(z)$. In order to find F such that the surface is isometric to the Poincaré disk, we must solve a second-order differential equation. Among its solutions, we find the *tractrix*, the curve followed by an object pulled by a rope of constant length and whose free end moves along a line (FIG. 1.5). Thus

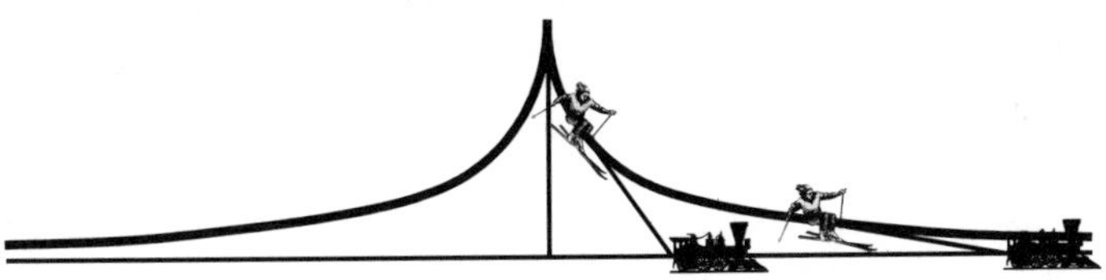

FIGURE 1.5. Tractrix.

the tractrix of revolution, often called a *pseudosphere* (FIG. 1.6), is a local model of the Poincaré disk [**58**].

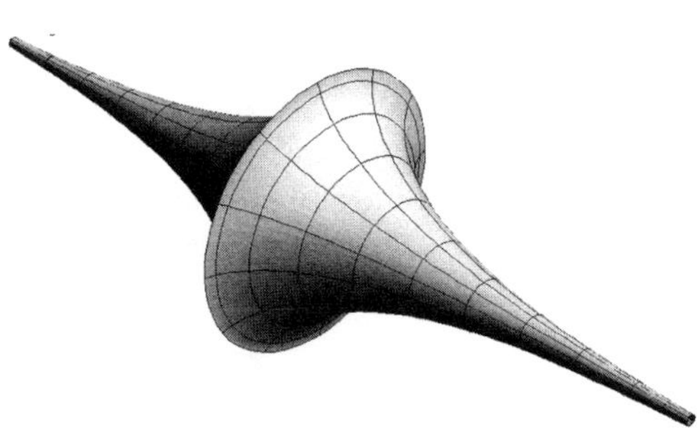

FIGURE 1.6. Pseudosphere.

[8]Thus the foregoing argument of Hilbert shows that the centroid of a triangle cannot be constructed using only a ruler; this is perhaps more surprising than for the center of a circle.

This tractrix contains a singular point (a cusp), however, in such a way that the surface of revolution is also singular along a cusp circle. Therefore we only obtain a piece of a surface isometric to a piece of the disk. The left part of FIG. 1.7 (inspired by [**66**]) shows a disk in $\mathbb{D}$ with a small radius and its isometric image in the pseudosphere. The right part shows a part of $\mathbb{D}$ (with area 2π) which is isometric to the complement of a generatrix in a half-pseudosphere.

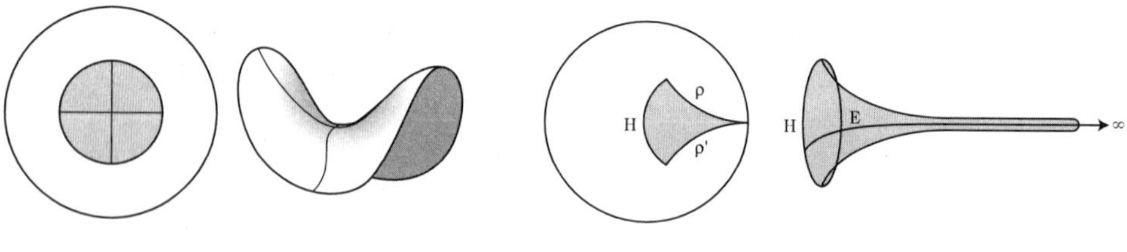

FIGURE 1.7. Pseudosphere and disk.

Moreover, we cannot escape these singularities: Hilbert proved[9] that there is no isometric embedding of the the Poincaré disk into 3-dimensional Euclidean space, such that the embedding is of class C^2. The proof is quite clever: it consists of a detailed analysis of two families of asymptotic lines[10] on a surface in $\mathbb{R}^3$ which is locally isometric to the disk. These curves form a *Chebychev net*, which was a notion introduced for a very concrete problem in the article entitled "Sur la coupe des vêtements" ("The cutting out of clothes") [**22**]. One considers a piece of fabric in the plane formed by the interlaced threads

$$x = i/N, \quad y = j/N, \qquad i, j = 0, \ldots, N \quad (N \text{ large}).$$

Next one deforms the fabric in space in such a way that the sides of each stitch, initially in the shape of a square, keep the same length. In other words, one considers the surfaces "clothed" in the fabric $u : [0, a] \times [0, b] \to \mathbb{R}^3$ such that $\partial u/\partial x$ and $\partial u/\partial y$ have norm 1 everywhere (but are not necessarily orthogonal). One finds that the asymptotic lines of a surface locally isometric to the Poincaré disk clothe this surface in the preceding sense (see FIG. 1.8).

This is the point at which Hilbert's proof begins. For the (interesting) continuation, see [**20, 30, 45**]. It is worth remarking that this problem of Chebychev continues to develop: apart from the fact that it is intimately linked to the so-called sine-Gordon partial differential equations, one uses this kind of ideas in the construction of certain contemporary materials (see, for example, [**64**]).

The proof of Hilbert's theorem could not be too elementary, because the theorem is false for class C^1. It follows from a theorem of Nash that there is a C^1 isometric embedding of the Poincaré disk into the space $\mathbb{R}^3$ [**37, 54**].

[9]In truth, Hilbert assumed the embedding to be analytic, and it was only much later that the C^2 version was proved.

[10]Let m be a point on a smooth surface S embedded in $\mathbb{R}^3$. Denote the tangent plane at m by Π and the normal line by N. Locally, S is the graph of a function u from a neighborhood of m in Π to N. The second derivative of u at m is the *second fundamental form* at the point m. The isotropic directions of this quadratic form, if they exist, are the *asymptotic directions*, and the curves that are everywhere tangent to one of these directions are the *asymptotic lines* of S [**20**].

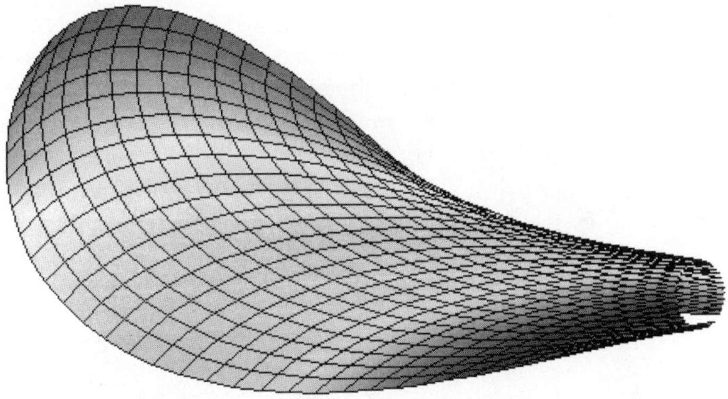

FIGURE 1.8. Clothing a piece of the pseudosphere.

What do these embeddings look like? It is very useful to construct concrete models of them, for example out of paper, as Beltrami did in the 19th century (FIG. 1.9).

FIGURE 1.9. Beltrami's model [**16**].

Take a large number of equilateral cardboard triangles, say a few centimeters to the side, and tape their edges together in such a way that each vertex is surrounded by 7 triangles: you will get an approximation of a piece of the Poincaré disk. Notice that if you replace 7 with 6, you will end up constructing a Euclidean plane with the usual tiling by equilateral triangles. By using 5 triangles around each vertex, one obtains a sphere, tiled like an icosahedron. (What happens with 4, 3, 2 triangles, with 8, 9, etc.?)

The remarkable internet site [**70**] contains a virtual exhibit that seeks to develop one's intuition for this type of objects. There is even a set of instructions for crocheting a Poincaré disk (FIG. 1.10)!

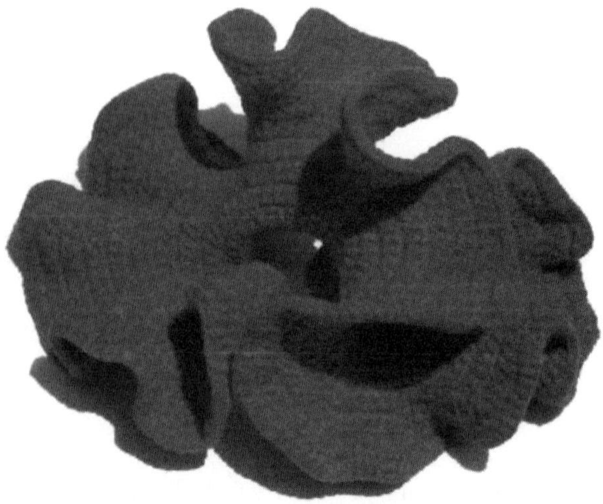

FIGURE 1.10. Hyperbolic crochet.[11]

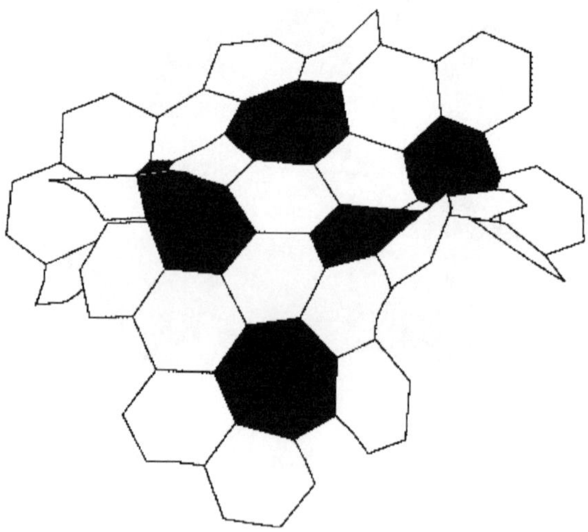

FIGURE 1.11. Hyperbolic football.[12]

In [**61**], the authors do not hesitate to evoke a "jupe à godets" ["gored skirt"] (FIG. 1.12) and compare the pseudosphere to a lily (FIG. 1.13).

Of course, it is possible to apply general theorems about C^∞ isometric embeddings of Riemannian manifolds into Euclidean spaces of sufficiently large dimension. Following Gromov [**37**] (generalizing work of Nash), a k-dimensional Riemannian

[11]Made by Dr. Daina Taimina. Image used courtesy of the Institute for Figuring [**70**].

[12]Drawn by Brian McMullen and Caitlin Berrigan of Cabinet Magazine. Image used courtesy of the Institute for Figuring [**70**].

FIGURE 1.12. Jupe à godets.

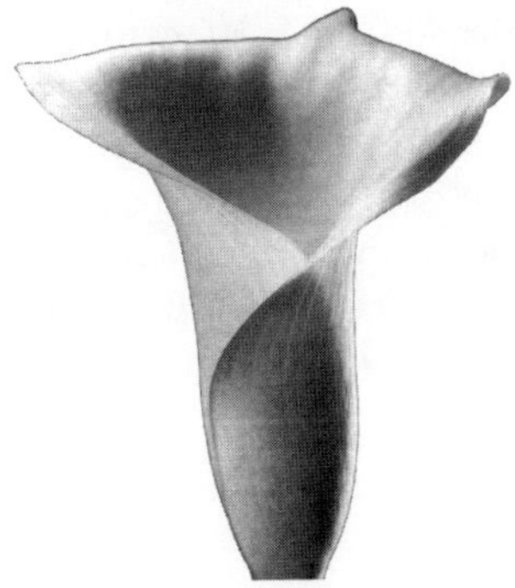

FIGURE 1.13. Calla lily.

manifold has a C^∞ isometric embedding into $(k+2)(k+3)/2$-dimensional Euclidean space. As for k-dimensional hyperbolic space (here we have only discussed the dimension 2 case), there is a C^∞ isometric embedding into $(5k-5)$-dimensional Euclidean space. In fact, Blanuša constructed in 1955 an *explicit* C^∞ embedding of the Poincaré disk into $\mathbb{R}^6$ [**14**]. Can the disk be isometrically immersed in $\mathbb{R}^4$? This seems to be an open question. Gromov showed, however, that any compact piece of the disk can be isometrically immersed in an arbitrary[13] (non-empty!) open subset of $\mathbb{R}^4$.

Such embeddings are in fact of little use, because they cannot be natural. A *natural* embedding would be an embedding $i : \mathbb{D} \to \mathbb{R}^N$ such that for any isometry f of $\mathbb{D}$, there would be an isometry $\bar{f}$ of $\mathbb{R}^N$ satisfying $i \circ f = \bar{f} \circ i$. Such an embedding cannot exist for purely algebraic reasons: there is no non-trivial homomorphism from the group of (orientation-preserving) isometries of the disk into the group of isometries of Euclidean space (exercise, not so easy!).

On the other hand, there do exist natural embeddings into *infinite*-dimensional space. Later I will show an explicit construction of an embedding $i : \mathbb{D} \to H$, where H is a Hilbert space, satisfying the following properties:

- $\text{dist}_{\text{hyp}}(x,y) = \|i(x) - i(y)\|^2$ for all $x, y \in D$;
- i is natural in the above sense.

1.6. A little bit of triangle geometry

The most famous theorem of hyperbolic geometry is due to Gauss [**33**]:

THEOREM. *The area of a triangle with angles* α, β, γ *is* $\pi - (\alpha + \beta + \gamma)$.

Before sketching two proofs of this theorem (there are plenty of others), we must explain the words "triangle, angle, area". The Poincaré disk is a Riemannian manifold whose metric $ds = |dz|/(1 - |z|^2)$ is *conformal* to the Euclidean metric $|dz|$; this means that the angle between two tangent vectors at the same point of the disk is the same regardless of whether one calculates with ds or $|dz|$. The hyperbolic area element is given by $d\,\text{area}_{\text{hyp}} = (1 - |z|^2)^{-2}\, d\,\text{area}_{\text{Eucl}}$.

[13]One might believe that the diameter of the open set would necessarily be larger than the diameter of the compact set, but nothing of the sort is true: an isometric embedding is not an isometry! For example, an isometric embedding of $\mathbb{R}$ into $\mathbb{R}^3$ is just an embedded curve, parametrized by arclength; it is not difficult to "wind up" such a curve into an arbitrarily small ball.

A triangle must of course be understood to be defined by three points and bounded by three geodesic segments; it therefore has three angles and an area.

It is easy to convince oneself that the sum of the angles of a triangle is less than π (put one of the vertices at the center of the disk and compare the angles of the hyperbolic triangle with those of the Euclidean triangle having the same vertices).

The first sketch of a proof of the theorem may require a few extra developments to confirm it, but it is rather intuitive. If P is a geodesic polygon with n sides and whose angles measure $\alpha_1, \ldots, \alpha_n$, we set

$$\mathcal{A}(P) = (n-2)\pi - \sum_{i=1}^{n} \alpha_i.$$

Then we remark that

- if we cut P along a geodesic to obtain polygons P_1 and P_2, we evidently have $\mathcal{A}(P) = \mathcal{A}(P_1) + \mathcal{A}(P_2)$ (see FIG. 1.14);
- if P is a triangle with small diameter, $\mathcal{A}(P)$ is small ("because" in a small neighborhood, a Riemannian metric is "almost" Euclidean: this is a point that must be made more precise);
- $\mathcal{A}$ is clearly invariant under isometries.

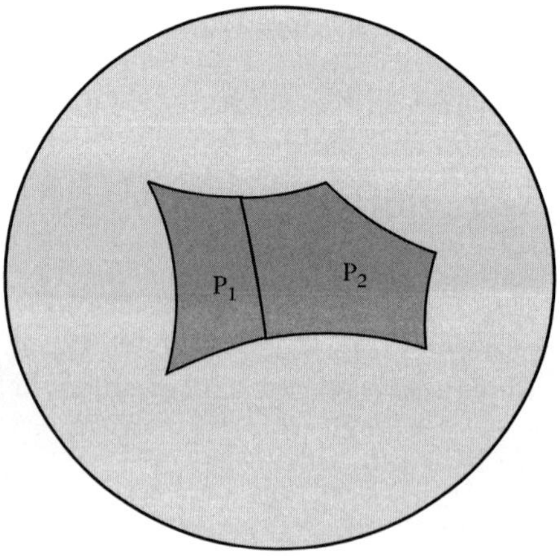

FIGURE 1.14. Cutting a polygon.

Thus we can mimic the classical construction of Lebesgue measure on the plane. We use $\mathcal{A}(P)$ as the measure of a polygon, and we define the measure of a Borel set using coverings by polygons. Thus, $\mathcal{A}$ defines a measure on the disk that is invariant by isometries. Homogeneity shows that $\mathcal{A}$ must be a constant multiple of the hyperbolic area. Therefore we have "established" that

$$\operatorname{area}(P) = c\left((n-2)\pi - \sum_{i=1}^{n} \alpha_i\right)$$

and all that is left is to determine the constant c. We will not do this here, partly because the calculation has no particular interest; but especially because in multiplying the Poincaré metric by a strictly positive constant λ, the constant c is divided by λ^2. In fact, the choice of the constant in the definition of the Poincaré metric is mostly dictated by the desire to normalize the constant c to 1, which is possible because we know that c is strictly positive.

Here is a second, more convincing proof. Three points on the boundary of $\mathbb{D}$ determine an *ideal triangle* whose "vertices" are "at infinity" (see FIG. 1.15).

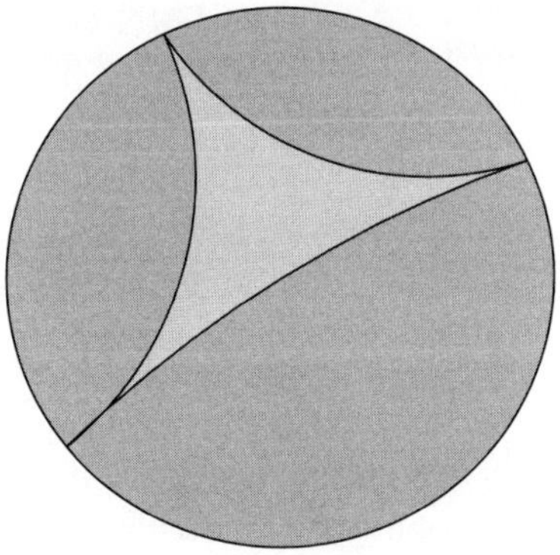

FIGURE 1.15. Ideal triangle.

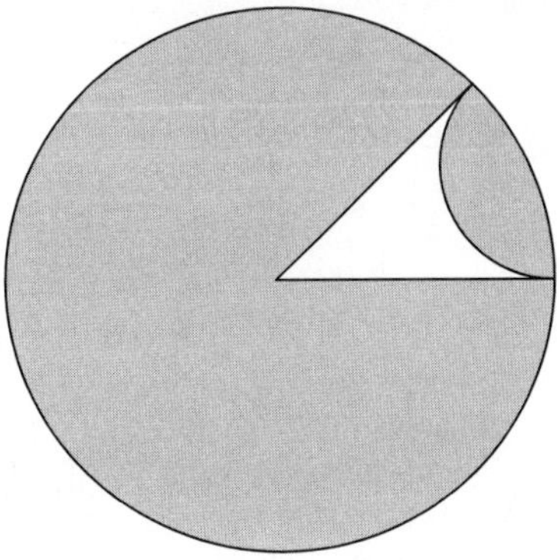

FIGURE 1.16. $T(\alpha)$.

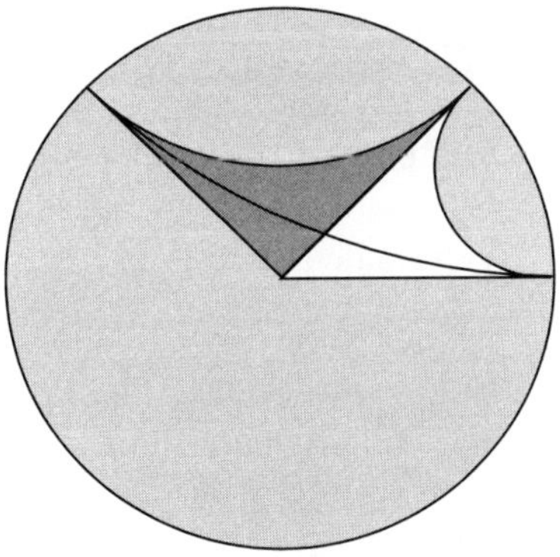

FIGURE 1.17. $T(\alpha+\beta)$.

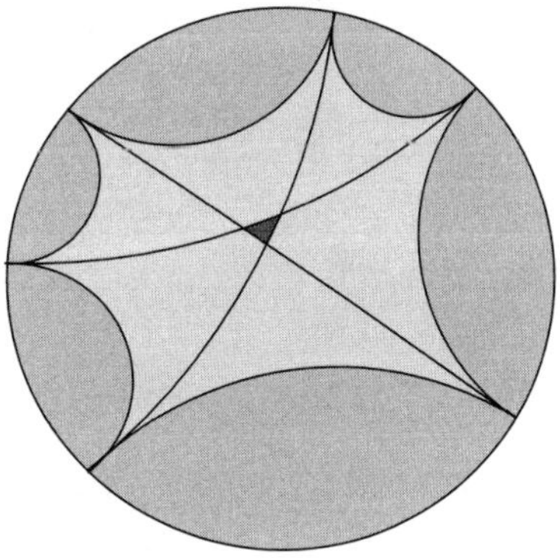

FIGURE 1.18. Ideal hexagon.

These ideal triangles, despite not being bounded, have bounded area. This follows from the fact that two geodesics converging to the same point on the boundary approach each other exponentially fast (in the hyperbolic sense, of course), so that the area integral converges. Since the group of isometries of the disk acts transitively on triplets of boundary points (exercise), all of these ideal triangles have the same area. One finds the value of this area to be π (calculate this!).

Now let us consider the triangle $T(\alpha)$ with two vertices at infinity and the remaining vertex having angle $\alpha \in (0, \pi)$ (FIG. 1.16). FIG. 1.17 shows that

$$\operatorname{area}(T(\alpha+\beta)) = \operatorname{area}(T(\alpha)) + \operatorname{area}(T(\beta)) - \pi,$$

from which it follows that $F(\alpha) = \pi - \operatorname{area}(T(\alpha))$ satisfies $F(\alpha+\beta) = F(\alpha) + F(\beta)$. Since F is continuous, there exists a constant c such that $F(\alpha) = c\alpha$. As $\operatorname{area}(T(\pi)) = 0$, we obtain $c = 1$ and $\operatorname{area}(T(\alpha)) = \pi - \alpha$.

Finally let us consider a "real" triangle $T(\alpha, \beta, \gamma)$ having all three vertices at "finite distance" (FIG. 1.18). By extending the sides to infinity, we obtain an ideal hexagon.

Each vertex of the triangle determines two isometric copies of $T(\alpha)$, $T(\beta)$, and $T(\gamma)$ respectively, one of which contains $T(\alpha, \beta, \gamma)$. The area of the hexagon is therefore

$$2[(\pi - \alpha) + (\pi - \beta) + (\pi - \gamma)] - 2\,\mathrm{area}(T(\alpha, \beta, \gamma)).$$

Moreover, as this hexagon can evidently be decomposed into four ideal triangles, its area is 4π. As announced before, we obtain

$$\mathrm{area}(T(\alpha, \beta, \gamma)) = \pi - (\alpha + \beta + \gamma).$$

I propose an "application exercise" to end this section. Euclid (almost) showed in the *Elements* that if two polygons P and Q in the Euclidean plane have the same area, one can cut each of them into a finite number of pieces $P_1, \ldots, P_n$ and $Q_1, \ldots, Q_n$ so that for all i the "pieces" P_i and Q_i are isometric [**32**, Book VI]. Show that the same is true in the Poincaré disk. This property is no longer true in higher dimensions and leads to fascinating developments concerning Hilbert's third problem: Can two polyhedra with the same volume in Euclidean space be cut into isometric pieces? See [**15**] for an elementary introduction and [**21**] for more information.

1.7. The disk is a tree

It follows from Gauss's formula that the area of a triangle is bounded by π. Strange geometry in which triangles can be arbitrarily large in size, but whose area is bounded!

The hyperbolic area of a disk of radius ρ is easy to calculate:

$$A(\rho) = \int_0^{\tanh \rho} \frac{2\pi t}{(1-t^2)^2}\, dt = \frac{\pi}{2}(\cosh \rho - 1).$$

We thus notice that this area tends to infinity as the radius tends to infinity (exponentially fast—we will return to this point). It follows from this that the radius ρ of the circle inscribed in a triangle is *bounded* independently of the triangle, since $\frac{\pi}{2}(\cosh \rho - 1) \leq \pi$ leads to $\rho \leq 1$ (FIG. 1.19). Consequently, in hyperbolic geometry, triangles are *thin*: each side is not very far away from the union of the two others.

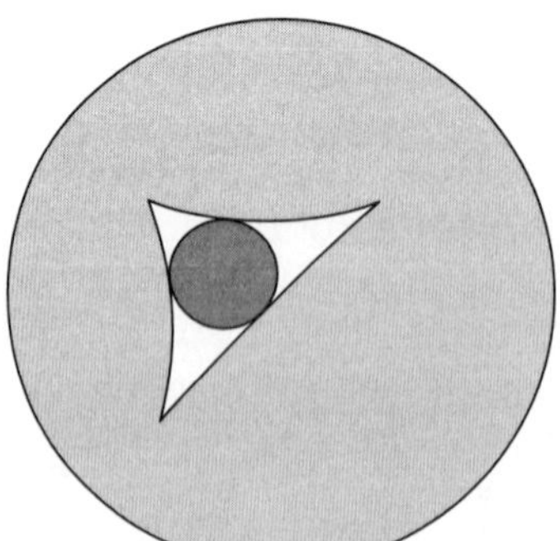

FIGURE 1.19. Inscribed circle.

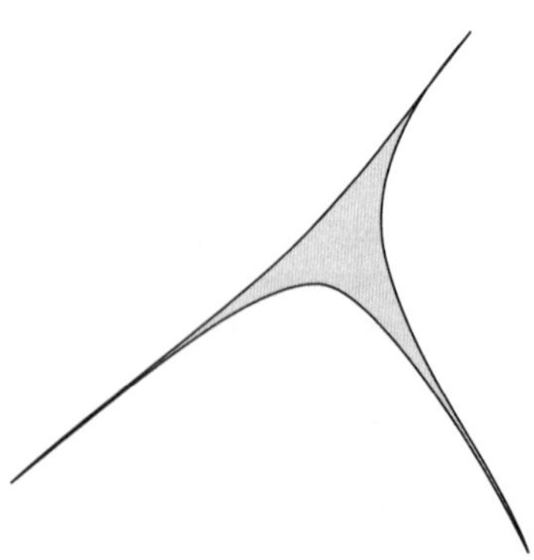

FIGURE 1.20. Thin triangle.

DEFINITION. We say that a geodesic metric space (X, d) is δ-hyperbolic if, for any triple of points (x, y, z) and any choice of geodesics $[x, y]$, $[y, z]$, $[z, x]$ pairwise connecting them, every point of $[x, y]$ is at a distance less than δ from a point of $[x, z]$ or of $[z, y]$.

Gromov extracted this geometric property from the Poincaré disk and recognized that this benign definition captures the essence of this geometry [**38**]. A bounded metric space (X, d) is of course δ-hyperbolic, with $\delta = \text{diam}(X, d)$, but this particular case evidently holds no interest. The theory is only interesting for unbounded spaces. We have seen, for example, that the Poincaré disk is δ-hyperbolic with $\delta = 2$. Numerous metric spaces are δ-hyperbolic and are therefore cousins of the disk. Regretfully, I will nonetheless remain in the disk!

Another way to express the δ-hyperbolicity of the Poincaré disk is to say that the union $[x, y] \cup [y, z] \cup [z, x]$ of the three sides of a triangle with vertices x, y, z is at a bounded distance from the union $[x, o] \cup [y, o] \cup [z, o]$, where o denotes the center of the inscribed circle: the union of the three sides is nearly a Y (see FIG. 1.20).

It turns out that this property generalizes to any finite subset of the disk—that it can be approximated by a finite tree,[14] with an error that depends only on the number of points.

Let us make this statement more precise. If we choose an arbitrary length for each edge of a finite tree, we can construct a metric realization of the tree: we attach Euclidean segments having the chosen lengths by their endpoints, following the combinatorics of the tree. By definition, the distance between two points is the length of the shortest path connecting them. We call a metric space constructed in this way a *metric tree* (see FIG. 1.21).

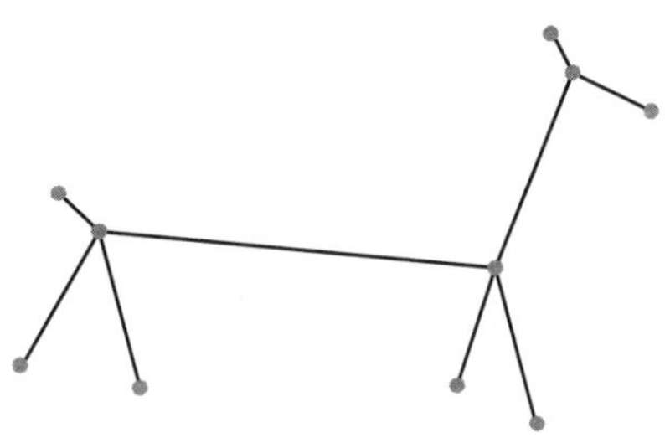

FIGURE 1.21. Metric tree.

PROPOSITION. *Consider n points $x_1, \ldots, x_n$ in a δ-hyperbolic metric space (X, d). Then there exists a metric tree (T, d_T) and n points $x'_1, \ldots, x'_n$ in T such that*

$$d(x_i, x_j) - 100\,\delta \log n \leq d_T(x'_i, x'_j) \leq d(x_i, x_j)$$

for all i, j.

In the photograph of "hyperbolic lettuce" (FIG. 1.22), one plainly sees arborescent nerves that approximate the whole leaf.

I reiterate that the "lack of isometry" $100\,\delta \log n$ depends (a little) on n but not on the points x_i, which could be very far away from each other.

[14] A tree is a connected graph with no cycles.

FIGURE 1.22. Hyperbolic lettuce.

The proof of this property is not difficult, but it is austere. The reader may begin by trying to construct it for himself, and then, in case of failure, he may go consult [**34, 38**] to appreciate the virtuosity with which Gromov manipulates the triangle inequality!

Thus, if we want to draw figures in the Poincaré disk formed from a large number of points very distant from each other, the result is nearly a tree. Very often, this gives a good intuition of hyperbolic geometry (FIG. 1.23).

FIGURE 1.23. Cactus approaching the geometry of the disk.

If (X, d) is a δ-hyperbolic metric space, one can divide the distance by a (large) constant $k > 0$ and set $d_k = d/k$. This amounts to looking at the space (X, d) "from far away", and the resulting metric space is of course (δ/k)-hyperbolic. If k approaches infinity, the defect $100(\delta/k)\log n$ of approximation by trees tends to 0 (n being fixed), that is to say "the space (X, d_k) tends to a tree." One can give a precise meaning to the previous sentence. I will not do so, however, because that would lead us to a discussion of Hausdorff–Gromov topology on the space of metric spaces, ultrafilters, etc. The interested reader may consult [**40**].

Whatever the case, let us retain that *the Poincaré disk, seen from far away, resembles a tree.*

1.8. Some examples of dendrologic intuition

Dendrologic[15] geometry is often intuitive. We shall see, using several simple examples, how it can guide our understanding of hyperbolic geometry.

The Pythagorean theorem. Let us compare right triangles in the Euclidean plane, in a tree, and in the disk.

When we say the triangle ABC "has a right angle at A", we mean that A is the foot of the altitude from C to the line (AB), or equivalently that A is the closest point to C and lying on (AB). In dendrologic geometry, right triangles are therefore those for which the shortest path from C to B passes through A, in other words those for which $BC = AB + AC$ (FIG. 1.24).

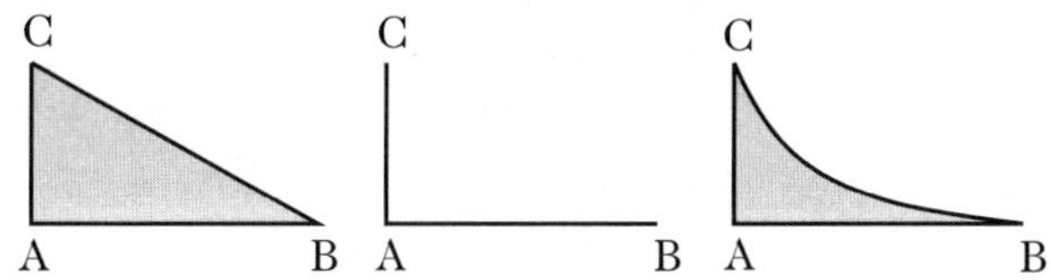

FIGURE 1.24. Right triangles.

In trees, the Pythagorean theorem has "lost its square": the hypotenuse is the sum of the sides. What is it for the Poincaré disk? One finds in books the hyperbolic Pythagorean theorem:

$$\cosh BC = \cosh AB \cdot \cosh AC.$$

If we take into account the fact that $\cosh x \simeq \exp(x)/2$ (for large real x), we find that $BC \simeq AB + AC$ as our dendrologic intuition had suggested.

Growth and transience. Because the Poincaré disk is homogeneous, we should compare it with a homogeneous tree. Let us consider, for example, the infinite homogeneous tree all of whose vertices have valence 3 (i.e., each is incident with three edges) (FIG. 1.25).

The ball of radius n centered at a point x_0 contains $3 \cdot 2^{n-1}$ vertices. We have rediscovered what we encountered before: the volume of a ball grows exponentially as a function of its radius.

Now let us consider a *random walk* in the tree. A point starts at the vertex x_0 and jumps randomly each second to one of the three neighboring vertices, with equal probability. Let us denote by $d(n)$ the distance between the point x_n at the moment n and the starting point x_0. If $d(n) \neq 0$, then $d(n+1) = d(n) + 1$ with probability $2/3$ and $d(n+1) = d(n) - 1$ with probability $1/3$. On the other hand, if $d(n) = 0$, then we have $d(n+1) = 1$. There is therefore a definite tendency to run off to infinity. It is not difficult to conclude that almost surely:

[15] DENDROLOGY from *dendro-* and *-logy*, n. The botanical study of trees and other woody plants. (*American Heritage Dictionary*) [TRANSLATOR'S NOTE: The original text refers to *Le Petit Robert*.]

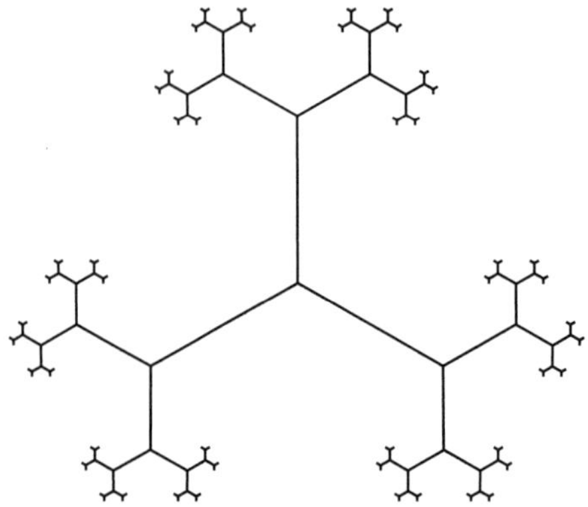

FIGURE 1.25. Homogeneous tree.

- $d(n)/n$ tends to $1/3$ (in particular $d(n)$ tends to infinity);
- the geodesic $[x_0, x_n]$ tends to an infinite geodesic $\gamma = [x_0, x_\infty)$ and the distance between x_n and γ is bounded by a constant multiple of $\log n$.

The same thing happens in the Poincaré disk for *Brownian motion*. For any Riemannian manifold M and for any point x_0, there exists a probability measure, called *Wiener measure*, on the space of paths $\gamma : \mathbb{R}^+ \to M$ starting from x_0 (see for example [**11, 28**]). For the Poincaré disk:

- almost every curve $\gamma : \mathbb{R}^+ \to \mathbb{D}$ starting from a fixed point converges to a point of the boundary $\omega(\gamma) \in \partial\mathbb{D}$;
- the hyperbolic distance between $\gamma(t)$ and the geodesic $[x_0, \omega(\gamma))$ is bounded by a constant multiple of $\log t$.

The proof is of course more difficult than in the case of trees, but the underlying idea is the same...

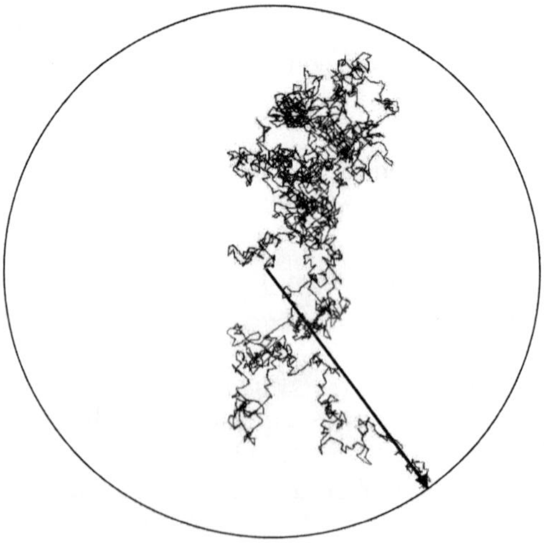

FIGURE 1.26. A Brownian path in the disk.

The fact that a random walk runs off to infinity can also be expressed in terms of the behavior of the solutions to the heat equation as time tends to infinity. D. Sullivan explained to me that it is almost impossible to heat houses in the Poincaré disk, because you cannot stop the heat from escaping to infinity!

Quasi-geodesics. One of the attractive features of hyperbolic geometry is its robustness. We shall illustrate it here in a simple but fundamental example.

A curve $\gamma : \mathbb{R} \to X$ in a metric space (X, d) is a *quasi-geodesic* if there exist a and $b > 0$ such that, for all t_1, t_2, we have:

$$a^{-1}|t_1 - t_2| - b \leq d(\gamma(t_1), \gamma(t_2)) \leq a|t_1 - t_2| + b.$$

THEOREM. *Every quasi-geodesic in a δ-hyperbolic metric space is within a bounded distance of a geodesic.*

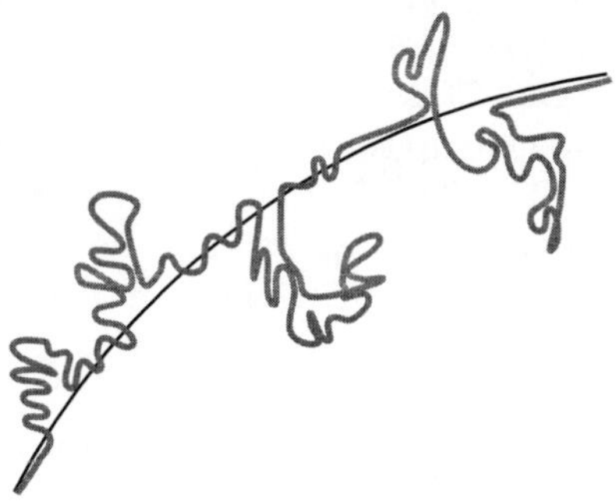

FIGURE 1.27. Quasi-geodesic.

The idea of the proof consists first in studying the case of a tree. A quasi-geodesic in a tree can cross several times over the same vertex, but these events cannot happen at very different times ($\gamma(t_1) = \gamma(t_2)$ leads to $|t_1 - t_2| \leq ab$). The result is that a quasi-geodesic in a tree consists in fact of a geodesic onto which are grafted some round trips of bounded length (FIG. 1.27). The general case is nothing but an adaptation of the particular case of the tree (see for example [**34**]).

Here is a sample application. We consider a Riemannian metric g on the disk whose ratio with the hyperbolic metric is bounded. This means that there exists a constant $\mathrm{Ct} > 1$ such that for any tangent vector v, the ratio between the g-norm and the hyperbolic norm of v is bounded by Ct^{-1} and Ct. Then a geodesic $\gamma : \mathbb{R} \to \mathbb{D}$ for the metric g is evidently a quasi-geodesic for the hyperbolic metric: it therefore remains within a bounded distance from a geodesic of the hyperbolic metric. This permits us a canonical way of associating a Poincaré geodesic to a geodesic[16] of g. This is the starting point for the phenomenon of *structural stability* of the geodesic flow of negatively curved compact manifolds: the qualitative behavior of geodesics does not depend (too much) on the choice of metric. Surrounding this is a long history which would carry us too far from the quasi-geodesic that we are trying to follow in this chapter, and would lead us from Hadamard to Gromov passing by Anosov [**5, 36, 43**]...

[16] In this text, a geodesic is by definition a curve that minimizes the length between any two points, even greatly distant ones. One normally speaks of *minimizing geodesics*.

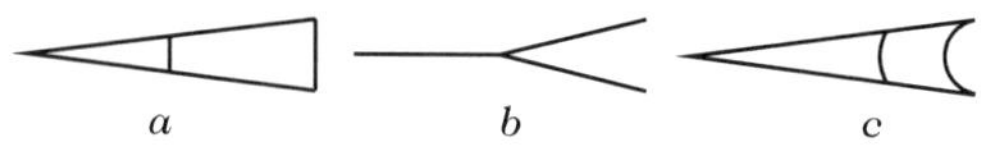

FIGURE 1.28. "Similar" triangles.

Sensitivity to initial conditions. Let us consider two geodesics $\gamma_1, \gamma_2 : \mathbb{R}^+ \to \mathbb{R}^2$ in the Euclidean plane (that is, two rays). Choose $\varepsilon > 0$ (small) and $T > 0$ (large). Suppose that γ_1 and γ_2 coincide at $t = 0$ and are close at time T, that is,

$$\mathrm{dist}_{\mathrm{Eucl}}(\gamma_1(T), \gamma_2(T)) \leq \varepsilon.$$

Then, by properties of similar triangles,[17]

$$\mathrm{dist}_{\mathrm{Eucl}}(\gamma_1(2T), \gamma_2(2T)) \leq 2\varepsilon.$$

In other words, after doubling the amount of time, the distance between $\gamma_1(t)$ and $\gamma_2(t)$ has only doubled: it therefore remains small (FIG. 1.28-a).

Now let us consider the case of a tree. FIG. 1.28-b shows two geodesics starting from the same point that coincide up to time T and diverge afterwards, in the sense that $\mathrm{dist}(\gamma_1(2T), \gamma_2(2T)) = 2T$. The fact that γ_1 and γ_2 remain close (and even coincide) on $[0, T]$ does not lead to any estimate for $\mathrm{dist}(\gamma_1(2T), \gamma_2(2T))$ (other than the triangle inequality). No similar triangles in dendrologic geometry!

The same phenomenon occurs in the disk. Let us consider two geodesics γ_1 and $\gamma_2 : \mathbb{R}^+ \to D$ such that $\gamma_1(0) = \gamma_2(0)$ and $\mathrm{dist}_{\mathrm{hyp}}(\gamma_1(T), \gamma_2(T)) = \varepsilon$. It is possible to calculate $F(\varepsilon, T) = \mathrm{dist}_{\mathrm{hyp}}(\gamma_1(T), \gamma_2(T))$ explicitly thanks to the formulas of hyperbolic trigonometry (see [**2, 24**]). One finds $F(\varepsilon, T) = 2\sinh^{-1}(2\sinh\frac{\varepsilon}{2}\cosh T)$. If ε and T are small, we recover similar triangles: $F(\varepsilon, T) \simeq 2\varepsilon$. In contrast, if $\varepsilon > 0$ is fixed and T tends to infinity, then $F(\varepsilon, T) \simeq 2T$, as for a tree (FIG. 1.28-c).

This is the simplest example of *sensitivity to initial conditions.* If one has a precise knowledge of a geodesic on an interval $[0, T]$ at one's disposal, it is impossible to deduce from this a precise knowledge on the interval $[T, 2T]$. The future seems to have forgotten the past. This is one of the most important ideas hiding behind the concept of *deterministic chaos.* The geodesics of the disk are deterministic in the sense that they are completely determined by their initial position and velocity, but their behavior is unpredictable in practice [**8, 26, 35, 43, 68**].

1.9. The disk is a curve

The unit disk in $\mathbb{C}$ has *complex* dimension 1 and the orientation-preserving isometries $f_{\alpha,a}$ which we encountered above are *holomorphic* in one complex variable. We find on the one hand that these $f_{\alpha,a}$ are the *only* orientation-preserving isometries, and on the other hand that they are equally the *only* holomorphic bijections of the disk. These two assertions are elementary but important: let us check them.

Let f be an orientation-preserving isometry of the Poincaré disk. By composing f with an appropriate $f_{\alpha,a}$, we may suppose that f leaves the origin fixed and that its derivative at this point is the identity. Being an isometry, f is also the

[17]TRANSLATOR'S NOTE: I have chosen to refer to similar triangles throughout this section where the original has "le théorème de Thalès", because the property being invoked is not commonly known by the name of Thales' theorem in English-speaking countries, at least in the U.S. and the U.K.

identity along the geodesic starting from 0 with an arbitrary initial direction v. Consequently, f is the identity everywhere, and the initial isometry was indeed of the form $f_{\alpha,a}$.

Let f be a holomorphic bijection of the disk. By composing f with an appropriate $f_{\alpha,a}$, we may suppose that f leaves the origin fixed. The classical Schwarz lemma then asserts that $|f(z)| \leq |z|$ for every point z of the disk [**3, 4, 53, 67**]. By considering the inverse f^{-1}, we obtain that in fact $|f(z)| = |z|$ and that f is a rotation. The initial holomorphic bijection was indeed of the form $f_{\alpha,a}$.

Thus, *the disk* $\mathbb{D}$ *can be endowed with an (oriented) metric structure and with a holomorphic structure whose automorphism groups coincide.* It is for this reason that the relationships between hyperbolic geometry (in real dimension 2) and holomorphic geometry (in complex dimension 1) are so close.

The omnipresence of the Poincaré disk as a metric space has a holomorphic counterpart: it concerns the famous *uniformization theorem*, probably one of the most beautiful mathematical jewels discovered in the 19th century, the result of efforts by Gauss, Riemann, Schwarz, Klein, Koebe, and Poincaré.

A *Riemann surface* is a holomorphic manifold of dimension 1. In other words, it is a topological space covered by open sets ("chart domains") homeomorphic to open sets in $\mathbb{C}$ so that the coordinate changes are holomorphic. Riemann surfaces are "curves" because they have complex dimension 1, but they are surfaces of real dimension 2. It is this curve-surface duality that gives the theory its flavor. The literature on this subject is immense, but I particularly recommend [**18, 23, 42, 50, 65**], among others...

A Riemann surface is *simply connected* if it is connected and if every closed curve on it can be continuously deformed to a point.

THEOREM (Uniformization theorem). *Let S be a simply connected Riemann surface. Then S is biholomorphically equivalent to the complex plane $\mathbb{C}$, the Riemann sphere $\mathbb{C} \cup \{\infty\}$, or the Poincaré disk $\mathbb{D}$.*

One must carefully distinguish this theorem from the particular case, taught in undergraduate courses, which asserts that a simply connected open set in the complex plane (non-empty and not all of $\mathbb{C}$) is biholomorphically equivalent to the disk. The uniformization theorem deals with an abstract Riemann surface, which cannot *a priori* be embedded in the plane. One may in particular apply it to the universal cover of an arbitrary Riemann surface, for example compact (that is, an algebraic curve, following Riemann). A proof can be found, for example, in [**42**]. This concerns one of Poincaré's initial motivations: a compact Riemann surface with genus greater than or equal to 2 can be identified with a quotient of the disk by a *Fuchsian group*, a discrete group of isometries.

FIG. 1.29 contains two illustrations of this kind of group, drawn by J. Leys [**27, 56**]; the second was inspired by the work of the artist M. C. Escher [**17, 25, 31**].

With regret, I will not discuss these groups, which however would deserve it. See for example [**25, 57, 60, 71**].

The connection between holomorphic structure and hyperbolic metric is summarized by Schwarz's lemma as expressed intrinsically by Pick. Every holomorphic function $f : \mathbb{D} \to \mathbb{D}$ is contracting for the hyperbolic metric: for any (z_1, z_2), we have $\mathrm{dist_{hyp}}(f(z_1), f(z_2)) \leq \mathrm{dist_{hyp}}(z_1, z_2)$. In truth, Pick's contribution was not limited to expressing Schwarz's lemma in an invariant way. He solved the difficult

FIGURE 1.29. Two Fuchsian groups.

problem of "holomorphic interpolation": If $z_1, \ldots, z_n$ and $w_1, \ldots, w_n$ are two n-tuples of points of the disk $\mathbb{D}$, under what condition does there exist a holomorphic map $f : \mathbb{D} \to \mathbb{D}$ such that $f(z_i) = w_i$ for all i? The necessary and sufficient condition is that the Hermitian matrix with coefficients $\frac{1-w_i\overline{w_j}}{1-z_i\overline{z_j}}$ be positive or zero. The necessity of this condition is not hard to see, but sufficiency is much more delicate (see for example [**1**]).

A *harmonic* function on the disk is a function $h : \mathbb{D} \to \mathbb{R}$ that is the real part of a holomorphic function [**3, 67, 72**]. Such a function is characterized by the fact that its value at a point is the average of its values on a circle centered at the point. The dendrologic analogue is a real function defined on the vertices of a tree, whose value at each vertex is the arithmetic mean of its values on the neighboring vertices. Let us take for example a harmonic function h on the infinite tree A all of whose vertices have valence 3. Let us now suppose that h takes positive values. Evidently, if a positive number a is the average of three positive numbers, each among them is at most equal to $3a$. In other terms, if $h : A \to \mathbb{R}^+$ is harmonic and if x and y are two neighboring vertices, one has $1/3 \leq h(x)/h(y) \leq 3$. We have just proved the dendrologic version of *Harnack's principle*: If $h : \mathbb{D} \to \mathbb{R}^+$ is harmonic, the hyperbolic norm of the gradient of $\log h$ is bounded by 1.

Now let us consider two points z_0 and z_1 in the disk. Let us define $\mathrm{dist}(z_0, z_1)$ as the upper bound of $\log(h(z_0)/h(z_1))$ over all positive harmonic functions h on the disk. Evidently, this defines a distance on the disk, homogeneous because the disk is holomorphically homogeneous. According to the characterization theorem, we find a constant multiple of the Euclidean plane or the hyperbolic plane. Of course, we have just discovered a new incarnation of the Poincaré distance! The reader may check for himself or consult [**9**].

1.10. Arrival at the boundary

We have seen that a random excursion in the disk ends almost surely on the boundary, and it is precisely on this boundary that we will end this chapter. The points of the boundary of the Poincaré disk are not *in* the disk, but the founders of hyperbolic geometry quickly took note of the importance of this boundary, which they christened the *absolute*. There are many intrinsic definitions of the absolute (which moreover generalize to δ-hyperbolic spaces). The simplest is the following: consider the set of *rays*—that is, isometric embeddings $\gamma : [0,\infty) \to \mathbb{D}$—and identify two such rays γ_1 and γ_2 if the distance $\mathrm{dist}_{\mathrm{hyp}}(\gamma_1(t), \gamma_2(t))$ is bounded. The quotient space is by definition the absolute $\partial\mathbb{D}$. It is not (too) difficult to endow the union $\mathbb{D} \cup \partial\mathbb{D}$ with a topology that makes it homeomorphic to a closed disk. The action of any isometry extends canonically to the boundary [**34**].

Lots of things happen at the boundary: I will content myself with illustrating them by keeping a promise made above and by naturally embedding the Poincaré disk in a Hilbert space. Given two distinct points u and v of $\partial\mathbb{D}$, there exists a unique oriented geodesic $\gamma \subset \mathbb{D}$ that tends at infinity towards u and v. Thus, the space $\mathcal{G}$ of pairs of distinct points of the boundary can be identified with the space of oriented geodesics, and the general theory of geodesics shows that this space possesses a natural volume (or rather an area in dimension 2); this is Liouville's theorem [**6, 7**]. In our case, it is not hard to identify this area. Consider the half-plane model, in which the boundary can be identified with $\mathbb{R} \cup \{\infty\}$. In these coordinates, the element of area is $\omega = du\,dv/(u-v)^2$, which one can also interpret as the cross ratio $-[u : v : u + du : v + dv]$. If I and J are two disjoint intervals of the boundary $\partial\mathbb{D}$, then the set of pairs (u, v) such that $u \in I$ and $v \in J$ has an area equal to the logarithm of the cross ratio of the four endpoints of the intervals.

This being posed, one may consider the Hilbert space $H = L^2(\mathcal{G}, \omega)$ of square-integrable functions on $\mathcal{G}$, on which the group of isometries of the disk acts isometrically. If z is a point of the disk, we define $\mathcal{G}_z \subset \mathcal{G}$ to be the set of pairs (u, v) such that the geodesic going from u towards v passes to the left of z. We remark that the indicator function $1_{\mathcal{G}_z}$ of $\mathcal{G}_z$ is not square-integrable. However, if z and z' are two points of $\mathbb{D}$, the difference $1_{\mathcal{G}_z} - 1_{\mathcal{G}_{z'}}$ is square-integrable (exercise). The L^2 norm of $1_{\mathcal{G}_z} - 1_{\mathcal{G}_{z'}}$ is ... $\mathrm{dist}_{\mathrm{hyp}}(z, z')^{1/2}$ (check this, without calculating!). There is then an obvious embedding of $\mathbb{D}$ into H. One chooses a base point z_0 in the disk and one sends the point z to $i(z) = 1_{\mathcal{G}_z} - 1_{\mathcal{G}_{z_0}} \in H$. It is now clear that $\|i(z) - i(z')\|^2 = \mathrm{dist}_{\mathrm{hyp}}(z, z')$ and that i is *natural*: any isometry f of $\mathbb{D}$ naturally defines an (affine) isometry $\bar{f}$ of H such that $i \circ f = \bar{f} \circ i$.

One inconvenience of the embedding i we have just constructed is that the equality $\|i(z) - i(z')\| = \sqrt{\mathrm{dist}_{\mathrm{hyp}}(z, z')}$ shows in particular that i is not differentiable, so that i is not an isometric embedding in the sense defined above! In 1932, Bieberbach constructed a *natural isometric embedding* of the disk into a Hilbert space [**13**]. Here is a modern presentation. One considers the Hilbert space H of holomorphic differential forms $\omega = f(z)\,dz$ on the disk, which are square-integrable: $\int_{\mathbb{D}} \omega \wedge \overline{\omega} < \infty$. If z, z' are two fixed points of the disk, the integration of holomorphic forms along a path joining z to z' defines a linear form on H and thus, by duality, a vector $V_{z,z'}$ in H. Let z_0 be a base point of the disk. The Bieberbach embedding consists in sending z to $j(z) = V_{z_0,z} \in H$. The naturality of j is evident because the holomorphic bijections of the disk evidently act by linear isometries of H. The fact that j is differentiable is an interesting exercise. An elementary (but not very interesting) calculation shows that $\|j(z) - j(z')\| = F(\mathrm{dist}_{\mathrm{hyp}}(z, z'))$ where $F(t) = \sqrt{2 \log \cosh t}$. In a neighborhood of 0, we have $F(t) \sim t$ so that the differential of j is indeed an isometry, as claimed. On the other hand, for large values of t, the difference between $F(t)$ and $\sqrt{2t}$ is small and the distortion of j for points distant from each other is of the same order of magnitude as for our first embedding i.

This property of the natural isometric embedding into a Hilbert space has important generalizations: a-T-menable groups, etc. (see [**10**] to learn more).

1.11. A few regrets...

Our excursion has not gone by so many places which would however have deserved the detour!

I could (should?) have explained what happens on the *inside* of the disk rather than limiting myself to a description of the disk, seen from the *outside*. The disk is in fact a special place in which one *does* functional analysis [**1**], complex analysis [**46, 67**], dynamical systems [**12**], number theory and modular forms on which one acts by Fuchsian groups [**57**], etc.

I should as well have gone farther. The disk has a numerous family. Of course, there are versions in all dimensions (hyperbolic balls) that have analogous properties. In addition, while searching for higher-dimensional Riemannian manifolds that have strong homogeneity properties, É. Cartan founded the theory of *symmetric spaces*, of which he gave a magnificent classification (see for example [**11, 29**]). Some of these spaces are moreover 2-homogeneous and deserve particular attention. The symmetric spaces also have combinatorial cousins: the *Bruhat–Tits buildings*, whose geometry contains just as many unbelievable riches [**19**].

And there are the spaces which are neighbors to the disk: negatively curved manifolds, δ-hyperbolic spaces and groups, etc. All this without forgetting the infinite-dimensional hyperbolic spaces about which Gromov made the following commentary in [**40**, p. 121]:

> *"These spaces look as cute and sexy to me as their finite-dimensional siblings but they have been for years shamefully neglected by geometers and algebraists alike."*

More excursions to come!

Appendix: Sketch of a proof of the characterization theorem

Let us a consider a homogeneous metric space (X, d) that is a surface, and let us denote by G its isometry group. This group has the topology of uniform convergence on compact subsets, which in fact makes it a locally compact group (by Ascoli's theorem).

Hilbert's fifth problem was solved in the middle of the 20th century by Montgomery and Zippin [**59**]. In its final form, the result established that a locally compact group that has a neighborhood of the identity without a non-trivial subgroup is a *Lie group*, which means that this group is a differentiable manifold and that the group structure $G \times G \to G$ is differentiable.

Replacing X by its universal cover, if necessary, we may begin by assuming that X is homeomorphic to the plane or the sphere (this uses the classification of surfaces, which was also a major event of 20th century mathematics).

I claim that G does actually contain a neighborhood of the identity without any non-trivial subgroup. For this, we use another difficult theorem, due to Kerékjártó [**47, 51**], according to which every compact group of homeomorphisms of the plane or the sphere is conjugate to a group of rotations (contained in O(2) or O(3) according to the case, the plane or the sphere). Since O(2) and O(3) obviously contain neighborhoods of the identity without non-trivial subgroups, the claim follows.

According to Montgomery and Zippin, the group G is therefore a Lie group, which we may assume to be connected. It acts transitively on X so that we can identify X with G/K, where K is the stabilizer of a point, compact of course, and therefore contained in O(2) in view of the previous result.

The group K may have dimension 0 or 1, and G has dimension 2 or 3. We are thus reduced to making a list of Lie groups of dimension 2 or 3, and in the second case looking for the compact subgroups isomorphic to O(2). This is not hard. Here are the possible results (always in the case where X is simply connected).

a) **G is two-dimensional and K is trivial**: the space X can be identified with $\mathbb{R}^2$ or with the affine group of transformations $x \mapsto ax + b$, $a > 0$.

b) **G is three-dimensional and K is isomorphic to O(2)**: the group G can be identified with the group of orientation-preserving isometries of the sphere (SO(3)), the Euclidean plane, or the Poincaré disk ($\mathrm{PSL}(2,\mathbb{R})$). In this case, the homogeneous space X can be identified with the sphere, the Euclidean plane, or the Poincaré disk.

If the metric space is 2-homogeneous, case **a)** cannot occur and we have indeed identified, not yet the metric space, but at least its group of isometries. It is not hard to show that, under the hypothesis that (X, d) is geodesic, it is in fact isometric to a constant multiple of the elliptic, Euclidean, or hyperbolic metric. To conclude, we must eliminate the hypothesis that X is simply connected. This is not hard once the universal cover has been identified. This ends the sketch of the proof of the characterization theorem which, as we have seen, costs a great deal in the sense that it uses many difficult things.

If we only keep the hypothesis of homogeneity (no longer assuming 2-homogeneity), we must also consider case **a)**. If $G \simeq \mathbb{R}^2$, it suffices to take a translation-invariant distance in the plane, for example any norm. In the same way, we may consider distances on the affine group that are invariant by translations on the left. These examples are well understood and clearly do not present the richness of the Euclidean plane, the sphere, and the Poincaré disk.

* * * * *

Thanks to Aurélien Alvarez, Pierre de la Harpe, Yann Ollivier, Martine Rufin and Bruno Sévennec for their detailed readings of this text as well as for their good advice.

Bibliography

[1] AGLER, J. & MCCARTHY, J.: *Pick interpolation and Hilbert function spaces.* Graduate Studies in Mathematics, **44**. American Mathematical Society, Providence, RI, 2002.

[2] ANDERSON, J.: *Hyperbolic geometry.* Springer Undergraduate Mathematics Series. Springer-Verlag London, Ltd., London, 1999.

[3] AHLFORS, L.: *Conformal invariants: topics in geometric function theory.* McGraw-Hill Series in Higher Mathematics. McGraw-Hill Book Co., New York-Düsseldorf-Johannesburg, 1973.

[4] AHLFORS, L.: *Complex analysis. An introduction to the theory of analytic functions of one complex variable*, third edition. International Series in Pure and Applied Mathematics. McGraw-Hill Book Co., New York, 1978.

[5] Anosov, D. V.: *Geodesic flows on closed Riemann manifolds with negative curvature.* Proceedings of the Steklov Institute of Mathematics, **90** (1967). American Mathematical Society, Providence, R.I., 1969.
[6] Arnold, V.: *Mathematical methods of classical mechanics*, Second edition. Translated from the Russian by K. Vogtmann. Springer Graduate Texts in Mathematics. Springer-Verlag, New York and Berlin, 1989.
[7] Arnold, V. & Avez, A.: *Problèmes ergodiques de la mécanique classique.* Monographies Internationales de Mathématiques Modernes, **9**. Gauthier-Villars, Éditeur, Paris, 1967.
[8] Barreau, H. & Noël, E.: *Le Hasard aujourd'hui.* Points Sciences, Seuil, 1991.
[9] Bear, H. S.: Part metric and Hyperbolic metric. *The American Mathematical Monthly*, **98**, 2 (1991), pp. 109–123.
[10] Bekka, B., de la Harpe, P., & Valette, A.: *Kazhdan's Property (T).* New Mathematical Monographs. Cambridge University Press, Cambridge, 2008.
[11] Berger, M.: *A panoramic view of Riemannian geometry.* Springer-Verlag, Berlin, 2003.
[12] Berteloot, F. & Mayer, V.: *Rudiments de dynamique holomorphe.* Cours Spécialisés, **7**. Société Mathématique de France, Paris, EDP Sciences, Les Ulis, 2001.
[13] Bieberbach, L.: Eine singularitätenfreie Fläche konstanter negativer Krümmung im Hilbertschen Raum. *Comm. Math. Helv.* **4** (1932), pp. 248–255.
[14] Blanuša, D.: Über die Einbettung hyperbolischer Räume in euklidische Räume. *Monatsh. Math.* **59** (1955), pp. 217–229.
[15] Boltianskiĭ, V.: *Hilbert's third problem.* Translated from the Russian by Richard A. Silverman. With a preface by Albert B. J. Novikoff. Scripta Series in Mathematics. V. H. Winston & Sons, Washington, D.C.; Halsted Press [John Wiley & Sons], New York-Toronto-London, 1978.
[16] Bonola, R.: *Non-Euclidean geometry, a critical and historical study of its developments.* Translation with appendices by H. S. Carslaw. Supplement containing the translation by G. B. Halsted of "The science of absolute space" by John Bolyai and "The theory of parallels" by Nicholas Lobachevsky. Dover Publications, Inc., New York, 1955.
[17] Bool, F. H., Kist, J. R., Locher, J. L., & Wierda, F.: *M. C. Escher, his life and complete graphic work.* Harry N. Abrahms, Inc., New York, 1982.
[18] Bost, J.-B.: *Introduction to compact Riemann surfaces, Jacobians, and abelian varieties. From number theory to physics.* (Les Houches, 1989), pp. 64–211, Springer, Berlin, 1992.
[19] Brown, K. & Abremenko, P.: *Buildings: Theory and Applications.* Graduate Texts in Mathematics. Springer-Verlag, New York, 2008.
[20] do Carmo, M.: *Differential geometry of curves and surfaces.* Translated from the Portuguese. Prentice-Hall, Inc., Englewood Cliffs, N.J., 1976.
[21] Cartier, P.: Décomposition des polyèdres: le point sur le troisième problème de Hilbert. Séminaire Bourbaki, Vol. 1984/85. *Astérisque* **133-134** (1986), pp. 261–288.
[22] Chebychev, P. L.: Sur la coupe des vêtements (1878), *Œuvres* II, Chelsea, 1962.
[23] Cohn, H.: *Conformal mapping on Riemann surfaces.* Reprinted from the 1967 edition. Dover Books on Advanced Mathematics. Dover Publications, Inc., New York, 1980.
[24] Coxeter, H. S. M.: *Non-Euclidean geometry*, sixth edition. MAA Spectrum. Mathematical Association of America, Washington, DC, 1998.
[25] Coxeter, H. S. M.: *Angels and devils. The Mathematical Gardner*, David A. Klarner, editor, Wadsworth International, 1981 (out of print). Reprinted in: *Mathematical Recreations: A Collection in Honor of Martin Gardner*, David A. Klarner, ed., Dover Publishers, 1998.
[26] Dahan-Dalmédico, A., Chabert, J.-L., & Chemla, K.: *Chaos et déterminisme.* Points Sciences, Seuil, 1992.
[27] Delahaye, J.-P.: Jos Leys, un artiste géomètre. *Pour la Science*, April 2006.
[28] Dynkin, E. & Yushkevich, A.: *Markov processes; Theorems and problems.* Translated from the Russian by James S. Wood. Plenum Press, New York, 1969.
[29] Eberlein, P.: *Geometry of nonpositively curved manifolds.* Chicago Lectures in Mathematics, University of Chicago Press, 1996.
[30] Efimov, N.: *Géométrie supérieure.* Translated from the Russian by E. Makho. Second edition. Mir, Moscow, 1985.
[31] Escher, M. C.: *Official M. C. Escher Web site*, published by the M. C. Escher Foundation and Cordon Art B.V. `http://www.mcescher.com/`

[32] EUCLID: *Elements.* Bilingual edition, with English translation (2007) by Richard Fitzpatrick, available for download at `http://farside.ph.utexas.edu/euclid.html`.
[33] GAUSS, K.: *Werke*, volume VII, Göttingen, 1900.
[34] GHYS, É. & DE LA HARPE, P.: *Sur les groupes hyperboliques d'après Mikhael Gromov.* (Bern, 1988), pp. 1–25, Progr. Math., **83**, Birkhäuser Boston, Boston, MA, 1990.
[35] GLEICK, J.: *Chaos. Making a new science.* Penguin Books, New York, 1987.
[36] GROMOV, M.: Three remarks on geodesic dynamics and fundamental group. Preprint SUNY (1976), reprinted in *L'Enseignement Mathématique* **46** (2000), pp. 391–402.
[37] GROMOV, M.: *Partial Differential Relations.* Springer-Verlag, 1986, Ergeb. der Math. 3. Folge, Bd. 9.
[38] GROMOV, M.: *Hyperbolic groups.* Essays in Group Theory, Mathematical Sciences Research Institute Publications **8** (1988), pp. 75–263, Springer-Verlag.
[39] GROMOV, M.: Sign and geometric meaning of curvature. *Rend. Semin. Mat. Fis. Milano* **61** (1991), pp. 9–123.
[40] GROMOV, M.: *Asymptotic invariants of infinite groups. Geometric group theory.* Volume 2, Proc. Symp. Sussex Univ., Brighton, July 14–19, 1991. *Lond. Math. Soc. Lecture Notes* **182**, Niblo and Roller ed., Cambridge Univ. Press, Cambridge, 1993, pp. 1–295.
[41] GROMOV, M.: Spaces and questions. GAFA 2000 (Tel Aviv, 1999). *Geom. Funct. Anal.* (2000), Special Volume, Part I, pp. 118–161.
[42] GUNNING, R. C.: *Lectures on Riemann surfaces.* Princeton Mathematical Notes. Princeton University Press, Princeton, N.J., 1966.
[43] HADAMARD, J.: Les surfaces à courbures opposées et leurs lignes géodésiques. *Journ. de Math.* **5** (4) (1898), pp. 27–73.
[44] DE LA HARPE, P.: On Hilbert's metric for simplices. *Geometric group theory.* Vol. 1 (Sussex, 1991), pp. 97–119, London Math. Soc. Lecture Note Ser., **181**, Cambridge Univ. Press, Cambridge, 1993.
[45] HILBERT, D.: *Grundlagen der Geometrie.* Thirteenth edition. Teubner Studienbücher Mathematik. B. G. Teubner, Stuttgart, 1987.
[46] HOFFMAN, K.: *Banach spaces of analytic functions.* Reprinted from the 1962 original. Dover Publications, Inc., New York, 1988.
[47] DE KERÉKJÁRTÓ, B.: Sur les groupes compacts de transformations topologiques des surfaces. *Acta Math.* **74** (1941), pp. 129–173.
[48] KLEIN, F.: *Vorlesungen über nicht-euklidische Geometrie.* Chelsea Publishing Co., New York, 1959.
[49] KLEIN, F.: *Development of mathematics in the 19th century.* With a preface and appendices by Robert Hermann. Translated from the German by M. Ackerman. Lie Groups: History, Frontiers and Applications, IX. Math Sci Press, Brookline, Mass., 1979.
[50] KLEIN, F.: *On Riemann's theory of algebraic functions and their integrals. A supplement to the usual treatises.* Translated from the German by Frances Hardcastle. Dover Publications, Inc., New York, 1963.
[51] KOLEV, B.: Note sur les sous-groupes compacts d'homéomorphismes de la sphère. Prépublication Marseille, 2005.
[52] KOLMOGOROV, A. N. & YUSHKEVICH, A. P.: *Mathematics in the 19th Century.* Geometry, Analytic function theory. Translated from the Russian, Birkhäuser Verlag, 1996.
[53] KRANTZ, S.: *Complex analysis: The geometric viewpoint.* The Carus Mathematical Monograph **23**, The Mathematical Association of America, 1990.
[54] KUIPER, N.: On C^1-isometric imbeddings I and II, *Indag. Math.* **17** (1955), pp. 545–556 and 683–689.
[55] LÉVY, P.: *Processus stochastiques et mouvement brownien.* Reprinted from the second edition (1955). Les Grands Classiques Gauthier-Villars. Éditions Jacques Gabay, Sceaux, 1992.
[56] LEYS, J.: Mathematical Imagery: `http://www.josleys.com/`
[57] MAGNUS, W.: *Noneuclidean tessellations and their groups.* Pure and Applied Mathematics, Vol. 61. Academic Press, New York-London, 1974.
[58] MCLACHLAN, R.: A gallery of constant-negative-curvature surfaces, *Mathematical Intelligencer* **16** (1994), pp. 31–37.
[59] MONTGOMERY, D. & ZIPPIN, L.: *Topological transformation groups.* Interscience Publishers, New York-London, 1955.

[60] MUMFORD, D., SERIES, C., & WRIGHT, D.: *Indra's pearls. The vision of Felix Klein.* Cambridge University Press, New York, 2002.
[61] NECHAEV, S. & VOITURIEZ, R.: On the plant leaf's boundary, "jupe à godets" and conformal embeddings. *J. Phys. A.* **34** (2001), no. 49, pp. 11069–11082.
[62] POINCARÉ, H.: Théorie des groupes fuchsiens, *Acta Mathematica* **1** (1882), pp. 1–62. *Œuvres*, tome II, pp. 108–168. Sur les fonctions fuchsiennes, *Acta Mathematica*, **1** (1882), pp. 193–294. *Œuvres*, tome II, pp. 169–257.
[63] POINCARÉ, H.: *La Science et l'hypothèse.* Paris, Flammarion, 1902.
[64] POPOV, E.: Geometric approach to Chebyshev net generation along an arbitrary surface represented by NURBS, `http://graphicon2002.unn.ru/demo/2002/`
[65] REYSSAT, E.: *Quelques aspects des surfaces de Riemann.* Progress in Mathematics, **77**. Birkhäuser Boston, Inc., Boston, MA, 1989.
[66] RUCKER, R.: *The fourth dimension: A guided tour of the higher universes.* Houghton Mifflin, 1984.
[67] RUDIN, W.: *Real and complex analysis.* Third edition. McGraw-Hill Science Engineering, 1986.
[68] RUELLE, D.: *Hasard et chaos.* Odile Jacob, 2000.
[69] SERRE, J.-P.: *Cours d'arithmétique.* Collection SUP: "Le Mathématicien" **2**, Presses Universitaires de France, Paris, 1970.
[70] The Institute For Figuring: `www.theiff.org/`
[71] THURSTON, W: *Three-dimensional geometry and topology*, Vol. 1. (Silvio Levy, ed.). Princeton Mathematical Series, **35**. Princeton University Press, Princeton, NJ, 1997.
[72] TSUJI, M.: *Potential theory in modern function theory.* Reprinted from the 1959 original. Chelsea Publishing Co., New York, 1975.

CHAPTER 2

Differential Equations with Algebraic Coefficients over Arithmetic Manifolds

NICOLAS BERGERON

"Hyperbolic geometry, founded on the hypothesis that the sum of the angles in a triangle is smaller than two right angles, at first seems to be nothing but a simple mental game that has no interest except for the philosopher, without being of any use to the mathematician. It is nothing of the sort. [...] I found myself, therefore, in the presence of a whole theory, conceived, it is true, for a philosophical goal, but whose every proposition, appropriately interpreted, provided me with a theorem applicable to ordinary Geometry. I found that by combining all of these theorems, I easily obtained the solution to the difficulty of which I spoke earlier."

Henri Poincaré
Analysis of the scientific work of Henri Poincaré by himself
Œuvres [**7**, vol. I, pp. X–XI]

2.1. Introduction

Pre-Poincaré, "non-Euclidean" geometries existed only as possibilities, unnecessary curiosities. Post-Poincaré, these geometries become necessities, they are involved in "normal" mathematics.

This question of existence in mathematics is obviously too rich for us to penetrate it more than superficially in this introduction. Let us specify, nonetheless, following the commentary of Lautman,[1] taken from [**6**] (by Joël Merker), that there are two meanings to the word **exist**, or more exactly to the question of existence, when applied to mathematics. There is, first of all, a "meta-theoretic" (metaphoric) sense that amounts to the possibility of an object's definition in connection with previous theories. The existence of hyperbolic geometry (due to Lobachevsky and Beltrami, at least for the construction of a model) is a typical example. As Merker emphasizes, "the thought of such a connection is indebted to a general philosophy of the possible." But there is also an "infra-theoretic" sense: defined within a theory, it applies to the objects of that theory. Merker's thesis is that the infra-theoretic meaning of existence of mathematical entities is not reducible to the concept of the possible. This is a first distinction from a **passage to the explicit** that can be assimilated to **a grasp of awareness on the order of concept** which one sees outlined in Lautman's genesis schemas. Even though the unthinkable passage from

[1]Albert Lautman, philosopher (1908–1944). "Essay on the notions of structure and existence in mathematics."

essence to existence is an essential philosophical challenge over which the metaphysicians have stumbled, Lautman presents it as a passage that does not truly pose any problem. For Lautman, this passage is indebted to the **understanding**, which becomes the source of the **genesis** of real theories. Lautman illustrates this genesis explicitly: his intent is to show that a being's achievement is asserted in his creative power. One aspect of this thesis that is most intimately related to the standard mathematical theories is the following: the essence of a material gives birth to the forms that its structure designs. As an example, Lautman notably addresses Riemann's theory of algebraic functions: in a way, the essence of algebraic functions of one variable constrains Riemann to associate to them a radically new object, which we call today the "Riemann surface" of the function.

In a similar class, we shall see that one of Poincaré's great merits is to have allowed hyperbolic geometry to exist in this infra-theoretic sense of which we have just spoken. The quote given in the epigraph shows clearly that, despite all his modesty, Poincaré is fully aware of this. He **derives** (in the sense of derive = provide a proof, demonstrate) the existence of hyperbolic geometry from the theory of algebraic differential equations. We therefore begin with a few reminders concerning integrals of algebraic functions.

When one goes around a singular point, the integrals of algebraic functions increase by a constant. For example, $\log x$, which is the integral of $1/x$, increases by 2π when one goes once around the singular point 0: this is the reason for the periodicity of the complex exponential (the inverse of the logarithm). But if the integral of $1/x$ or even the integral

$$\int_0^1 (1-x^2)^{-1/2}\,dx = \int_0^{\pi/2} (1-\sin^2\theta)^{-1/2}\cos\theta\,d\theta = \pi/2$$

are elementary, the equally innocent-appearing integral

$$\int_0^1 (1-x^4)^{-1/2}\,dx$$

(length of the lemniscate) is not. This latter is an elliptic integral. Nevertheless, as in the case of $\log x = \int dy/y$, it may be fruitful to consider the inverse functions of the integrals

$$\int_0^x (1-y^2)^{-1/2}\,dy \quad \text{and} \quad \int_0^x (1-y^4)^{-1/2}\,dy.$$

The elementary integral

$$\arcsin x = \int_0^x (1-y^2)^{-1/2}\,dy$$

maps the upper half-plane injectively into the shaded band in FIG. 2.1; in fact, by reflection in the punctured plane $\mathbb{C} - \{\pm 1\}$, we obtain a tiling of the complex plane in the range by non-overlapping images of the lower $(-)$ and upper $(+)$ half-planes. One thus rediscovers that the function defined by the integral has as an inverse a **sine** function $\sin x$, which is a uniform function of $x \in \mathbb{C}$ with period

$$2\pi = 4 \times \text{the complete integral} \int_0^1 \frac{dy}{\sqrt{1-y^2}},$$

which can therefore be viewed as a function defined on the complex cylinder $X = \mathbb{C}/L$, with $L = 2\pi\mathbb{Z}$:

$$x = \int_0^{\sin x} (1 - y^2)^{-1/2}\, dy,$$

up to periods.

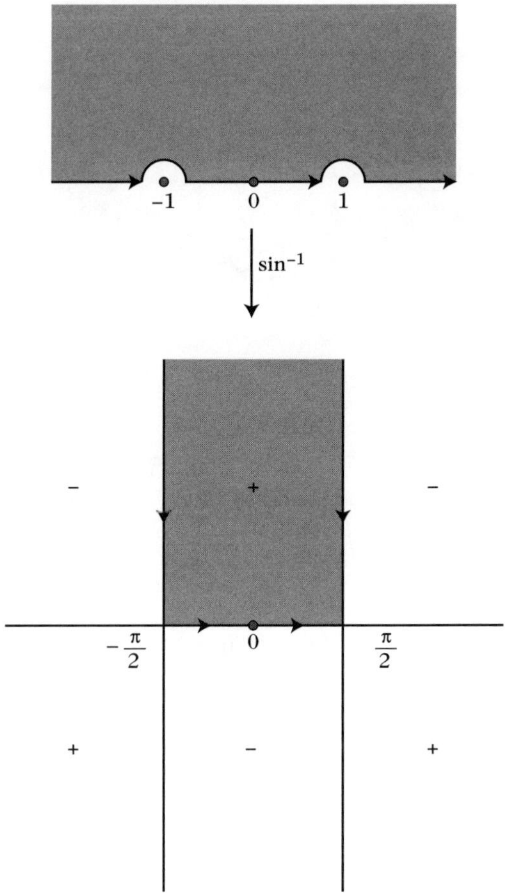

FIGURE 2.1. The arcsin function.

It is remarkable that even though the integral $\int (1 - y^4)^{-1/2}\, dy$ is not elementary, a similar phenomenon (discovered by Gauss and Abel) persists for the elliptic integrals

$$x \mapsto \int_0^x [(1 - y^2)(1 - k^2y^2)]^{-1/2}\, dy, \quad \text{with } 0 < k < 1 \text{ fixed.}$$

Here the square root appearing in the integral is chosen to be positive for $-1 < x < 1$ and extended by analytic continuation to the entire closed upper half-plane, avoiding the singular (branch) points ± 1, $\pm 1/k$ by infinitesimal half-circles as shown in FIG. 2.2. The expression under the integral is multiplied by i as one passes 1 or

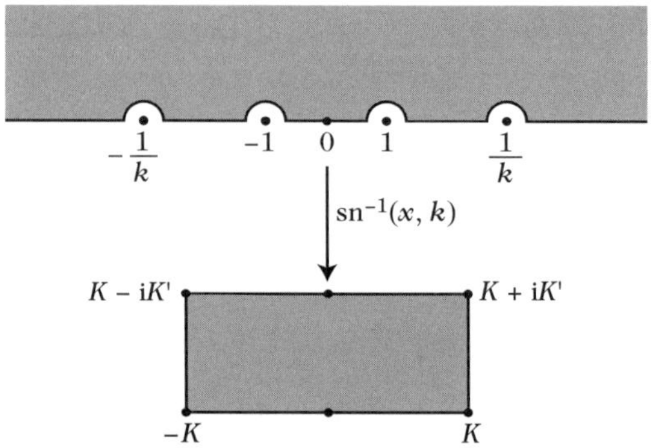

FIGURE 2.2. The function $y = \mathrm{sn}^{-1}(x, k)$ (defined in the text).

$1/k$ from the left to the right; as x runs along the real axis from 0 to infinity, its image begins by increasing from 0 to

$$K = K(k) = \int_0^1 [(1-x^2)(1-k^2x^2)]^{-1/2}\,dx,$$

then makes a $(+\pi/2)$-angle turn to climb up to the point $K + iK'$ with

$$K' = K'(k) = \int_1^{1/k} [(x^2-1)(1-k^2x^2)]^{-1/2}\,dx.$$

Finally, the image of x makes another $(+\pi/2)$-angle turn and returns to the left until it reaches the point iK' as indicated in FIG. 2.2. The image of the negative real axis is obtained by reflection of the image of $[0, +\infty)$ in the imaginary axis. The remarkable phenomenon is that these two images close up at $x = \pm\infty$, in such a way that the real axis forms exactly the perimeter of a rectangle. The integral therefore maps the (open) upper half-plane injectively onto the interior of this rectangle. By reflection in the punctured plane $\mathbb{C} - \{\pm 1, \pm 1/k\}$, we obtain a tiling of the complex plane in the domain by (isometric and non-overlapping) rectangles, the images of the lower $(-)$ and upper $(+)$ half-planes; cf. FIG. 2.3.

Let us now consider the inverse function, which Jacobi called the *sinus amplitudinus*, and which we denote $\mathrm{sn}(x, k)$. We therefore have

$$x = \int_0^{\mathrm{sn}(x,k)} [(1-y^2)(1-k^2y^2)]^{-1/2}\,dy,$$

up to periods. Four copies of the original rectangle from FIG. 2.2 are joined and shaded in FIG. 2.3; these large rectangles form a tiling of the complex plane. The sinus amplitudinus function maps each of these large rectangles in the same way on the complex projective line $\mathbb{P}^1 = \mathbb{C} \cup \{\infty\}$. The function $\mathrm{sn}(x, k)$ therefore admits two independent complex periods $\omega_1 = 4K(k)$ and $\omega_2 = 2iK'(k)$; it defines a uniform function with values in $\mathbb{P}^1$ on the torus $\mathbb{C}/L$, with $L = \mathbb{Z}\omega_1 + \mathbb{Z}\omega_2$; this is an *elliptic function*. This function is a degree 2 covering ramified over the points ± 1, $\pm 1/k$.

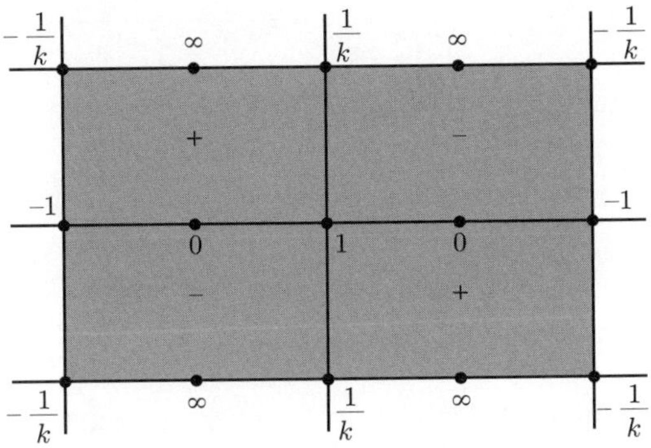

FIGURE 2.3. The four rectangles.

1. Elliptic functions

We have recalled in the Introduction that inverting **elliptic integrals of the first kind**

$$x \mapsto \int_0^x [(1-y^2)(1-k^2y^2)]^{-1/2}\,dy \quad \text{with } 0 < k < 1 \text{ fixed}$$

leads to considering doubly periodic meromorphic functions. Thus the sinus amplitudinus function defines a doubly periodic function whose periods are in the ratio $iK/2K$, where we always denote by

$$K = K(k) = \int_0^1 [(1-x^2)(1-k^2x^2)]^{-1/2}\,dx$$

and

$$K' = K'(k) = \int_1^{1/k} [(x^2-1)(1-k^2x^2)]^{-1/2}\,dx$$

the complete elliptic integrals of the first kind. It is not *a priori* evident that an arbitrary ratio can be thus obtained by an appropriate choice of modulus $k \in \mathbb{C} - \{0, \pm 1\}$. One may show, however, that this is indeed the case. Recall, more simply, how Weierstrass constructed a doubly periodic meromorphic function, having period lattice $\mathbb{L} = \omega_1\mathbb{Z} + \omega_2\mathbb{Z}$ (with ω_1 and ω_2 independent) and possessing a unique *double* pole modulo periods. The idea is simple: one places the pole at zero by considering the function x^{-2} and makes it periodic by summing its translates by the period lattice. This idea is too naïve because such a sum diverges; one may nonetheless save it by taking

$$\wp(x) = x^{-2} + \sum_{\omega \in \mathbb{L} - \{0\}} [(x-\omega)^{-2} - \omega^{-2}].$$

The subject of elliptic functions is riddled with marvelous identities, delicate to prove. The simplest example is perhaps the differential equation satisfied by the $\wp$-function:

$$(\wp')^2 = 4(\wp - e_1)(\wp - e_2)(\wp - e_3),$$

where $e_1 = \wp(\omega_1/2)$, $e_2 = \wp(\omega_1/2 + \omega_2/2)$, and $e_3 = \wp(\omega_2/2)$. Therefore the $\wp$-function inverts the integral

$$\frac{1}{2}\int_{\infty}^{x} [(y-e_1)(y-e_2)(y-e_3)]^{-1/2}\, dy.$$

This latter expression is another example of an **elliptic integral**, the notion under which one gathers the non-elementary integrals $\int F(x,y)\,dx$, where F is a rational function in x and $y = \sqrt{Q(x)}$, with $Q \in \mathbb{C}[x]$ of degree ≤ 4. Remark that in our case $Q(x) = (x-e_1)(x-e_2)(x-e_3)$ with e_1, e_2, and e_3 distinct, and the change of variable $x \mapsto x^2 + e_1$ transforms the differential dx/y into $2\,dx/\sqrt{(x^2+e_1-e_2)(x^2+e_1-e_3)}$. It therefore reduces the study of the elliptic integral above to the case where Q has degree 4, which itself reduces to the case of elliptic integrals of the first kind by mapping, via a projective transformation, the various roots of Q to ± 1 and $\pm 1/k$, for a certain $k \in \mathbb{C} - \{0, \pm 1\}$. One can thus verify that

$$\wp(x) = e_3 + \frac{e_1 - e_3}{\operatorname{sn}^2(x\sqrt{e_1-e_3}, k^2)},$$

where $k^2 = (e_2-e_3)/(e_1-e_3)$. This makes k^2 a function of the ratio of periods $\tau = \omega_2/\omega_1 = iK'/K$. In [**1**], Fuchs described a new method for studying K and K': if we set $w = k^2 = k^2(\tau)$, then the functions $\eta_1 = K/\sqrt{w}$ and $\eta_2 = iK'/\sqrt{w}$ in the variable w satisfy the *hypergeometric equation*

$$2w(w-1)\frac{d^2\eta}{dw^2} + 2(2w-1)\frac{d\eta}{dw} + \frac{1}{2}\eta = 0.$$

When $k^2 \in \mathbb{C}-\{0,1\}$, the ratio $\tau = \eta_2/\eta_1$ has a strictly positive imaginary part. We therefore obtain the function $\tau \mapsto k^2(\tau)$ as the inverse of the ratio η_2/η_1 of two independent solutions to the hypergeometric equation above. This is a single-valued function of τ, defined and holomorphic on the upper half-plane, invariant under the subgroup $\Gamma(2)$ of the modular group $\mathrm{SL}(2,\mathbb{Z})$ consisting of matrices congruent to the identity modulo 2.

Finally, let us remark that since $\operatorname{Im}(\tau) > 0$, the variable $q = e^{i\pi\tau}$ has modulus < 1, and the entire series in q converges rapidly. Such series appear in the theory of *theta-functions*, introduced by Jacobi in 1829. A typical theta series is

$$\vartheta_3(z,q) = 1 + 2\sum_{n=1}^{+\infty} q^{n^2}\cos(2nz),$$

which is an entire function of z. Jacobi showed that $\sqrt{k}$ is the quotient of two theta series at the point $z = 0$:

$$\sqrt{k} = \frac{\vartheta_2(0,q)}{\vartheta_3(0,q)}.$$

In an analogous manner, and following Fuchs, Poincaré considered the ratio ρ_1/ρ_2 of two independent solutions of an order 2 differential equation with algebraic coefficients. When one travels around a singular point, this quotient no longer undergoes a simple translation, but a projective transformation: inverting the function, one obtains, under certain conditions, a uniform (i.e., single-valued) function, as were the exponential or the elliptic functions, but this time invariant under a group of projective transformations (exactly those by which the quotient ρ_1/ρ_2 transforms when one goes around singular points). We will see that in order to study these functions (and, in the first place, even their existence), it is as though Poincaré were constrained to introduce a new concept of length in the upper half-plane. The very essence of the subject "order 2 differential equations with algebraic coefficients" thus gives birth to a new form of geometry, hyperbolic geometry, and its structure sketches this geometry. Although, in the case of elliptic integrals, Gauss and Abel had recognized a Euclidean symmetry (tiling the Euclidean plane by rectangles), Poincaré derived from this new problem an extremely rich (and surprising) symmetry, indebted to hyperbolic geometry. Our goal here is to follow Poincaré's own description of this genesis schema of hyperbolic geometry as he relates it in [**4**], while inserting the mathematics that are behind it. This will guide us to the study of arithmetic groups whose theory, at least that which concerns the geometric aspect, was born with the work of Poincaré.

2.2. Algebraic linear differential equations

We begin by describing Poincaré's motivations to study the discrete subgroups of the group $\mathrm{PGL}(2,\mathbb{C})$ of complex projective transformations. Consider a second-order linear differential equation

$$\eta'' + A(w)\eta' + B(w)\eta = 0, \tag{2.1}$$

where A and B are holomorphic functions of the variable w in a region S of the Riemann sphere $\mathbb{P}^1$. Given two linearly independent solutions $\eta_1(w)$ and $\eta_2(w)$ of this equation at a point w_0, any other solution is of the form

$$\eta(w) = a\eta_1(w) + b\eta_2(w), \tag{2.2}$$

where a and b are constants. Now consider a loop L in S based at the point w_0; if we analytically continue the functions η_1 and η_2 along L, we obtain a new pair (η_1', η_2') of solutions (called *monodromic*[2]) at w_0. This pair is, in general, different for the same reason that the integral of $1/x$ increases by 2π when we go once around the singular point 0 (cf. box 2, p. 54, for more details). According to (2.2),

$$\eta_1'(w) = a\eta_1(w) + b\eta_2(w),$$
$$\eta_2'(w) = c\eta_1(w) + d\eta_2(w),$$

with $ad - bc \neq 0$ (if η_1' and η_2' were linearly dependent, by going around the loop in the opposite direction to come back to the initial determinations η_1 and η_2 one would obtain that these, too, were linearly dependent). The ratio η_1/η_2 is therefore transformed into

$$\eta_1'/\eta_2' = \frac{a(\eta_1(w)/\eta_2(w)) + b}{c(\eta_1(w)/\eta_2(w)) + d},$$

that is, by a projective transformation. (In Poincaré's time, one said a "fractional linear substitution", or even simply a "linear substitution".[3]) The set of all transformations obtained in this way, i.e., by considering all the loops based at w_0, forms a group Γ that Poincaré called the *group of the differential equation*. If one begins

[2]Caution: in (η_1', η_2'), the prime is not the derivative!

[3]TRANSLATOR'S NOTE: The terminology of "fractional linear" maps persists in English usage.

with a different pair of solutions or base point, one obtains a conjugate of the group Γ by a fractional linear transformation.

If the region S is simply connected, then any solution of (2.1) at w_0 is analytic on all of S, and the group Γ is trivial. In general, the group Γ can be finite, infinite, or even a continuous subgroup of $\mathrm{PGL}(2, \mathbb{C})$.

2. Fuchsian differential equations

Consider equation (2.1). The solutions of such an equation are locally holomorphic in any plane domain in $\mathbb{C}$ where A and B are holomorphic. In a series of fundamental works, Fuchs studied such equations in a neighborhood of an **isolated** singular point $w = a$ of the equation. More precisely, he studied how a system of solutions is transformed by analytic continuation along a small loop L (oriented counterclockwise) around the singularity. We have recalled that such a transformation is governed by a fractional linear transformation and therefore by a matrix in $\mathrm{GL}(2, \mathbb{C})$. This matrix depends *a priori* on the choice of a system of solutions from which one begins, but it is not hard to verify that its conjugacy class in $\mathrm{GL}(2, \mathbb{C})$ is well defined. In particular, its eigenvalues are well-defined; let us denote them ω_1 and ω_2. We then distinguish two cases according to whether ω_1 and ω_2 are distinct or not.

In the first case there exists a fundamental system of solutions (η_1, η_2) transformed into $(\omega_1\eta_1, \omega_2\eta_2)$ by analytic continuation along L. Let us then set

$$\rho_1 = \frac{\log \omega_1}{2i\pi}, \qquad \rho_2 = \frac{\log \omega_2}{2i\pi},$$

where we do not fix the determination of the logarithm and therefore consider ρ_1 and ρ_2 modulo 1. The solutions η_1 and η_2 are thus of the form

$$\eta_1(w) = (w - a)^{\rho_1} V_1(w), \qquad \eta_2(w) = (w - a)^{\rho_2} V_2(w),$$

where V_1 and V_2 are two functions defined and single-valued in a neighborhood of the point $w = a$; they can be represented by convergent Laurent series in $w - a$.

In the second case there exists a fundamental system of solutions (η_1, η_2) transformed into $(\omega_1\eta_1, \beta\omega_1 + \omega_2\eta_2)$ by analytic continuation along L, where $\beta \in \mathbb{C}$. The solutions η_1 and η_2 are then of the form

$$\eta_1(w) = (w - a)^{\rho_1} V_1,$$
$$\eta_2(w) = (w - a)^{\rho_1} \left[V_3(w) + \frac{\beta}{\omega_1} \frac{1}{2i\pi} V_1(w) \log(w - a) \right],$$

where $V_3 = V_1 V_2$ and the V_j are convergent Laurent series in $w - a$.

It is natural to ask under what conditions the Laurent series V_j can be replaced by entire series. We say that an isolated singular point a is *regular singular* if the functions $w \mapsto V_j(w)$ $(j = 1, 2)$ are holomorphic at $w = a$ for an appropriate choice of ρ_j. The question is then to find a necessary and sufficient condition at $w = a$ so that it is a regular singular point. The following theorem, due to Fuchs, answers this question.

THEOREM. *A point $w = a$ of the equation* (2.1) *is a regular singular point if and only if*

(1) *the point a is an isolated singularity of A or B,*
(2) *the function $A(w)$ is regular or admits a simple pole at $w = a$, and*
(3) *the function $B(w)$ is regular or admits a pole of order ≤ 2 at $w = a$.*

To show that the conditions of the theorem are sufficient, we write the equation (2.1) in the form

$$(w-a)^2\eta''(w) + (w-a)p(w)\eta'(w) + q(w)\eta(w) = 0,$$

where $p(w)$ and $q(w)$ are holomorphic in a neighborhood of $w = a$. The second degree equation

$$s(s-1) + p(a)s + q(a)s = 0$$

is called the *indicial equation*: its roots are precisely ρ_1 and ρ_2.

We say that the equation (2.1) is *Fuchsian* if it has only a finite number of singular points and they are all (including the point ∞) regular singular points.

Poincaré's goal was to solve the general equation (2.1) when A and B are rational functions. As he mentioned in a letter to Klein, dated March 30, 1882 (cf. [**7**, volume II, pp. 106–107]), an important step was his discovery of Fuchs's work:

> *"As for M. Fuchs, in his Memoirs in Volumes 83 and 89 of Crelle's Journal,*[4] *he raised himself to a new point of view and shone the light on the narrow link that unites the theory of differential equations with that of certain uniform functions. It was the reading of these Memoirs that became the point of departure for my research."*

Fuchs, an analyst specializing in differential equations, applied himself in this second memoir to generalize the Jacobi inversion and thus to invert the integrals of a given differential equation. In this article, he considered in particular (and very briefly) the inverse function of the ratio $z = \eta_1(w)/\eta_2(w)$ of two solutions: $w = F(z)$. This function is in general multivalued. Fuchs considered more particularly a **Fuchsian** equation (2.1) (cf. note on Fuchsian differential equations) with A and B rational (every singularity of equation (2.1) is then a pole of order ≤ 1 for A or a pole of order ≤ 2 for B). We can thus take S to be the Riemann sphere $\mathbb{P}^1$ with the poles of A and B removed. We shall assume in what follows that the equation is in fact given in its canonical form

$$\frac{d^2\eta}{dw^2} = Q(w)\eta \tag{2.3}$$

with Q a **rational** function.[5]

3. Complex projective structures

Let S be a smooth surface without boundary. We recall that a *complex projective structure* σ on S is a maximal atlas such that the coordinate changes are realized by transformations in $\mathrm{PGL}(2,\mathbb{C})$. The Riemann sphere $\mathbb{P}^1 = \mathbb{C} \cup \{\infty\}$

[4] The first, in French, is the one in which Fuchs described his new method for studying K and K' (cf. box 1, p. 51); the second, in German, contains the "theorem" of inversion which we discuss in the following.

[5] We could just as well suppose that Q is only algebraic; we would then have to take for S the Riemann surface associated to Q with the singular points of Q removed.

is a trivial example of a complex projective structure: consider the coordinate z, defined everywhere except at the north pole, and the coordinate $1/z$, defined everywhere except at the south pole.

Every complex projective structure on S defines a local diffeomorphism dev from the universal cover $\tilde{S}$ to the Riemann sphere $\mathbb{P}^1$. Locally, the map dev is a complex projective diffeomorphism with respect to the complex projective structures on $\tilde{S}$ and $\mathbb{P}^1$. The map dev is called the *developing map* of σ; it is uniquely defined up to post-composition by a projective transformation. Consider the action of the fundamental group $\pi_1(S)$ on $\tilde{S}$ as a group of covering transformations. Then there exists a morphism $\rho : \pi_1(S) \to \mathrm{PGL}(2, \mathbb{C})$ such that dev is ρ-equivariant (that is, $\mathrm{dev}(\gamma \cdot x) = \rho(\gamma) \cdot \mathrm{dev}(x)$). The representation ρ is called the *monodromy representation* of the structure σ. This representation is unique up to conjugation in $\mathrm{PGL}(2, \mathbb{C})$. The group $\Gamma = \rho(\pi_1(S))$ is called the *monodromy group*. Conversely, if one has a local ρ-equivariant homeomorphism $\mathrm{dev} : \tilde{S} \to \mathbb{P}^1$, one can pull back by ρ the canonical complex projective structure of $\mathbb{P}^1$ to obtain a complex projective structure on $\tilde{S}$. The group $\pi_1(S)$ preserves this complex projective structure, which therefore descends to a complex projective structure σ on $S = \pi_1(S)\backslash\tilde{S}$. The map dev is the developing map of this structure.

An important class of complex projective structures are those obtained by *uniformization*. Suppose that $\Gamma \subset \mathrm{PGL}(2, \mathbb{C})$ is a discrete subgroup acting freely and properly on a domain $D \subset \mathbb{P}^1$. The canonical complex projective structure on D projects to a complex projective structure on $S = \Gamma\backslash D$, which is *uniformized* by Γ. In this case, the monodromy representation is a surjective morphism $\rho : \pi_1(S) \to \Gamma$ with kernel $\pi_1(D)$. The developing map is the covering $\tilde{S} \to D$.

There exist complex projective structures that are not of this type: notably, it could happen that the monodromy representation is not discrete, or that the image of the developing map is the entire Riemann sphere. To any complex projective structure σ on S corresponds a complex structure (on S). We only consider the projective structures σ that correspond to a Riemann surface of *finite type*, that is, a compact surface from which is removed a finite set of points, the *punctures*. We will say that σ is of *finite type* if for any puncture q of S and for a small disk U around q in S, the restriction of the developing map to U is either an everywhere-defined injective holomorphic map, or a map equivalent (that is, equal up to composition by diffeomorphisms in the domain and the range) to the logarithm map

$$\log : \{z \in \mathbb{C} : 0 < |z| < 1\} \to \mathbb{P}^1,$$

or a map equivalent to the map $z^{1/n}$, with n an integer such that $n \geq 2$ (conical singularity).

Let us return to the differential equation (2.3). Let S be the Riemann sphere $\mathbb{P}^1$ with the poles of Q removed. In a neighborhood of a point w_0 in S, the quotient η_1/η_2 defines a projective coordinate, and since the coordinate changes are realized by projective transformations, the equation (2.3) endows the surface S with a complex projective structure σ (cf. note above). The multivalued function

$w \mapsto \eta_1(w)/\eta_2(w)$ lifts (while always fixing the base point w_0) to the universal cover of S and thus defines a developing map of the complex projective structure σ. One can therefore think of the multivalued map $w \mapsto \eta_1(w)/\eta_2(w)$ as the developing map of the complex projective structure σ and of Γ as its monodromy group. A simple calculation shows that the function

$$z = \eta_1(w)/\eta_2(w) \tag{2.4}$$

satisfies the equation

$$\{z, w\} = -2Q(w), \tag{2.5}$$

where

$$\{z, w\} = \frac{z'''(w)}{z'(w)} - \frac{3}{2}\left(\frac{z''(w)}{z'(w)}\right)^2$$

is the *Schwarzian derivative.*[6]

If S is a Riemann surface of finite type (cf. box 2, p. 55), σ a complex projective structure of finite type on S, and $f = \mathrm{dev}_\sigma$, the developing map of σ, then the Schwarzian derivative of f defines what is now called a holomorphic quadratic differential on S. The fact that σ is of finite type implies that this quadratic differential has poles of order ≤ 2 at the punctures of S.

In a somewhat obscure article [**3**], Fuchs considers the equations (2.3) such that the roots of the indicial equation (see box 2 on p. 54) of each pole of Q are the inverses of integers. In modern terms, the rational function $-2Q(w)$ determines a holomorphic quadratic differential on S that (since the equation (2.3) is Fuchsian) has poles of order ≤ 1 at the punctures of S. Solving the equation (2.3) therefore allows us to solve the Schwarzian differential equation (2.5). However, every solution z of (2.5) lifts to a locally injective holomorphic map of the universal cover $\widetilde{S}$ to $\mathbb{P}^1$, which is unique up to post-composition by a projective transformation, and which is therefore ρ-equivariant (see box 2, p. 55) with respect to a morphism $\rho : \pi_1(S) \to \mathrm{PGL}(2, \mathbb{C})$ with image Γ. Fuchs's condition on the roots of the indicial equation at each pole of Q then immediately implies that the projective structure σ defined by the equation (2.3) is **of finite type**. And Fuchs then "shows" that w (otherwise called F) is a meromorphic function of x.

Poincaré found this result so intriguing that he devoted a memoir to it [**7**, vol. I, pp. 336–373], which he submitted to the Academie des Sciences for its Grand Prix. A corollary of Fuchs's result is that in fact the function F must be single-valued. For

---

[6]Recall that given four points x, y, z, t on the real projective line $\mathbb{P}^1(\mathbb{R}) = \mathbb{R} \cup \{\infty\}$, their cross-ratio $[x, y, z, t] = (x - z)(y - t)/(x - y)(z - t)$ is a "projective invariant", meaning that it is invariant when one applies the same transformation $u \mapsto (au + b)/(cu + d)$ (with $a, b, c, d \in \mathbb{R}$ and $ad - bc \neq 0$) to x, y, z, t. The Schwarzian derivative is a "local" projective invariant, obtained in the following manner. We take a small parameter ε, and for example R_ε the rotation by angle ε on the circle (to which we identify the real projective line). Then we take some x, followed by $y = R_\varepsilon x$, $z = R_\varepsilon y$, and $t = R_\varepsilon z$ (that is, x and its first three iterates). We consider the cross-ratio $[x, y, z, t]$: how does it change by a diffeomorphism f of the real projective line? The answer is that

$$[f(x), f(y), f(z), f(t)] = [x, y, z, t] - 2\{f, x\}\varepsilon^2 + o(\varepsilon^2),$$

which defines the Schwarzian derivative and gives its intuitive interpretation: it is a kind of projectively invariant second derivative. The Schwarzian derivative was first introduced in the theory of conformal transformations: it was discovered by Lagrange in his research on the construction of geographic maps. Schwarz knew this, but Cayley was unaware of Lagrange's discovery and gave it the name of "Schwarzian derivative".

Fuchs, who was more interested in local analytic problems (such as the condition on the indicial equation), this was not central. For Poincaré, who was interested in global properties of solutions to differential equations, the question:

"Is the function F single-valued?"

was primordial (it is a first form of uniformization of the projective structure). With this in view, Poincaré remarked that if (η_1', η_2') is a pair of monodromic solutions to (2.3) obtained by analytically continuing η_1 and η_2 along a certain loop, then

$$F(\eta_1'/\eta_2') = F(\eta_1/\eta_2);$$

in other words, F is invariant under the group action of Γ (i.e., F is a Γ-automorphic function). If we suppose that F is single-valued, then it is a (non-constant) holomorphic function on the image $\Omega \subset \mathbb{P}^1$, invariant under the action of Γ, with developing map $w \mapsto \eta_1(w)/\eta_2(w)$. The orbit of a point of Ω under Γ cannot have accumulation points in Ω, otherwise the function F would be constant (F is analytic). The group Γ must therefore act *properly* on Ω.

Such a group is discrete in $\mathrm{PGL}(2, \mathbb{C})$.[7] This point of view allowed Poincaré to distinguish three types of points:

(1) the points reached by $z = \eta_1(w)/\eta_2(w)$ when w follows a finite arc from the base point w_0 (the points inside Ω),
(2) the points reached by z when w follows an infinite arc, and
(3) the points that are never reached.

As Poincaré remarked, Fuchs's proof only works for the points in the closure D of Ω in the domain of discontinuity of Γ.

As we have said, although Fuchs was only looking to locally invert the integrals of a differential equation, Poincaré was interested in the global aspect of the solutions. Poincaré began by studying the case where D is the entire Riemann sphere and remarked that in this case the function F is rational.

Elliptic (doubly periodic) functions are important examples of Γ-automorphic functions for which $D = \mathbb{C}$ and Γ is isomorphic to the group $\mathbb{Z}^2$. Poincaré then wondered which elliptic functions could be obtained from a linear differential equation of the type (2.3). (In other words, which elliptic curves can one uniformize by the above method?) He showed that all elliptic functions can be obtained in this manner. His proof arises from the following way in which every elliptic function F allows the integration of a linear equation of order 2 with rational coefficients. If we set

$$w = F(z), \quad \eta_1 = \sqrt{\frac{dF}{dz}}, \quad \eta_2 = z\sqrt{\frac{dF}{dz}}, \tag{2.6}$$

then η_1 and η_2 satisfy the differential equation (2.3) for a certain rational function Q of w. It is an instructive exercise to verify all of this for the Weierstrass $\wp$-function (see box 1 on p. 51) and therefore to conclude the following theorem.

[7] In modern terms, Ω is therefore a Γ-invariant subset of the domain of discontinuity of Γ.

THEOREM. *Fix a lattice in $\mathbb{C}$. Let α, β, and γ be be the values of the Weierstrass $\wp$-function on the half-periods of this lattice. Then the differential equation*

$$\frac{d^2\eta}{dw^2} = \left[\frac{2\alpha-\beta-\gamma}{8(w-\alpha)} + \frac{2\beta-\alpha-\gamma}{8(w-\beta)} + \frac{2\gamma-\alpha-\beta}{8(w-\gamma)} - \frac{3}{16(w-\alpha)^2} - \frac{3}{16(w-\beta)^2} - \frac{3}{16(w-\gamma)^2}\right]\eta$$

(which has four singular points $\alpha, \beta, \gamma, \infty$) can be integrated (in the above sense) with the aid of the Weierstrass $\wp$-function.

It can happen that the domain D is not analytically isomorphic to either the Riemann sphere or the complex plane: for example, this is the case when D is the open unit disk in $\mathbb{C}$, or rather the Poincaré half-plane viewed as an open disk in the Riemann sphere. The set of projective transformations preserving the Poincaré half-plane is the group $\mathrm{PSL}(2,\mathbb{R})$. Poincaré therefore returns to the question of the existence of functions F as above, single-valued with D equal to the upper half-plane and invariant under a discrete subgroup of $\mathrm{PSL}(2,\mathbb{R})$. In honor of Fuchs, Poincaré would call these *Fuchsian* functions. As Poincaré remarked, these new transcendental functions would allow (in a manner analogous to elliptic functions) the integration of new linear differential equations of order 2. But Poincaré quickly realized that these transcendental functions (like the elliptic functions) were promises of an even brighter future, and his next articles would thenceforth start from them (and no longer from differential equations).

But before all that, he first had to construct them. As he relates in [**4**]:

> *"For fifteen days, I sought to prove that no function analogous to what I have since called Fuchsian functions could exist; I was thereby greatly ignorant; every day I sat at my work table, I spent an hour or two on it, I tried a large number of combinations, and I failed to arrive at any result."*

His first intuition was that Fuchsian functions could not exist. This was the beginning of a long series of false intuitions tied to the surprising symmetries of hyperbolic geometry.[8]

2.3. The step of an omnibus at Coutances

Very soon an agitated night came to completely upend Poincaré's first intuition:

> *"One evening, I had a cup of coffee, contrary to my usual habits, and I could not sleep: the ideas crowded around me; it felt as though they were beating against me, until two of them clung together, that is to say, they formed a stable combination. In the morning, I had established the existence of a class of Fuchsian functions, those which are derived from hypergeometric series; I had nothing left to do but write up the results, which only took me a few hours."* [**4**]

[8]In the same vein, before proving the first part of his famous *hyperbolization theorem* (for fibered manifolds), Thurston attempted to prove that a manifold that fibers over the circle can never be hyperbolic.

Recall that the *hypergeometric equation* (cf. box 1 on p. 51) considered by Euler and Gauss is the equation

$$w(1-w)\eta'' + [\gamma - (\alpha+\beta+1)w]\eta' - \alpha\beta\eta = 0, \tag{2.7}$$

where α, β, γ are real constants. The ratio z of two linearly independent solutions to (2.7) therefore satisfies the Schwarzian equation

$$\{z, w\} = \frac{1-\lambda^2}{2z^2} + \frac{1-\mu^2}{2(1-z^2)} + \frac{1-\lambda^2-\mu^2+\nu^2}{2z(1-z)}, \tag{2.8}$$

where the relations between the two sets of parameters are

$$\lambda = 1-\gamma, \quad \mu = \gamma - \alpha - \beta, \quad \nu = \alpha - \beta.$$

A solution to (2.8) maps the upper and lower half-planes onto two adjacent triangles whose sides are either straight lines or arcs of circles and whose angles are $\lambda\pi$, $\mu\pi$, and $\nu\pi$. Poincaré studied how these triangles were copied by analytic continuation around the singular points and succeeded in creating a satisfactory description when $1/\lambda$, $1/\mu$, and $1/\nu$ are integers. This hypothesis ensures that the triangles do not line up incorrectly, so that the inverse function $F(z)$ is single-valued. The most interesting case is where $\lambda + \mu + \nu < 1$. Poincaré showed that, in this case, in fact the triangles fill up the disk bounded by the circle orthogonal to the sides of the triangle. He called such figures "mixtilinear" and proved that they do not overlap by tracing out the figure on another disk where they have straight sides.[9]

One easily checks that if F is a Fuchsian function, the relations (2.6) permit the integration of an equation of type (2.3). Thus the Fuchsian function deduced from the hypergeometric equation enables the integration of this latter. But evidently in order to say that this function solves the equation, it must be independently defined.

> *"I wanted next to represent these functions as the quotient of two series; this idea was perfectly conscious and reflective; the analogy with elliptic functions guided me. I wondered what properties these series should have, if they existed, and I arrived without difficulty at forming the series that I called theta-Fuchsian."* [**4**]

The quotient of two theta-Fuchsian series is immediately a Fuchsian function. Poincaré therefore demonstrated the existence of these functions, found their analytic expression, and reduced the problem of constructing Fuchsian functions to the construction of discrete subgroups of finite covolume in $\mathrm{PSL}(2,\mathbb{R})$. The first proofs of these results are a bit obscure, but Poincaré was ripe for the intuition that was

[9]This case had previously been treated by Schwarz, as Klein later informed Poincaré. It was probably in reading Fuchs's article [**1**] (briefly described in the box on p. 51) that Poincaré was led to rediscover Schwarz's results. In fact, the proof that Poincaré had in his head at this point in his research does not seem to have been very rigorous—for that, he would need the help of hyperbolic geometry. Moreover, in a letter to Mittag-Leffler dated July 27, 1882, it was questionable what was the manner by which Schwarz proved this result: *"M. Schwarz therefore stated in this memoir a result of the highest importance, and it is this which I have cited. He gives no proof. There is in the proof of this result a very delicate point, a difficulty of a special nature; I am unaware of how M. Schwarz overcame it."* The problem to which Poincaré alluded founded the use of geometric techniques in the study of Fuchsian functions. He showed that this problem is equivalent to the determination of Fuchsian groups, which is to say, to finding a regular subdivision of the unit disk into infinitely many congruent regions. We shall see that it was by using hyperbolic geometry that Poincaré uncovered the notion of a generating polygon and brought to light the duality between polygons with certain properties and Fuchsian groups.

going to completely overturn the theory and whose revelation he described in the following famous passage [**4**]:

> *"At that time, I left Caen, where I was living, to take part in a geological course undertaken by the École des Mines. The events of this trip made me forget my mathematical work; once we arrived at Coutances, we boarded an omnibus for some excursion or another; the moment that I set foot on the footboard, the idea came to me, apparently without anything in my previous thoughts having prepared me for it, that the transformations I had used to define the Fuchsian functions were identical to non-Euclidean geometry. I did not carry out the verification; I would not have had time to do so, since, once I was barely seated on the omnibus, I took up again the conversation that had been started, but right away I was completely certain. Returning to Caen, I verified the result at my leisure in order to set my mind at rest."*

We have said it before: **this discovery completely changed the status of hyperbolic geometry.** It was no longer only logically possible, but equally natural, familiar, and useful in the theory of linear differential equations. Poincaré's marvelous intuition was in fact that the transformations in $\mathrm{PSL}(2,\mathbb{R})$ can be used to **define** a new concept of length, for which the different triangles obtained via the developing map that solves (2.8) have the same size. **The corresponding geometry is hyperbolic geometry.**

4. Hyperbolic geometry

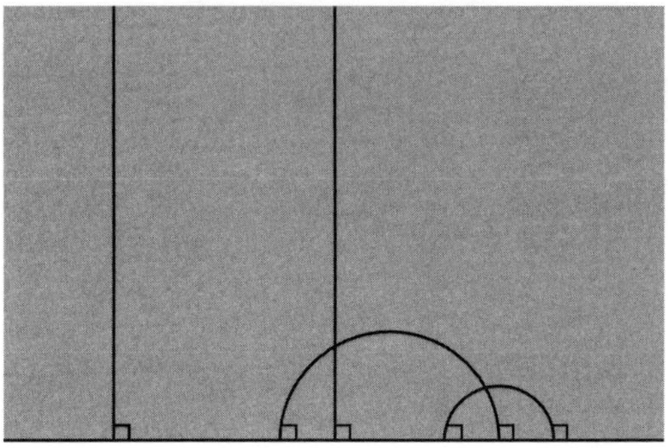

FIGURE 2.4. Geodesics of the hyperbolic plane.

The mathematical literature concerning non-Euclidean, or hyperbolic, geometry begins in 1829 with several articles by Lobachevsky. Nonetheless, during the first forty years of its history, hyperbolic geometry was a curiosity with no apparent connection to the rest of mathematics, and even lacking solid foundations. An important event in its history arrived in 1868 with the discovery by Beltrami (and starting from previous work by Riemann) of several *models* for hyperbolic geometry, such as the *upper half-plane* model, which permitted Lobachevsky's geometry to rest on a firm mathematical basis. The **hyperbolic**

plane can thus be thought of as the upper half-plane

$$\mathbb{H} = \{z = x + iy \in \mathbb{C} : y = \mathrm{Im}\,(z) > 0\}$$

endowed with the element of infinitesimal length

$$ds = \frac{\sqrt{dx^2 + dy^2}}{y}.$$

The lines (of shortest length) in this model are the usual vertical lines and the half-circles meeting the real axis $\mathbb{R}$ at a right angle (cf. FIG. 2.4).

Ten years later, Poincaré reintroduced Beltrami's half-plane model in his work on Fuchsian functions and used it to identify the group of orientation-preserving isometries of the hyperbolic plane with the group, now denoted $\mathrm{PSL}(2,\mathbb{R})$, of projective transformations

$$z \mapsto (az + b)/(cz + d)$$

with real coefficients and determinant $ad - bc = 1$. Poincaré reduced the construction of Fuchsian functions to the construction of discrete (sufficiently large: of "finite covolume") subgroups of $\mathrm{PSL}(2,\mathbb{R})$, or rather to the construction of tilings of the hyperbolic plane $\mathbb{H}$ by pairwise congruent (that is, all translations by a projective transformation in $\mathrm{PSL}(2,\mathbb{R})$) finite polygons (perhaps with vertices "at infinity" $\mathbb{R} \cup \{\infty\}$). Poincaré remarked that such examples of discrete groups had already been considered, notably by Fuchs in [**1**]. FIG. 2.5 represents the tiling associated to the Fuchsian function k^2 (see box 1 on p. 51): one can compare it to FIG. 2.2 to measure the richness of the symmetry that Poincaré derived from his work on differential equations.

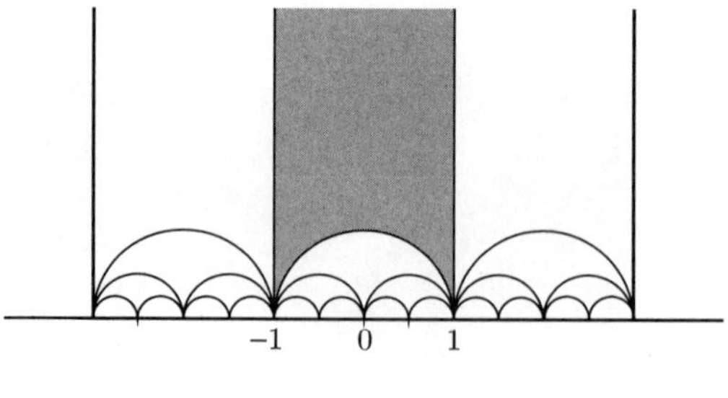

FIGURE 2.5. Tiling of the hyperbolic plane by ideal triangles.

Once guessed, this fact is easily verified, particularly in the half-plane model, due to Beltrami (see box 4 above). Poincaré immediately used hyperbolic geometry to give a rigorous proof that the triangular decomposition of the disk is possible, by showing that each point of the disk is really in a triangle. Recall that he did not know Schwarz's memoir on the hypergeometric equation at the time.

He also showed that the theta-Fuchsian series associated to a discrete subgroup of finite covolume in $\mathrm{PSL}(2,\mathbb{R})$ are indeed convergent, by an ingenious reasoning concerning hyperbolic area and the hyperbolic perimeter of the region composed of transforms of the triangles in a hyperbolic circle of increasing radius. Now the task was to construct other **Fuchsian groups** than those associated to the hypergeometric equation.

2.4. Quadratic forms and arithmetic surfaces

"I then put myself to the task of studying arithmetic questions without any great apparent result and without suspecting that this could have the least relation with my previous research. Disgusted with my lack of success, I went to spend a few days on the seaside, and I thought about other things. One day, while I was walking along the sea cliff, the idea came to me, again with the same characteristics of brevity, suddenness, and immediate certitude, that the arithmetic transformations of indefinite ternary quadratic forms were identical to those of non-Euclidean geometry." [**4**]

The number-theoretic questions that thus interested Poincaré dealt with the theory of quadratic forms and traced back essentially to two sources. The first is a theorem surely known to Archimedes (cf. the problem of the oxen), partially known to the Indian mathematician Brahmagupta (7th century), and whose first written proof is due to another Indian mathematician: Bhaskara (12th century). It is the solution to the so-called Pell's equation. But it is to Fermat (who could not have known the solutions of his predecessors) that modern Western mathematicians owe the recognition of the importance of arithmetically studying quadratic forms and the solution of Pell's equation.

THEOREM 2.4.1 (Fermat). *Pell's equation*

$$x^2 - ay^2 = 1 \quad \textit{with } a \in \mathbb{N},\ \sqrt{a} \notin \mathbb{N}$$

has infinitely many solutions $(x, y) \in \mathbb{Z}^2$.

Let us remark that this theorem is really non-trivial: for example, the smallest non-trivial ($y \neq 0$) integer solution to the equation $x^2 - 94y^2 = 1$ is

$$(2143295, 221064).$$

The other source is Gauss's *theory of reduction of quadratic forms*, developed by Lagrange and then Hermite. It consists of the study of orbits of the group $\mathrm{GL}(n, \mathbb{Z})$ on quadratic forms in n variables.[10] Let X be the set of positive definite quadratic forms; we then look for a fundamental set $D \subset X$ for the action of $\mathrm{GL}(n, \mathbb{Z})$ on X, that is, a "small" subset D of X whose orbit under the action of $\mathrm{GL}(n, \mathbb{Z})$ is X (in other words, $\mathrm{GL}(n, \mathbb{Z}) \cdot D = X$).

In his *Disquisitiones Arithmeticae*, Gauss considered the case $n = 2$. Every positive definite quadratic form can be written in a unique way as

$$q(x, y) = a(zx + y)(\overline{z}x + y), \tag{2.9}$$

where $a > 0$ and z belongs to the Poincaré half-plane $\mathbb{H} = \{z : \mathrm{Im}\,(z) > 0\}$. The action of positive scalars on X commutes with that of $\mathrm{GL}(2, \mathbb{Z})$, and (2.9) implies that the quotient $X/\mathbb{R}_+^*$ is isomorphic to the half-plane $\mathbb{H}$. The group $\mathrm{SL}(2, \mathbb{Z})$ acts by projective transformations on $\mathbb{H}$, and Gauss proved the following famous theorem.

[10]For the action given by $(A \cdot q)(x) = q({}^tAx)$. This is a natural action, from the point of view of number theory, since two quadratic forms in the same orbit have the same integer values.

THEOREM 2.4.2 (Gauss). *Let* $D = \{z \in \mathbb{H} : |z| \geq 1$ *and* $|\mathrm{Re}\,(z)| \leq 1/2\}$. *Then*

$$\mathbb{H} = \mathrm{SL}(2,\mathbb{Z}) \cdot D.$$

The forms corresponding to $z \in D$ are called *reduced.*

More generally, when $n \geq 2$, every positive definite quadratic form q in n variables can be written in a unique way in the following form:

$$q(x) = t_1(x_1 + u_{12}x_2 + u_{13}x_3 + \cdots + u_{1n}x_n)^2 + t_2(x_2 + u_{23}x_3 + \cdots + u_{2n}x_n)^2 + \cdots + t_n{x_n}^2, \tag{2.10}$$

where $t_1, \ldots, t_n$ are positive real numbers and $u_{ij} \in \mathbb{R}$.

THEOREM 2.4.3. *After replacing* q *with* $A \cdot q$ *for some matrix* $A \in \mathrm{GL}(n,\mathbb{Z})$, *we may assume that*

$$|u_{ij}| \leq \frac{1}{2} \quad \textit{when} \quad 1 \leq i < j < n$$

and

$$t_i \leq \frac{4}{3}t_{i+1} \quad \textit{when} \quad 1 \leq i \leq n-1.$$

The subset $\mathfrak{S}$ of X defined by the inequalities of Theorem 2.4.3 is called the *Siegel set.* When $n = 2$, Theorem 2.4.3 follows from Gauss's theorem.

Gauss's theory of reduction does not immediately extend to indefinite quadratic forms due to the fact that the group $\mathrm{SL}(n,\mathbb{Z})$ does not act properly on the space of indefinite quadratic forms. Nonetheless, using an ingenious technique, Hermite developed an analogous theory for $n = 3$ that applies to indefinite forms in three variables, such as $x^2 + y^2 - z^2$, as soon as one restricts to transformations that preserve this quadratic form. More precisely, let q be an indefinite quadratic form in three variables with integer coefficients and the associated bilinear form φ_q. To every point ξ of the upper sheet $\mathcal{H}_q$ of the hyperboloid $q = -1$, Hermite associates the positive definite quadratic form

$$q[\xi](\,\cdot\,) = q(\,\cdot\,) + 2\varphi_q(\xi,\,\cdot\,)^2.$$

(As ξ varies, one obtains the set of Hermite upper bounds on the form q.) Gauss's reduction theory thus applies to each of the quadratic forms $q[\xi]$. The quadratic form q is called *reduced* (in the sense of Hermite) if one of the forms $q[\xi]$ is reduced in the sense of Gauss. This definition is motivated by the following finiteness lemma, proved by Hermite.

LEMMA 2.4.4. *The set of reduced integral quadratic forms with non-zero determinant is finite.*

The special nature of indefinite forms in three variables led Poincaré to discover a new model of the hyperbolic plane, the *hyperboloid model.*[11] As he described in the above quote, Poincaré had the intuition that arithmetic transformations of indefinite quadratic forms in three variables are identical to the transformations of non-Euclidean geometry. The subject thus takes on a new dimension by opening the way to a geometric study of what are now called *arithmetic groups.*

Let us return to the indefinite quadratic form q with integer coefficients in three variables. Denote by $G = O(q)$ the subgroup of $\mathrm{GL}(3,\mathbb{R})$ consisting of matrices that stabilize the quadratic form q, and by $\Gamma = G \cap \mathrm{GL}(3,\mathbb{Z})$ the corresponding

---

[11]The only one really due to Poincaré!

arithmetic group. The group Γ is a Fuchsian group[12] which Poincaré proposed to study.

Theorem 2.4.3 provides a fundamental set Σ for the action of the group $\mathrm{GL}(n,\mathbb{Z})$ on the space of positive definite quadratic forms. If $A \in \mathrm{GL}(n,\mathbb{Z})$, we denote by $A\cdot\Sigma$ the image of the fundamental domain Σ under A. The geometric representation that Poincaré made of Hermite's results is formulated in the following theorem (cf. [**7**, vol. II: *Les fonctions fuchsiennes et l'arithmétique*, p. 474]).

THEOREM 2.4.5. *There exist m transformations $A_1, \dots, A_m \in \mathrm{GL}(3,\mathbb{Z})$ such that the subset $\Omega \subset \mathcal{H}_q$ consisting of the points ξ satisfying*

$$q[\xi] \in \bigcup_{i=1}^{m} A_i \cdot \Sigma$$

is a fundamental set for the action of the group Γ on $\mathcal{H}_q$.

PROOF. Since q is integral and all the forms in the orbit of q by $\mathrm{GL}(3,\mathbb{Z})$ have the same determinant, by lemma 2.4.4 there is only a finite number among them which are reduced, say $A_1^{-1}\cdot q, \dots, A_m^{-1}\cdot q$. We will show that the subset $\Omega \subset \mathcal{H}_q$ defined by

$$q[\Omega] = \left(\bigcup_{i=1}^{m} A_i \cdot \Sigma\right) \cap q[\mathcal{H}_q]$$

is a fundamental set.

First let us show that every orbit of Γ in $\mathcal{H}_q$ meets Ω. Let ξ be a point of $\mathcal{H}_q$. To it corresponds the positive definite quadratic form $q[\xi]$. According to the theory of reduction for definite forms, there exists $A \in \mathrm{GL}(n,\mathbb{Z})$ such that $(A\cdot q)[{}^tA^{-1}\xi] = A\cdot(q[\xi]) \in \Sigma$. The quadratic form $A\cdot q$ is therefore reduced in the sense of Hermite, and hence equal to $A_i^{-1}\cdot q$ for some i. Thus $A_iA \in O(q)\cap \mathrm{GL}(3,\mathbb{Z}) = \Gamma$, and $q[{}^tA^{-1}\,{}^tA_i^{-1}\,\xi] = (A_iA\cdot q)[{}^tA^{-1}\,{}^tA_i^{-1}\,\xi] = A_iA\cdot(q[\xi]) \in A_i\cdot\Sigma$: the orbit of ξ under Γ does indeed meet Ω.

Finally, the fact that Σ is itself a fundamental set for $\mathrm{GL}(3,\mathbb{Z})$ implies that Ω is "small". □

One can prove that the volume of such a fundamental set is finite, which is what Poincaré did. We will content ourselves with proving the following theorem, also due to Poincaré.

THEOREM 2.4.6. *Suppose that the quadratic form q does not represent* 0 *over* $\mathbb{Q}$. *Then the group Γ is identified with a discrete subgroup of* $\mathrm{PSL}(2,\mathbb{R})$ *(a Fuchsian group) that is* **cocompact**.[13]

PROOF. The image $q[\mathcal{H}_q]$ in X is a (natural) orbit of G in X. In modern terminology, we see that the quotient $\Gamma\backslash G$ is immersed in the space $\mathrm{GL}(3,\mathbb{Z})\backslash\mathrm{GL}(3,\mathbb{R})$ (which is also the space of lattices in $\mathbb{R}^3$) by looking at the orbit of the standard lattice $\mathbb{Z}^3$ under G. Theorem 2.4.5 implies that this image is closed.[14] The goal then is to show that the image is relatively compact. But according to Theorem 2.4.3,

[12] In order to avoid heavy notation, we will also denote by Γ the corresponding subgroup of $\mathrm{PSL}(2,\mathbb{R})$, which acts on $\mathbb{H}$.

[13] In other words, the quotient $\Gamma\backslash\mathbb{H}$, equipped with the quotient topology, is compact.

[14] This could moreover easily be shown directly.

for a sequence Λ_i of lattices in $\mathbb{R}^n$ there are only two ways to go to infinity:

- either by allowing the volume of Λ_i tend to infinity,
- or by allowing the *size*[15] of Λ_i tend to 0.

The first possibility is ruled out since every element of G has determinant ± 1. Finally, since the form q does not represent 0 over $\mathbb{Q}$, it has a restriction to $\mathbb{Z}^3 - \{0\}$ with non-zero integer values. Since q is continuous and $q(0,0,0) = 0$, the neighborhood $U = \{(x,y,z) : |q(x,y,z)| \leq 1/2\}$ of the origin contains no point of $\mathbb{Z}^3 - \{0\}$. But G leaves the form q invariant; therefore there does not exist any point of the orbit of $\mathbb{Z}^3$ under G inside U, and the size is uniformly bounded below on the image of $\Gamma\backslash G$. This concludes the proof of Theorem 2.4.6. □

We remark that this theorem implies in particular that the set Γ of integral transformations preserving the quadratic form q is infinite (Hermite's theorem). The case of quadratic forms in two variables $x^2 - ay^2$, with a square-free, implies Fermat's theorem mentioned above.[16] Poincaré breathed a geometric point of view into these arithmetic questions. The surface (*a priori* having conical singularities) $\Gamma\backslash\mathbb{H}$ is an *arithmetic surface.*

Poincaré studied in detail the fundamental domains of the arithmetic groups constructed above (the sets Ω). It is interesting to note that Poincaré introduced to this end what we now call the *trace fields.*[17] Poincaré in particular obtained the following theorem (cf. [**7**, vol. II: *Les fonctions fuchsiennes et l'arithmétique*, VII]).

THEOREM 2.4.7. *No torsion element (that is, no element of finite order) in* $\Gamma \subset \mathrm{GL}(3,\mathbb{Z})$ *is congruent to the identity modulo* 3.

PROOF. We shall show more generally the following lemma, due to Minkowski.

LEMMA 2.4.8. *Let F be a finite subgroup of* $\mathrm{GL}(n,\mathbb{Z})$. *Then F injects into* $\mathrm{GL}(n,\mathbb{Z}/l\mathbb{Z})$ *for all* $l \geq 3$.

This lemma is proved by first reducing to the case of a subgroup F that acts *irreducibly* on $\mathbb{R}^n$. Then the idea is to introduce an F-invariant norm such that the shortest vector in $\mathbb{Z}^n - \{0\}$ has norm 1. Now the orbit of such a vector is a generating system of the vector space $\mathbb{R}^n$. But these vectors are pairwise at a distance ≤ 2 and therefore cannot be identified modulo $l \geq 3$. □

In this spirit, Poincaré introduced the *congruence subgroups* of Γ where the nine coefficients of the matrices in $\Gamma \subset \mathrm{GL}(3,\mathbb{Z})$ are subject to certain congruences modulo some integer. We denote

$$\Gamma(N) = \ker(\Gamma \to \mathrm{GL}(3,\mathbb{Z}/N\mathbb{Z})).$$

A congruence subgroup in Γ contains a subgroup $\Gamma(N)$ for some N; it is therefore a finite-index subgroup in Γ. In addition, Theorem 2.4.7 implies that $\Gamma(3)$ (and in fact $\Gamma(N)$, for $N \geq 3$) is torsion-free. Therefore the arithmetic surfaces $\Gamma(N)\backslash\mathbb{H}$ have no singularities.

[15] In other words, the shortest distance to the origin from a non-zero point of the lattice.

[16] The integer matrices preserving the quadratic form are in fact of the form $\left(\begin{smallmatrix} m & an \\ n & m \end{smallmatrix}\right)$, where (m,n) is a solution to Pell's equation.

[17] A fundamental invariant in the work of Takeuchi and Reid on surfaces and 3-dimensional arithmetic hyperbolic manifolds.

"Having returned to Caen, I reflected on this result and derived its consequences. The example of quadratic forms showed me that there were Fuchsian groups other than those corresponding to the hypergeometric series; I saw that I could apply to them the theory of theta-Fuchsian series and that, consequently, there existed Fuchsian functions other than those derived from the hypergeometric series, which were the only ones I knew up until then." [**4**]

Thus Poincaré came to discover the first examples of discontinuous groups and automorphic functions defined by number-theoretic means. (This work would become the object of a vast extension: the theory of automorphic functions in several variables, an extension which is essentially the work of Siegel, then Borel and Harish-Chandra.) Relying on these examples (as well as those which he had extracted from studying the hypergeometric series), Poincaré took little time to construct a general theory, with a particularly geometric flavor, of all Fuchsian groups and the corresponding automorphic functions. (There is nothing analogous in the case of functions of several variables, with the notable exception of hyperbolic geometry in dimension 3: work of Thurston. In most cases, we even know from the work of Margulis that all discontinuous groups giving birth to such automorphic functions are those provided by number theory.)

But the interest of the Fuchsian functions which are tied to the theory of indefinite quadratic forms in three variables is not merely historical and anecdotal. These functions have a property that, when discovered by Poincaré ([**7**, vol. II: *Les fonctions fuchsiennes et l'arithmétique*, IX]), greatly struck him (indeed, he mentioned it on several occasions in his works): it is that they provide a generalization of the theory of the transformation of modular functions. This is a point worth dwelling on.

We again denote by Γ an arithmetic Fuchsian group associated to an indefinite quadratic form q in three variables with integer coefficients. Poincaré distinguished two cases (A) and (B) according to whether or not q represents zero over $\mathbb{Q}$.

In case (A), the group Γ is commensurable with the modular group $\mathrm{SL}(2,\mathbb{Z})$, if we agree with Poincaré to say that two groups are *commensurable* if their intersection has finite index in both. We then return, in principle, to the theory of modular functions. It should be noted that in this case the group Γ necessarily contains parabolic substitutions (that is, translations $z \mapsto z + a$) and therefore the corresponding Fuchsian functions admit Fourier series expansions. These expansions are fundamentally important in the theory of modular functions up through its most number-theoretic aspects: work of Hecke, Ramanujan's conjecture... [18]

In case (B), the group Γ is, according to Theorem 2.4.6, cocompact and therefore cannot contain parabolic substitutions. From the point of view of pure function theory, in principle this makes the study of Γ simpler. On the other hand, one can no longer find a Fourier series expansion; for a long time, this is what made the automorphic functions corresponding to such groups more mysterious.

In both cases, let g be a substitution belonging to $O(q) \cap \mathrm{GL}(3,\mathbb{Q})$; there exists an integer a such that $a^{-1}g$ belongs to $\mathrm{GL}(3,\mathbb{Z})$. If γ is an arbitrary substitution with integer coefficients, in order for $g^{-1}\gamma g$ to also have integer coefficients it suffices that γ be congruent to the identity modulo D, the determinant of $a^{-1}g$.

[18] EDITOR'S NOTE: Chapter 3 deals with the applications of case (A) in analytic number theory.

As Poincaré observed, an immediate result of this remark is that the groups Γ and $g^{-1}\Gamma g$ are commensurable for all $g \in O(q) \cap \mathrm{GL}(3, \mathbb{Q})$ (they both contain $\Gamma(D)$). The application of general principles from the theory of automorphic functions shows immediately that if f is a Fuchsian function of the group Γ, then *there is an algebraic relation between* f *and its transform by any substitution* g *in* $O(q) \cap \mathrm{GL}(3, \mathbb{Q})$. The group Γ that leaves f invariant is therefore contained in a much larger group, one which is even dense in G, that transforms f into a function connected with f by an algebraic relation. In case (A), the result is essentially the one that gives rise to the theory of transformations of modular functions and of "modular correspondences"; on this rests, for example, the entire theory of Hecke. In case (B), it took a long time to understand the significance of these algebraic relations; thus Weil wrote in 1954, in a text ([**8**]: *Poincaré et l'arithmétique*) which we have largely plundered in the above paragraphs:

> *"Is it the fault of the tool provided by Fourier series that the analogous problems for case (B) have not flowered up until now, in spite of Poincaré's insistence to draw attention to them? It is certainly to be wished that they be attacked as soon as possible..."*

Since then, the work of Shimura and then of Langlands (in much greater generality and at the same time introducing the fertile point of view of representation theory) have allowed the realization of Weil's wish and the development of the general notion of "Hecke correspondence".

Given a compact manifold M, a *non-singular correspondence* on M is a submanifold $C \subset M \times M$ such that the projections $C \to M$ onto each of the factors are finite (non-ramified) covering maps. Poincaré's observation comes back to the construction of an infinite set of non-singular correspondences on arithmetic hyperbolic surfaces: to each element $g \in O(q) \cap \mathrm{GL}(3, \mathbb{Q})$ there corresponds an isometric transformation of the hyperbolic plane $\mathbb{H}$, and the map

$$\begin{aligned} \mathbb{H} &\to \mathbb{H} \times \mathbb{H}, \\ z &\mapsto (z, gz) \end{aligned}$$

induces on the quotient a so-called "Hecke" correspondence

$$(g\Gamma g^{-1} \cap \Gamma)\backslash\mathbb{H} \to \Gamma\backslash\mathbb{H} \times \Gamma\backslash\mathbb{H}.$$

Let us conclude this matter by remarking that Poincaré thought of his transformation theory for arithmetic Fuchsian functions as an analogue to the addition theorem for elliptic functions, and, moreover, he implied that this property would characterize arithmetic groups. This remarkable intuition is now a theorem of Margulis, as long as we extend the notion of arithmetic group to other number fields: a hyperbolic surface of finite volume is arithmetic if and only if it admits infinitely many non-singular correspondences. Finally, we remark that Poincaré's theorem on transformations of arithmetic Fuchsian functions, in analogy with the addition theorem for elliptic functions, is the first in a sequence of results (which were developed in recent years) concerning Shimura varieties (or, more generally, arithmetic varieties) and analogous to theorems concerning abelian varieties.

Discovered before Poincaré's famous theorem on tilings of the hyperbolic plane, arithmetic groups therefore played a key role in Poincaré's progression towards the study of Fuchsian groups, and then of the problem of uniformization. He writes [**4**]:

> *"Naturally I posed myself the problem of forming all of these functions; I made upon it a systematic siege, and one after the other I pulled off all of the leading fortifications. There was however one that still held and whose fall would lead to that of the whole guard. But all my efforts at first only served to improve my knowledge of the difficulty, which was already something. All of this work was perfectly conscious.*
>
> *Upon that, I left for Mont Valérien, where I had to do my military service; I therefore had very different preoccupations. One day, as I was crossing the boulevard, the solution to the difficulty that had blocked me appeared all of a sudden. I did not seek to immediately deepen the idea, and it was only after my service that I again took up the question. I had all of the elements; I had only to assemble and organize them. I therefore composed my definitive memoir at one stroke and without trouble."*

The last step of which Poincaré spoke here is his famous theorem on tilings of the hyperbolic plane, and the memoir is the one from 1882 entitled "Theory of Fuchsian groups" ([**4**, vol. 2, p. 108]). Between his first research on Fuchs's theorem (end of May 1880), his recognition of the role of hyperbolic geometry (June 1880), and the construction of arithmetic groups (end of August 1880), only three months had run by. Less than one year later (February 1881), Poincaré's theorem on tilings of the hyperbolic plane was proved and the theory of Fuchsian functions was ready. It was announced in two notes of the *Comptes Rendus de l'Académie des Sciences*.

As we have tried to emphasize, the geometric basis of Poincaré's work is evident from the beginning (convergence of theta-Fuchsian series, fundamental sets, and arithmetic groups...). Moreover, Poincaré very quickly drew near to the hypothesis that every Riemann surface could be obtained as a quotient of the hyperbolic plane by a Fuchsian group. He advanced this theory (the uniformization theorem) in April and May of the following year (1881). Let us then conclude this article with a brief description of his first (incomplete) proof of the uniformization theorem, via what Poincaré called the method of continuity.

2.5. The question of uniformization

(For more details on this section, we refer to Kapovich's book [**5**].)

We return to the case of the differential equation (2.3). The surface S is then the Riemann sphere with some finite number n of points removed. Let us denote by Σ the underlying topological surface and consider the moduli space $\mathcal{M}(\Sigma)$ of complex structures on Σ. This is the quotient of

$$C_{n-3} = \{(z_1, \ldots, z_{n-3}) \in \mathbb{C}^{n-3} : z_j \neq 0, 1 \text{ and } z_i \neq z_k \quad \text{for } j \neq k\}$$

by the action of the permutation group of a set of $n-3$ elements. The space $\mathcal{M}(\Sigma)$ is an *orbifold*; it is more convenient to work with the complex manifold C_{n-3} (the Teichmüller space of Σ). The problem of uniformization then consists in associating to each point $(z_1, \ldots, z_{n-3}) \in C_{n-3}$ a torsion-free Fuchsian group Γ such that

$$\mathbb{P}^1 - \{0, 1, \infty, z_1, \ldots, z_{n-3}\} \cong \Gamma \backslash \mathbb{H}$$

(as complex manifolds).

Poincaré's approach to proving the uniformization theorem was composed of the following three steps (which are not without their echoes in the approach Thurston would develop, almost a century later, to prove his hyperbolization theorem for Haken manifolds).

The first step consists in finding a surface $S_0 \in C_{n-3}$ (viewed as Teichmüller space) that is uniformizable, i.e., such that $S_0 = \Gamma \backslash \mathbb{H}$, where Γ is a torsion-free Fuchsian group. Such a surface is easy to find: take a cyclic covering of the sphere with three points removed (which is itself uniformizable via the solution $= k^2$ of the equation (2.8), with $\lambda = \mu = \nu = 0$; see box 1 on p. 51).

In the second step, we consider $S \in C_{n-3}$ and the Schwarzian differential equation $\{f, z\} = \varphi$, where φ is a holomorphic quadratic differential with at most simple poles at the punctures. This equation admits a solution f, which is unique up to post-composition by transformations of $\mathrm{PGL}(2, \mathbb{C})$ and is a multivalued locally injective holomorphic function. The solution f defines a complex projective structure σ of finite type on S (without conical singularities because φ has simple poles at the punctures) and depends holomorphically on φ and the position of the punctures. Let $[\rho]$ be the conjugacy class of the monodromy representation of the complex projective structure σ. This defines the *monodromy map*

$$\mu_S : \mathbb{C}^{n-3} \to \mathcal{R}_{\mathrm{par}}(\Gamma, \mathrm{PGL}(2, \mathbb{C})),$$

where $\mathbb{C}^{n-3}$ is identified with the space of holomorphic quadratic differentials on S that have at most simple poles at the punctures and $\mathcal{R}_{\mathrm{par}}(\Gamma, \mathrm{PGL}(2, \mathbb{C}))$ designates the space of representations of Γ into $\mathrm{PGL}(2, \mathbb{C})$ sending parabolics onto parabolics and considered modulo conjugations. The μ_S do not assemble to give a map $C_{n-3} \times \mathbb{C}^{n-3} \to \mathcal{R}_{\mathrm{par}}(\Gamma, \mathrm{PGL}(2, \mathbb{C}))$; in order to do this we must pass to the universal cover of C_{n-3}. Nevertheless, if $\gamma \subset C_{n-3}$ is a smooth curve immersed in C_{n-3} starting from S_0, we obtain a well-defined smooth map $\mu_\gamma : \gamma \times \mathbb{C}^{n-3} \to \mathcal{R}_{\mathrm{par}}(\Gamma, \mathrm{PGL}(2, \mathbb{C}))$.

Now the third step is to apply Poincaré's method of continuity. Let us consider $S \in C_{n-3}$; we want to show that there exists φ such that if $[\rho] = \mu_S(\varphi)$, we have $\rho(\Gamma) \subset \mathrm{PSL}(2, \mathbb{R})$ and $\rho(\Gamma) \backslash \mathbb{H} = S$. In other words, S is uniformizable and $[\rho]$ is the uniformizing representation. The method of continuity consists in showing that if γ is a smooth immersed path going from S_0 to S in C_{n-3}, the subset U of points of γ that constitute uniformizable Riemann surfaces is open and closed. The fact that U is open seemed to flow from the source for Poincaré. The reason is surely that the fact that U is open is a consequence of the following two intuitive (but non-trivial) results.

(1) For all $S \in C_{n-3}$, the map μ_S is transverse to the space of discrete, faithful representations of Γ into $\mathrm{PSL}(2, \mathbb{R})$ (this result is now a theorem of Faltings).
(2) The set of representations in $\mathcal{R}_{\mathrm{par}}(\gamma, \mathrm{PSL}(2, \mathbb{R}))$ that are faithful and discrete is an open subset of $\mathcal{R}_{\mathrm{par}}(\gamma, \mathrm{PSL}(2, \mathbb{R}))$ (which now can easily be deduced from a classical theorem of Ehresmann).

The "proof" that U is closed occupies the major part of [**7**, vol. II, p. 300, *Sur les groupes des équations linéaires*]. Its only tool for working with discrete subgroups of $\mathrm{PSL}(2, \mathbb{R})$ is the notion of fundamental polygon: it seems to be insufficient to obtain a rigorous proof.

Nonetheless, since C_{n-3} is arcwise connected, Poincaré's method is intuitively convincing and constitutes an important historical step towards the proof[19] of the uniformization theorem by Poincaré and Koebe.

---

[19]By completely different and more analytic methods, thereby reflecting the recent history of the geometrization conjecture (or Perelman's theorem: see Chapter 12).

Bibliography

[1] L. Fuchs. *Sur quelques propriétés des intégrales des équations différentielles, auxquelles satisfont les modules de périodicité des intégrales elliptiques des deux premières espèces*, J. für die reine und angewandte Mathematik (J. de Crelle) **83** (1877), pp. 13–38.

[2] L. Fuchs. *Ueber die linearen Differentialgleichungen zweiter Ordnung, welche algebraische Integrale bestizen*, J. für die reine und angewandte Mathematik (J. de Crelle) **85** (1878), pp. 1–26.

[3] L. Fuchs. *Ueber eine Klasse von Funktionen mehrerer Variabeln, welche durch Umkehrung der Integrale von Lösungen der linearen Differentialgleichungen mit rationalen Coefficienten entstehen*, J. für die reine und angewandte Mathematik (J. de Crelle) **89** (1880), pp. 151–169.

[4] Jacques Hadamard, Henri Poincaré. *Essai sur la psychologie de l'invention dans le domaine mathématique* [Hadamard]. *L'invention mathématique* [Poincaré]. Les Grands Classiques Gauthier-Villars, éditions Jacques Gabay, Sceaux, 1993.

[5] Michael Kapovich. *Hyperbolic manifolds and discrete groups. Progress in Mathematics*, Vol. 183. Birkhäuser Boston Inc., Boston, MA, 2001.

[6] J. Merker. *L'ontologie explicite des théorèmes d'existence en mathématiques*, Séminaire de Philosophie et Mathématiques de l'École normale supérieure, Paris (under the direction of Martin Andler, Pierre Cartier, Maurice Loi, Hourya Sinaceur): session of December 18, 1996.

[7] Henri Poincaré. *Œuvres*. Les Grands Classiques Gauthier-Villars. Éditions Jacques Gabay, Sceaux.

[8] André Weil. *Scientific works. Collected papers.* Vol. II (1951–1964). Springer-Verlag, New York, 1979.

CHAPTER 3

Poincaré and Analytic Number Theory

EMMANUEL KOWALSKI

"One area of arithmetic where unity seems to be absolutely lacking is the theory of prime numbers; we have only found asymptotic laws, and this is all we should hope for; but these laws are isolated and one can only arrive at them by different paths that seem incapable of communicating with each other. I believe I can glimpse where the desired unity will come from, but I glimpse only vaguely; everything will doubtless reduce to the study of a family of transcendental functions that will make it possible, from the study of their singular points and the application of Darboux's method, to calculate asymptotically certain functions of very large numbers."

H. Poincaré, *The future of mathematics* [**5**]

3.1. Introduction

Analytic number theory is a relatively young branch of arithmetic, even though the natural impulse that determines its essential questions is as old as the theorem saying that there are infinitely many prime numbers.

Poincaré did not really practice analytic number theory directly. If one consults his complete works, in particular volume V, which is devoted to number theory, only two articles can be tied to it. One [**7**] considers prime numbers of the form $4n+1$ or $4n+3$, and extends to these residue classes the methods used by Chebychev to give upper and lower bounds for $\pi(x)$, the number of prime numbers $\leq x$; this is not one of Poincaré's greatest works! The second [**8**], in the volume of *Crelle's Journal* in honor of Dirichlet, presents different analytic functions and obtains the analytic class-number formula for quadratic forms with a given discriminant, by a different approach than that of Dirichlet.

As the citation at the beginning of this text shows, it was not a disinterest in (let us say, for simplicity) asymptotic properties of prime numbers that explains such a modest contribution. It is not obvious exactly what analytic functions he was thinking about as he wrote these lines. Without a doubt, his thoughts dealt with what we now call Dirichlet series, or more precisely, L-functions.

And yet, even though it was not discovered until later and somewhat slowly, one of the relatively little-known revelations of 20th century number theory was the breakthrough of these automorphic, or Fuchsian, functions (dear to Poincaré) into analytic number theory. By a path that is no less magnificent for having been a bit off the beaten track, using along the way a great number of different insights, an entire area in this theory of prime numbers was found to be completely overturned (the essential steps were the works of Hecke and Petersson, then of Maass and

Selberg, then of Kuznetsov, and finally of Iwaniec and Deshouillers). And, again very recently, a connection (still tenuous) has appeared between these results and the ancient (and still open) *Twin Primes Problem*—that is, to know if there are infinitely many prime numbers p such that $p+2$ is also prime, and if that is the case, to specify their distribution.

In what follows, we will obviously not attempt to handle proofs. Those desirous to know more will find complementary information, often with detailed proofs, in the texts [**2, 4, 9**], which are accessible at the level of first- or second-year graduate students.

3.2. Poincaré series and Kloosterman sums

The functions that are the subject of this discussion are among those justly called "Poincaré series". Let us first recall their definition. The general principle is the following. Suppose we are given a group Γ operating on a set X. We wish to exhibit (complex-valued) functions on X that are invariant under the action of Γ, i.e., such that $f(\gamma \cdot x) = f(x)$ for all $\gamma \in \Gamma$ and $x \in X$. In general this is a very delicate problem; but if one has at one's disposal a function F on X that turns out to be invariant under a *subgroup* B of Γ, the brilliant idea of Poincaré series is that, at least formally, the average of F over the cosets B_γ of B in Γ provides such a function f:

$$f(x) = \sum_{g \in B\backslash\Gamma} F(g \cdot x).$$

In fact, it suffices to rearrange the corresponding expression for $f(\gamma \cdot x)$ by setting $g\gamma = g'$ (which permutes the classes modulo B) to reveal that f is invariant... if it exists, because the indicated sum may well be divergent.

In the case that interests us, this question of existence is indeed non-trivial, although relatively easy to solve. Let us put ourselves in a relatively elementary frame, sufficient for the number-theoretic applications that will follow: here the set X will be what we nowadays call the *Poincaré half-plane*

$$\mathbb{H} = \{z = x + iy \in \mathbb{C} \mid y > 0\}$$

and Γ will be what we call a "Fuchsian subgroup of the first kind" of $\mathrm{SL}(2,\mathbb{R})$: it will suffice to know that the Hecke congruence subgroups $\Gamma_0(q)$, defined by

$$\Gamma_0(q) = \left\{ g = \begin{pmatrix} a & b \\ c & d \end{pmatrix} \mid a,b,c,d \in \mathbb{Z},\ ad - bc = 1,\ c \equiv 0 \bmod q \right\},$$

are examples of these groups (without a doubt atypical, because they have an arithmetic nature that is not present in general). Even the simplest case with $q = 1$, for which $\Gamma = \Gamma_0(1)$ is the group $\mathrm{SL}(2,\mathbb{Z})$, already reveals a good portion of the theory's aspects.

The action of such a group on $X = \mathbb{H}$ is given by

$$g \cdot z = \frac{az+b}{cz+d}.$$

The *automorphic functions* are complex-valued functions defined on $\mathbb{H}$ and satisfying

$$f\left(\frac{az+b}{cz+d}\right) = f(z) \tag{3.1}$$

for $\gamma = \left(\begin{smallmatrix} a & b \\ c & d \end{smallmatrix}\right) \in \Gamma$. We also impose on them other regularity conditions, which are important, but about which we will here say only a few things: *grosso modo*, f must be holomorphic (in other words, it must preserve angles and orientation), and as z approaches the "boundary" of $\mathbb{H}$, the function f must have "moderate" growth. (For example, if $\Gamma = \mathrm{SL}(2, \mathbb{Z})$, $|f(z)|$ must be bounded by a power of $\mathrm{Im}(z)$ for $\mathrm{Im}(z) \geq 1$.)

In addition to *automorphic functions*, one classically considers *automorphic forms* of weight k, where k is a positive integer (which is necessarily even in the case under consideration if the form is non-zero): these are functions f, again defined on $\mathbb{H}$ and complex-valued, with similar regularity conditions to those mentioned above, such that

$$f\left(\frac{az+b}{cz+d}\right) = (cz+d)^k f(z) \tag{3.2}$$

for $\gamma = \left(\begin{smallmatrix} a & b \\ c & d \end{smallmatrix}\right) \in \Gamma$. One of Poincaré's intuitions was, precisely, that such forms are easier to construct directly than automorphic functions, and that if f_1 and f_2 are automorphic forms of the same weight (and f_2 is non-zero...), then by setting $f = f_1/f_2$ one obtains an automorphic function.

The principle of Poincaré series applies here by taking $\Gamma = \Gamma_0(q)$ and, for the subgroup B, the subgroup of "unipotent" matrices of $\Gamma_0(q)$:

$$\Gamma = \Gamma_0(q), \qquad B = \left\{ g = \begin{pmatrix} 1 & n \\ 0 & 1 \end{pmatrix} \mid n \in Z \right\} \subset \Gamma_0(q).$$

The matrices of B operate on $\mathbb{C}$ by $z \mapsto z + n$; the idea is therefore, in order to find a function F *ad hoc*, to choose an arbitrary 1-periodic function defined on $\mathbb{H}$. It is then natural to consider

$$F(z) = \exp(2i\pi mz) = \exp(-2\pi my)\exp(2i\pi mx) \qquad (z = x + iy)$$

for $m \in \mathbb{Z}$, which certainly satisfies this property. The corresponding Poincaré series are then the expressions, which at first are only formally defined,

$$P_m(z) = \sum_{\gamma \in B\backslash\Gamma_0(q)} \exp(2i\pi m\gamma \cdot z),$$

$$P_m^{(k)}(z) = \sum_{\gamma \in B\backslash\Gamma_0(q)} (cz+d)^{-k} \exp(2i\pi m\gamma \cdot z);$$

the second definition corresponds to the modification of the above principle that is necessary in order to obtain an automorphic form rather than an automorphic function.

It is not very hard to establish that the first series *diverges* for all values of m and all z; as soon as $k > 2$, however, the second converges and defines an automorphic form of weight k.

In the case of $\mathrm{SL}(2, \mathbb{Z})$, these series are explicitly written by Poincaré in an article published after his death[1] [**6**], where, in the very last pages, he carried out the next step in analyzing their properties: to calculate the *Fourier expansion* of $P_m^{(k)}(z)$. In fact, relation (3.2), when applied to a matrix belonging to B, implies

[1] EDITOR'S NOTE: Poincaré mailed this article the same day as he entered the clinic in which he would die ten days later.

that the Poincaré series are again periodic with period 1. Fourier's theory implies that they have a series expansion of the type

$$P_m^{(k)}(z) = \sum_{h\in\mathbb{Z}} p_m^{(k)}(h)\exp(2i\pi hz)$$

for some *Fourier coefficients* $p_m^{(k)}(h) \in \mathbb{C}$. For $m \neq 0$, the general formula, now classical, is of the form

(3.3)

$$p_m^{(k)}(h) = \left(\frac{h}{m}\right)^{(k-1)/2}\left\{\delta(m,h) - 2\pi i^{-k}\sum_{c\geq 1,\, q\mid c} c^{-1}S(m,h;c)J_{k-1}\left(\frac{4\pi\sqrt{mn}}{c}\right)\right\}.$$

(We recall that the notation $q|c$ signifies that q divides c.) This expression, with a fairly impressive appearance, brings into play a Bessel function, J_{k-1}, whose power series expansion (convergent on all of $\mathbb{C}$) can easily be given,

$$J_{k-1}(z) = \left(\frac{z}{2}\right)^{k-1}\sum_{n\geq 0}\frac{(-1)^n}{2^{2n}n!(n+k-1)!}z^{2n},$$

and above all a *Kloosterman sum*

$$\text{(3.4)} \qquad S(m,h;c) = \sum_{1\leq x<c,\,(x,c)=1}\exp\left(\frac{mx+h\overline{x}}{c}\right), \qquad \text{with} \quad x\overline{x}\equiv 1 \bmod c$$

(where (x,c) denotes the GCD of x and c). In FIG. 3.1, we reproduce the part of the calculation where Poincaré writes such a sum, which he denotes $\sum E$; note that he says nothing more about it than the fact that it "is in general non-zero" [*"n'est pas nulle en général"*].

These innocent expressions were headed for a brilliant future. They reappeared first of all, independently of these calculations of Poincaré, in an important article by H. Kloosterman, published in 1926 [**3**] (FIG. 3.2).

There, amply justifying that these sums bear his name rather than Poincaré's, Kloosterman extracts a remarkable arithmetic application: he succeeds in obtaining an asymptotic formula for the number of representations of a large integer $n \gg 1$ by a quadratic form in four variables, positive definite and with integer coefficients, that is, for the number of solutions $(x_1, x_2, x_3, x_4) \in \mathbb{Z}^4$ to an equation

$$\text{(3.5)} \qquad n = a_1x_1^2 + a_2x_2^2 + a_3x_3^2 + a_4x_4^2,$$

where a_i, $1 \leq i \leq 4$, are fixed integers ≥ 1. The context of this result is the *circle method* of Hardy, Ramanujan, and Littlewood, of which Kloosterman invented a new, greatly refined version. This method had allowed Hardy and Littlewood to give a new solution to Waring's problem (the original one being due to Hilbert): for a fixed integer $k \geq 1$, there exists an integer $g = g_k \geq 1$ such that every integer $n \geq 1$ can be written as the sum of g kth powers of non-negative integers:

$$n = x_1^k + \cdots + x_g^k, \qquad \text{with } x_i \geq 0.$$

For $k = 2$, the circle method provides only the upper bound $g_2 \leq 5$ (at least for the representation of "sufficiently large" integers), even though it had been known since Lagrange that every integer is the sum of four squares (that is, $g_2 = 4$). Thus, Kloosterman's theorem permits the analytic approach of Hardy and Littlewood to be extended and made just as effective as the more algebraic methods that provide Lagrange's theorem.

The crucial point of Kloosterman's proof was to obtain a non-trivial estimate for the size of a sum $S(m,n;c)$ when m, n, and c are relatively prime. The case of prime $c = p$ is the most important, to which the general case can easily be reduced. In this case, we note that according to definition (3.4), $S(m,n;p)$ is the sum of $p-1$ terms all of which have modulus 1, but with (apparently) random arguments, because when x ranges among integers from 1 to $p-1$, the inverse $\overline{x}$ of x modulo p varies in a chaotic manner. Apart from the obvious symmetries, one can therefore see the calculation of the Kloosterman sum as the realization of a "random walk" in the plane. FIG. 3.3 reproduces such a path for the sum $S(1,1;1021)$: starting from the origin, segments are drawn connecting the intermediary sums leading to the calculation of the Kloosterman sum.

According to this description, it is therefore natural to expect that $S(m,n;p)$ has a modulus substantially smaller than what the obvious upper bound—the triangle inequality—provides, which is $|S(m,n;p)| \leq p-1$. And indeed, in the present

148 H. POINCARÉ.

c'est une fonction qui se déduit immédiatement des fonctions de Bessel et que j'écrirai $J(m, G)$; on aura donc

$$\omega(\gamma, \delta) = \sum \mu_j E q^j J(m, G).$$

Nous allons maintenant grouper ensemble les termes qui correspondent aux diverses valeurs de δ non congrues entre elles suivant le module γ. Si nous appelons $\omega(\gamma)$ la somme de ces termes, le coefficient de q^j dans $\omega(\gamma)$ sera

$$\mu_j J(m, G) \sum E.$$

Il faut donc calculer $\sum E$, c'est-à-dire

$$\sum e^{\frac{2i\pi}{\gamma}(j\delta - p\alpha)}.$$

Les entiers j, p et γ sont donnés; mais on donne à α toutes les valeurs entières premières avec γ et incongrues entre elles par rapport au module γ, et à δ les valeurs correspondantes, de telle façon que

$$\alpha\delta \equiv 1 \pmod{\gamma}.$$

Je me bornerai à constater que $\sum E$ n'est pas nul en général. Il reste à sommer par rapport à γ et notre coefficient s'écrit :

$$\sum_\gamma \mu_j \left[\sum E\right] J\left(m, \frac{4pj\pi^2}{\gamma^2}\right).$$

Il n'y a aucune raison pour qu'il y ait des relations linéaires entre les valeurs des fonctions de Bessel $J\left(m, \frac{4pj\pi^2}{\gamma^2}\right)$ correspondant aux différentes valeurs de γ. Il n'y a donc aucune raison pour que ce coefficient s'annule.

FIGURE 3.1. Poincaré's article making use of a Kloosterman sum.

from which the statement follows.

2. 4. The sum $S(u, v; \lambda, \Lambda; q)$.

We shall show afterwards, that the approximation for large values of q of the sum occurring on the right hand side of the formula of lemma 3^*, can be reduced to the calculation for large values of q of the sum

$$S(u, v; \lambda, \Lambda; q) = \sum_{p \equiv \lambda \ (\mathrm{mod}\ \Lambda)}' \exp\left(\frac{2\pi i u p}{q} + \frac{2\pi i v p'}{q}\right).$$

But before performing the reduction, we shall first consider this sum S. The object of this section is the proof of lemma 4. The lemmas 4 *b*—4 *e* are special cases of lemma 4, from which the general lemma 4 will be deduced.

FIGURE 3.2. The Kloosterman sum in Kloosterman's article.

case, Kloosterman obtains the following upper bound:

$$|S(m, n; p)| < 2p^{3/4}.$$

The readers keen on probability will doubtlessly wonder at this point whether the exponent 3/4 is the best possible, and we might suspect that it could be replaced by 1/2. This is indeed the case, but the proof of the following bound,

$$|S(m, n; p)| \leq 2\sqrt{p} \qquad \text{(if } p \text{ does not divide } nm), \tag{3.6}$$

which is due to A. Weil, is much more difficult: it comes as a corollary to some particular cases of the Riemann Hypothesis for algebraic curves over finite fields, whose general proof Weil obtained between 1940 and 1948.

It is, in contrast, easy to check that this exponent 1/2 is in fact the best possible if we vary m and n: precisely, there always exists m such that $|S(1, m; p)| > \sqrt{2p - 2}$.

Given this, an interesting point already appears in this first number-theoretic application of Kloosterman sums: it is not a single sum that plays a part in the

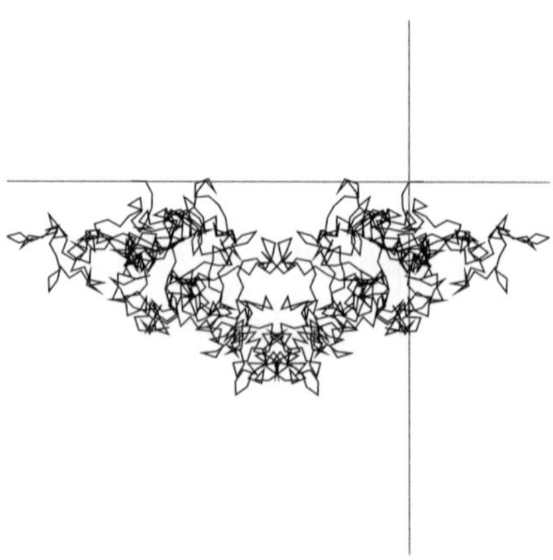

FIGURE 3.3. The Kloosterman sum $S(1, 1; 1021) = -18.608411\ldots$

problem, but a *sum* of Kloosterman sums, eventually weighted by a "test function". One can therefore look at (3.3) with a different view, recalling that equality is a symmetric relation, and see the formula in question as expressing (in the case $q = 1$ to clarify the discussion) a weighted sum

$$\sum_{c \geq 1} c^{-1} S(m, n; c) f\left(\frac{4\pi\sqrt{mn}}{c}\right) \tag{3.7}$$

(where $f = J_{k-1}$) in terms of Poincaré series, and therefore of automorphic forms (of weight k). We note that by linearity, it is then possible to obtain such a "summation formula" for every function f that can be expressed as a linear combination (or infinite series) of Bessel functions J_{k-1}, with consideration for various questions of convergence:

$$f(x) = \sum_{k \geq 4,\, k \text{ even}} a_k J_{k-1}(x).$$

However, the sums that appear "in (arithmetic) nature" have hardly any reason to be of this form (even allowing infinite series). For example, one could see this immediately by observing that as soon as the series above converges, we have

$$\int_0^{\infty} f(x)\,(J_{it}(x) - J_{-it}(x))\,\frac{dx}{x} = 0$$

for *every* Bessel function with pure imaginary index J_{it}, $t \in \mathbb{R}$ being fixed. Therefore the functions in the space generated by these Bessel functions J_{it} can *never* be expressed as combinations of the J_{k-1}...[2]

Could we complete the stock of summation formulas in a way that incorporates the "missing" functions? This is indeed the case, but this discovery was not made without difficulty...

Following the thread of our history, the necessary ingredient appeared in the work of A. Selberg. He studied (for relatively disjoint reasons) automorphic functions satisfying (3.1), not necessarily holomorphic, but solutions to the eigenvalue equation for the hyperbolic Laplacian: for some constant λ (which is necessarily ≥ 0 in the relevant cases due to growth conditions), we have

$$-y^2\left(\frac{\partial^2 f}{\partial x^2}(z) + \frac{\partial^2 f}{\partial y^2}(z)\right) = \lambda f(z)$$

for all $z \in \mathbb{H}$. (Such functions were originally considered by H. Maass.) Selberg introduced into this study a new type of Poincaré series that allow one to compensate for the divergence of those we saw previously once $k = 0$. Depending on one "auxiliary" complex variable s, the Poincaré series defined by Selberg[3] is

$$P_m(z; s) = \sum_{\gamma \in B \backslash \Gamma_0(q)} \operatorname{Im}(\gamma \cdot z)^s \exp(2i\pi m \gamma \cdot z)$$

(here again, the restriction to the groups $\Gamma_0(q)$ is a simplification). As for the previous series, it is possible to determine a Fourier expansion because these functions

[2] Some authors discourage the use of the (innumerable) integral formulas and various transformations of Bessel functions; be careful not to give in to this temptation: by refusing to plunge into the integral tables, one risks passing over a large number of discoveries...

[3] The addition of such a factor depending on s to "regularize" a divergent series was inaugurated by E. Hecke.

remain periodic of period 1. We obtain the relation

$$P_m(z;s) = \sum_{h\in\mathbb{Z}} p_m(h;y,x)\exp(2i\pi hz),$$

where y denotes as before the imaginary part of z and where

$$p_m(h;y,s) = \delta(m,h) + \sum_{c\geq 1,\, q|c} c^{-2s} S(m,n;c) B(m,h,c,y,s)$$

and

$$B(m,h,c,y,s) = y^s \int_{-\infty}^{+\infty} (x^2+y^2)^{-s} \exp\left(-2i\pi\left(hx + \frac{m}{c^2(x+iy)}\right)\right) dx,$$

which is perhaps not so surprising any longer (the dependence of the coefficient on y is more complicated than in the previous case, because in the present case the Poincaré series are not holomorphic)!

We now have an additional parameter (that is to say, s) and many new sums of Kloosterman sums. Are they enough to express the functions at play in the applications, by weighted average (integral) over certain values of s? The answer is affirmative, because once $s = 1+it$, the function B (generally inextricable) becomes close to the functions J_{it}, and by invoking an old theorem of Sears and Titchmarsh, one can see that every "reasonable" function on $[0,+\infty)$ can be expressed as a sum of a linear combination of the J_{k-1} and a weighted integral of the $J_{it} - J_{-it}$.

This step (which in order to be complete requires deep developments and turns out to be very delicate analytically) was first reached by N. Kuznetsov (and independently R. Bruggeman) in 1977, 1980, for the value $q = 1$, and by Deshouillers and Iwaniec in 1982 for general q. The *Kuznetsov formula* then expresses that (3.7) can be represented as a mix of integrals and sums of the type

$$\sum_j \lambda_j(m)\overline{\lambda_j(n)}\widehat{f}(t_j) + \int_{-\infty}^{\infty} \rho(m;\tfrac{1}{2}+it)\,\overline{\rho(n;\tfrac{1}{2}+it)}\,\widehat{f}(t)\,dt,$$

bringing into play a certain integral transform $\widehat{f}$ of f (more complicated than the Fourier transform) and the Fourier coefficients $\lambda_j(n)$, $\rho(n;\frac{1}{2}+it)$ of automorphic functions (forms) chosen to constitute an orthonormal basis of the space of automorphic functions and forms.

The number-theoretic interest of such a formula brings us back to Kloosterman: in applications (such as those already described), it is more often required to consider a *sum* of Kloosterman sums rather than an individual sum. Even if very interesting results (for example, that of Kloosterman) can emerge from individual study (in particular from Weil's bound (3.6)), the very existence of such formulas leaves the hope of deeper developments, which were previously out of reach, by opening the door to the analysis of the oscillations of $S(m,n;c)$ as c varies... with the precondition of understanding the coefficients $\lambda_j(n)$ and $\rho(n;\frac{1}{2}+it)$, that is, understanding well the space of automorphic functions and forms.

The interest in understanding these oscillations for natural number-theoretic problems was pointed out by Yu. Linnik in a prescient exposition to the International Congress of Mathematicians of Stockholm in 1962. He had formulated a conjecture on the matter (see FIG. 3.4), which is often stated in the following simplified form: For fixed m and n, there exists for all $\varepsilon > 0$ a constant $C(\varepsilon)$, depending on m and n, such that

$$\Big|\sum_{c\leq X} c^{-1}S(m,n;c)\Big| \leq C(\varepsilon)X^{\varepsilon} \quad \text{for all } X \geq 1;$$

here again, if one thinks of $S(m,n;c)/\sqrt{c}$ as a "random variable", essentially bounded but with variable sign, this prediction can be justified by a probabilistic analogy. This conjecture, at least in a "smoothed-out" form,[4] follows easily from Kuznetsov's method.[5]

or the corresponding values for $y \leqslant \sqrt{n}\ln n$ will be of importance. Hence, in the problem (3.1) we can effect the levelling with the error $O(n^{(\frac{1}{4}2k-\frac{1}{2})+\varepsilon})$.

The levelling in the problem involving a general quadratic form with two or more variables is a little more sophisticated.

The possibility of levelling in the problem (3,1) with the error term $O(n^{(\frac{1}{4}2k-\frac{1}{2})+\varepsilon})$ leads to the fact that the problem of the optimal estimate of the error term in this problem is connected with the summation of the Kloosterman sums (2,1). The following hypothesis arises in this connection:

Hypothesis on Kloosterman's sums.

Let N be a large number;

$$T(N,g) = \sum_{x \bmod g} \exp\frac{2\pi i}{g}(x' + Nx) \quad \text{the Kloosterman sum,}$$

$$g_1 > N^{1/2-\varepsilon_0} \quad (\varepsilon_0 > 0 \text{ arbitrarily small}).$$

Then

$$\sum_{g\leqslant g_1} T(N,g) = O(g_1^{1+\varepsilon}) \quad \text{for each } \varepsilon > 0. \tag{3.6}$$

This hypothesis implies the estimate of the remainder term in the problem (2.2): $F(x_1,\dots,x_k)=n$ for even k's: $|\sigma_F(n)|\gamma_F(\varepsilon)n^{(\frac{1}{4}k-\frac{1}{2})+\varepsilon}$. Moreover, the proof of this hypothesis would lead to considerable advances in the problem of counting the points inside the circle or in general an ellipsoid of even dimension.

FIGURE 3.4. Linnik's conjecture.

It would now be possible to describe numerous applications of these summation formulas for Kloosterman sums. Just to give one, before concluding this section: it is the crucial ingredient in the proof, due to B. Conrey, that at least 60% of the zeroes of the Riemann zeta function are located on the critical line.

But we shall now return to the source of analytic number theory, the study of prime numbers...

[4]That is, by replacing the sum over $c \leq X$ by a sum of the type $\sum c^{-1}S(m,n;c)\exp(-c/X)$, which attenuates the extreme terms and avoids the purely analytic problems.

[5]It is interesting to note, however, that only very recently É. Fouvry and P. Michel succeeded in showing that this resolution is indeed a confirmation of a randomness in the signs of the Kloosterman sums, and not the effect of a hypothetical estimate of the type

$$|S(m,n;c)| \leq C(m,n)c^{1/2-\gamma}$$

once $c \geq 1$, with $\gamma > 0$, *for fixed m and n.* (If m and n vary, Weil's bound cannot be improved.)

3.3. Prime numbers, arithmetic progressions, and a message

Let us return to the distribution of prime numbers, and again denote by $\pi(x)$ the number of integers $p \leq x$ such that p is prime. The most fundamental result in analytic number theory is the *prime number theorem*, which determines the *asymptotic* behavior of $\pi(x)$ as $x \to +\infty$:

THEOREM (Hadamard, de la Vallée-Poussin, 1895). *We have*

$$\lim_{x \to +\infty} \frac{\pi(x)}{x/\ln x} = 1,$$

which we also write as

$$\pi(x) \sim \frac{x}{\ln x} \quad \textit{as} \quad x \to +\infty.$$

This result was conjectured by Legendre and Gauss, made plausible by Chebychev, and the methods of Hadamard and de la Vallée-Poussin are fundamentally based on Riemann's revolutionary point of view, that is, on the introduction of functions of one complex variable. "Elementary" proofs are now known; this does not mean that they are simple.

It was quickly realized that the result, in this form, is insufficient for many applications. More precisely, a basic question quickly poses itself: can one, similarly, "count" the prime numbers satisfying additional conditions? For example, count the $p \leq x$ such that the last digit of p equals 1, or more generally such that the remainder from the division of p by a fixed integer $q \geq 1$ equals a remainder a, also fixed (the previous case corresponds to a remainder of 1 following division by 10); or again, can one count the $p \leq x$ such that $p - 1$ is a square, in other words $p = n^2 + 1$ for an integer $n \geq 1$; or perhaps, can one count the $p \leq x$ such that $p + 2$ is also prime (the "twin prime numbers")?

Only the case of constraints on the remainder by division is solved at this time. Precisely, taking as a base the ideas of Dirichlet (just as revolutionary as those of Riemann, since they are the basis of the theory of *group representations*), one can adapt the method of proof of the prime number theorem to deduce the following:

THEOREM. *Let $q \geq 1$ be a fixed integer, and let $a \in \mathbb{Z}$ be a non-zero integer relatively prime to q. Then the number $\pi(x; q, a)$ of prime numbers $p \leq x$ such that $p - a$ is divisible by q (i.e., such that $p \equiv a \mod q$) satisfies*

$$\pi(x; q, a) \sim \frac{1}{\varphi(q)} \frac{x}{\ln x} \sim \frac{1}{\varphi(q)} \pi(x) \quad \textit{as} \quad x \to +\infty,$$

where $\varphi(q)$, the Euler function, is the number of integers a such that $0 \leq a \leq q$ and a is relatively prime to q.

For example, for $q = 10$, $\varphi(10) = 4$ (the allowed values of a being $1, 3, 7, 9$), and this result, compared with the prime number theorem, says that the last digit (the units) has a value of 1 for approximately 25% of prime numbers, 3 for approximately 25%, and likewise 7, 9 for approximately 25%, with the awareness of course that if the last digit is even, the integer is divisible by 2 and therefore can only be prime if it equals 2.

But here again, as soon as one wants to apply this result in practice—that is, to other problems that reduce, more or less easily, to considerations about $\pi(x; q, a)$—it turns out almost always that this result is dramatically insufficient. More precisely, the crucial question is the *uniform* dependence on q especially and on a (less crucially... although: see further down!) of the error term $E(x;q,a)$ appearing when we write

$$\pi(x;q,a) = \frac{1}{\varphi(q)}\mathrm{li}(x) + E(x;q,a), \text{ where } \mathrm{li}(x) = \int_2^x \frac{dt}{\ln t}.$$

This latter function, the "logarithmic integral", satisfies for $x \geq 2$ the inequality

$$\left|\mathrm{li}(x) - \frac{x}{\ln x}\right| \leq \frac{12x}{(\ln x)^2},$$

and the prime number theorem can also be written in the form

$$\pi(x) \sim \mathrm{li}(x) \quad \text{as} \quad x \to +\infty.$$

This form was moreover the one predicted by Gauss, who interpreted $1/\ln x$ as the "density" of prime numbers around x, or as the "probability" that a number the size of x would be prime.

The utilization of $\mathrm{li}(x)$ allows one to render the "error term" $E(x;q,a)$ much smaller (that is, it gives a much better approximation to $\pi(x;q,a)$ than the elementary function $x/\ln x$). In particular, it is relatively easy to show that the famous Riemann Hypothesis, generalized to Dirichlet's L-functions, is *equivalent* to the following estimate: There exists a constant $C \geq 0$ such that

$$|E(x;q,a)| \leq Cx^{1/2}\ln x \tag{3.8}$$

for all $q \geq 1$, any a relatively prime to q, and all $x \geq 2$. What must be remembered from this bound is that the term $E(x;q,a)$ is *negligible* next to the principal term $\mathrm{li}(x)/\varphi(q)$, essentially if q is a little smaller than $\sqrt{x}/\ln x$.

A simple and rather natural problem, which leads directly to the consideration of the uniform dependence of $E(x;q,a)$ on q, is the asymptotic estimation of the sum

$$S(x) = \sum_{p\leq x} d(p-1),$$

where the sum is restricted to prime numbers and $d(n)$ denotes the number of divisors $d \geq 1$ of an integer $n \geq 1$. Such sums may appear in applications, but especially we can view the question as a simple test example of what we know concerning the interaction between the multiplicative properties of integers (which define the prime numbers) and their additive properties (revealed by the passage from p to $p-1$); in other words, do there exist particular multiplicative properties for numbers of the form $p-1$?

This exact problem, which goes back to Titchmarsh, reduces to the study of $E(x;q,1)$ by writing

$$d(p-1) = \sum_{\substack{q,q'\geq 1\\ p-1=qq'}} 1$$

and by inverting the two sums:

$$\begin{aligned} S(x) &= \sum_{q\leq x-1} \sum_{\substack{p\leq x\\ p\equiv 1 \bmod q}} 1 = \sum_{q\leq x-1} \pi(x;q,1) \\ &= \mathrm{li}(x)\left(\sum_{q\leq x-1}\frac{1}{\varphi(q)}\right) + \sum_{q\leq x-1} E(x;q,1). \end{aligned}$$

The first term is easy to manage, and the difficulty is in dealing with the remaining sum of remainder terms, where q can be very large. A trick of Dirichlet (if $ab = n$, then one of a or b has size at most $\sqrt{n}$) easily allows one to reduce to $q \leq \sqrt{x}$ instead of $q \leq x$. But, by any method, *even under the (Generalized) Riemann Hypothesis*, that is, even using (3.8), this sum is not sufficiently controlled to conclude that the first term dominates.[6]

This is where Kloosterman sums can make their reappearance. By using *sieve methods*, it is possible to analyze very precisely the prime numbers $p \leq x$ by reducing their study to that of sequences dd' of multiples of an integer d, *from the moment* that this study is done uniformly with respect to d as large as possible with respect to x, ideally $d \asymp \sqrt{x}$ (which indicates that d and $\sqrt{x}$ are "of the same order of magnitude": more precisely, that $d/\sqrt{x}$ and $\sqrt{x}/d$ are bounded). However, the equation $p \equiv a \bmod q$, which can be rewritten as $p - 1 = qq'$, becomes, for these sequences of multiples of d, the simple equation

$$dd' - 1 = qq', \text{ or } dd' - qq' = 1.$$

The objective is then to count the number of solutions to such an equation with maximal uniformity and precision when $dd' \leq x$. Why should we hope that Kloosterman sums will come into play? Let us remark that $dd' - qq' = 1$ means that d' is the inverse of d modulo q, which recalls the definition (3.4). It is not too surprising that after the application of different forms of harmonic analysis to "detect" the solutions to this equation, one brings into play sums like the Kloosterman sums... Such an explanation is evidently only suggestive (and hardly that...). But it is in fact by reducing to sums of Kloosterman sums, and by invoking the results that Deshouillers and Iwaniec deduced from the Kuznetsov formula generalized to $\Gamma_0(q)$, that Fouvry and Iwaniec first, then more efficiently Bombieri, Friedlander, and Iwaniec, succeeded in *getting around* the Riemann Hypothesis. The best result currently known takes the following form: for any fixed integer $a \neq 0$, any $A \geq 1$, and any $\varepsilon > 0$, there exists a constant $C(a, A, \varepsilon)$ such that for all "well-factorable"[7] coefficients γ_q and every $x \geq 2$, we have

$$\left| \sum_{q \leq x^{4/7-\varepsilon}} \gamma_q E(x; q, a) \right| \leq C(a, A, \varepsilon) \frac{x}{(\ln x)^A}. \tag{3.9}$$

Let us repeat: *even* under the Generalized Riemann Hypothesis, this result is out of reach of an individual estimate of $E(x; q, a)$. Definitively, it depends on what we know about automorphic functions of (high) level q, and in particular the properties (and even the existence!) of Poincaré series.

Let us end with a very recent discovery by Goldston, Pintz, and Yıldırım. Using a completely new way of analyzing the small differences between prime numbers, they succeeding in proving the following remarkable result: *if* there exists a $\delta > 0$

[6]Note a certain analogy with the problem of Kloosterman sums: the Riemann Hypothesis is optimal when q and a vary; but for an average of remainder terms, as in the present case, one may hope for compensations from changes of size and (especially) of sign.

[7]This notion, introduced into sieve methods by Iwaniec, is relatively technical and we omit the definition; however, the essential thing to know is that it is in fact in the form written in (3.9), with such weights, that the averages of $E(x; q, a)$ appear most often in applications.

such that for all $A \geq 1$ and $x \geq 2$ one has[8]

$$\sum_{q \leq x^{1/2+\delta}} \max_{y \leq x} \max_{(a,q)=1} |E(y;q,a)| \leq D(A) \frac{x}{(\ln x)^A} \tag{3.10}$$

for some constant $D(A) \geq 0$, then there exists at least one integer $k \geq 2$ such that infinitely many (consecutive) prime numbers differ by k. It is hoped that this latter statement holds for all even k, the case $k = 2$ being the famous twin prime conjecture, but up to the present clear connections with well-established results from the analytic theory of prime numbers (other than sieve methods) have been lacking.

Alas, (3.9) is not the same thing as (3.10) for $\delta = 4/7 - 1/2$! In fact, in (3.10) one must be able to deal with variable (not fixed) a, and every interval $p \leq y \leq x$, not just $p \leq x$; finally, the implicit coefficient $\gamma_q = \operatorname{sgn}(E(y;q,a))$ is *not* well-factorable...

However, these differences will doubtlessly be the object of particular attention from the number-theorists already mentioned and many others! Perhaps new surprises will emerge from this attention...

Bibliography

[1] J. M. Deshouillers and H. Iwaniec, *Kloosterman sums and Fourier coefficients of cusp forms*, Invent. Math. **70** (1982), pp. 219–288.

[2] H. Iwaniec and E. Kowalski, *Analytic Number Theory*, A.M.S. Colloquium Publ. **53** (2004).

[3] H. D. Kloosterman, *On the representation of numbers in the form $ax^2 + by^2 + cz^3 + dt^2$*, Acta Math. **49** (1926), pp. 407–464.

[4] E. Kowalski, *Un cours de théorie analytique des nombres*, Cours Spécialisé **13**, Soc. Math. France, 2004.

[5] H. Poincaré, *L'avenir des mathématiques*, Atti IV Congr. Internaz. Matematici, Roma, 1908 (*Œuvres*, vol. 5, p. 22).

[6] H. Poincaré, *Fonctions modulaires et fonctions fuchsiennes*, Ann. Fac. Sci. Toulouse 3rd series, vol. **3** (1911), pp. 125–149; available at `http://numdam.org` or in *Œuvres*, vol. 2, pp. 592–618.

[7] H. Poincaré, *Extension aux nombres premiers complexes des théorèmes de M. Tchebicheff*, Journal de Math. 4th series, vol. **8** (1891), pp. 25–28; or in *Œuvres*, vol. 5, pp. 442–479.

[8] H. Poincaré, *Sur les invariants arithmétiques*, J. für die reine und angew. Math. (1905), pp. 89–150; or in *Œuvres*, vol. 5, pp. 203–266.

[9] P. Sarnak: *Some applications of modular forms*, Cambridge Tract. Math. **99**, Cambridge Univ. Press 1990.

[8]For $\delta = -\varepsilon$, $\varepsilon > 0$ as small as one likes, this statement already has a meaning and is known: it is the famous Bombieri–Vinogradov theorem; it is not stronger than the Riemann Hypothesis.

CHAPTER 4

The Theory of Limit Cycles

JEAN-PIERRE FRANÇOISE

4.1. Introduction

The notion of limit cycle (an isolated periodic orbit in the set of periodic orbits of a vector field; see the precise definition in Section 4.2) was introduced by H. Poincaré in his article [**25**]. The importance of this notion did not reveal itself at the time, however, perhaps because applications were most often, in Newtonian tradition, described by conservative (and even Hamiltonian) systems as opposed to dissipative systems. The school of Andronov would play, in what followed, an important role in showing the great interest in this notion for applications (cf. [**1**]). Nowadays, the theory of limit cycles plays a fundamental role in modeling, for example, life rhythms (cf. [**7, 16**]).

This chapter seeks to emphasize the advancement of mathematical research on limit cycles, which involves varied domains, including the theory of bifurcations (also formed by Poincaré), complex analysis, and analytic geometry. We begin by defining the Poincaré map of a periodic orbit in arbitrary dimensions and briefly presenting the notion, fundamental for applications, of frequency locking. Then we restrict ourselves to the case of dimension 2. First we mention Poincaré's contributions (the Poincaré–Bendixson theorem, compactification for polynomial systems). The essential problem, which is still open today, is to control the number of limit cycles as a function of the degree of the equation. The local version of this problem and its relation with the center-focus problem posed by Poincaré led to Bautin's theory. This theory now has a quantitative version founded on the methods of complex analysis. This way of approaching the study of limit cycles has given rise to more recent studies on Abel equations, thanks to which we have been able to better understand the structure of first return maps.

4.2. Periodic orbits

The notion of limit cycle is now mainly reserved for dimension 2, but it is not necessary to distinguish this notion from that of an isolated periodic orbit that, in any dimension, can represent the asymptotic state of a system for some initial conditions.

DEFINITION 4.2.1. Let X be a complete vector field[1] on an open subset of $\mathbb{R}^n$ (or, more generally, of a manifold...). A periodic orbit of the vector field X is an orbit $t \mapsto x(t)$ passing through a point x_0, which is not a singular point (i.e.,

[1]Meaning that the orbits $x(t)$ are defined on the entire interval $(-\infty, +\infty)$.

$X(x_0) \neq 0$), and for which there exists a number $T > 0$ (called the *period*) that satisfies $x(T) = x(0)$.

We specify the minimal period as the smallest positive real number T that satisfies this condition. The multiples of the minimal period are also periods. Unless otherwise specified, for a periodic orbit of period T, we always mean the minimal period.

When, for a vector field X of class C^k $(k \geq 1)$, we have a periodic orbit Γ, a solution to the differential equation $\dot{x} = X(x)$ passing through the point x_0, and a germ of a hyperplane[2] Σ transverse to Γ at x_0, then, for every point $x \in \Sigma$ sufficiently close to x_0, the solution to the differential equation passing through x at the instant $t = 0$ will be denoted $\varphi_t(x)$.

THEOREM 4.2.2. *We denote by*

$$\Gamma = \{x \mid x = \varphi_t(x_0),\ 0 \leq t \leq T\},$$

a periodic orbit of period T. Let Σ be the hyperplane orthogonal to Γ at x_0,

$$\Sigma = \{x \mid (x - x_0) \cdot X(x_0) = 0\}.$$

Then there exists $\delta > 0$ and a unique C^k-function $x \mapsto \tau(x)$ defined for

$$x \in \Sigma, \qquad |x - x_0| < \delta,$$

such that

$$\tau(x_0) = T \qquad \text{and} \qquad \varphi_{\tau(x)}(x) \in \Sigma.$$

The function $x \mapsto \tau(x)$ is called the "first return time" function of the periodic orbit passing through x_0. The proof of this theorem is very simple: it suffices to apply the implicit function theorem to the function $(t, x) \mapsto (\varphi_t(x) - x_0) \cdot X(x_0)$.

DEFINITION 4.2.3. The map $P : x \mapsto P(x) = \varphi_{\tau(x)}(x)$ is called the Poincaré first return map associated to the periodic orbit Γ.

The implicit function theorem shows that it has the same differentiability class as the vector field.

DEFINITION 4.2.4. In dimension 2, we call a limit cycle any periodic orbit that is isolated in the set of periodic orbits.

In what follows, we consider m oscillators:

$$\frac{dx_i}{dt} = f(x_i, y_i), \qquad \frac{dy_i}{dt} = g(x_i, y_i), \qquad i = 1, \dots, m. \tag{4.1}$$

The term "oscillator" here is taken in the sense of a dissipative oscillator, that is, each of the m oscillators of the system (4.1) has an attracting limit cycle. The vector field obtained by considering the set of variables (x_i, y_i) $(1 \leq i \leq m)$ therefore has a torus of periodic orbits, obtained by taking the product of the attracting limit cycle, and denoted in the following by $T^m(0)$.

Now assume that these m oscillators are weakly coupled:

$$\frac{dx_i}{dt} = f(x_i, y_i) + \varepsilon F_i(x, y, \varepsilon), \qquad \frac{dy_i}{dt} = g(x_i, y_i) + \varepsilon G_i(x, y, \varepsilon), \tag{4.2}$$

where ε is also assumed to be as small as desired.

[2]That is, a piece of a hyperplane in an unspecified, arbitrarily small neighborhood of x_0.

DEFINITION 4.2.5. We say that the system (4.2) demonstrates frequency locking if, for every sufficiently small value of ε, it has a periodic orbit γ_ε that, as $\varepsilon \to 0$, tends (in the sense of Hausdorff distance[3]) toward a periodic orbit contained in the periodic torus $T^m(0)$.

The simplest example is the "van der Pol" oscillator, forced by a sinusoidal oscillation of frequency $1/T$, which develops oscillation with multiple frequencies: see for example [**7**] or [**28**].

Let U be a neighborhood of an attracting periodic orbit $t \mapsto \gamma(t)$, such that $\gamma(0) = 0$.

LEMMA. *There exists a constant* $\alpha > 0$ *such that, for every* $x \in U$, *there exists* t_0 *(called the* asymptotic phase*) such that for every* $t \geq 0$,

$$|\varphi_t(x) - \gamma(t - t_0)| < Ke^{-\alpha t/T}.$$

PROOF. Let Π be a section transverse to the flow passing through $\gamma(0)$. The first return map can be written as

$$P : x_0 \mapsto x_1 = A \cdot x_0 + D(x_0), \qquad x_0 \in \Pi,$$

where $a = \|A\| < e^{-\alpha} < 1$ and D vanishes at 0 along with all of its derivatives. Therefore, if ε is sufficiently small and $\|x_1\| < \varepsilon$, we have $\|x_1\| < e^{-\alpha}\|x_0\|$ and, more generally, all of the $x_n = P^n(x_0)$ satisfy $\|x_n\| < e^{-\alpha n}\|x_0\|$. This shows in particular that x_n tends to 0, which expresses the orbital stability. Given an arbitrary ε, by continuity of the flow along a compact invariant set, there exists $\delta = \delta(\varepsilon)$ such that if the distance $d(\gamma, x_0) < \delta$, there exists a smaller strictly positive value $\mathcal{T}_0 = \mathcal{T}_0(x_0)$ for which $\varphi_{\mathcal{T}_0}(x_0) \in \Pi$ and $\|\varphi_{\mathcal{T}_0}(x_0)\| < \varepsilon$. Let $\tau(x)$ be the first return time function (here $\tau(0) = T$). We introduce $\tau_0 = 0$, $\tau_1 = \tau(x_0)$, $\tau_n = \tau_{n-1} + \tau(x_{n-1})$ for $n \geq 1$. Note that $\varphi_{\tau_n}(x_0) = x_n$.

Because the function τ is of class C^1 and x_n tends to 0, there exists L_0 such that

$$|\tau(x_n) - T| < L_0\|x_n\| < L_0 e^{-\alpha n}\|x_0\|.$$

The series whose general term is $(\tau_n - nT) - (\tau_{n-1} - (n-1)T)$ is therefore normally convergent. It follows that the sequence $\tau_n - nT$ converges, and we denote by t_0 its limit. By taking a partial sum of the series, we obtain in addition the upper bound

$$|\tau_n - (nT + t_0)| \leq L_1 e^{-\alpha n}\|x_0\|,$$

with $L_1 = L_0/(1 - e^{-\alpha})$. We obtain the following upper bound:

$$\|\varphi_{t+\tau_n}(x_0) - \varphi_{t+nT+t_0}(x_0)\| \leq L_2 e^{-\alpha n}\|x_0\|.$$

We have in addition

$$\|\varphi_{t+\tau_n}(x_0) - \gamma(t)\| = \|\varphi_t(x_n) - \varphi_t(0)\| \leq L_3\|x_n\| \leq L_3 e^{-\alpha n}\|x_0\|.$$

This makes it possible to write the upper bound

$$\|\varphi_{t+nT+t_0}(x_0) - \gamma(t)\| \leq (L_2 + L_3)e^{-\alpha n}\|x_0\|,$$

and by replacing $t + nT$ with t, we obtain the desired result:

$$\|\varphi_{t+t_0}(x_0) - \gamma(t)\| \leq (L_2 + L_3)e^{-\alpha t/T}\|x_0\|.$$

□

[3]The Hausdorff distance between two compact sets K, K' in a metric space (E, d) is, by definition, the smallest real number $r > 0$ such that $d(x, K') < r$ for all $x \in K$ and $d(y, K) < r$ for all $y \in K'$.

For a planar system, it is thus possible to define a system of "amplitude-phase" coordinates by completing the *asymptotic phase* function, defined thanks to the lemma with, for example, the square of the distance function to the periodic orbit.

Now we assume that the system (4.2) has frequency locking, and we assume that the periodic orbit $\Gamma(t)$ that defines this locking is attracting. We consider the projections $\Gamma_i(t)$ of this cycle onto the coordinate planes (x_i, y_i), $i = 1, \ldots, m$. We assume that ε is sufficiently small for this projection to lie in the open set U_i on which are defined the "amplitude-phase" coordinates of the system (4.2). This system is defined up to a choice of an origin on the periodic orbit. The differences of the phases between the different projections are therefore defined unambiguously. This determines a system of "amplitude-phase difference" coordinates. We can therefore write the system when restricted to the open set $U = \prod_{i=1}^{m} U_i$:

$$\frac{d\rho_i}{dt} = f_i(\rho, \alpha, \varepsilon), \qquad \frac{d\alpha_i}{dt} = \Phi_i(\rho, \alpha, \varepsilon), \qquad i = 1, \ldots, m. \tag{4.3}$$

DEFINITION 4.2.6. We say that the system (4.3) demonstrates phase locking if the system induced on $\Gamma(t)$ by the phase differences:

$$\frac{d(\alpha_i - \alpha_1)}{dt} = \Phi_i(0, (\alpha - \alpha_1), \varepsilon) \tag{4.4}$$

has an attracting singular point.

In what follows, we will often consider the case where the (non-perturbed) oscillators are identical and in particular where they have the same frequency ω. The frequency locking of the coupled system can therefore be written on a frequency vector

$$\underline{\omega} = \underline{k}\omega, \qquad \underline{k} \in \mathbb{Z}^N.$$

DEFINITION 4.2.7. A phase locking that occurs with the frequency vector such that $\underline{k} = (1, \ldots, 1)$ is called a synchronization.

The notion of frequency locking is discussed by Poincaré within the framework of existence theorems for periodic orbits of Hamiltonian systems. His method allows the search for periodic orbits to be reduced to that of critical points of functions[4] (Jacobi generating functions, secular functions). If these critical points are of Morse type,[5] the signature of the Hessian at these points determines the stability of the periodic orbit [**8**, **27**]. Poincaré originated a method [**27**], later developed by Malkin [**22**] and Roseau [**28**], which is now largely used to prove, for example, the existence of synchronization (neurosciences, cardiac physiology [**21**]).

4.3. Limit cycles. Poincaré–Bendixson theorem

The Poincaré–Bendixson theorem[6] is the most fundamental result that distinguishes 2-dimensional systems. Its statement is the following:

THEOREM (Poincaré–Bendixson). *We consider a C^1-vector field $X = (f, g)$ in the plane. Given a point m, let $\gamma_m = \{\varphi(t, m),\ t \in \mathbb{R}\}$ be an integral curve of X, defined for all t, such that the positive orbit $\gamma_m^+ = \{\varphi(t, m),\ t \geq 0\}$ passing through*

[4] A critical point of a function is a zero of the gradient.

[5] A function has a Morse critical point if the determinant of its Hessian (matrix of second derivatives) at this critical point is non-zero.

[6] Poincaré [**25**] for polynomial fields; Bendixson [**2**] for fields of class C^1.

the point m is contained in a compact set K. We define the ω-limit set of γ_m, denoted by $\Omega(m)$, as

$$\Omega(m) = \{q \in K \mid \text{there exists a sequence } (t_n) \text{ such that } t_n \to \infty \text{ as } n \to \infty \text{ and } \lim_{n\to\infty} \varphi(t_n, m) = q\}.$$

We assume that the field X has finitely many singularities in $\Omega(m)$. There are three possibilities:

(a) *If $\Omega(m)$ does not contain any singular points, it is a periodic orbit.*
(b) *If $\Omega(m)$ contains both singular and regular points, $\Omega(m)$ is composed of a set of orbits, each of which tends toward one of the singular points as $|t| \to \infty$. In this case, $\Omega(m)$ is called a graph.*
(c) *If $\Omega(m)$ does not contain any regular points, then it is a singular point.*

(A proof may be found in [**7**], for example.)

In the same article, Poincaré establishes, for a polynomial vector field in the plane, the existence of a compactification that makes it possible to read the phase portrait of the vector field over a disk by extending the orbits to the boundary of the disk. This allows the Poincaré–Bendixson theorem to be extended to infinity. For polynomial vector fields of degree less than a fixed integer n, one can bound the number of singular points and the number of graphs by a constant that only depends on n. It is natural to ask if one can also uniformly bound the number of limit cycles as a function of the degree, and this question is precisely what is now called part (B) of Hilbert's 16th problem. The question initially posed by Hilbert in 1900 was phrased in vaguer terms ("to find the number and location of the Poincaré limit cycles"). In 1923, H. Dulac announced the proof of the fact that for a fixed polynomial field, the number of limit cycles is finite. His "proof" is based on an asymptotic analysis of the Poincaré first return map in a neighborhood of the graphs. It was considered correct until Yu. Ilyashenko noticed, in 1982, that one of the key lemmas of the proof was false! Following that, J. Ecalle [**5**] and Ilyashenko independently corrected the proof and showed new properties of the asymptotic expansion classes involved [**17**]. In the rest of this chapter, we will focus on a local variant of the problem that was discovered by N. N. Bautin. Bautin belonged to the research group founded by the students of Andronov at Nizhny Novgorod.

4.4. The local Hilbert's 16th problem, Bautin's theory

We consider the polynomial vector field:

$$\dot{x} = -y + cx + P(x,y) = -y + cx + \sum_{i,j \mid i+j \geq d} a_{i,j} x^i y^j, \tag{4.5}$$

$$\dot{y} = x + cy + Q(x,y) = x + cy + \sum_{i,j \mid i+j \geq d} b_{i,j} x^i y^j. \tag{4.6}$$

We write this system in polar coordinates (r, θ):

$$x = r\cos\theta, \qquad y = r\sin\theta.$$

This leads to:

$$\begin{aligned} r\dot{r} &= x\dot{x} + y\dot{y} = cr^2 + xP + yQ = cr^2 + r^{d+1}A(r,\theta), \\ \dot{\theta} &= (x\dot{y} - y\dot{x})/(x^2+y^2) = 1 + r^{d-1}B(r,\theta), \end{aligned}$$

where $A(r,\theta)$ and $B(r,\theta)$ are two polynomials in $(r, \cos\theta, \sin\theta)$, linear in the parameters a, b. This gives

$$\frac{dr}{d\theta} = \frac{cr + r^d A(r,\theta)}{1 + r^{d-1}B(r,\theta)}, \tag{4.7}$$

and therefore

$$\frac{dr}{d\theta} = cr\left[\sum_{k=0}^{\infty}(-1)^k r^{k(d-1)}B(r,\theta)^k\right] + \sum_{k=0}^{\infty}(-1)^k r^{k(d-1)+d}A(r,\theta)B(r,\theta)^k. \tag{4.8}$$

This equation can be written as

$$\frac{dr}{d\theta} = \sum_{k=0}^{\infty} r^k R_k(r,\theta), \tag{4.9}$$

where the coefficients $R_k(r,\theta)$ are polynomials in $(r, \cos\theta, \sin\theta)$ that depend polynomially on the parameters a, b. Bautin proposed looking for the solution $r = r(\theta)$ to (4.9) such that $r(0) = r_0$ as a series:

$$r = v_1(\theta)r_0 + v_2(\theta)r_0^2 + \cdots + v_k(\theta)r_0^k + \cdots. \tag{4.10}$$

Comparing (4.9) and (4.10) produces:

$$\begin{aligned} &v_1'(\theta) = c, \quad v_2(\theta) = \cdots = v_{d-1}(\theta) = 0, \quad v_d'(\theta) = R_d(\theta), \\ &v_k'(\theta) = \sum_{i=2}^{k} B_{ik}[v_d(\theta), \ldots, v_{k-1}(\theta)]R_i(\theta), \qquad k \geq d+1. \end{aligned} \tag{4.11}$$

The polynomial $B_{ik}[a_d, \ldots, a_{k-1}]$ has integer coefficients and coincides with the coefficient of X^{k-i} in $(X + a_d X^d + \cdots + a_p X^p + \cdots)^i$. The relation (4.11) makes it possible to construct the functions $v_k(\theta)$ recursively:

$$v_k(\theta) = \int_0^\theta \sum_{i=2}^{k} B_{ik}[v_d(\varphi), \ldots, v_{k-1}(\varphi)]\, R_i(\varphi)\, d\varphi. \tag{4.12}$$

This construction makes two facts apparent.

(i) The coefficient $v_k(\theta)$ is a polynomial in θ and $\exp(c\theta)$ (of degree less than k) and in $(\sin(\theta), \cos(\theta))$.
(ii) The coefficient $v_k(\theta)$ is a polynomial in the vector field parameters (a, b).

Therefore, in particular the coefficients $v_k(2\pi)$ of the first return map are polynomials in the parameters (a, b). We obtain the analyticity of the Poincaré first return map expressed by the convergent series

$$r = v_1(2\pi)r_0 + v_2(2\pi)r_0^2 + \cdots + v_k(2\pi)r_0^k + \cdots. \tag{4.13}$$

If we add as a hypothesis that the first non-linear term of the equation is such that $v_2(2\pi) < 0$, we can easily deduce from Bautin's analysis the "Hopf bifurcation theorem" (in this particular context), that is, *the existence of a limit cycle that appears from the origin once the parameter c becomes strictly positive* (and here we get in addition the stability of the limit cycle). In fact, this is equivalent to the existence of a positive root $r = r(c)$ of the equation $L(r, c) - r = 0$, which has the form:

$$f(r, c) = (\exp(2\pi c) - 1) + v_2(2\pi)r + \cdots = 0.$$

The implicit function theorem implies that there exists a solution $r = r(c)$ such that $r(0) = 0$, since $f'_r(0,0) = v_2(2\pi) \neq 0$, and this solution is positive because it is close to $-(\exp(2\pi c) - 1)/v_2(2\pi)$.

If the second coefficient of the Poincaré map is zero, it is necessary to transfer the discussion to the third, and so on. In general, one has a bifurcation of greater order that is supercritical or subcritical according to the sign of the first non-zero coefficient of the Poincaré map.

In what follows, we assume that $c = 0$, and we denote by $\lambda = (a, b)$ the set of coefficients of the perturbation. The most important point of Bautin's theory is the idea of replacing the Weierstrass preparation theorem,[7] which does not apply here (because we are in the case where $f(x, \lambda_0)$ is identically zero), with the following theorem (stated and proved in full generality for the first time in [**10**]):

THEOREM. *Let $f(x, \lambda) = \sum_{i=0}^{\infty} f_i(\lambda)x^i$ be a germ of an analytic function whose coefficients depend polynomially on λ. Let λ_0 be a value of λ for which $f(x, \lambda_0) = 0$ for all x. Consider the increasing sequence of ideals generated by the coefficients of $f(x, \lambda)$ in the ring of polynomials in λ. This increasing sequence stabilizes, and we call the limit of these ideals the* Bautin ideal. *Let l be the number of elements of a generating set of the Bautin ideal. There exists a neighborhood U of the origin in $\mathbb{R}$ and a neighborhood V of λ_0 in $\mathbb{R}^d$ such that for all $\lambda \in V$, the number of $x \in U$ that are isolated zeroes of $f(x, \lambda)$ is smaller than l.*

When this theorem is applied to the first return map, it leads to an upper bound for the number of limit cycles that can appear in a neighborhood of the origin.

Bautin's theory gave rise to recent developments [**12, 13**] with a quantitative version based on methods of complex analysis. Not only do they lead to precise bounds on the number of limit cycles, but they also give estimates on their location.

4.5. The Poincaré center-focus problem

Let us again take the system (4.5)–(4.6) with $c = 0$. For $P = Q = 0$, the orbits are concentric circles. *A priori,* for a small perturbation (P, Q), two cases are possible in a neighborhood of the origin: either the orbits are again periodic (we then say that the origin is a *center*), or they wind around the origin like spirals (we then say that the origin is a *focus*, stable if the trajectories approach it, or unstable if they move away from it).

In 1885, Poincaré posed the problem of characterizing the coefficients P and Q of the perturbation for which the origin remains a center (cf. [**26**]). This question is now called the Poincaré *center-focus problem.*[8] The case that has given the most developments is the one where the perturbation is homogeneous. In this case, it amounts to the same thing as requiring that all the orbits of the system

$$\dot{x} = y + \varepsilon P(x, y), \qquad \dot{y} = -x + \varepsilon Q(x, y) \tag{4.14}$$

[7] If $f(x, \lambda)$ is a germ of an analytic function at the point $(x_0, \lambda_0) \in \mathbb{R} \times \mathbb{R}^n$, such that $f(x, \lambda_0)$ is not identically zero, then there exists a unique germ of an analytic function $a(x, \lambda) \neq 0$ and a unique monic polynomial in x, $\Pi(x, \lambda) = (x - x_0)^m + b_{m-1}(\lambda)(x - x_0)^{m-1} + \cdots + b_0(\lambda)$ (with $m \geq 0$, and where the $b_j(\lambda)$ are germs of analytic functions at λ_0 that are zero at this point), such that $f(x, \lambda) = a(x, \lambda)\Pi(x, \lambda)$.

[8] EDITOR'S NOTE: Or simply the *center problem.* We will see it again in Chapter 5.

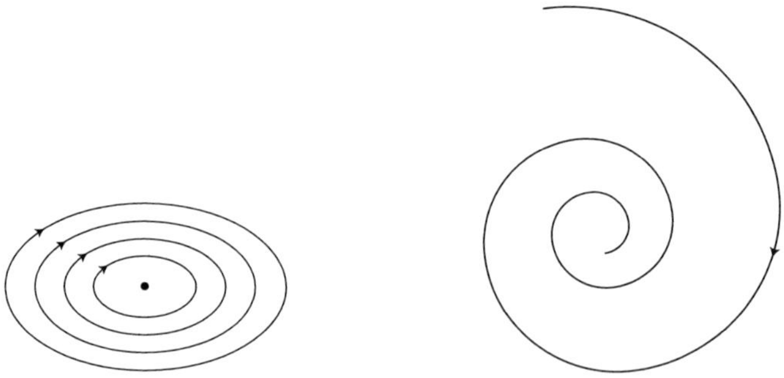

FIGURE 4.1. A center and a focus.

be periodic in a neighborhood of the origin and for every ε. This allows us to make an expansion in powers of ε and to look for an expression for the successive derivatives of the Poincaré map with respect to ε in a neighborhood of the origin.

The vector field defines a foliation of the plane, to which we can associate the differential 1-form

$$\Omega = dH + \varepsilon\omega,$$

where

$$H(x,y) = \frac{1}{2}(x^2 + y^2) \qquad \text{and} \qquad \omega = P(x,y)\,dy - Q(x,y)\,dx.$$

The first derivative of the Poincaré map (with respect to ε) is given by a formula of Pontryagin:

$$L_1(t) = \int_{H=t} \omega.$$

Work on the succeeding derivatives began (cf. [**6, 23**]) with the observation that, if we assume that L_1 is identically zero, then there exist polynomials g and R such that

$$\omega = g\,dH + dR.$$

In this case, the second derivative is given by

$$L_2(t) = \int_{H=t} g\omega.$$

And so forth; more generally, we obtain an expression for the first derivative that is not identically zero.

Subsequently, this approach underwent several developments. It was possible, for example, to extend this process to a more general class of functions H. In the case where one cannot apply variants of the process, it was shown that the first non-zero derivative is always a solution to a Fuchs-type equation (an equation with regular singularities); there exist cases where it is not given by an abelian integral.[9] More recently, several authors have independently remarked upon the connection between these successive derivatives and iterated integrals [**6, 9, 14, 15, 18, 19, 20, 29**].

[9] An integral of a polynomial differential form over an oval of an algebraic curve.

4.6. Abel's equations and Hilbert's 16th problem in dimension 1

In 1976, L. Cherkas [**4**] became interested in the case where P and Q are homogeneous of the same degree (and $c = 0$), and he proposed transforming the equation (4.7), which in this case becomes

$$\frac{dr}{d\theta} = \frac{r^d A(\theta)}{1 + r^{d-1}B(\theta)}, \tag{4.15}$$

by setting

$$\rho = \frac{r^d}{1 + r^{d-1}B(\theta)}. \tag{4.16}$$

We obtain the equation

$$\frac{d\rho}{d\theta} = q(\theta)\rho^3 + p(\theta)\rho^2, \tag{4.17}$$

where $p(\theta) = (d-1)f(\theta) + g'(\theta)$ and $q(\theta) = -(d-1)f(\theta)g(\theta)$ are trigonometric polynomials of degree $d+1$ and $2d+2$, respectively.

This type of equation is connected with polynomial Abel equations:

$$\frac{dy}{dx} = q(x)y^3 + p(x)y^2, \tag{4.18}$$

and it is natural to call (4.17) a trigonometric Abel equation. The Poincaré center-focus problem and Hilbert's 16th problem naturally extend to this context, and this has given rise to the development of an entirc field of recent research, which we shall summarize in what follows.

The common context that surrounds the two previous situations is that of a complex differential equation with rational function coefficients:

$$\frac{dz}{dw} = q(w)z^3 + p(w)z^2, \tag{4.19}$$

which defines an analytic complex foliation of $\mathbb{C} \times \mathbb{C}$. Given a path γ in the w-plane that avoids the singularities (both fixed and moving) of the equation and goes from w_0 to w_1, we can define (at least for z_0 near 0) an analytic map $P_\gamma : z_0 \mapsto z_1$. In the special case where this path is a loop, we recover the case of the usual (trigonometric) center, and this situation is also more general that that envisaged above in the particular case where p and q are polynomials and the foliation is real. This generalization is in the spirit of an approach, proposed by Petrovskii and Landis [**24**], to Hilbert's 16th problem. It may be picked up again after a detour by Abel's equations, which should certainly shed a new light on this way of attack.

4.7. Complex Poincaré map

We will not go deep into the subject here, because it is currently the object of very active research. We limit ourselves to an example that was studied in detail in [**11**]. The idea is to consider the following Abel equation:

$$\frac{dy}{dx} = cxy^3 + y^2. \tag{4.20}$$

In the case where the parameter c equals

$$c = \frac{1}{4} - \frac{1}{4}\left[\frac{1}{2n+1}\right]^2,$$

the equation admits the first integral

$$H(x,y) = \frac{y(1 - xy/v_1)^n}{(1 - xy/v_2)^{n+1}},$$

where $v_1 = -(2n+1)/n$ and $v_2 = -(2n+1)/(n+1)$. It is shown in [**11**] that the curves $H(x,y) = c$ are irreducible (for non-critical c). The equations (4.20) thus provide an example of a family of polynomial equations of bounded degree with a first integral whose degree is unbounded. As $n \to \infty$, $c \to \frac{1}{4}$, and one can show that equation (4.20) has, for $c = \frac{1}{4}$, infinitely many complex limit cycles ($y(1) = y(0)$) given as the complex solutions to the equation

$$\frac{y_0}{y_0 + 2} e^{\frac{2}{y_0+2}} = \frac{e}{2} y_0.$$

Bibliography

[1] A. Andronov, *Sur les cycles limites de Poincaré et la théorie des oscillations auto-entretenues*, Note in the Comptes Rendus de l'Acad. des Sci. presented by J. Hadamard, October 14, 1929.

[2] I. Bendixson, *Sur les courbes définies par des équations différentielles*, Acta Math. **24** (1901), pp. 1–88.

[3] A. Brudnyi, *An explicit expression for the first return map in the center problem*, J. Diff. Equations **206** (2004), pp. 306–314.

[4] L. A. Cherkas, *Number of limit cycles of an autonomous second-order system*, Diff. Eq. **5** (1976), pp. 666–668.

[5] J. Ecalle, *Introduction aux fonctions analysables et preuve constructive de la conjecture de Dulac*, Hermann, Paris, 1992.

[6] J.-P. Françoise, *Successive derivatives of a first return mapping: application to quadratic vector fields*, Ergodic Th. & Dynamical Systems **16** (1996), pp. 87–96.

[7] J.-P. Françoise, *Oscillations en biologie*, Collection "Mathématiques & Applications", Springer, Berlin, 2005.

[8] J.-P. Françoise, V. Guillemin, *On the period spectrum of a symplectic map*, Journal of Functional Analysis **100** (1991), pp. 317–358.

[9] J.-P. Françoise, M. Pelletier, *Iterated integrals, Leray residue and first return map*, J. of Dynam. Control Systems **12**, 3 (2006), pp. 357–369.

[10] J.-P. Françoise, C. C. Pugh, *Keeping track of limit cycles*, J. Differential Equations **65** (1986), pp. 139–157.

[11] J.-P. Françoise, N. Roytvarf, Y. Yomdin, *Analytic continuation and fixed points of the Poincaré mapping for a polynomial Abel equation*, J. Eur. Math. Soc. **10** (2008), pp. 543–570.

[12] J.-P. Françoise, Y. Yomdin, *Bernstein inequality and applications to analytic geometry and differential equations*, J. Funct. Analysis **146** (1997), pp. 185–205.

[13] J.-P. Françoise, Y. Yomdin, *Projection of analytic sets and Bernstein inequalities*, Singularities Symposium-Lojasiewicz 70, B. Jacubczyk, W. Pawlucki, Y. Stasica, eds., Banach Center Publications (Warsaw) **44** (1998), pp. 103–108.

[14] L. Gavrilov, *Higher order Poincaré–Pontryagin functions and iterated path integrals*, Ann. Fac. Sci. Toulouse Math. (6) **14** (2005), no. 4, pp. 663–682.

[15] L. Gavrilov, I. D. Iliev, *The displacement map associated to polynomial unfoldings of planar Hamiltonian vector fields*, Amer. J. Math. **127** (2005), no. 6, pp 1153–1190.

[16] A. Golbeter *Biochemical oscillations and cellular rhythms*, Cambridge University Press, Cambridge, 1996.

[17] Yu. S. Ilyashenko, *Finiteness theorems for limit cycles*, Russian Math. Surveys **45** (1990), pp. 129–203.

[18] I. D. Iliev, *The cyclicity of the period annulus of the quadratic Hamiltonian triangle*, J. Differential Equations **128** (1996), pp. 309–326.

[19] A. Jebrane, P. Mardešić, M. Pelletier, *A generalization of Françoise's algorithm for calculating higher order Melnikov functions*, Bulletin des Sciences Mathématiques **126** (2002), pp. 705–732.

[20] A. Jebrane, P. Mardešić, M. Pelletier, *A note on a generalization of Françoise's algorithm for calculating higher order Melnikov functions*, Bulletin des Sciences Mathématiques **128** (2004), pp. 749–760.

[21] J. Keener, J. Sneyd, *Mathematical Physiology*, Interdisciplinary Applied Mathematics, Springer, New York, 1998.

[22] I. G. Malkin, *Certain problems in the theory of nonlinear oscillations*, Moscow, English translation in U.S. Atomic Energy Commission, 1959.

[23] L. Perko, *Differential Equations and Dynamical Systems*, 3rd edition, Texts in Applied Mathematics **7**, Springer, 2001.

[24] I. Petrovskii, E. Landis, *On the number of limit cycles of the equation* $dy/dx = P(x,y)/Q(x,y)$ *where* $P(x,y)$ *and* $Q(x,y)$ *are polynomials of* 2nd *degree*, Mat. Sb. **37** (1955), pp. 209–250.

[25] H. Poincaré, *Mémoire sur les courbes définies par une équation différentielle*, Chapter VI (Cycles limites), Journal de Mathématiques, 3rd series **7** (1881), pp. 375–422; **8** (1882), pp. 251–296.

[26] H. Poincaré, *Sur les courbes définies par les équations différentielles*, Chapter XI (Théorie des centres), Journal de Mathématiques pures et appliquées, 4th series **1** (1885), pp. 167–244.

[27] H. Poincaré, *Les méthodes nouvelles de la mécanique céleste*, vol. 3, Gauthier-Villars, Paris, 1892. Reprintings: Dover Publications Inc., New York, 1957, and Blanchard, Paris, 1987.

[28] M. Roseau, *Vibrations non linéaires et théorie de la stabilité*. Springer Tracts in Natural Philosophy, vol. 8, Berlin-Heidelberg-New York, 1966.

[29] M. Uribe, *Principal Poincaré–Pontryagin function of polynomial perturbations of the Hamiltonian triangle*, J. Dyn. Control Systems **12** (2006), no. 1, pp. 109–134.

CHAPTER 5

Singular Points of Differential Equations: On a Theorem of Poincaré

DOMINIQUE CERVEAU

Introduction

The study of singular points of differential equations, especially in two real or complex variables, for more than a century was the exclusive domain of the Russian and French schools. From very early on (Briot and Bouquet for example, 1856), a privileged place was given to the search for special solutions (called *separatrices*), which are germs of curves γ (holomorphic in the complex case, real analytic in the real case) passing through the singular point 0, and such that $\gamma - \{0\}$ is a solution to the differential equation in question; then one leaps into the study of first integrals "à la Liouville", that is, bringing into the picture formulas that are algebraic in the elementary functions (typically the logarithm).

In order to construct such curves, one uses techniques from the mid-19th century for the resolution of singularities (blow-ups, Newton polygons), even though the final result on the existence of reduction of singularities for singular points of ordinary differential equations was not formally established until much later [**9**]. In a similar way, well before Zariski and Hironaka, Picard used the resolution of singularities on surfaces as an obvious fact!

In the holomorphic (i.e., complex analytic) case, the existence of a separatrix, which also appears to have been a tautology to early researchers, was rigorously proved only in the 1980s by Camacho and Sad in a famous article, one of the beautiful results of the Brazilian school, which has been very active since that same decade. Here is the statement:

THEOREM ([**4**]). *Let $\Omega = A\,dz + B\,dw$ be a holomorphic 1-form defined in a neighborhood of the point 0 in $\mathbb{C}^2$. There exists a holomorphic curve γ passing through 0 such that $\gamma - \{0\}$ is a solution to the differential equation $\Omega = 0$.*

Let us be more explicit. Here A and B are given by convergent series in z and w. If one presents γ in the parametrized form $t \mapsto (z(t), w(t))$, the theorem simply says that $z'(t)\,A(z(t), w(t)) + w'(t)\,B(z(t), w(t))$ is identically zero; in other words, $t \mapsto (z(t), w(t))$ is a parametric solution to the ordinary differential equation:

$$w' = -\frac{A}{B}(z, w).$$

Of course this kind of statement fails in the real case. The simplest example is given by the differential equation $x\,dx + y\,dy = 0$, whose solutions are concentric circles centered at 0: evidently no analytic curve passing through the singular point 0 can be a solution to this equation.

The study of holomorphic differential equations is still alive and well. Besides the Camacho–Sad theorem, it uses "holonomic" techniques, products of the theory of global foliations, whose very essence goes back to Poincaré (the famous first return map[1]) and the resolution of singularities mentioned above. We will come back to it.

In parallel, one finds an abundant literature concerning the description of real differential equations, motivated in large part by Hilbert's 16th problem: we are given an equation $\Omega = P\,dx + Q\,dy$, where P and Q are Nth degree polynomials in two real variables; the idea is to provide an upper bound for the number of isolated closed curves (limit cycles) that solve $\Omega = 0$, as a function of the degree N. Nothing is known today, even in degree 2 (the lowest interesting degree). All the same we know that this number is finite! Announced by Dulac at the beginning of the 20th century [**6**], this fact was only proved recently by Ecalle [**7**] and Ilyashenko [**10**]. The two proofs are difficult, conceptually and technically; Dulac's "proof" was only recognized as incomplete in the 1980s and was taught for many years in prestigious universities... As a result of the arrival of Hilbert's 16th problem there were developments, under the impetus of the Russian school—in particular Ilyashenko and his students—in the study of bifurcations of "phase portraits" (i.e., drawings of the solutions), in general in a neighborhood of "Hamiltonian" equations, the case where Ω has a first integral, for example polynomial, or more generally of type $\prod_i f_i^{\lambda_i}$, where the f_i are polynomials and the λ_i are real numbers. This approach led to problems in analysis known to be difficult; the idea was to estimate the number of roots of functions φ constructed from abelian integrals $\varphi(t) = \int_{f=t} \eta$, where η is a certain 1-form and f is the first integral of Ω.

As for the study of singular points of real analytic equations $\Omega = 0$, the "topological" description of solutions is complete. This means that we know how to predict the phase portraits produced by such equations. The same is not true for the analytic description; it must be said that we desire the impossible: a list of "invariants" that would allow us to predict that two equations $\Omega_i = 0$, $i = 1, 2$, can be exchanged by a coordinate transformation once they have the same invariants.

Ever since the work of the pioneers, the study of the real case has often been rendered opaque by the abundant research concerning the complex domain, the more-or-less general feeling being that it is sufficient to take "the real part of complex results" and that consequently there is no reason to dwell upon it, the power of complex analysis as developed in the 19th century taking care of the rest. Poincaré's approach did not emerge from this point of view. He was without a doubt the first to extract concepts of a global nature—i.e., not resting on arguments calculated in coordinates—thus leading to theorems also of a global nature: one inevitably thinks of the *Poincaré–Bendixson theorem*[2] describing the limit sets of trajectories in dimension 2.

It is of course out of the question to present what would only be a summary of Poincaré's contributions in the realm of differential equations. We will only touch upon one theme that he mentions as important and that deals with periodic solutions, more precisely with "stable" configurations.

After several pictures with commentary to prepare the reader for the use of complex fields, we will begin, in order to fix some basic ideas and techniques, by

[1] EDITOR'S NOTE: See p. 88.

[2] EDITOR'S NOTE: See p. 90.

stating and recalling a proof of the Poincaré–Lyapunov center theorem; the idea of the proof that we present is not Poincaré's; it was suggested by G. Reeb and carried out by R. Moussu. We will give several generalizations of it in connection with reversibility and integrability in the sense of Liouville. We will finish with an anecdotal result, a kind of generalization of the center theorem in the hyperbolic case.

5.1. The Poincaré–Lyapunov center theorem

We consider an ordinary differential equation

$$E = \begin{cases} x' = A(x, y), \\ y' = B(x, y), \end{cases}$$

where A and B are analytic functions in a neighborhood of the origin 0 in $\mathbb{R}^2$. We assume that A and B vanish at 0 (the equation E is said to be singular) but do not simultaneously vanish on the same curve. The solutions to E are the trajectories of the vector field X_E,

$$X_E = A\frac{\partial}{\partial x} + B\frac{\partial}{\partial y},$$

and they parametrize the integral curves of the differential form

$$\omega_E = B\,dx - A\,dy.$$

When the Jacobian matrix J_E,

$$J_E = \begin{pmatrix} \partial A/\partial x & \partial A/\partial y \\ \partial B/\partial x & \partial B/\partial y \end{pmatrix},$$

is invertible at the origin, E is called non-degenerate. Under the hypothesis of non-degeneracy, Poincaré distinguished the following *topological* types:

(1) The **nodes**, whose phase portrait (up to orientation) is of the following type:

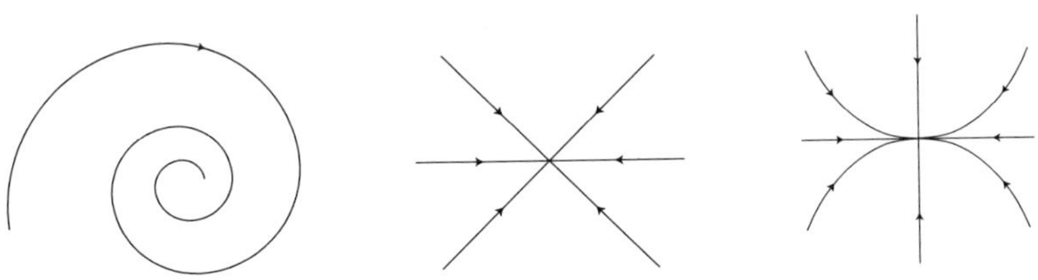

All of the trajectories contain the singular point in their closures. Note that the above pictures are all "topologically" equivalent (i.e., identical *up to homeomorphism*).

(2) The **saddles**, with phase portrait:

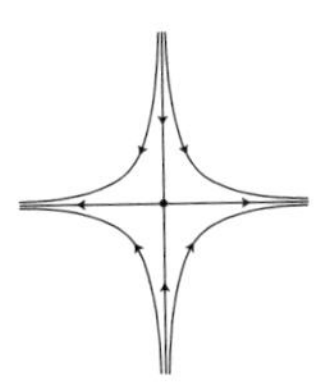

Here only four trajectories are attached to the singular point.

(3) The **centers**, the case where all the solutions of E are periodic:

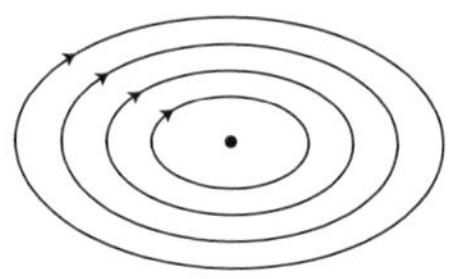

For example, if one observes the solutions to the differential equation associated with the simple pendulum:

$$x' = y,$$
$$y' = \sin x,$$

one observes the existence of singular points of center type and saddle type.

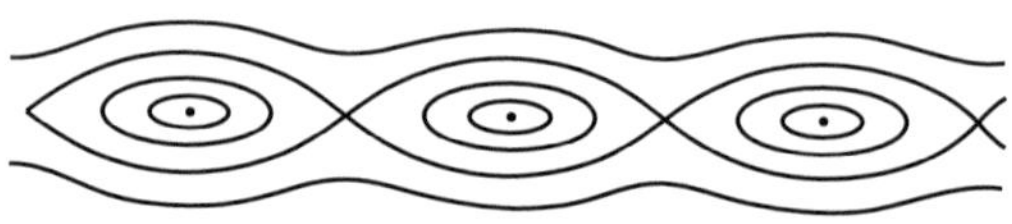

FIGURE 5.1. Phase portrait of a simple pendulum.

The introduction of friction leads to the disappearance of the centers, which become sinks, as well as that of the "saddle connections", although the saddles themselves persist.

Poincaré mentions the rarity of centers but underlines their importance: these are the only cases where the singular point is "stable" (in the sense of Lyapunov). He proposes, at least in the non-degenerate case, to decide, starting from the knowledge of the coefficients of the series A and B, if the singular point is a center or not. Poincaré [**8**] remarked that a center can only occur if infinitely many conditions are satisfied; more precisely, he observed that the non-vanishing of certain "quantities that one cannot help comparing with the secular-correction terms" dear to astronomers, permits the construction of closed algebraic curves C, encircling the singular point, which are transverse to the field X_E:

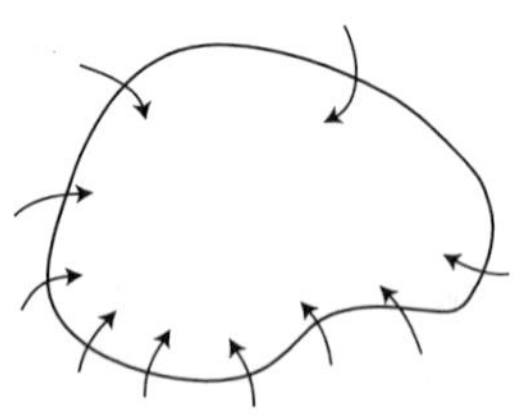

This constitutes an obstruction to the singular point being a center. The idea here is in fact a remarkable geometric point of view on what we today call the theory of "normal forms".

Of course, for a non-degenerate field X_E to be a center, it is necessary that its linear part (the degree 1 part in the Taylor expansion) be, up to linear isomorphism, of the following form:

$$L_E = y\frac{\partial}{\partial x} - x\frac{\partial}{\partial y}.$$

We remark that the linear field L_E has the first integral $x^2 + y^2$. Poincaré proved that a non-degenerate field of center type has an analytic first integral f of Morse type (i.e., of type $x^2 + y^2$ for a good choice of coordinates). His proof is difficult to read, in particular the part dealing with the convergence of f, but the approach is not without recollections of that of B. Malgrange in "Frobenius with singularities, I: codimension 1". This result is known today by the name of the Poincaré–Lyapunov center theorem:

THEOREM ([**8**]). *If the equation E is non-degenerate, then E is of center type if and only if X_E has an analytic first integral of Morse type.*

To the extent that the arguments we present use and abuse complex fields, here is a section where we try to give a description of the solutions of a linear vector field on the space $\mathbb{C}^2$.

5.2. A few pictures with commentary

We will here attempt to visualize the complex trajectories of linear vector fields:

$$X_\lambda = z\frac{\partial}{\partial z} + \lambda \cdot w\frac{\partial}{\partial w}, \quad \text{where} \quad \lambda \in \mathbb{C} - \{0\}.$$

The flow of X_λ is given by

$$\varphi_T(z, w) = (ze^T, we^{\lambda T}), \quad T \in \mathbb{C},$$

and this flow parametrizes the aforementioned trajectories. We note that the multivalued function $\lambda \log z - \log w$ is a first integral of X_λ, and we therefore propose to draw its level curves. As we are in four dimensions, we will proceed by *tomography*, that is, cutting by real submanifolds of $\mathbb{C}^2$.

Before this let us make two observations. The first is obvious: the two coordinate axes (with the origin removed) are trajectories of X_λ. The second rests on the following fact: for (z, w) not belonging to the axes, the map $T \mapsto \varphi_T(z, w)$ is injective if and only if λ is not rational ($\lambda \in \mathbb{C} \setminus \mathbb{Q}$). This has the following geometric consequence: if (z, w) is not situated on one of the two axes, the trajectory passing through (z, w) can be identified with $\mathbb{C}$ (it is therefore a plane, from the topological point of view) when λ is non-rational and with $\mathbb{C} - \{0\}$ (an annulus, for the topologists) when λ is rational.

One last remark before setting the tomography on its way: by a clever trick of homothety, it suffices for us to study the trajectory $\mathcal{L}_\lambda$ from the point $(1, 1)$, which is what we will do.

5.2.1. Complex tomography. We want simply to describe the intersection S_λ of $\mathcal{L}_\lambda$ with the complex line $z = 1$:

$$S_\lambda = \{(1, e^{2i\pi k\lambda}),\ k \in \mathbb{Z}\}.$$

This is child's play: if λ is rational, $\lambda = p/q$, then S_λ is nothing other than the set of qth roots of unity. When λ is irrational ($\lambda \in \mathbb{R} \setminus \mathbb{Q}$), the points of S_λ form a dense subset of the unit circle, so well that our eye will tell us that S_λ is a circle! Only the generic case, where λ is not real, remains: we observe that the points of S_λ are distributed on spirals going from the origin to infinity.
This elementary tomography already indicates to us that, when λ is not real, $\mathcal{L}_\lambda$ contains the axis $w = 0$ in its closure; it is of course the same for the axis $z = 0$.

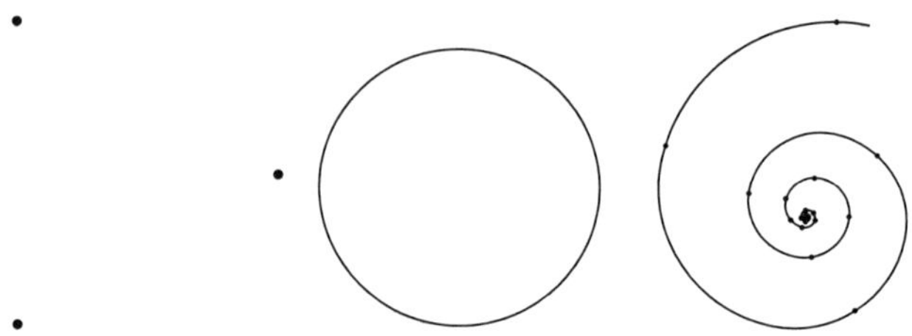

FIGURE 5.2. The cases of $\lambda = 1/3$, $\lambda = \sqrt{2}$ and generic λ, respectively.

5.2.2. Tomography by real 3-spaces. This time we intersect $\mathcal{L}_\lambda$ with the "cylinder":

$$\mathcal{C} = \{(z, w),\ |z| = 1\},$$

a space containing the line $z = 1$. This is the product of a circle and a plane. Everyone knows that the plane is homeomorphic to the open disk. One can therefore represent $\mathcal{C}$ as the product of a circle and a disk Δ, a product that is nothing other than the interior of a filled torus having as its core the circle $S^1 = \mathcal{C} \cap \{w = 0\}$:

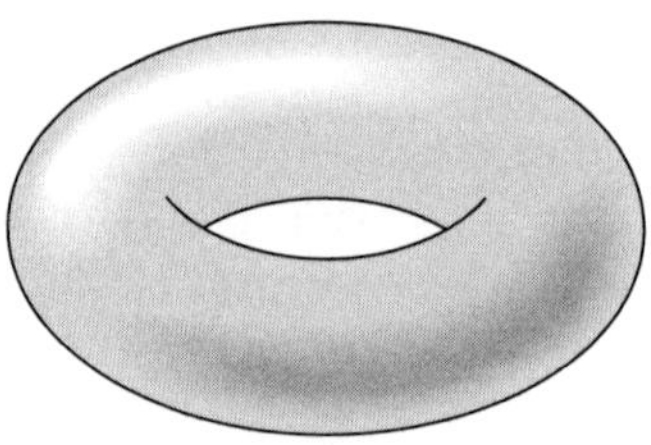

Let us now examine, according to the value of λ, the intersection $\mathcal{C}_\lambda$ of $\mathcal{L}_\lambda$ and $\mathcal{C}$:

$$\mathcal{C}_\lambda = \{(e^{2i\pi\theta}, e^{2i\pi\lambda\theta}),\ \theta \in \mathbb{R}\}.$$

We begin with the case, *a priori* easy but not actually so, where $\lambda = 1$. The trajectories of X_1 are lines in $\mathbb{C}^2$ passing through 0 (with 0 removed). We note that $\mathcal{C}_1$ is traced on the two-dimensional torus $\mathbb{T}$,

$$\mathbb{T} = \{|z| = |w| = 1\},$$

and is precisely the "primary bisector" of this torus: "the thread that goes around once in both directions". We observe that the configuration of the two circles $\mathcal{C}_1$ and $\mathcal{C} \cap \{w = 0\}$ is the following:

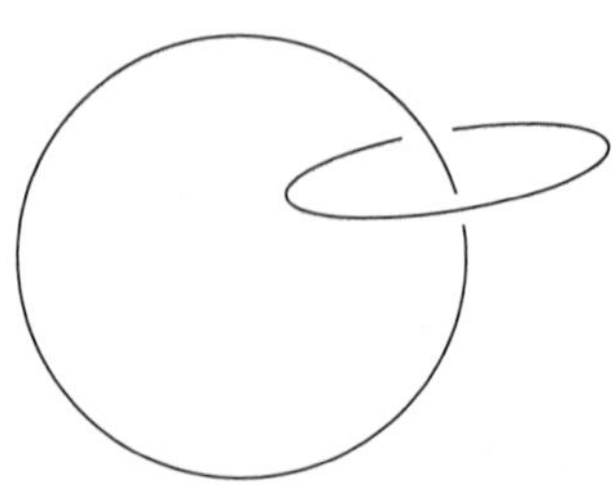

Imagine a chain with two links... If we now take the intersection of three lines with

$\mathcal{C}$, we do not obtain a chain with three links... but the **Borromean rings**:[3]

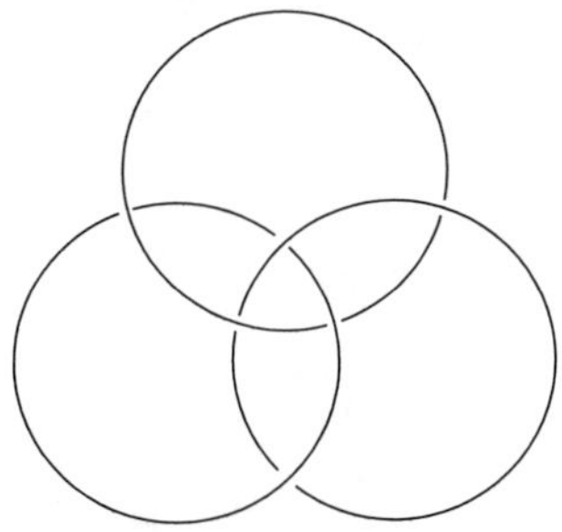

We can moreover easily convince ourselves that if, instead of cutting through the cylinder $\mathcal{C}$, we cut through the sphere S^3, the boundary of the unit Euclidean ball in $\mathbb{C}^2$, we will observe an analogous phenomenon (we could use the fact that S^3 and the boundary of the polydisk $\{|z| \leq 1, |w| \leq 1\}$ are homeomorphic via radial projection).

We are now ready to study the case where λ is an arbitrary real number: here again $\mathcal{C}_\lambda$ is contained in the torus $\mathbb{T}$ and, according to whether λ is rational or not, $\mathcal{C}_\lambda$ is either a closed line or dense:

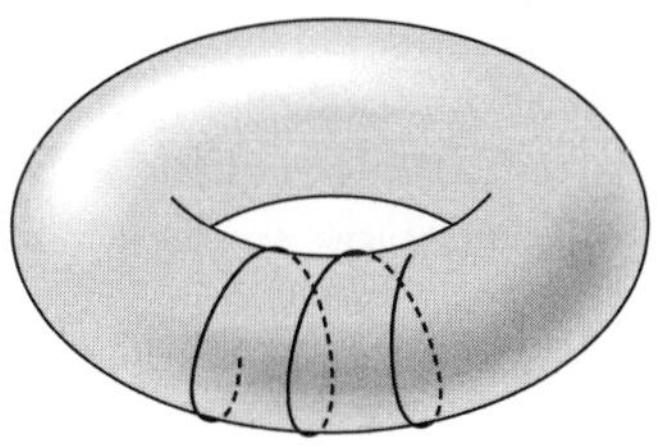

The remaining case is where $\lambda = a + ib$ is not real, i.e. $b \neq 0$. We observe that

$$\mathcal{C}_\lambda = \{M(\theta) = (e^{2i\pi\theta}, e^{-2\pi b\theta} e^{2i\pi a\theta}),\ \theta \in \mathbb{R}\}.$$

As θ varies we observe that the first coordinate of $M(\theta)$ runs along a circle at uniform speed while the second draws a spiral in the plane $\mathbb{C} \simeq \Delta$:

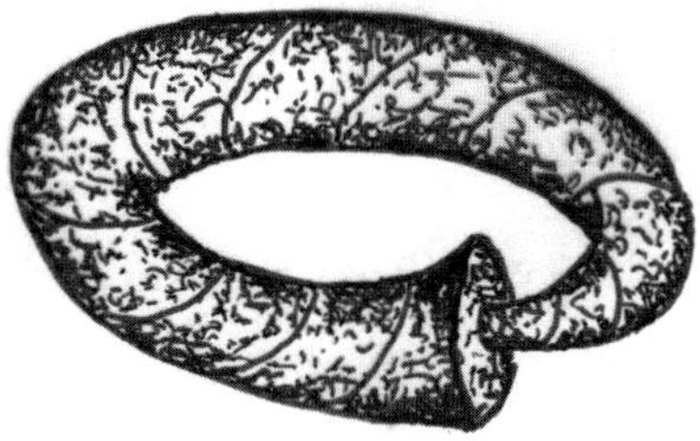

[3]TRANSLATOR'S NOTE: This configuration differs from the one usually called the "Borromean rings", in which no two of the circles are linked, and only by taking all three together is the object inseparable; here, each pair of rings is linked together.

Obviously our way of drawing obscures the axis w around which a similar phenomenon takes place. We are more used to making tomography via the sphere S^3 where, as we have said, the trace of our two coordinate axes consists of 2 interlaced circles. The curve $\mathcal{C}_\lambda$ spirals "into the past" around one, "into the future" around the other:

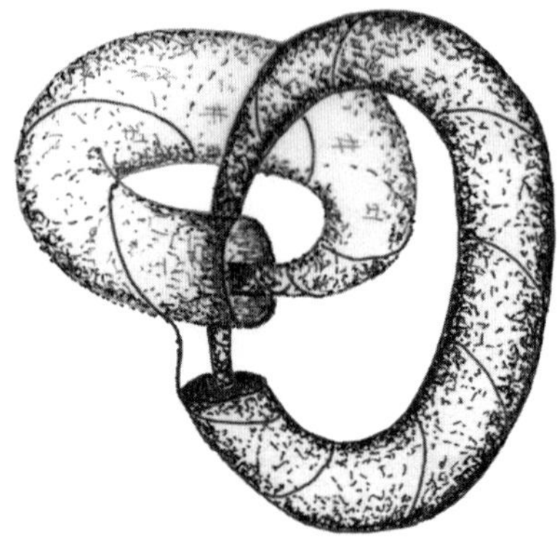

Tomography is very pretty, but how can we have a more precise idea, if possible in dimension 4, of the nature of $\mathcal{L}_\lambda$? This is not easy. When λ is not real, one often uses the following subterfuge. One looks at the function $\varphi = |z|^2 + |w|^2$ whose level sets are spheres centered at the origin; one then remarks that apart from 0 the level sets of φ are all transverse to $\mathcal{L}_\lambda$. This indicates that one can see $\mathcal{L}_\lambda$ as a cone over $\mathcal{L}_\lambda \cap S^3$, but a cone whose generatrices are certainly not lines... Do you know how to find them? To finish, here is what the image of the map $T \mapsto (e^T, |e^{\lambda T}|)$ in $\mathbb{R}^3 = \mathbb{C} \times \mathbb{R}$ might look like:

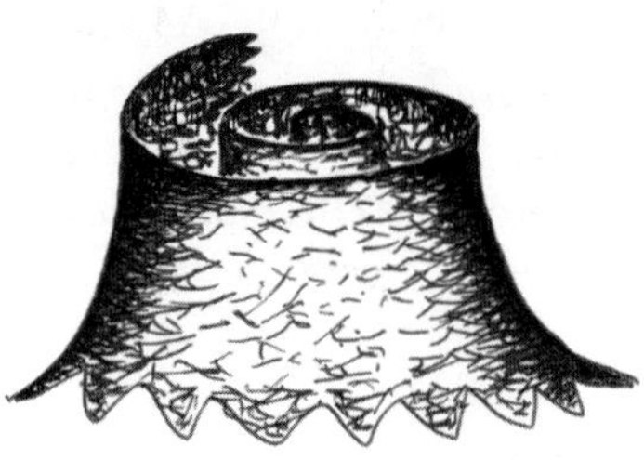

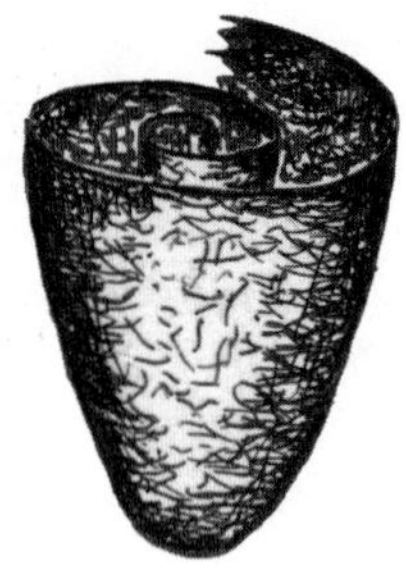

These figures cannot be distinguished topologically.

5.3. Towards a proof of the center theorem

The most naive—but entirely correct—approach is the following. Suppose that E defines a center; the flow φ_t of the field X_E,

$$\frac{\partial \varphi_t}{\partial t} = X_E \circ \varphi_t,$$

has the property that for each initial condition m near the origin, the map $t \mapsto \varphi_t(m)$ is periodic. Let us denote by $T(m)$ the smallest positive period of $t \mapsto \varphi_t(m)$. If T is non-constant, the function $m \mapsto T(m)$ is a first integral of the field X_E, but is it analytic at the origin? If T is constant (we then say that the field X_E is isochronous), the flow φ_t induces an action of the circle S^1 in a neighborhood of the origin. We can then take, for example, the average of the function $x^2 + y^2$ on the trajectories of X_E and obtain a first integral, here a truly analytic one.

The second approach is due to Georges Reeb. After passing to polar coordinates the field X_E no longer has singularities. One constructs a first integral f of X_E by extending any function defined, for example, on the x-axis in a constant way along the trajectories. The problem is essentially the same as before: how to justify the analyticity of f at 0 (moreover, for this to happen it will be necessary to choose a function along the x-axis *ad hoc*). But the idea is indeed present, as long as we exploit it well and proceed by complexification. *Grosso modo* here is how we proceed. The equation E induces on $\mathbb{C}^2$ a differential equation $E^{\mathbb{C}}$ of the same kind and a vector field $X_E^{\mathbb{C}}$ defined by

$$X_E^{\mathbb{C}} = A(z,w)\,\frac{\partial}{\partial z} + B(z,w)\,\frac{\partial}{\partial w},$$

where A and B are the initial series that are applied to the variables z and w, this time complex. In place of passing to polar coordinates, we substitute a complex blow-up (this is nothing but complex formalization of the magic: we set $y = tx$). To be more accurate, we construct a surface $\widetilde{\mathbb{C}}^2$, the blow-up of $\mathbb{C}^2$ at 0, and a so-called blow-up map $\pi : \widetilde{\mathbb{C}}^2 \to \mathbb{C}^2$: here, $\pi^{-1}(0)$ is identified with the complex projective line $\mathbb{CP}(1)$, which must be thought of as the space of complex lines through the origin in $\mathbb{C}^2$, and the restriction of π to $\widetilde{\mathbb{C}}^2 - \pi^{-1}(0)$ is a holomorphic isomorphism onto $\mathbb{C}^2 - \{0\}$. It turns out that we can describe $\widetilde{\mathbb{C}}^2$ by means of two superimposed "coordinate spaces" $\{(z,t)\} \simeq \mathbb{C}^2$ and $\{(s,w)\} \simeq \mathbb{C}^2$, the coordinates s and t being added so that they satisfy the relation $st = 1$. In coordinates, the map π is simply defined by

$$\pi(z,t) = (z,tz) \qquad \text{and} \qquad \pi(s,w) = (sw,w).$$

The equation $E^{\mathbb{C}}$ thus lifts to $\widetilde{\mathbb{C}}^2$; the blow-up procedure produces, in a rather artificial way, a new solution, in this case $\pi^{-1}(0)$. On $\pi^{-1}(0)$, once L_E is normalized, there are two singular points i and $-i$ (viewed as the slopes of the lines $w = iz$ and $w = -iz$), and through each of these singular points passes a complex separatrix γ_i, γ_{-i} (this is the theorem of Briot and Bouquet, a preliminary version of the statement of Camacho–Sad); see FIG. 5.3. In the space $\widetilde{\mathbb{C}}^2$, the inverse image of the real plane $\mathbb{R}^2$ (viewed as a subspace of $\mathbb{C}^2$) is a **Möbius strip** $\mathcal{M}$, whose intersection with $\pi^{-1}(0)$ is a circle $\mathcal{C}$. On this Möbius strip are drawn the pre-images by π of the solutions to the real equation E; see FIG. 5.4. There is the circle $\mathcal{C}$ (which is the core curve of $\mathcal{M}$), and "neighboring circles" (a bit deformed, but circles in the topologist's sense) that "pass twice" near $\mathcal{C}$.

Specialists in foliations have introduced the notion of holonomy, a concept that generalizes the first return map of Poincaré. Let us describe this on our example. Our complex surface $\widetilde{\mathbb{C}}^2$ is foliated (think of layers of sediment) by the solutions (of the lift of $E^{\mathbb{C}}$), which have complex dimension 1, and are therefore of real dimension 2. Away from the singular points i and $-i$, the solutions (corresponding to initial conditions near 0) are neighboring to the artificial solution $\pi^{-1}(0)$. Imagine that each solution is materialized by a plate of glass, and that from infinity comes a radiating light whose (complex!) rays are carried by the lines of $\mathbb{C}^2$ passing through 0... or rather their lifts to $\widetilde{\mathbb{C}}^2$. As we walk along a solution curve $\widetilde{\gamma}$, our shadow

FIGURE 5.3

FIGURE 5.4

then traces a curve γ on $\pi^{-1}(0)$ (which, although we have not yet said so, is a sphere); see FIG. 5.5. Conversely, if we are given a curve drawn on $\pi^{-1}(0)$, say for example the circle $\mathcal{C}$, we can carry out a walk along a curve $\widetilde{\mathcal{C}}$ in such a way that the projected shadow gives us back $\mathcal{C}$. Starting from a point z situated on the ray of the z-axis, we come back, having closed our path, over $\mathcal{C}$ at a point of the same axis, denoted $\tau(z)$. We remark, apart from the fact that τ defines a holomorphic function from the axis to itself, that if z is on $\mathcal{M}$, then $\tau(\tau(z)) = z$: this is the center condition. It follows that the map τ satisfies $\tau \circ \tau = \text{id}$: this is the isolated zero theorem.

It is easy to construct non-constant τ-invariant holomorphic functions (for example $z + \tau(z)$); then we can extend such a function by deciding that it will be constant along trajectories of $X_E^{\mathbb{C}}$ (lifted to $\widetilde{\mathbb{C}}^2$). That this construction is consistent comes from the fact that $\pi^{-1}(0)$ with the two singular points removed is identified with $\mathbb{C} - \{0\}$ and that the circle $\mathcal{C}$ generates its Poincaré group (the first homotopy group). A small technical refinement near the points i and $-i$ allows us to define a holomorphic function $\widetilde{f} : \widetilde{\mathbb{C}}^2 \to \mathbb{C}$, a non-constant first integral of our equation. Using the fact that π is an isomorphism from $\widetilde{\mathbb{C}}^2 - \pi^{-1}(0)$ onto $\mathbb{C}^2 - \{0\}$, we obtain on the initial space $\mathbb{C}^2$ a first integral f, which is certainly defined away

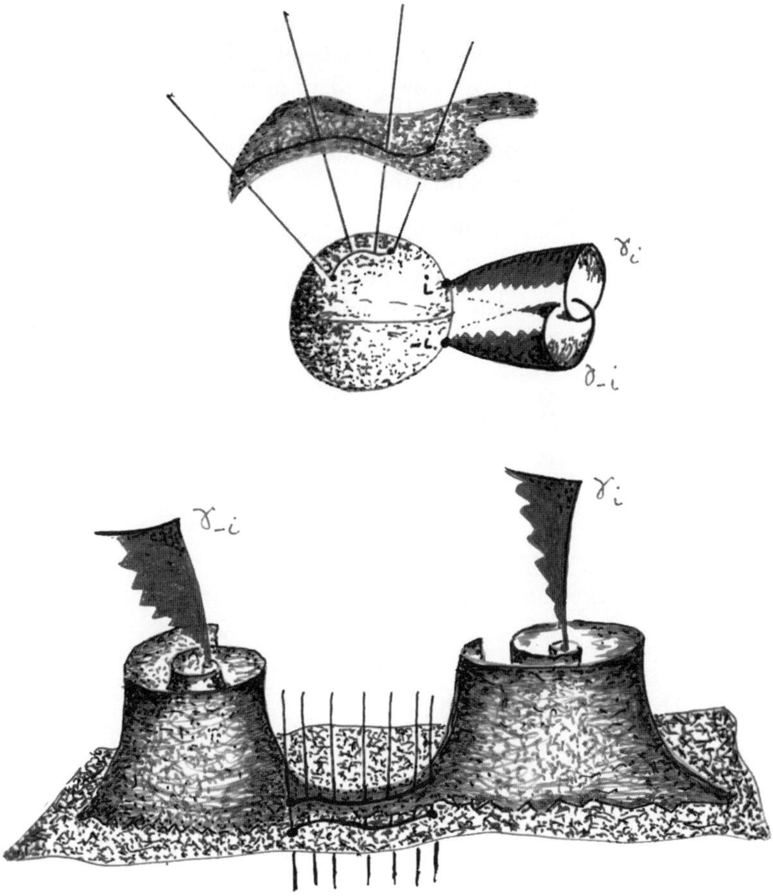

FIGURE 5.5

from 0. In fact, this extends to the origin thanks to a famous theorem of Hartogs, which is truly a theorem of the complex world. It is easy to see that the function f thus constructed is real on the real plane (... otherwise take the real part); in addition, f is a Morse function, which can be checked by a harmless calculation. Thus Reeb's method, thanks to passing to the complex world where we can use the Hartogs theorem, is made consistent.

5.4. Examples of centers and connections with "integrability"

Obviously, the presence of a center does not necessitate non-degeneracy. For example, if $f : \mathbb{R}^2 \to \mathbb{R}$ is a non-constant positive analytic function with a critical point at 0, the Hamiltonian field X_f of f, defined by

$$X_f = \frac{\partial f}{\partial y}\frac{\partial}{\partial x} - \frac{\partial f}{\partial x}\frac{\partial}{\partial y},$$

of course has f as a first integral and satisfies the center condition. As soon as f is not a Morse function, one can show that the period function T tends to infinity as m tends to 0. One can in the same way construct centers associated to functions of type $f_1^{\lambda_1} \cdots f_P^{\lambda_P}$, where the f_i are now positive analytic functions and the λ_i are positive real numbers.

Here is a another way to construct centers in connection with the phenomenon of "reversibility". Consider the folding map:

$$p : (x, y) \mapsto (x^2, y).$$

On the space $\mathbb{R}^2$, we give a differential equation $\Omega = 0$ (where $\Omega = A\,dx + B\,dy$), "monodromic" with respect to the y-axis, which means that the phase portrait in the half-plane $x \geq 0$ is of the following type:

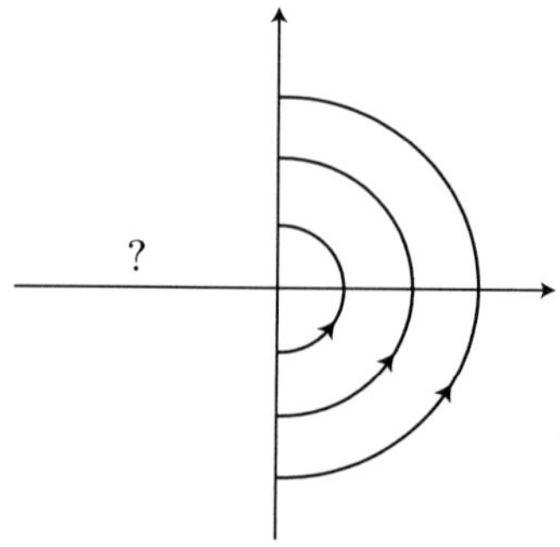

For example, a spiraling node is monodromic:

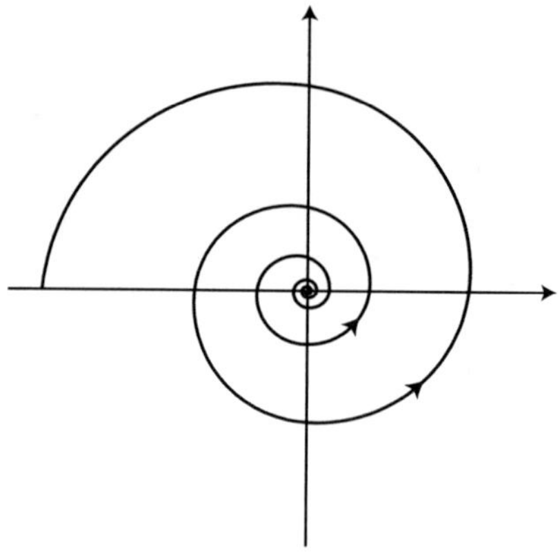

The differential equation associated to

$$p^*\Omega = 2xA(x^2, y)\,dx + B(x^2, y)\,dy$$

therefore defines a reversible center, i.e., one whose phase portrait is invariant under the involution $(x, y) \mapsto (-x, y)$. We remark that this construction can be generalized by making finitely many folds along distinct lines. By juxtaposing the work of M. Brunella [**1**], M. Berthier, and R. Moussu [**3**], we obtain the following very pretty result concerning "nilpotent" centers, those for which the matrix J_E is nilpotent, which is a degenerate case.

THEOREM. *Every nilpotent center is reversible, i.e., invariant for an involution of type $(-x, y)$, after choosing an appropriate set of coordinates.*

Obviously it is to be understood that such a center can be obtained by the preceding folding construction.

Even the statement of the Poincaré–Lyapunov theorem allows an extension of the concept of the center to the complex case: we will say that a differential equation $E^{\mathbb{C}}$ is of center type at a singular point if it has a holomorphic local first integral at this point. Henri Dulac [**5**] at the beginning of the 20th century became interested in polynomial differential equations having a non-degenerate center at the origin of $\mathbb{C}^2$. Here is one of his results, slightly surprising.

Dulac considers differential equations of the following type:

$$E(Q_1, Q_2) : \begin{cases} x' = -y + Q_1, \\ y' = \quad x + Q_2, \end{cases}$$

where the Q_i are quadratic forms. The set of all the $E(Q_1, Q_2)$ is identified with the space of coefficients of the Q_i, i.e., with $\mathbb{C}^6$. In $\mathbb{C}^6$, the $E(Q_1, Q_2)$ of center type at 0 form a proper algebraic variety C, a direct consequence of a finiteness theorem due to Hilbert. Dulac gave the explicit description: C has five irreducible components, for which he gave parametrizations (we say that these are unirational varieties). To do so, Dulac showed that if E belongs to C, then E is integrable in the sense of Liouville. Without going too much into details, this means that the differential form $\Omega_E^{\mathbb{C}}$ has an "integrating factor" P_E: this is a polynomial such that $\Omega_E^{\mathbb{C}}/P_E$ is closed. The "Liouvillian" function $\int \Omega_E/P_E$ is then a first integral of E. Dulac's proof, which is a computational *tour de force*, has undeniable similarities with that of Poincaré regarding the problem of the center. Although it is not lacking in concept, today it is hard to read, the deep "meaning" of the calculation being outdated and truly doomed to obsolescence:[4] it would be interesting to propose a geometric alternative.

It had been conjectured that a differential equation E of center type is either reversible or integrable in the sense of Liouville. In fact, nothing of the sort is true: in [**2**], counterexamples to this conjecture are constructed, using complex techniques and a hint of non-linear Galois theory.

5.5. The "hyperbolic" Poincaré–Lyapunov theorem

The center theorem gives necessary and sufficient conditions on the phase portrait of a differential equation, obtained by perturbing the one corresponding to the Morse function $x^2 + y^2$, in order to insure the existence of an analytic first integral. What is it for a perturbation of the Hamiltonian function xy?

Consider such a perturbation, i.e., a vector field:

$$X_E = (x + \cdots)\frac{\partial}{\partial x} - (y + \cdots)\frac{\partial}{\partial y} = L + \cdots$$

and its complexification $X_E^{\mathbb{C}}$. The flow of the linear part $L^{\mathbb{C}}$ of $X_E^{\mathbb{C}}$ is given by

$$\varphi_\tau(z, w) = (e^\tau z, e^{-\tau} w), \qquad \tau \in \mathbb{C}.$$

[4]TRANSLATOR'S NOTE: The original text has *voué aux gémonies*, a reference to the Gemonian stairs of ancient Rome; this phrase in French has come to mean, since the 19th century, extreme public discrediting and humiliation.

This flow obviously parametrizes the level curves of the first integral zw. We remark that the intersection of a trajectory of $L^{\mathbb{C}}$ with the real plane $\mathbb{R}^2$ gives two trajectories of the real linear field $L = x\frac{\partial}{\partial x} - y\frac{\partial}{\partial y}$. This is of course tied to the fact that the complex level curves of zw are connected, while those of the real function xy are not. Thus, to each real solution of the linear field L we can associate another solution, which we will call its companion. Observe that each real solution can be slid along the complex solution to be wedded to its companion. By using arguments similar to those of the proof of the Poincaré–Lyapunov theorem, we obtain the following fact: *If every solution of X_E has a companion* (in fact, if the real trace of the complex solutions is not connected), *then the field X_E has a first integral that is a Morse function of type xy.* Obviously, one may object that this phenomenon is not visible to the naked eye in the phase portrait of the field X_E. This is true, but all the same this is the hyperbolic version of the center theorem! Here again, when we consider the degenerate cases, the presence of companions is not directly tied to Liouvillian integrability any more than it is to reversibility. This property of companionship is intriguing, however, and deserves to be studied in full generality.

Bibliography

[1] M. Brunella, *Vanishing holonomy and monodromy of certain centres and foci*, Astérisque No. 222 (1994), pp. 37–48.

[2] M. Berthier, D. Cerveau, A. Lins Neto, *Sur les feuilletages analytiques réels et le problème du centre*, J. Diff. Eq. **131** (1996), no. 2, pp. 244–266.

[3] M. Berthier, R. Moussu, *Réversibilité et classification des centres nilpotents*, Ann. Inst. Fourier **44** (1994), no. 2, pp. 465–494.

[4] C. Camacho, P. Sad, *Invariant varieties through singularities of holomorphic vector fields*, Ann. of Math (2) **115** (1982), no. 3, pp. 579–595.

[5] H. Dulac, *Détermination et intégration d'une certaine classe d'équations différentielles ayant pour point singulier un centre*, Bulletin des Sci. Math. **32** (1908).

[6] H. Dulac, *Sur les cycles limites*, Bull. Soc. Math. France **51** (1923), pp. 45–188.

[7] J. Ecalle, *Introduction aux fonctions analysables et preuve constructive de la conjecture de Dulac*, coll. "Travaux en cours", Hermann, 1992.

[8] H. Poincaré, *Mémoire sur les courbes définies par une équation différentielle*, Journal de Mathématiques, 3rd series, **7** (1881), pp. 375–422, and **8** (1882), pp. 251–296.

[9] A. Seidenberg, *Reduction of singularities of the differential equation $A\,dx + B\,dy$*, Amer. J. Math. **90** (1968), pp. 248–269.

[10] Y. Ilyashenko, *Finiteness theorems for limit cycles*, Russian Math. Surveys **45** (1990), no. 2, pp. 123–203.

CHAPTER 6

Periodic Orbits of the Three-Body Problem: Early History, Contributions of Hill and Poincaré, and Some Recent Developments

Michael Nauenberg

A brief history is given of the early development of periodic orbits of the restricted three-body problem, correcting a long standing historical error that attributes the first application of this concept to Euler or Jacobi rather than to Newton: in fact, Newton introduced this approximation in a treatment of lunar motion in the third book of *Principia*. Hill's seminal work on this problem, 190 years later, greatly influenced Poincaré who extended it further giving existence proofs for a large class of periodic orbits when two of the three masses are relatively small. Recently there has been renewed interest in periodic orbits after Cristopher Moore's remarkable discovery in 1993 of a periodic orbit with three equal masses moving on a single trajectory in the shape of a figure eight. The last section describes some new results obtained in collaboration with Moore.

Introduction

In the *Principia* [**33**], Newton laid the foundations of orbital dynamics and the universal theory of gravitation, describing in mathematical form the physical origin of the empirical laws of planetary motion discovered by Kepler [**3**]. Although this was in itself an enormous achievement, Newton demonstrated that these laws are justified only when the interplanetary gravitational forces are neglected. As he remarked in the Preface to his masterpiece,

> *But after I began to work on the inequalities of the motion of the Moon, and ... the motions of several bodies with respect to one another ... I thought that publication should be put to another time so that I might investigate these other things...*

Remarkably, a large part of these investigations took place at the same time that Newton was composing the *Principia*, and it led to his development of methods to calculate the perturbations due to interplanetary gravitational forces, and to the effect of the solar gravitational force on the lunar orbit. In particular, he gave approximate solutions of the three-body problem which stimulated the work of mathematicians and astronomers for the next two centuries, and remains a challenge to the present time. For example, in Corollaries 3 and 4 of Proposition 17, Newton gave a succinct description of the method of variation of orbital parameters, which later was developed by Euler and Lagrange and is usually credited to them [**8**]. In the first edition of the *Principia* Newton claimed to have calculated correctly the rate of precession of the lunar apogee,[1] but by the third edition he apparently changed his mind stating that *"the [advance of the] apsis of the Moon is about twice as swift"* as he had calculated it.[2] The resolution to this discrepancy was not found until 60 years later when Alexis Clairaut and Jean d'Alembert carried out Newton's calculation to a higher order approximation [**46, 47**]. Euler was greatly impressed with this result, and declared [**46**]

> *For it is certain that it is only since this discovery that one can regard the law of attraction reciprocally proportional to the square of the distances as solidly established, and on this depends the entire theory of astronomy.*

However, it is generally forgotten that in treating the lunar problem Newton also obtained the first approximation of a periodic orbit for the *restricted* three-body problem:[3] indeed, in Book 3, Prop. 28, he calculated a fictitious lunar orbit, which is periodic in a *uniformly rotating* frame centered on the Earth with the Sun fixed, taking also its gravitational force into account (this solution is discussed below, p. 115). Newton's solution is not well known, and the first treatment of this problem is attributed to Euler in 1772 [**11, 45**], with later contributions by Jacobi [**20**] and by George W. Hill [**17**]. It was Hill's work which had a fundamental influence on Poincaré's research, particularly in connection with his theory of periodic orbits. The analysis of the stability of these orbits for the restricted three-body problem ultimately led Poincaré to his ground breaking discovery of chaotic motion in this dynamical system, which first appeared in a revised version of his prize winning essay for the King Oscar II competition [**37**]. The original manuscript contained a fundamental error that was pointed out by Edvard Phragmén, and its correction led Poincaré to his discovery of the homoclinic tangle.[4]

Poincaré's great admiration for Hill can be gleaned from a lengthy Introduction he wrote for Hill's *Collected Mathematical Works*, which began with the words: *"M. Hill est une des physionomies les plus originales du monde scientifique américain."* (*M. Hill is one of the most original faces in the American scientific world.*) [**17**]. When Poincaré came to the United States for the 1904 St. Louis Congress and had the opportunity to meet Hill, he greeted him with the words *"you are the one man I came to America to see"*.[5]

It should be mentioned that in 1772 Lagrange published a remarkable solution of the three-body problem [**21**], where each body moves in accordance to Kepler's law on a separate elliptic orbit with the same eccentricity and a common focus at the center of mass. The motion takes place in such a manner that each body lies on the vertex of an equilateral triangle with the sides corresponding to the relative distances, see FIG. 6.1. For the case that one of the three masses vanishes, the center of mass lies on the line joining the massive bodies, and when the orbits are circles, this is a special solution of the restricted three-body problem. In a rotating frame in which the massive bodies are fixed, the massless body is then at one of

---

[1]See *Principia* 1687, Book 3, Prop. 35. The details of Newton's calculation are described in unpublished manuscripts in the Portsmouth collection [**8**].

[2]See *Principia*, Book 1, Prop. 45, Cor. 2.

[3]The expression *"problème restreint"* was introduced by Poincaré to describe the three-body problem in a plane, when one of the masses is neglected and the two others move along circular orbits (necessarily around their common center of mass).

[4]See Chapter 8 in this book. For further details of this intriguing story see the book [**1**] by June Barrow-Green.

[5]Quoted in [**1**], p. 28.

two equilibrium points forming with the two massive bodies an equilateral triangle. Shortly before, Euler [**10**] had discovered another remarkable solution (rediscovered by Lagrange), where the three bodies are permanently aligned; in particular, there are three additional equilibrium points, situated on the axis between two massive bodies: one between the two bodies and the other two on the outside. In the limiting case treated by Hill, only the one in between the masses remains, and Poincaré demonstrated the existence of periodic orbits centered at this equilibrium point (see Section 6.3).

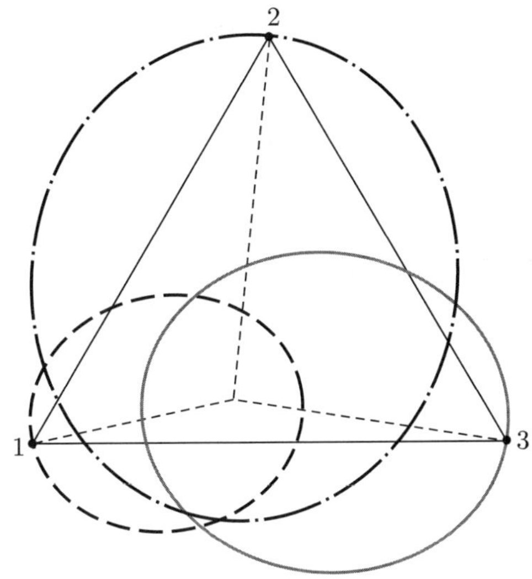

FIGURE 6.1. Lagrange's 1772 three-body periodic orbit with masses in the ratio 1 : 2 : 3. The vertex of the equilateral triangle locates the position of the bodies at apogee, and the intersection of the dashed lines along the axis of the ellipses is the center of mass.

In the next section Newton's periodic lunar orbit theory is presented, and the further development of this theory by Hill is given in Section 6.2, as well as its extension by Poincaré. Section 6.3 describes several aspects of Poincaré's treatment of periodic orbits of the three-body problem: proof of existence when two of the masses are relatively small, using the method of analytic extension; possibility of periodic orbits that in the limit of vanishingly small masses do not correspond to Keplerian ellipses; periodic orbits near an equilibrium point. Poincaré's geometrical proofs of existence of infinite families of periodic orbits of the three-body problem are sketched in Section 6.4, written by Éric Charpentier. In Section 6.5 some recently discovered periodic orbits are described, which owe much to the inspiration provided by Poincaré's work.

6.1. Newton's periodic lunar orbit[6]

In Prop. 28, Book 3 of the *Principia*, Newton considered a fictitious lunar orbit which is periodic in a uniformly rotating frame. He assumed that the orbit of the

[6]This section is based on my article "Newton's perturbations methods for the three-body problem and their application to the lunar problem", in: *Isaac Newton's Natural Philosophy* edited by J. Z. Buchwald and I. B. Cohen (MIT Press, Cambridge, 2001).

Moon was approximated by an ellipse of small eccentricity with the Earth located at its *center*, in such a way that its major axis remains normal to the Earth-Sun radius.

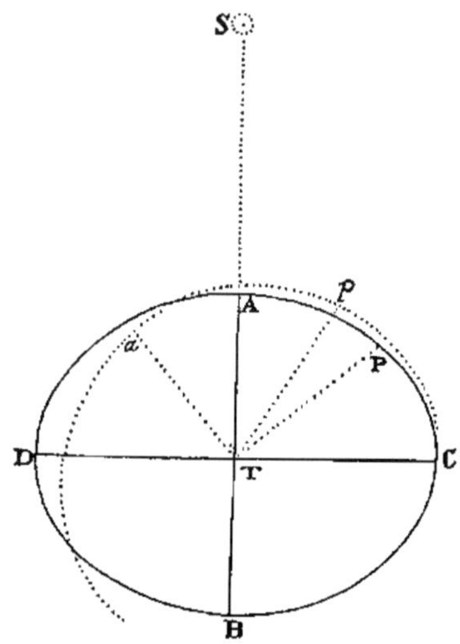

FIGURE 6.2. Newton's figure for the lunar orbit in Prop. 28, Book 3 of the *Principia* (third edition, 1726). The oval $ADBC$ represents the lunar orbit in a frame rotating about the Earth at T, such that the Sun remains fixed at S.

He obtained this solution by applying a curvature method, which he had developed in the 1670s [**28**], [**4**], but he did not discuss this approach in Book 1 of the first edition (1687) of the *Principia*, a situation which must have thoroughly mystified his readers.

To understand Newton's curvature method and his periodic lunar solution it is useful to present it first in modern mathematical form, and afterwards turn to the presentation given in the *Principia*. The earliest approach developed by Newton to calculate orbital motion for general central forces was based on the extension of the concept of centrifugal force or acceleration, which he and Huygens [**19**] had developed independently for the case of uniform circular motion [**3**]. If a body is moving with constant velocity on a circular orbit with radius ρ, then the acceleration a along the radial direction is given by

$$a = v^2/\rho. \tag{6.1}$$

This acceleration is also a measure of the force required to keep a body in such a motion. For motion along a general orbit, this relation remains valid for the normal component of the acceleration, provided v is the instantaneous velocity, and ρ is the radius of curvature at the corresponding point of the orbit. In Book 3, Prop. 28 starts with the statement that

> *The curvature of the orbit which a body describes, if attracted in lines perpendicular to the orbit, is directly as the force of attraction and inversely as the square of the velocity.*[7]

The term curvature, however, was nowhere defined in the 1687 *Principia*, but Newton's calculation reveals that it is indeed proportional to $1/\rho$. Assuming that the force acts in a plane and choosing a fixed origin, the normal component a_n of the acceleration can be written in terms of the radial component a_r and transverse component a_θ of the acceleration a,

$$a_n = a_r \sin\alpha + a_\theta \cos\alpha, \tag{6.2}$$

where α is the angle between the radius vector and the velocity vector. Expressing the velocity v in terms of the angular momentum h, called the *moment of areas* in the *Principia*, we have

$$v = \frac{h}{r\sin\alpha}, \tag{6.3}$$

and substituting (6.2) and (6.3) in (6.1) (with a replaced by a_n) leads to a general equation for orbital motion,

$$\frac{1}{\rho\sin^3\alpha} = \frac{r^2}{h^2}[a_r + a_\theta \cot\alpha]. \tag{6.4}$$

In the second edition of the *Principia*, Newton added several new corollaries to Prop. 6 of Book 1, which deals with the measure of central forces, and in particular in Cor. 3 one finds the first explicit statement of the curvature measure,

> ... *the centripetal force will be inversely as the solid* SY^2PV ...,

where in our notation $PV = 2\rho\sin\alpha$, and $SY = r\sin\alpha$. This measure corresponds to equation (6.4) for the special case that $a_\theta = 0$, apart from the overall constant of proportionality h. In this case h is a constant of the motion expressing the area law or conservation of angular momentum proved in Book 1, Prop. 1. However, if the transverse component of the force, a_θ, does not vanish, then

$$\frac{dh}{dt} = ra_\theta. \tag{6.5}$$

Setting

$$h = r^2\frac{d\theta}{dt}, \tag{6.6}$$

where θ is the polar angle, and eliminating the time variable t, equality (6.5) becomes

$$\frac{dh}{d\theta} = \frac{r^3}{h}a_\theta. \tag{6.7}$$

While Newton did not explicitly write this relation in the *Principia*, Book 3, Prop. 26, entitled *"to find the hourly increment of the area which the Moon, by*

---

[7] It is interesting to compare this statement with a cryptic comment in Newton's 1664 Waste Book:

> *If the body b moves in an Ellipsis, then its force in each point (if its motion in that point be given) may be found by a tangent circle of equal crookedness with that point of the Ellipsis.*

The term *crookedness* was Newton's earlier word for curvature, and *tangent circle* is now known as the osculating circle. For further details see [**28**].

a radius drawn to the Earth, describes in a circular motion", contains a calculation for the change in h at the syzygies and quadratures[8] obtained by integrating (6.7) approximately for the case that a_θ is the transverse component of the solar perturbation.

In fact, (6.4) and (6.7) correspond precisely to the differential equations of motion obtained by Alexis Clairaut and Jean d'Alembert [**47, 44**], who applied Leibniz's form of the calculus to Newton's laws of motion written in the *Principia* in an equivalent geometrical form. This correspondence can be seen from the differential equation in polar coordinates for the projection of the curvature ρ along the radial direction. Newton [**34**] obtained this relation around 1671 in the form

$$\rho \sin\alpha = r\,\frac{(1+z^2)}{(1+z^2-dz/d\theta)}, \tag{6.8}$$

where $z=\cot\alpha=(1/r)\,dr/d\theta$. It follows that

$$\frac{1}{\rho\sin^3\alpha}=\left(\frac{d^2}{d\theta^2}+1\right)\frac{1}{r}, \tag{6.9}$$

and substituting this relation in the left hand side of (6.4) yields the more familiar equation

$$\left(\frac{d^2}{d\theta^2}+1\right)\frac{1}{r}=\frac{r^2}{h^2}\left[a_r+\frac{a_\theta}{r}\frac{dr}{d\theta}\right] \tag{6.10}$$

as was first written by Clairaut [**47, 44**], with

$$h^2=h_0^2+2\int d\theta\, r^3 a_\theta \tag{6.11}$$

according to (6.7). In this form, Newton's geometrical relation, (6.4), becomes a differential-integral equation, with θ as an independent variable. Except for special cases, such an equation can generally be solved only approximately or by numerical integration.

Up to the first order in r/R, where R is the mean solar distance, Newton derived in Prop. 25 and 26, Book 3, the expressions for the radial and angular component of the solar perturbing force

$$a_r-\mu/r^2=-\frac{g}{2}r(1+3\cos 2\psi), \tag{6.12}$$

and

$$a_\theta=-\frac{3g}{2}r\sin 2\psi, \tag{6.13}$$

where ψ is the angle between the Sun and the Moon as measured from the Earth, and μ and g are constants ($\mu=GM$, $g=GM_0/R^3$, M is the sum of the mass of the Earth and the Moon, M_0 is the mass of the Sun, and G is Newton's gravitational constant).

In the limit of a large number of lunar cycles per year or small values of the parameter $m=T_L/T_E$, where T_L is the sidereal lunar period and T_E the sidereal period of the Earth, Newton conjectured that a solution of (6.4) has the form

$$r\approx r_0\,(1-\epsilon\cos 2\psi), \tag{6.14}$$

[8] At the syzygies, the Sun, the Earth and the Moon are aligned, and at quadrature, the line between the Earth and the Moon is perpendicular to the line between the Earth and the Sun.

approximating an ellipse with eccentricity ϵ and $\psi = (1-m)\theta$, with the Earth at the center. The form of this solution is suggested by the requirement that it satisfies Newton's curvature relation (6.4) with a_r and a_θ given by (6.12) and (6.13). The radial component a_r of the perturbing force, which appears on the right hand side of this equation, contains a term proportional to $\cos 2\psi$, while the contribution of the transverse force component a_θ is multiplied by $\cot\alpha = (1/r)dr/d\theta$, which is of order ϵ and can be neglected in the lowest order approximation. However, a_θ contributes to the variation in the angular momentum h, and neglecting the variation in r on the right hand side of (6.11) yields

$$h^2 = h_0^2 + \frac{3m^2\mu r_0}{2(1-m)}\cos 2\psi, \tag{6.15}$$

where h_0 is the average value of the angular momentum. Substituting (6.14) in the curvature equation (6.10) and neglecting terms of order m^2 and $m^2\epsilon$, we obtain

$$r_0 = (h_0^2/\mu)(1+m^2/2) \tag{6.16}$$

and

$$\epsilon = \frac{3}{2}m^2\frac{(2-m)}{(1-m)[4(1-m)^2-1]}. \tag{6.17}$$

This analytic form for ϵ was obtained by Laplace [**22**] and by Tisserand [**44**, p. 39]. Evidently ϵ is of order m^2, which justifies the approximations made previously. However, as terms of order m^4 have been neglected, (6.17) is valid only up to cubic terms in m, leading to

$$\epsilon \approx m^2 + \frac{19}{6}m^3. \tag{6.18}$$

This result corresponds to the first two terms of an expansion of Hill's solutions [**17**] in powers of m. In Book 3, Prop. 28, Newton gave only a numerical result for $\epsilon = 0.00719$, which is in excellent accord with the result obtained from (6.17); substituting Newton's value for $m^2 = 1/178.725$ yields $\epsilon = 0.00720$.

The relation between time t and polar angle θ can be obtained by integrating (6.6),

$$t = \int d\theta\frac{r^2}{h}. \tag{6.19}$$

Substituting (6.14) and (6.15) for r and h, respectively, we find

$$\omega t = \theta - \xi\sin(2\psi), \tag{6.20}$$

where $\omega = 2\pi/T_L = h_0/r_0^2$ and

$$\xi = \frac{1}{(1-m)}\left[\epsilon + \frac{3m^2}{8(1-m)}\right] \approx \frac{11}{8}m^2 + \frac{59}{12}m^3. \tag{6.21}$$

The parameter ξ is the maximum variation from the mean angular motion of the Moon, ωt. This variation was discovered by Tycho Brahe after he missed observing an eclipse of the Moon which he had expected to occur at the time predicted by the lunar mean motion.

In the *Principia*, Book 3, Prop. 29, Newton first calculated the maximum variation $\bar{\xi}$ in the angular variable $\psi = (1-m)\theta$ corresponding to the angular position of the Moon, relative to that of the Sun in his geometrical approximation. In this case

$$\psi = \bar{\omega}t + \bar{\xi}\sin(2\psi), \tag{6.22}$$

where $\bar{\omega} = 2\pi/\bar{T}_L = (1-m)\omega$ is the mean angular frequency of the Moon in the rotating frame, $\bar{T}_L$ is called the *synodic period* of the Moon, and $\bar{\xi} = (1-m)\xi$. Numerically, Newton found $\bar{\xi} = 32'32''$ in good agreement with the value $\bar{\xi} = 32'34''$, found by using Newton's value for his parameter m in (6.21). Aware, however, that astronomers measure this variation with respect to the inertial frame, Newton remarked:

> *Thus it would be* [that the greatest variation is 32′32″] *if the Moon in passing from the quadrature to the syzygy describes an angle* CTA (see FIG. 6.2) *of* 90° *only.*[9] *But because of the motion of the Earth, by which the Sun is transferred forward in its apparent motion, the Moon before it reaches the Sun describes an angle* CTa (see FIG. 6.2) *greater than a right angle, in the time of the synodic revolution of the Moon to the time of its periodic revolution* [sidereal period], *that is, in the ratio of* $29^d12^h44^m$ *to* $27^d7^h43^m$. *And in this way all the angles about the centre* T *are dilated in the same ratio and the greatest variation, which otherwise would be* but 32′32″, *now augmented in the said proportion, becomes* 35′10″.

Newton's "ratio" $\bar{T}_L/T_L$ equals $1/(1-m)$, and his "proportion" refers to the relation $\xi = \bar{\xi}/(1-m)$. This gives $\xi = 35'12''$ in good agreement with his quoted result.[10]

We now turn to Newton's method as described in Book 3, Prop. 28. In this proposition Newton evaluates ϵ by considering the curvature equation only at two special points: the syzygies $\psi = 0, \pi$ and the quadratures $\psi = \pi/2$ and $3\pi/2$. This does not suffice to show that Newton's revolving ellipse (6.14) is actually a solution of the curvature equation of motion, but as Laplace commented ([**22**], p. 419):

> *This consideration is exact, but it required a proof... These calculational hypotheses, founded on likely insights, are permitted to inventors, in such difficult researches...*[11]

At these special points $\alpha = \pi/2$, according to (6.4),

$$\frac{\rho_s}{\rho_q} = \frac{h_s^2 r_q^2 a_q}{h_q^2 r_s^2 a_s}, \tag{6.23}$$

where the subscripts s, q refer to the value of the corresponding quantities at the syzygies and at the quadratures respectively. Substituting for the angular momentum h, radius r and radial force a at these special points the values given by (6.15),

---

[9] This remark seems to indicate that Newton erroneously believed that in the rotating frame the maximum variation is $\bar{\xi}$.

[10] Newton's geometrical methods in the *Principia*, where he eliminated the time variable by relating areal changes to time intervals, led him to approximate the angular position of the Sun by the relation $\theta' = m\theta$. But according to (6.21), $\theta' - \Omega t = m\xi\sin(2\psi)$, where $\Omega = m\omega$ is the mean angular rotation rate of the Earth around the Sun, which leads to a *spurious* solar variation. Hence, his method leads to results which are correct in an expansion in powers of m only up to and including cubic terms.

[11] *Cette considération est exacte, mais elle exigeait une démonstration... Ces hypothèses de calcul, fondées sur des aperçus vraisemblables, sont permises aux inventeurs, dans des recherches aussi difficiles...*

(6.14) and (6.12) respectively leads to the relation[12]

$$\frac{\rho_s}{\rho_q} = \frac{[1 + 3m^2/2(1-m)](1+m^2)}{[1 - 3m^2/2(1-m)](1-2m^2)}, \tag{6.24}$$

while the curvature relation (6.9) implies that[13]

$$\frac{\rho_s}{\rho_q} = \frac{1 + [4(1-m)^2 - 1]\epsilon}{1 - [4(1-m)^2 - 1]\epsilon}. \tag{6.25}$$

Equating these two relations and solving for ϵ gives again the result in (6.17).

6.2. Hill's periodic lunar orbits and Poincaré's loops

In 1877, Hill [**17**] published his ground breaking paper on lunar theory. Like Newton, he considered the motion of the Moon in a uniformly rotating frame centered on the Earth with the Sun fixed, obtaining periodic solutions of the restricted three-body problem. Previously, the usual approach had been to start with a Keplerian orbit of the Moon, and to incorporate afterwards the effect of the solar force by various approximations. Hill's work had a major influence on Poincaré's research, and in the Preface of *New methods of celestial mechanics* [**38**], Poincaré acknowledged this declaring that

> *In this work, unfortunately incomplete, it is possible for us perceive the germ of most of the progress that Science has since made.*[14]

In a uniformly rotating frame centered on the Earth with the Sun-Earth axis fixed (taken as x axis), and up to the lowest order in the expansion of the solar gravitational force in powers of r/R, where r is the distance to the Moon and R is the distance to the Sun, Hill's scaled equations take the form

$$\begin{aligned} \ddot{x} + \frac{x}{r^3} - 3m^2x - 2m\dot{y} &= 0, \\ \ddot{y} + \frac{y}{r^3} + 2m\dot{x} &= 0, \end{aligned} \tag{6.26}$$

where $m = T_L/T_E$ is now the ratio of the *synodic* lunar period[15] to the sidereal period T_E of the Earth. Here the radius r is in units of the mean Earth-Moon

---

[12] In Book 3, Prop. 28, Newton evaluated the dynamical condition (6.23) for the curvature ratio ρ_s/ρ_q by substituting for the ratio of the radial forces $a_s/a_q = [u(1/AT^2) - (2000/w)AT]/[u(1/CT^2) + (1000/w)CT]$, where $u = 178725$ corresponds to $u = 1000/m^2$, $AT = 1-x$, $CT = 1+x$ and $w = AT \times CT = 1 - x^2$. Actually w should be unity, but the additional term x^2 is fourth order in m and consequently negligible. The ratio h_s/h_q, evaluated in Prop. 26, is given as $(y+50)/(y-50) = 11073/10973$ where $y = 11023$ corresponds to $y = 200(1-m)/(3m^2)$. For $w = 1$, Newton's result agrees with (6.24).

[13] The evaluation of the curvature ratio from the assumed shape of an elliptical orbit rotating about its center is given in Book 3, Prop. 28 in the form $(AT^3 + zCT^2AT)/(CT^3 + zAT^2CT)$, with $z = 16824/100000$ corresponding to $z = 1/(1-m)^2 - 1$. This result agrees with (6.25) neglecting terms of order m^2 and higher. In Cajori's revision of Motte's translation of Newton's *Principia* (University of California Press, 1934), this relation appears on page 447 with a sign error.

[14] *Dans cette œuvre, malheureusement inachevée, il est permis d'apercevoir le germe de la plupart des progrès que la Science a faits depuis.*

[15] Recall that in Newton's treatment, T_L was the *sidereal* lunar period. Thus the parameter m in Hill's equations differs from Newton's (Section 6.1).

distance and the time is scaled so that 2π corresponds to the lunar period. These equations have a constant of the motion, C, introduced by Jacobi [**20**],

$$-\frac{1}{2}(\dot{x}^2 + \dot{y}^2) + \frac{1}{r} + \frac{3}{2}m^2x^2 = C. \tag{6.27}$$

For each value of the parameter m, there exists a unique periodic solution of these equations with period 2π, which then determines Jacobi's constant C as a function of m. For values of $m \leq 1/3$, Hill obtained this solution by expanding the Cartesian coordinates in a Fourier series. By symmetry considerations, taking $t = 0$ at a point where $y = 0$, we have

$$x(-t) = x(t), \quad y(-t) = -y(t), \tag{6.28}$$

and

$$x(t+\pi) = -x(t), \quad y(t+\pi) = -y(t), \tag{6.29}$$

which implies that this solution has the form

$$x = \sum_k a_k \cos(kt), \qquad y = \sum_k b_k \sin(kt), \tag{6.30}$$

where the sum over k is restricted to odd integers. It can be readily verified that Newton's approximate solution corresponds to the first two terms of this expansion with $a_1 = 1 - (1/2)(\epsilon + \bar{\xi})$, $b_1 = 1 + (1/2)(\epsilon + \bar{\xi})$ and $a_3 = b_3 = (1/2)(\bar{\xi} - \epsilon)$, where ϵ and $\bar{\xi}$ are the parameters defined by Newton's solution, (6.14) and (6.22).[16]

For larger values of m, Hill resorted to numerical integration and interpolation to obtain periodic solutions, and for convenience he rescaled his equations to eliminate the parameter m in (6.26). The corresponding solutions, $x'(t)$, $y'(t)$, are related to the previous ones according to

$$x'(t) = m^{2/3}x(t/m), \quad y'(t) = m^{2/3}y(t/m), \tag{6.31}$$

and have period $T_m = 2\pi m$. Indeed, this is how the value of m associated with a periodic solution is determined from its period in Hill's rescaled equations which do not contain this parameter explicitly. The corresponding Jacobi constant is $C' = m^{-2/3}C$.

Some periodic lunar orbits that Hill obtained from his analysis are shown in FIG. 6.3

In particular, one of these orbits has two cusps on the symmetry axis which is normal to the Earth-Sun axis: they correspond to the two points where the Moon comes to rest in the rotating frame when the combined gravitation force of the Earth and the Sun is balanced by the centrifugal force. This orbit is the one having the maximum value of m among all those given by Hill, and he calculated this periodic solution to great accuracy by numerical integration and interpolation by the requirement that the velocity vanish on the symmetry axis.[17] Hill wrote:

---

[16]To compare the expansions of these parameters (which are valid up to cubic powers of m) with the results obtained from the solution of Hill's equations, recall that in Newton's equations $m = T_L/T_E$, where T_L was the sidereal lunar period, while in Hill's equations it corresponds to the synodic period: thus one has first to substitute $m \to m/(1+m)$ in (6.14) and (6.20). I obtain $\epsilon = m^2 + (7/6)m^3$ and $\bar{\xi} = (11/8)m^2 + (13/6)m^3$, which are in agreement with the results obtained by using the expansion for Hill's coefficients given on p. 317 of his *Collected Works* [**17**].

[17]Hill determined that a quarter of the period T is 0.88116. This result gives $m = T/2\pi = 0.56096$, which I verified is correct to 5 decimal figures.

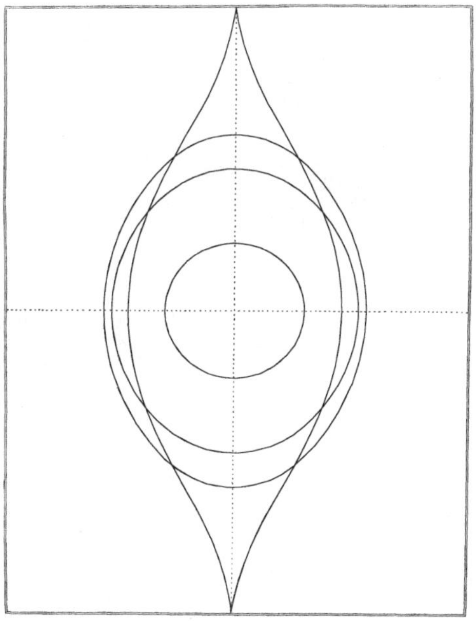

FIGURE 6.3. Hill's solutions for several periodic lunar orbits, from *Collected mathematical works of Hill*, vol. 1. Earth is at the center of coordinates; the Sun is at infinity on the horizontal axis (x axis).

> *The Moon of the last line* [the solution with two cusps] *is, of the class of satellites considered in this Chapter, that which, having the longest lunation, is still able to appear at all angles with the Sun and thus undergo all possible phases. Whether this class of satellites is properly to be prolonged beyond this Moon, can only be decided by further employment of mechanical quadratures* [integrations]. *But it is at least certain that the orbits, if they do exist, do not intersect the line of quadratures*[18] *and that the Moons describing them would make oscillations to and for, never departing as much as 90° from the points of conjunction or of opposition.*

But in the first volume of his *New Methods of Celestial Mechanics* [**38**] (1892), Poincaré pointed out by simple analytic considerations that Hill's assumption was incorrect,[19] because the success of Hill's numerical interpolation technique to calculate periodic orbits implied the existence of solutions near the orbit with a cusp.[20] These solutions can be obtained analytically by solving Hill's equations (6.26) for values of m near the critical value $m_c = 0.56096$ (corresponding to the cusp orbit) in a power series in time t, taking $t = 0$ when the Moon is at $x = 0$, $y = y_0$, where y_0 is the maximum value of y. Expanding the solution of Hill's equations (6.26) up

[18]The vertical axis in FIG. 6.3, perpendicular to the line Earth-Sun.

[19]In a footnote in his *Collected Works*, p. 328, Hill mentioned that *"the attribution of the maximum lunation to this moon is erroneous as was first pointed out to me by J. C. Adams and afterwards by M. Poincaré."*

[20]However, Poincaré showed that periodic solutions that never intersect the line of quadratures, as imagined by Hill, *do exist*: see below, p. 129. But these solutions are *not* the continuation of Hill's family.

to cubic powers in time t, we obtain

$$x(t) = \dot{x}_0 t + m\dot{y}_0 t^2 - \left(\frac{m}{3y_0^2} + \frac{\dot{x}_0}{6}\left(\frac{1}{y_0^3} + m^2\right)\right) t^3, \tag{6.32}$$

$$y(t) = y_0 + \dot{y}_0 t - \left(\frac{1}{2y_0^2} + m\dot{x}_0\right) t^2 + \frac{1}{3}\left(\frac{1}{y_0^3} - 2m^2\right)\dot{y}_0 t^3, \tag{6.33}$$

where $\dot{x}_0$ and $\dot{y}_0$ are the components of velocity at $t = 0$. In these equations, terms with $\dot{x}_0$ can be neglected, except for the first term in (6.32). For a periodic solution, $\dot{y}_0 = 0$ and the quadratic term in the expansion for $x(t)$ and the cubic term in the expansion for $y(t)$ must vanish. Poincaré pointed out that this result is also a consequence of the symmetry condition of Hill's periodic solutions, which in this case is[21]

$$x(-t) = -x(t); \quad y(-t) = y(t). \tag{6.34}$$

For values of $m < m_c$, there is only a single crossing with axis $x = 0$, and since Hill's periodic orbits rotate counterclockwise, $\dot{x}_0 < 0$. Therefore, for small t the cubic term in (6.32) can be neglected, and the orbital curve is approximately a parabola

$$y = y_0 - \frac{1}{2y_0^2\dot{x}_0^2} x^2. \tag{6.35}$$

For $m = m_c$, defined to be the value of m for which both $\dot{x}_0 = 0$ and $\dot{y}_0 = 0$, the cubic term in (6.32) becomes the dominant term, leading to an orbital curve in the shape of a cusp,

$$y = y_0 - \frac{1}{2}\left(\frac{3}{y_0 m}\right)^{2/3} |x|^{2/3}. \tag{6.36}$$

This gives a mathematical proof that Hill's critical periodic orbit has a cusp at the maximum value of y with exponent $2/3$, but Poincaré neglected to call attention to this proof.

For values of $m > m_c$, Poincaré assumed that $\dot{x}_0 > 0$, and setting $x(t) = 0$ in (6.32), he found that a second crossing occurs at time $t = t_0$, where

$$t_0 = \sqrt{\frac{3\dot{x}_0 y_0^2}{m}}. \tag{6.37}$$

And he gave a sketch of a periodic orbit with $m > m_c$, which can be compared to a solution obtained numerically:

In fact, to demonstrate that the resulting loop satisfies the requirement for a periodic orbit, it is necessary to show that at this second crossing with $x = 0$, $y(t_0)$ lies *below* y_0, and that both $\dot{x}(t_0)$ and $\dot{y}(t_0)$ are negative. That is the case, since equations (6.32)–(6.33) give

$$y(t_0) = y_0 - \frac{3}{2m}\dot{x}_0$$

and

$$\dot{x}(t_0) = -2\dot{x}_0, \quad \dot{y}(t_0) = -\frac{\sqrt{3\dot{x}_0/m}}{y_0}.$$

[21] It differs from (6.28) because now we take $t = 0$ when the Moon is at $x = 0$, $y = y_0$, where y_0 is the maximum value of y.

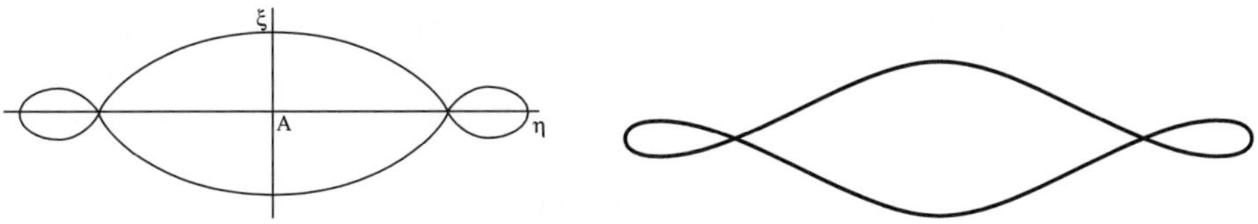

FIGURE 6.4. LEFT: Poincaré's sketch of a periodic lunar orbit with two additional loops, given in *Les méthodes nouvelles de la mécanique céleste*, vol. 1 (here, the Sun is at infinity on the vertical axis, the ξ axis, corresponding to Hill's x axis). RIGHT: a solution numerically calculated by the author with the value of $m = 0.65$.

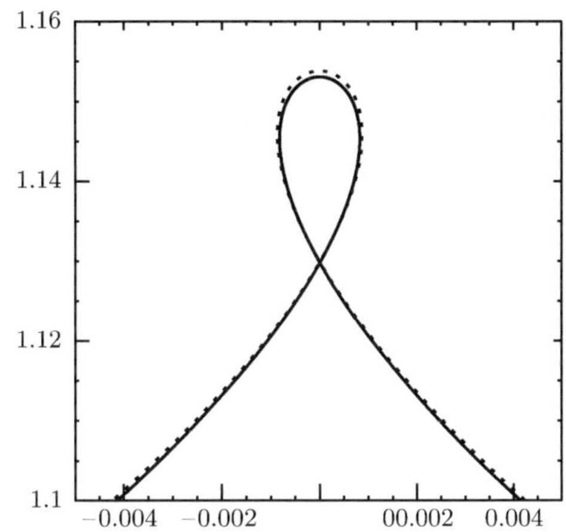

FIGURE 6.5. A loop of Poincaré's orbit (see FIG. 6.4), calculated numerically (dashed curve) and analytically (solid curve) by a power expansion in time with $t = 0$ at the self-intersection (here $m = 0.565$).

(These relations were not given by Poincaré.) At the second crossing we also find that

$$\dot{x}(t_0) = -\frac{2}{3}\, m\, \dot{y}(t_0)^2 y(t_0)^2. \tag{6.38}$$

Still for $m > m_c$ it is also possible to carry out the power series expansions (6.32)–(6.33), taking the origin in time at the crossing of the curve with initial values $\dot{x}_0' < 0$ and $\dot{y}_0' > 0$. In that case, however, the symmetry condition (6.34) does not apply. The condition for periodicity now requires that at a second crossing with the y axis, $x(t_0') = 0$, one has $\dot{y}(t_0') = 0$ and $\dot{x}(t_0') < 0$. According to (6.33), this crossing occurs at the time

$$t_0' = \dot{y}_0'\, {y_0'}^2, \tag{6.39}$$

and the condition $x(t_0') = 0$ requires that the initial parameter satisfy the relation

$$\dot{x}_0' = -\frac{2}{3}\, m\, (\dot{y}_0' y_0')^2. \tag{6.40}$$

When $\dot{y}'_0 = -\dot{y}(t_0)$ and $y'_0 = y(t_0)$, we have $t'_0 = t_0$ and $\dot{x}'_0 = \dot{x}(t_0)$. In this case, these two expansions are equivalent, but in comparison with the actual orbit they give different result because higher order terms in these expansions have been neglected.

In FIG. 6.5 we illustrate our results by a curve obtained analytically for $m = 0.565$ and show that it is in good agreement with a corresponding segment of the orbit obtained by numerical methods (dashed curve), with the values of y'_0 and $\dot{y}'_0$ from this calculation substituted for the initial values in the analytic power series expansions (6.32) and (6.33), with $\dot{x}'_0$ given by the consistency relation (6.40).

6.3. Poincaré's treatment of periodic orbits of the three-body problem

The importance of periodic orbits of the three-body problem in Poincaré's work in celestial mechanics cannot be overestimated. In his discussion of these orbits, in vol. 1 of *New methods of celestial mechanics* [**38**], Poincaré commented:

> *Moreover, what makes these periodic solutions so precious to us is that they are, so to speak, the sole breach from which we can try to penetrate into a place heretofore reputed to be inaccessible.*[22]

Later, in a tribute to Poincaré, Hadamard wrote [**15**]:

> *It is to Poincaré that the credit belongs for having shown that within the periodic solutions lay an instrument, one of the most powerful we can use in searching for and studying other solutions.*[23]

Poincaré conjectured that in the neighborhood of any bounded orbit there exists a periodic solution which could then serve as a starting point to investigate its properties, and he gave an existence proof for a large class of periodic solutions of the three-body problem for the case that two of the masses are very small. In the limit that these two masses vanish, the solution of the three-body problem reduces to two Keplerian orbits rotating around the massive body, and Poincaré recognized that there are numerous such orbits which are periodic when viewed from a suitable rotating frame. For example, all circular orbits with angular frequencies ω and ω', where $\omega > \omega'$ (say), are periodic solutions when viewed in a frame rotating uniformly with a certain angular frequency Ω: the condition for periodicity with a period T in such a rotating frame is that there exists integers n and n' $(n > n')$ such that

$$(\omega - \Omega)T = n2\pi \qquad \text{and} \qquad (\omega' - \Omega)T = n'2\pi$$

and that gives

$$T = (n - n')2\pi/(\omega - \omega') \tag{6.41}$$

and

$$\Omega = (n'\omega - n\omega')/(n' - n). \tag{6.42}$$

For example, taking $n = 1$ and $n' = 0$, we have $T = 2\pi/(\omega - \omega')$, $\Omega = \omega'$: this is the rotating frame where the body with smaller rotation frequency ω' is fixed. Taking

---

[22] *D'aillers, ce qui nous rend ces solutions périodiques si précieuses, c'est qu'elles sont, pour ainsi dire, la seule brèche par où nous puissions essayer de pénétrer dans une place jusqu'ici réputée inabordable.*

[23] *... c'est à Poincaré qu'il appartient d'avoir montré dans les solutions périodiques un instrument, l'un de plus puissants dont on dispose, pour la recherche et l'étude des autres solutions.*

$n = 2$ and $n' = 1$, we again have $T = 2\pi/(\omega - \omega')$, but now $\Omega = -(\omega - 2\omega')$: in this case, when $\omega > 2\omega'$, the rotating frame is retrograde.

Poincaré showed that when masses become positive, but remain small enough, almost all these periodic orbits subsist (they are distorted but do not cease to exist): he called them *"periodic solutions of the first kind"* (*"solutions périodiques de la première sorte"*). In addition, Poincaré considered periodic *"solutions of the second kind"* corresponding to a distortion of two Keplerian ellipses having finite eccentricities with the major axis aligned relative to each other, and *"solutions of the third kind"* where in the limit of vanishing masses we have two circular orbits with different angles of inclination. For "solutions of the second kind" the frequency ratio ω/ω' must be a rational number.[24]

It is less known that Poincaré recognized the unexpected fact that there exist also periodic orbits that in the limit of vanishingly small masses do not correspond to Keplerian ellipses. These are orbits where the minimum distance between the two masses also become vanishingly small in such a manner that their interaction remains finite. Hence, in this limit the massless bodies appear to collide with each other and abruptly change momenta and energy; these limit periodic orbits consist of arcs of elliptical orbits with endpoints at the collision point. In Chapter 32 of his *New methods* ([**38**], vol. 3, pp. 362–371), Poincaré explained these "solutions périodiques de deuxième espèce" as follows:

> *Let us assume, in fact, that the two planets, at first very far from each other, each describe a Keplerian ellipse. It could happen that these two ellipses meet, or pass very close to each other, and do so in such a way that at a certain moment the distance between the two planets becomes very small; at this moment, the action of their mutual perturbation could become detectable, and the two orbits would undergo significant perturbations. Then the planets, having once again become very distant from each other, would describe new Keplerian ellipses. Only these new ellipses would differ greatly from the old ones; the perihelions and the nodes would have undergone considerable variations.*
>
> *I denote this phenomenon by the name of* collision, *even though there is no collision in the proper sense of the word, since the two planets do not come into contact, and it suffices for their distance to become small enough for their attraction to become detectable, despite the smallness of their masses.*[25]

[24]It is interesting to note that several recently discovered extrasolar planetary systems satisfy such a resonance criteria to a good approximation. (A continuously updated reference is found at `http://exoplanet.edu`. For a detailed discussion of the 2 : 1 resonance in GJ876 see [**30**].)

[25]*Supposons, en effet, que les deux planètes, d'abord très éloignées l'une de l'autre, décrivent l'une et l'autre une ellipse képlérienne. Il pourra arriver que ces deux ellipses se rencontrent, ou passent très près l'une de l'autre, et cela de telle façon qu'à un certain moment la distance des deux planètes devienne très petite; à ce moment, leur action perturbatrice mutuelle pourra devenir sensible et les deux orbites subiront des perturbations importantes. Puis les planètes, s'étant de nouveau éloignées l'une de l'autre, décriront de nouveau des ellipses képlériennes. Seulement ces nouvelles ellipses différeront beaucoup des anciennes; les périhélies et les nœuds auront subi des variations considérables.*

Je désignerai sous le nom de choc *ce phénomène, bien qu'il ne s'agisse pas d'un choc au sens propre du mot, puisque les deux planètes ne viennent pas au contact et qu'il suffit que leur distance devienne assez petite pour que l'attraction soit sensible malgré la petitesse des masses.*

In his famous 1890 essay in Acta Mathematica on the three-body problem [**37**] Poincaré concluded with the comment:

> *If $\mu = 0$, this means that the masses of the two planets are infinitely small, and they cannot acts on each other in a detectable manner,* unless they are at an infinitely small distance from each other. *But if these planets pass infinitely close to each other, their orbits will be roughly modified as though they had collided. We may establish initial conditions in such a way that these collisions occur periodically, and we thus obtain discontinuous solutions that are truly periodic solutions to the problem of Keplerian motion, and* which have no right to set aside.[26]

Poincaré did not study these solutions in any detail because *these solutions deviate too much from the orbits actually traversed by celestial bodies.* During the past years, however, these periodic orbits, which have particular interest in space navigation when probes approach close to planets, have been investigated by Michel Hénon [**16**], who also considered the more familiar periodic orbits discussed by Poincaré.

To give a flavor of the various techniques developed by Poincaré to treat periodic orbits of the three-body problem (and first to prove their existence), we describe here his method of analytic continuation in terms of a small parameter μ (see [**38**], Vol. 1, Section 38). Setting $m = a\mu$ and $m' = a'\mu$ for the small masses, where a and a' are constants, the Hamiltonian equations of motion for the coordinates and canonical conjugate momenta $x_i(t, \mu)$ can be written in the general form

$$\dot{x}_i = X_i(x, \mu), \tag{6.43}$$

for $i = 1, 2, \ldots, n$, where the X_i's are analytic functions of μ near $\mu = 0$. Assuming that for $\mu = 0$ there are periodic solutions $x_i(t, 0) = \phi_i(t)$ of these equations with period T, i.e.

$$\phi_i(t + T) = \phi_i(t), \tag{6.44}$$

Poincaré established the conditions for the existence of nearby periodic solutions for small values of the parameter μ. Let

$$x_i(0, \mu) = \phi_i(0) + \beta_i \tag{6.45}$$

be the initial displacements from the periodic solution for $\mu \neq 0$, where the β_i's are small. At a later time $T + \tau$, where τ is a small time interval, Poincaré set

$$x_i(T + \tau, \mu) = \phi_i(0) + \beta_i + \psi_i \tag{6.46}$$

where the ψ_i's are additional small displacements which depend on the initial displacements β_i and on τ and μ. Hence, the condition that for $\mu \neq 0$, $x_i(t, \mu)$ be a periodic function with period $T + \tau$ is

$$\psi_i(\beta, \tau, \mu) = 0 \tag{6.47}$$

---

[26] *Si $\mu = 0$, c'est que les masses des deux planètes sont infiniment petites et qu'elles ne peuvent agir l'une sur l'autre d'une manière sensible,* à moins d'être à une distance infiniment petite l'une de l'autre. *Mais si ces planètes passent infiniment près l'une de l'autre, leurs orbites vont être brusquement modifiées comme si elles s'étaient choquées. On peut disposer des conditions initiales de telle façon que ces chocs se produisent périodiquement et on obtient ainsi des solutions discontinues qui sont de véritables solutions périodiques du problème du mouvement képlérien et* que nous n'avons pas le droit de laisser de côté.

for $i = 1, 2, \ldots, n$, where $\beta = (\beta_1, \ldots, \beta_n)$. This is a system of n nonlinear equations that determines the initial displacements β_i in terms of τ and μ.

Thus, Poincaré reduced the problem of establishing the existence of periodic solutions of the equations of motion for the three-body problem for small masses of two of the bodies to the existence of zeroes of certain functions. For scale invariant forces one of the β_j's can be set equal to zero, say β_1, which then determines τ as a function of μ. When the Jacobian determinant of ψ_i's with respect to β_j's for $j = 2, \ldots, n$ and τ does not vanish for $\mu = 0$, it follows from the implicit function theorem, which Poincaré discussed in Sections 30 and 31 of his *New Methods of Celestial Mechanics* [**38**] (vol. 1), that the β_j's and τ can be expanded in a power series in μ. To calculate this Jacobian it is sufficient to determine ψ_i to first order in the parameters μ and τ. Setting

$$x_i(t, \mu) = \phi_i(t) + \xi_i(t)\mu \tag{6.48}$$

and substituting this form in (6.43) gives a linear inhomogeneous differential equation for $\xi_i(t)$,

$$\dot{\xi}_i = \frac{\partial X_i}{\partial \phi_j}\xi_j + \frac{\partial X_i}{\partial \mu}, \tag{6.49}$$

where the partial derivates of X_i are calculated at $\mu = 0$. According to (6.45), the initial condition for $x_i(t, \mu)$ is

$$\xi_i(0) = \beta_i/\mu, \tag{6.50}$$

and we have (at lowest order)

$$\psi_i = \xi_i(T)\mu + \dot{\phi}_i(0)\tau - \beta_i, \tag{6.51}$$

where according to (6.43)

$$\dot{\phi}_i(0) = X_i(\phi(0), 0). \tag{6.52}$$

Further discussion of the solutions of the linear equation (6.49) is given in Poincaré's prize essay *Sur le problème des trois corps et les équations de la dynamique* [**37**]. Hence, the entries of the Jacobian determinant are determined by

$$\frac{\partial \psi_i}{\partial \beta_j} = \frac{\partial \xi_i(T)}{\partial \xi_j(0)} - \delta_{i,j}, \tag{6.53}$$

$$\frac{\partial \psi_i}{\partial \tau} = X_i(\phi(0), 0). \tag{6.54}$$

Due to the degeneracy of the Kepler problem, the Jacobian determinant vanishes for the three-body problem when $\mu = 0$, which requires a more subtle treatment of this limit described in Sections 30–34 of Poincaré's *New methods of celestial mechanics* [**38**] (vol. 1).

In the last section of Chapter 3 of his book, Poincaré demonstrated also the existence of a different class of periodic orbits that are characterized by the property that these orbits oscillate in the neighborhood of equilibrium points. As an example he discussed the general solutions of Hill's equations (6.26) near $x_0 = 1/(3m^2)^{1/3}$, $y_0 = 0$, which correspond to one of the equilibrium points of the three-body problem discovered by Euler and Lagrange.

Setting $x = x_0 + \delta x$ and $y = \delta y$, Hill's equations of motion become, up to the first order in the displacements δx and δy,

$$\delta\ddot{x} = 9m^2\delta x + 2m\delta\dot{y}, \tag{6.55}$$

$$\delta\ddot{y} = -3m^2\delta y - 2m\delta\dot{x}. \tag{6.56}$$

Setting

$$\delta x = a\cos(\omega t), \quad \delta y = b\sin(\omega t) \tag{6.57}$$

leads to two linear equations for the coefficients a and b,

$$(\omega^2 + 9m^2)a + 2m\omega b = 0, \tag{6.58}$$

$$2m\omega a + (\omega^2 - 3m^2)b = 0. \tag{6.59}$$

Solutions of these equations require that the determinant vanishes, which leads to a fourth order algebraic equation for the frequencies ω,

$$(\omega^2 + 9m^2)(\omega^2 - 3m^2) - 4m^2\omega^2 = 0. \tag{6.60}$$

Two solutions are real,

$$\omega_\pm = \pm m\sqrt{\sqrt{28} - 1}, \tag{6.61}$$

and the other two are imaginary,

$$\omega'_\pm = \pm im\sqrt{\sqrt{28} + 1} \tag{6.62}$$

corresponding to solutions of the form

$$\delta x = a\cosh(|\omega'|t), \quad \delta y = b\sinh(|\omega'|t). \tag{6.63}$$

The existence of imaginary frequencies indicates that the periodic solutions are unstable, because a small deviation grows exponentially in time according to the dominant term $e^{|\omega'|t}$. Poincaré's periodic orbit corresponds to an orbit conjectured by Hill (see quotation on p. 123 above) which does not intersect the axis of quadrature.

In Chapter 4, Poincaré generalized the linear stability analysis for equilibrium points[27] to the stability of periodic orbits of a dynamical system. He proved that in

---

[27] Concerning Euler and Lagrange equilibrium points, let us mention that Joseph Liouville ([**23**], 1842) showed that the Euler points (on the axis of two massive bodies) are linearly unstable, and that Gabriel Gascheau ([**12**], 1843) showed that the Lagrange points (forming an equilateral triangle with the massive bodies) are linearly stable if and only if the masses m_A, m_B of the two massive bodies A, B satisfy $\frac{m_A}{m_B} + \frac{m_B}{m_A} > 25$. (More generally, Gascheau showed that Lagrange's triangular solution for three bodies A, B, C with arbitrary masses is linearly stable if and only if $(m_A + m_B + m_C)^2 > 27(m_A m_B + m_B m_C + m_C m_A)$.) In particular, for the Sun, Jupiter and an asteroid, the two Lagrange points are linearly stable; and, indeed, one finds there two clouds of asteroids ("Greeks" at one point, and "Troyans" at the other). Gascheau's criterion is often credited to Edward J. Routh, who rediscovered it in 1875. This linear stability analysis was restricted to circular orbits, but it has been extended to finite eccentricities (which requires an application of Floquet's analysis: see footnote below) by J. M. A. Danby [**9**]. The linear stability analysis may indicate, however only a bifurcation to a more complex but still stable orbit. This turns out to be the case for the Gascheau–Routh critical point. For a recent numerical analysis which determines the actual boundary for unstable orbits for general masses and finite eccentricities, see [**31**].

this case there exist corresponding characteristic exponents[28] S_i which are defined by solving the motion equations near a periodic orbit $x_i^0(t)$, where $i = 1, \dots, n$: keeping only linear deviations $\delta x_i(t)$ Poincaré showed (see *Les Méthodes Nouvelles*, p. 176) that in this case

$$\delta x_i(t) = e^{S_i t}\phi_i(t), \tag{6.64}$$

where the $\phi_i(t)$ are periodic functions with the same period as $x_i^0(t)$. Thus, the condition for linear stability, as before, is that the characteristic exponents S_i are imaginary.

6.4. Poincaré's geometrical proofs for the existence of periodic orbits[29]

Let us outline here two other proofs of Poincaré for the existence of periodic orbits in the 3-body problem. They are of a geometrical nature. It is striking that they constitute the first (1883) and the last (1912) contributions of Poincaré to celestial mechanics.

The first proof (1883).
The first proof was announced in a note to *Comptes Rendus* of the French Academy of Sciences in 1883 [**35**], and the details were given in the *Bulletin astronomique* in 1884 [**36**]. Poincaré starts from the basic observation that any function satisfying two parity relations $f(-t) = f(t)$ and $f(\tau - t) = f(\tau + t)$ is periodic, with period $T = 2\tau$. Thus one has just to prove the parity relations.

Here, we consider only the case of a plane motion. Let A be a massive body and B, C two small masses bodies. Take a frame where A is at rest with the origin of coordinates at A. Poincaré says that there is a "symmetrical conjunction" at $t = 0$ if the three bodies are on the same line, with the velocity vectors of B, C perpendicular to this line. Thus, the mutual distances of the three bodies are the same at times t and $-t$ (they are even functions of time). If there is again a symmetrical conjunction at another time τ, the mutual distances of the three bodies will be the same at times $\tau + t$ and $\tau - t$; thus they will be periodic: at time $T = 2\tau$ we shall have the same relative configuration as at time 0; the line of conjunction will not be the same in general, but in a suitable rotating frame the motion will be periodic. So the problem is reduced to the existence of two symmetrical conjunctions.

Since we have chosen a frame where A is at rest and taken the origin of coordinates at A, we have 8 initial conditions: for example, the two coordinates and the two components of the velocity vectors for B and for C. But an orbit "coincides" at a given time with an (osculating) ellipse, and it is convenient to choose a set of parameters attached to these ellipses as initial data; say, the longitudes λ_B, λ_C of B and C, the longitudes of perihelia, the mean angular frequencies ω_B, ω_C, and eccentricities e_B, e_C. Assume that at $t = 0$ the longitudes of B, C and of their perihelia are zero: so the three bodies are on a same line, and the velocities of B, C are perpendicular to this line; this is a "symmetrical conjunction". We have spent four initial conditions, we have still four: ω_B, ω_C, e_B, e_C. Now, we have to

[28]The introduction of characteristic exponents for the solution of linear differential equations with coefficients that are periodic functions was first carried out by Gaston Floquet. He published his results in several articles starting with his thesis, *Sur la théorie des équations différentielles linéaires*, submitted to the Faculty of sciences in Paris on 8 April 1879.

[29]This section is written by Éric Charpentier.

prove that for masses m_B, m_C small enough, we can choose these parameters in such a way that a second symmetrical conjunction occurs. Here Poincaré uses a topological theorem:

THEOREM. *Let $X_1, \ldots, X_n$ be n continuous functions of n variables $x_1, \ldots, x_n$. Assume that $X_i > 0$ at $x_i = a_i$ and $X_i < 0$ at $x_i = -a_i$ $(1 \le i \le n)$. Thus there exists at least one point $x = (x_1, \ldots, x_n)$ such that $-a_i < x_i < a_i$ $(1 \le i \le n)$ and $X_1(x) = X_2(x) = \cdots = X_n(x) = 0$.*[30]

It is a natural generalization of the intermediate value theorem (which corresponds to $n = 1$). The proof is just an application of Kronecker's theorem (a higher-dimensional generalization of Cauchy's index theorem). Poincaré writes:

> *We see immediately the advantage that this theorem's application presents. To rigorously prove that certain equations can be satisfied, it suffices to study the sign of certain functions.*[31]

This very simple (but powerful) method gives to Poincaré the existence of a second symmetrical conjunction. Indeed, let X be the difference of longitudes of B and C at a time $t = \tau > 0$ (arbitrary); let $Y = \dot{r}_B$, $Z = \dot{r}_C$ be the time derivatives of the distances of B, C from A, at $t = \tau$, and X_0, Y_0, Z_0 the values of X, Y, Z in the limit $m_B = m_C = 0$. All are functions of the four arbitrary initial data ω_B, ω_C, e_B, e_C (recall that the four initial longitudes are zero). Fix arbitrarily $\omega_B + \omega_C$ and consider the three variables e_B, e_C and $\omega_B - \omega_C$. In the case of $e_B = e_C = 0$ and $\omega_B - \omega_C = \pi/\tau$, we have $X_0 = Y_0 = Z_0 = 0$. Assume temporarily that the Jacobian determinant of X_0, Y_0, Z_0 with respect to $e_B, e_C, \omega_B - \omega_C$ is non-vanishing at $e_B = e_C = 0$, $\omega_B - \omega_C = \pi/\tau$. Thus, by the local inversion theorem, we can take X_0, Y_0, Z_0 as the new variables, i.e. consider X, Y, Z as functions of X_0, Y_0, Z_0, for (X_0, Y_0, Z_0) in a small neighborhood of 0, containing a cube $[-a_1, a_1] \times [-a_2, a_2] \times [-a_3, a_3]$; but, for m_B, m_C small enough, $X_0 = -a_1$ implies $X < 0$, $X_0 = a_1$ implies $X > 0$, and the same with Y, Z, so the sign conditions of the topological theorem above are fulfilled. Then there exists (X_0, Y_0, Z_0) in the cube above (and thus, e_B, e_C near 0 and $\omega_B - \omega_C$ near π/τ) such that $X = Y = Z = 0$. This means that there is again a symmetrical conjunction at time $t = \tau$, and that proves the existence of periodic solutions, which (among other properties) take place in a fixed plane and are almost circular. Poincaré calls them the periodic solutions of the *"first kind"*. For fixed masses, they depend on four continuous parameters; for example, the times t_0, $t_0 + \tau$ of the first and the second symmetrical conjunctions (we remove the hypothesis $t_0 = 0$), and the longitudes α_0, $\alpha_0 + \varphi$ of B at t_0 and at $t_0 + \tau$ (we remove the hypothesis $\alpha_0 = 0$). We have assumed that the Jacobian determinant of X_0, Y_0, Z_0 with respect to $e_B, e_C, \omega_B - \omega_C$ did not vanish at $e_B = e_C = 0$, $\omega_B - \omega_C = \pi/\tau$. A calculation shows that it vanishes at that point if and only if φ $(= \omega_B \tau)$ is a multiple of π; thus one has just to exclude these isolated values. If we are interested only in the *shape* of orbits, we can fix the

---

[30] In other words, *if a continuous vector field $X = (X_1, \ldots, X_n)$ is pointing out on all faces of the cube $[-a_1, a_1] \times \cdots \times [-a_n, a_n]$, then it has at least one zero inside this cube.* Notice that the celebrated Brouwer fixed point theorem of 1909 can be immediately deduced from this, as will be remarked by Hadamard in 1910. In fact, these two theorems are equivalent.

[31] *On voit immédiatement l'avantage que peut présenter l'application de ce théorème. Pour démontrer rigoureusement que certaines équations peuvent être satisfaites, il suffit d'étudier le signe de certaines fonctions.*

origins of times and of angles by $t_0 = 0$ and $\alpha_0 = 0$ (as above) and take τ for the unit of time: there remains the single parameter φ.

As mentioned in Section 6.3, Poincaré proves also the existence of periodic solutions of *"second kind"* (planar but with finite eccentricities, and other features) and of *"third kind"* (with small eccentricities but finite inclination angles). He indicates that he was able to prove the existence of periodic solutions of *"fourth kind"* (finite eccentricities, finite inclination angles), but only for special values of the ratio m_B/m_C.

In fact, the use of the topological theorem above was not necessary, and Poincaré would no more mention it when he would explain this method again in [**38**] (Vol. 1, Chapter 3, n° 40): indeed, one just has to apply the implicit function theorem to X, Y, Z as functions of $e_B, e_C, \omega_B - \omega_C$, with parameters m_B, m_C.

The last proof (1912).

At the end of his life, Poincaré ([**41**], 1912) proves the existence of an infinity of periodic solutions of *fixed energy* and he asks which ones subsist and in which order they disappear when masses increase; he says that the answer depends on a certain geometric theorem, at least in the case of the restricted 3-body problem, and, more generally, of dynamical problems with only two degrees of freedom:

> *I have proved, a long time ago now, the existence of periodic solutions to the three-body problem. The result still left much to be desired, however; for, although the existence of each kind of solution was established for small values of the masses, one could not see what should happen for larger values, which were the ones among these solutions that would persist, and to what order they would disappear. Reflecting on this question, I convinced myself that the response must depend on the correctness or falsehood of a certain geometric theorem, whose statement is very simple, at least in the case of the restricted problem and dynamical problems where there are only two degrees of freedom.*[32]

This "geometric theorem" is the following:

THEOREM. *Let S be an annulus in the plane, and let f be a one-to-one and bicontinuous mapping (i.e., a homeomorphism) from S onto itself, preserving area, moving the points of one boundary circle clockwise and the points of the other one counterclockwise.*[33] *Then f has at least two fixed points.*

---

[32] *J'ai démontré, il y a longtemps déjà, l'existence des solutions périodiques du problème des trois corps; le résultat laissait cependant encore à désirer; car, si l'existence de chaque sorte de solution était établie pour les petites valeurs des masses, on ne voyait pas ce qui devait arriver pour des valeurs plus grandes, quelles étaient celles de ces solutions qui subsistaient et dans quel ordre elles disparaissaient. En réfléchissant à cette question, je me suis assuré que la réponse devait dépendre de l'exactitude ou de la fausseté d'un certain théorème de géométrie dont l'énoncé est très simple, du moins dans le cas du problème restreint et des problèmes de dynamique où il n'y a que deux degrés de liberté.*

[33] This condition may seem ambiguous, because rotation by angle α in one direction ($0 \leq \alpha \leq 2\pi$) has the same effect, geometrically, as rotation by angle $2\pi - \alpha$ in the opposite direction, so *a priori* we cannot say that f moves a point of the boundary in a specific direction. In fact, it is assumed that f acts on the *universal covering* of the annulus, that is, on the pairs (r, θ), where θ (the argument) is a real number (and not an "angle" defined modulo 2π); thus one can speak of a rotation in one direction or in the other (corresponding to a displacement of the argument to the right or to the left on the real axis).

Poincaré could prove only special cases. When he did not manage to solve a problem, he let it mature; but this time, fearing not having the possibility to come back to it, he published his work unfinished. A few months later, he died. Shortly after, a complete proof of the "last geometric theorem of Poincaré" was given by George David Birkhoff [**2**],[34] with a different approach. In 1990, Christophe Golé and Glen Hall [**14**] gave a proof following the original approach of Poincaré.

Let us come back to Poincaré's article. We consider the restricted 3-body problem.[35] Namely, one considers three bodies A, B, C, with masses $m_A \geq m_B$ and m_C negligible ($m_C = 0$); the motion takes place in a plane, A and B move along circular orbits (necessarily around their common center of mass). Taking the frame which rotates with them, we are now only interested in the motion of C. Let x, y be its coordinates. The quadruples $(x, y, \dot{x}, \dot{y})$ represent the points of the so-called phase space (space of states of the system), here $\mathbb{R}^4$. The solutions of equations of motion define curves in the phase space, called *phase trajectories.* A closed phase trajectory corresponds to a periodic solution. Conservation of energy induces in the *rotating* frame a constraint (Jacobi) of the form $\frac{1}{2}(\dot{x}^2 + \dot{y}^2) - F(x, y) = -c$, which defines a (three-dimensional) hypersurface in the phase space. This constraint implies $F(x, y) - c \geq 0$, determining an area in the plane; Poincaré assumes that this area has a bounded connected component and he restricts his study to motions confined in this bounded area. I shall denote by $\mathcal{E}_c$ the corresponding three-dimensional subset of the phase space.

For example, when $m_B = 0$, Jacobi's constraint is $\frac{1}{2}(\dot{x}^2+\dot{y}^2) - \frac{K}{r} - \frac{1}{2}\omega_B^2 r^2 = -c$, where ω_B is the angular frequency of the rotating frame (= that of B in the fixed plane) and K is the mass of the massive body A multiplied by Newton's constant of gravitation; then one has $\frac{K}{r} + \frac{1}{2}\omega_B^2 r^2 - c \geq 0$. When $c > \frac{3}{2}(K\omega_B)^{2/3}$, the equation $\frac{1}{2}\omega_B^2 r^3 - cr + K = 0$ has two positive roots, $r_1 < r_2$, and we can restrict ourselves to motions confined in the disc of radius r_1. In this case a point of $\mathcal{E}_c$ is determined by a point of this disc and an angle giving the direction of the velocity vector (its norm is given by Jacobi's constraint); on the boundary of the disc, this angle is undetermined (like longitude at Earth's poles).

The goal of Poincaré is to prove the existence of an infinite family of periodic solutions in $\mathcal{E}_c$, as rich as possible, and to determine in which order they disappear when the mass m_B increases from 0. For that, he assumes that we have a *Poincaré section* in $\mathcal{E}_c$: that is, a (two-dimensional) surface S whose boundary is a closed phase trajectory Γ_0 and whose interior is transverse to the flow (i.e., no phase trajectory is tangential to it), and for every point P on S, the phase trajectory passing through P will pass again through S. We denote by $T(P)$ its point of first return.[36] The hypothesis of transversality to the flow implies that T is a

[34]In fact, Birkhoff's proof provides the existence of at least *one* fixed point; for a second fixed point, his argument was not conclusive; in 1925 he proposed a modification but his new argument has been contested, until Morton Brown and Walter Neumann [**5**] gave, in 1977, a transparent exposition of it, showing clearly that it was correct.

[35]Poincaré had in mind, more generally, *dynamical problems with only two degrees of freedom* and, simultaneously to the restricted 3-body problem, he treated the problem of existence of an infinity of closed geodesics on a distorted sphere, a problem which would be developed later by Birkhoff: see Chapter 7 in this book (p. 153).

[36]In the case of $m_B = 0$ one can explicitly construct such a section S and the corresponding map T; as m_B increases, S subsists (undergoing some distortion), at least when m_B remains small enough and for generic values of c.

homeomorphism from S onto itself. Poincaré assumes further that S is topologically equivalent to a disc, and he identifies S with the disc $r \leq a$ in $\mathbb{R}^2$ (and Γ_0 with the circle $r = a$). Assuming, besides, that the periodic trajectory corresponding to the phase trajectory Γ_0 is *stable*, Poincaré shows that as a point P on S approaches the boundary, the difference of arguments $\arg T(P) - \arg P$ tends to a constant $2\pi\theta$. Then T induces on the boundary a rotation by angle $2\pi\theta$ (more precisely, a translation of $2\pi\theta$ on the argument). This fact implies (from the same index theorem of Cauchy and Kronecker which already implied the "topological theorem" in the proof of 1883) that in S there exists at least one fixed point for T, and even at least one *stable* fixed point. By distorting the system of coordinates in S, one can assume that this is the point 0. Thus, as above, as a point P on S approaches 0, the difference $\arg T(P) - \arg P$ tends to a constant $2\pi\varphi$. That is to say, T induces a rotation of angle $2\pi\varphi$ in the limit at 0. Besides, Poincaré proved (see *Méthodes nouvelles*, Vol. 3, Chap. 27) that T preserves "area", or more exactly some positive measure on S which will work as well: thus one could apply the geometric theorem if arguments $2\pi\theta$ and $2\pi\varphi$ were of opposite signs. But, first, there is no reason why this should be the case, and second, Poincaré wants to obtain an infinity of periodic solutions, with arbitrarily large periods (because he hopes to prove someday that one can approximate all bounded solutions in the phase space by periodic solutions, on arbitrarily long times). Thus he replaces T by T^p, $p \in \mathbb{N}$, and he performs after T^p the rotation of angle $2\pi q$, $q \in \mathbb{Z}$ (more precisely the translation of $2\pi q$ on the argument): the resulting map, f, induces on the boundary circle (corresponding to) Γ_0 a rotation of angle $2\pi(p\theta + q)$ and, in the limit at 0, a rotation of angle $2\pi(p\varphi + q)$. These angles (these arguments) will be of opposite signs if $\varphi < -q/p < \theta$ or $\theta < -q/p < \varphi$. Except for the special case of $\varphi = \theta$, this condition is satisfied by an infinity of pairs (p, q), and to each pair the "geometric theorem" associates at least two fixed points for f, and hence at least two periodic orbits. Moreover, the preceding argument allows us to know which of these solutions subsist and in which order they disappear when m_B increases from 0: as long as S exists (undergoing an adequate distortion), θ and φ vary continuously and the periodic solutions corresponding to a given (p, q) will disappear only when θ or φ will come to coincidence with $-q/p$.

6.5. Some recent developments in periodic orbit theory

Poincaré's work has been a source of inspiration for some recent work in finding new periodic solutions of the n-body problem for equal masses. In the determination of periodic orbits with masses not necessarily small, Poincaré [**39**] used the principle of least action[37] (let us recommend, in passing, his excellent popular discussion of this principle in Chapter 12 of [**40**]). His idea is to consider classes of virtual periodic motions ("virtual" means that they do not necessarily satisfy the equations of motion), and then to prove that in each of these classes the action integral has an extremum: motions corresponding to these extrema are periodic solutions of the equations of motion. This type of argument has recently been essential in establishing mathematically existence proofs for such orbits, and also it has been applied to calculate periodic orbits without solving Newton's equations of motion, as will be discussed below.

[37]More properly, this principle consists in finding an *extremum* of the action.

In 1993, Cristopher Moore [**26**] discussed a classification of periodic orbits of the n-body problem for equal masses confined in a plane according to their knot and braid types [**13, 18**]. Moore searched for such orbits by starting with an approximate periodic orbit with the desired symmetry and topology, and then minimized the action integral using a gradient descent procedure with respect to the Cartesian coordinates of the orbit. In this manner he found a number of new periodic 3-body orbits with a variety of braid types and symmetries. In particular, he discovered the celebrated orbit in the shape of a figure eight, FIG. 6.6, and found that it is stable.

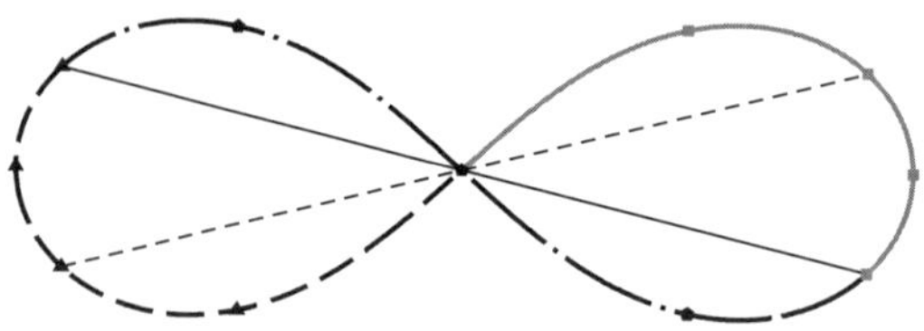

FIGURE 6.6. Moore's figure-8 orbit for three equal masses shown at 5 equal time intervals during a third of a period. The two straight lines show symmetrical conjunctions separated by half a period.

FIG. 6.6 shows two consecutive "symmetrical conjunctions" with the same body in between; contrary to Poincaré's situation (see Section 6.4), the velocities of the end masses are not perpendicular to the conjunction lines, but symmetry is restored if the two end masses are also interchanged under the time reversal transformation $t \to -t$.

Six years after Moore's discovery, Alain Chenciner and Richard Montgomery [**6**] gave a rigorous mathematical existence proof for this orbit by applying a variational method based on the principle of least action, and since then more precise numerical work has been done describing this orbit [**42, 29**]. Moreover, many new periodic orbits with equal masses following each other along a fixed trajectory have been identified[38] [**42**], and additional numerical evidence that the figure-8 orbit is stable under perturbations in the plane has been provided [**43**]. These results have given an impetus to the search of new periodic orbits of the n-body problem, and to the development of mathematical proofs, which continues unabated to the present time.

It turns out that Moore's gradient descent procedure for calculation of periodic orbits has some undesirable instabilities, and a better approach [**29**] is to capitalize on the periodicity of the orbits by expanding the Cartesian coordinates in a Fourier series and finding the extrema of the action integral directly as a function of the Fourier coefficients of the trajectory. The convergence of the Fourier expansion to a given accuracy can be readily tested by comparing the accelerations calculated from this expansion with the corresponding forces, thus verifying the equations of motion. Using this approach, I found extensions of the planar figure-8 orbit in three dimensions, which have non-zero angular momenta about the two axis of symmetry in the original plane of the figure-8, and the axis perpendicular to this plane [**29**].

[38]One can download a number of such "choreographies" from the web site of Carles Simó: `http://www.maia.ub.es/dsg/nbody.html`. A selection of them is given at `http://melusine.eu.org/syracuse/flash/nbody/`.

The case where the axis of rotation is along the horizontal symmetry axis of the figure-8 is shown in FIG. 6.7.

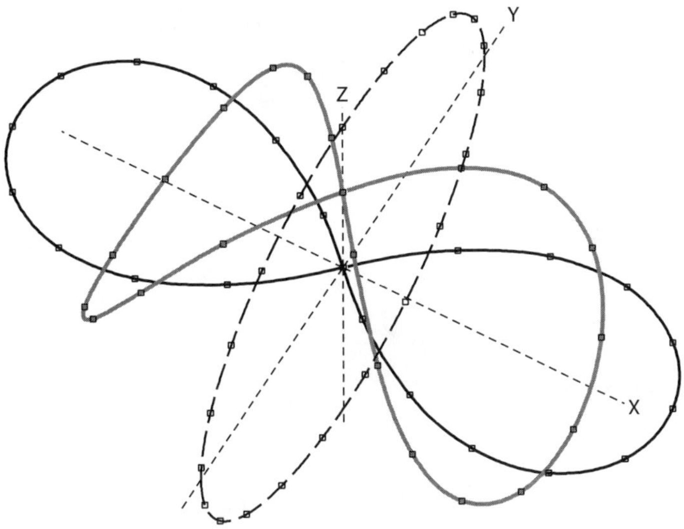

FIGURE 6.7. A three-dimensional figure-8 orbit rotating about the horizontal symmetry axis (x-axis), which interpolates between the planar figure-8 orbit and the Lagrange circular orbit (also shown). There are 21 masses moving on each of the orbits.

A proof for the existence of these orbits has been given by Christian Marchal [**24, 25**], who also showed that these orbits form a family which interpolates between the planar figure-8 orbit, which has zero angular momentum, and the (generalized) Lagrange solution;[39] see also references [**7, 32**].

Until now most of the newly discovered periodic orbits have been either planar orbits or were obtained by an extension of these orbits. Recently, however, Moore and I have found a number of new three-dimensional periodic n-body orbits with equal masses which have cubic symmetry [**27**].[40] An example for the case of $n = 12$ is shown in FIG. 6.8.

This orbit was conjectured by considering the faces of a *cuboctahedron*,[41] with its 24 edges as composing 4 hexagonal trajectories, which can be oriented in such a way that half of the triangular faces are oriented counterclockwise and the others are oriented clockwise. One assumes that the masses can start at the corners of the cuboctahedron and then travel along these trajectories in these directions: first orbiting around the counterclockwise triangles and then around the clockwise ones. Starting with the approximation where the hexagonal trajectories are replaced by circles, the actual orbits were then obtained numerically by the Fourier gradient

[39] In the case of three equal masses and zero eccentricity, Lagrange's solution ([**21**], 1772) is a circular orbit with the masses rotating on the vertex of an equilateral triangle.

[40] The web sites of Cris Moore and Michael Nauenberg show a number of animations of recently found periodic orbits of the n-body problem: `http://www.santafe.edu/~moore/gallery.html` and `http://physics.ucsc.edu/~michael`

[41] A polyhedron obtained by intersection of a cube and an octahedron. It is composed of six squares and eight equilateral triangles.

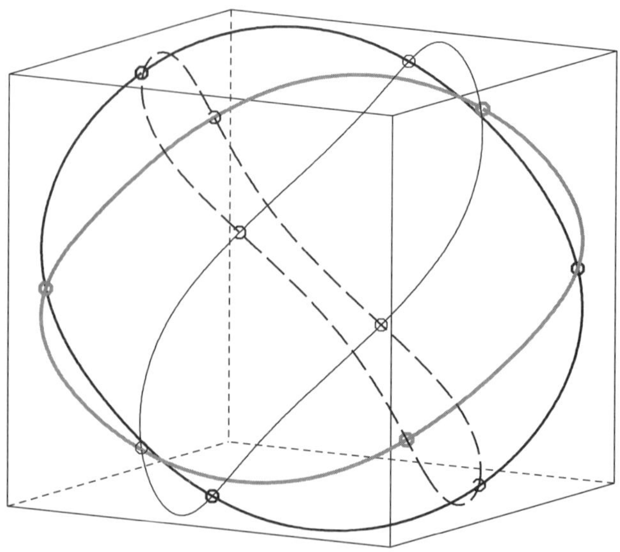

FIGURE 6.8. Periodic orbit with 12 equal masses and cubic symmetry.

descent method. It can be shown that on each of the four orbits there can be any odd number m of masses, and we have calculated the Fourier coefficients of such orbits up to $m = 121$ [**27**].

A simple way to understand these periodic orbits and other recently found n-body periodic orbits in three dimensions is to start by considering a fixed number of equal masses arranged periodically on each of several circles of equal radius, which rotate on these circles at constant velocity. These circles are arranged in a symmetric fashion around a common center, and intersect each other at various discrete points dictated by the symmetry. An essential restriction for constructing a periodic orbit is that at the intersections of the orbits the masses should not collide with each other, and that their center of mass remains fixed. Starting with an allowed configuration, the total action integral is calculated, and then the shape and orientation of the orbits is deformed to change the magnitude of the action until an extrema under these changes is found, which can be obtained iteratively by the Fourier gradient method [**29**].

An elementary example with two orbits and three masses on each is illustrated in FIG. 6.9.

Here the initial orbit consisted of two equal circular orbits with a common center, and the planes of the orbits were inclined relative to each other at some arbitrary but finite angle, while the three masses on each orbit were located symmetrically around the circle. Starting with a circular orbit whose plane makes an angle θ with the z axis, the Cartesian coordinates of the three masses on this orbit are given by

$$\begin{aligned} x_j(t) &= \sin\theta \, \sin t_j, \\ y_j(t) &= \cos t_j, \\ z_j(t) &= \cos\theta \, \sin t_j, \end{aligned} \tag{6.65}$$

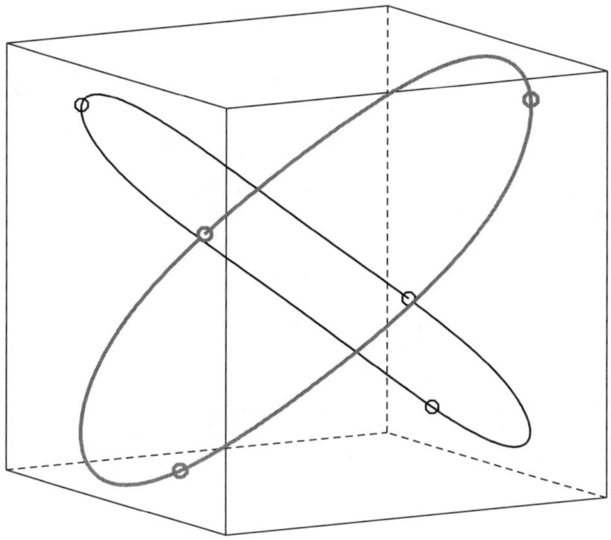

FIGURE 6.9. Three-dimensional periodic orbit consisting of two distorted circular orbits with three masses on each orbit rotating in opposite directions.

where $t_1 = t$, $t_2 = t+2\pi/3$, and $t_3 = t-2\pi/3$. The second circular orbit is obtained by the symmetry transformation $\theta \to -\theta$, $t \to -t+\pi$. In this case there are no collisions at the intersection points of these two orbits, which occur at $x = z = 0$ and $y = \pm 1$.

Minimizing the action integral results in Fourier expansion for $j = 1, 2, 3$ given by

$$\begin{aligned} x_j(t) &= \sum_k a(k)\sin(kt_j), \\ y_j(t) &= \sum_k b(k)\sin(kt_j), \\ z_j(t) &= \sum_k a(k)\sin(kt_j), \end{aligned} \tag{6.66}$$

where the sum is over odd values of k. The corresponding expansion for the second orbit is obtained by the transformation $a(k) \to -a(k)$, $b(k) \to -b(k)$, $c(k) \to c(k)$. The resulting Fourier coefficients calculated by the gradient method up to five decimal digits are given in the following table, and the resulting orbit is shown in FIG. 6.9.

Fourier Coefficients			
k	$a(k)$	$b(k)$	$c(k)$
1	1.06371	0.94684	0.90762
3	0.04244	0.02491	0.00000
5	0.00110	−0.00002	−0.00321
7	−0.00028	−0.00024	−0.00037
9	−0.00005	−0.00003	0.00000

Bibliography

[1] J. Barrow-Green, *Poincaré and the Three Body Problem*, American Mathematical Society (Providence, RI), 1996.

[2] G. D. Birkhoff, *Proof of Poincaré's geometric theorem*, Trans. Amer. Math. Soc. **14** (1913), pp. 14–22. *Démonstration du dernier théorème de géométrie de Poincaré*, Bull. Soc. math. France **42** (1914), pp. 1–12.

[3] J. B. Brackenridge, *The Key to Newton's Dynamics: the Kepler Problem and the Principia*, University of California Press (Berkeley, CA), 1995.

[4] J. B. Brackenridge and M. Nauenberg, *Curvature in Newton's Dynamics*, in: *The Cambridge Companion to Newton*, edited by I. B. Cohen and G. E. Smith, Cambridge Univ. Press, 2002, pp. 85–137.

[5] M. Brown and W. D. Neumann, *Proof of the Poincaré-Birkhoff fixed point theorem*, Michigan Math. J. **24** (1977), pp. 21–31.

[6] A. Chenciner and R. Montgomery, *A remarkable periodic solution of the three-body problem in the case of equal masses*, Ann. of Math. **152** (2001), pp. 881–901.

[7] A. Chenciner, J. Féjoz, and R. Montgomery, *Rotating eights I: the three* Γ_i *families*, Nonlinearity **18** (2005), pp. 1407–1424.

[8] *The Foundations of Newtonian Scholarship*, edited by R. Dalitz and M. Nauenberg, World Scientific (Singapore), 2000, p. 167.

[9] J. M. A. Danby, *Stability of the Triangular Points in the Elliptic Restricted Problem of Three Bodies*, Astrophysical Journal **69** (1964), pp. 165–172.

[10] L. Euler, *De motu rectilineo trium corporum se mutuo attrahentium* (1765): *Opera* (ser. 2), vol. 25, pp. 281–289.

[11] L. Euler, *Theoria Motuum Lunae, Nova Methodo Pertractata* (1772): *Opera* (ser. 2), vol. 22, pp. 1–41.

[12] G. Gascheau, *Examen d'une classe d'équations différentielles et application à un cas particulier du problème des trois corps*, Comptes Rendus Acad. Sci. Paris **16** (1843), pp. 393–394.

[13] R. W. Ghrist, P. J. Holmes, and M. Sullivan, *Knots and Links in Three-Dimensional Flows*, Springer Lecture Notes in Mathematics, volume 1654, Springer (Heidelberg), 1997.

[14] C. Golé and G. R. Hall, *Poincaré's proof of Poincaré's last geometric theorem*, pp. 135–151 of: *Twist mappings and their applications, Proc. Workshop, Minneapolis/MN (USA) 1989/90*, IMA Vol. Math. Appl. **44** (1992).

[15] J. Hadamard, *L'Œuvre mathématique de Poincaré*, reproduced in: *Œuvres de Jacques Hadamard*, Tome 4, Éditions du Centre National de la Recherche Scientifique (Paris), 1968, p. 1967.

[16] Michel Hénon, *Generating Families in the Restricted Three-Body Problem*, Lecture Notes in Physics, Vol. 1, Springer, 1977, and Vol. 2, Springer, 2001.

[17] G. W. Hill, *Researches in the Lunar Theory*, American Journal of Mathematics **1** (1878), pp. 5–26, 129–147, 245–260. Reprinted in *Collected Mathematical Works of G. W. Hill*, Vol. 1 (published by the Carnegie Institution of Washington, June, 1905), pp. 284–335.

[18] P. J. Holmes and R. F. Williams, *Knotted periodic orbits in suspensions of Smale's horseshoe: torus knots and bifurcation sequences*, Arch. Rat. Mech. Anal. **90** (1985), pp. 115–194.

[19] C. Huygens, *Horologium Oscillatorium*, Paris, 1673, Reproduced in vol. 18 of *Œuvres complètes de Christiaan Huygens* (OC), Société Hollandaise des Sciences (The Hague, Martinus Nijhoff 1888–1950). Translated into English by R. J. Blackwell: *The Pendulum Clock*, Iowa State University Press, 1966. See also *De Vi Centrifuga* (1659, publ. 1704): OC, Vol. 16, pp. 256–328.

[20] C. G. J. Jacobi, *Sur le mouvement d'un point et sur un cas particulier du problème des trois corps*, Comptes Rendus Acad. Sci. Paris **3** (1836), pp. 59–61.

[21] J. L. Lagrange, *Essai sur le Problème des Trois Corps* (1772): *Œuvres*, vol. 6, pp. 229–331.

[22] P. S. Laplace, *Sur la théorie lunaire de Newton*, in: *Mécanique Céleste*, Tome V, Chapitre II, Imprimerie Royale (Paris), MDCCCXLVI.

[23] J. Liouville, *Sur un cas particulier du problème des trois corps*, Comptes Rendus Acad. Sci. Paris **14** (1842), pp. 503–506. *Mémoire sur un cas particulier du problème des trois corps*, read at Académie des Sciences, 4 April 1842; J. Math. pures et appl. (2) **1** (1856), pp. 248–264.

[24] C. Marchal, *The family P_{12} of the three-body problem—the simplest family of periodic orbits, with twelve symmetries per period*, Celestial Mechanics and Dynamical Astronomy **78** (2001), pp. 279–298.

[25] C. Marchal, *How the method of minimization of action avoids singularities*, Celestial Mechanics and Dynamical Astronomy **83** (2002), pp. 325–353.

[26] C. Moore, *Braids in classical gravity*, Physical Review Letters **70** (1993), pp. 3675–3679.

[27] C. Moore and M. Nauenberg, *New orbits for the N-body problem*, in a Festschrift for Philip Holmes, Journal of Computational and Nonlinear Dynamics **1** (2006), pp. 307–311.

[28] M. Nauenberg, *Newton's Early Computational Method for Dynamics*, Arch. for Hist. Exact. Sci. **46** (1994), pp. 221–252.

[29] M. Nauenberg, *Periodic orbits for three particles with finite angular momentum*, Physics Letters **292** (2001), pp. 93–99.

[30] M. Nauenberg, *Determination of masses and other properties of extra-solar planetary systems with more than one planet*, Astrophysical Journal **568** (2002), pp. 369–376.

[31] M. Nauenberg, *Stability and eccentricity for two planets in a* $1:1$ *resonance, and application to extrasolar planetary systems*, The Astronomical Journal **124** (2002), pp. 2332–2338.

[32] M. Nauenberg, *Continuity and stability of families of figure-8 orbits with finite angular momentum*, Celestial Mechanics and Dynamical Astronomy **97** (2007), pp. 1–15.

[33] I. Newton, *Mathematical Principles of Natural Philosophy*, third edition translated by I. B. Cohen and Anne Whitman, University of California Press (Berkeley, CA), 1997.

[34] *The Mathematical Papers of Isaac Newton*, vol. 3; ed. by D. T. Whiteside, Cambridge University Press, 1969, p. 170.

[35] H. Poincaré, *Sur certaines solutions particulières du problème des trois corps*, Comptes Rendus Acad. Sci. Paris **97** (1883), pp. 251–252. Reproduced in: *Œuvres d'Henri Poincaré*, Tome VII, Gauthier-Villars (Paris), 1952, pp. 251–252.

[36] H. Poincaré, *Sur certaines solutions particulières du problème des trois corps*, Bulletin Astronomique **1** (1884), pp. 65–74. Reproduced in: *Œuvres d'Henri Poincaré*, Tome VII, Gauthier-Villars (Paris), 1952, pp. 253–261.

[37] H. Poincaré, *Sur le problème des trois corps et les équations de la dynamique*, Acta Math. **13** (1890), pp. 1–270. Reproduced in: *Œuvres d'Henri Poincaré*, Tome VII, Gauthier-Villars (Paris), pp. 262–490. (This is the revised form of the famous essay which was awarded the King Oscar II Prize in 1889.)

[38] H. Poincaré, *Les Méthodes Nouvelles de la Mécanique Céleste*, Gauthier-Villars (Paris), Vol. 1 (1892), Vol. 2 (1893), Vol. 3 (1899). Reproduced by Dover publications (New York, 1957) and Blanchard (Paris, 1987). Translated into English as *New Methods of Celestial Mechanics*, edited and introduced by Daniel L. Goroff, American Institute of Physics, 1993.

[39] H. Poincaré, *Sur les solutions périodiques et le principe de moindre action*, Comptes Rendus Acad. Sci. Paris **123** (1896), pp. 915–918. Republished in: *Œuvres d'Henri Poincaré*, Tome VII, Gauthier-Villars (Paris), 1952, pp. 224–226.

[40] H. Poincaré, *Science and Hypothesis* (1903), Dover, 1952, pp. 217–221.

[41] H. Poincaré, *Sur un théorème de géométrie*, Rend. Circ. Mat. Palermo **33** (1912) pp. 375–407. Reproduced in: *Œuvres d'Henri Poincaré*, Tome VI, Gauthier-Villars (Paris), 1952, pp. 499–538.

[42] C. Simó, *New families of solutions in n-body problems*; in: *Proc. ECM 2000*, Barcelona, July 10–14.

[43] C. Simó, *Dynamical properties of the figure eight solution of the three-body problem*, in: *Celestial Mechanics* (Evanston, IL, 1999), Contemp. Math. **292**, Amer. Math. Soc. (Providence, RI), 2002, pp. 209–228. Preprint at `http://www.maia.ub.es/dsg/2001`

[44] F. Tisserand, *Théorie de la Lune de Newton*, in: *Traité de Mécanique Céleste*, Tome III, Chapitre III, pp. 27–45, Gauthier-Villars (Paris), 1894.
[45] F. Tisserand, *Première et Seconde Théories de la Lune d'Euler*, dans: *Traité de Mécanique Céleste*, Tome III, Chapitres V–VI, pp. 65–88, Gauthier-Villars (Paris), 1894.
[46] C. B. Waff, *Clairaut and the motion of the lunar apse: the inverse square undergoes a test*, in: *Planetary Astronomy from the Renaissance to the rise of astrophysics, The General History of Astronomy* 2B, edited by René Taton and Curtis Wilson, Cambridge University Press, 1995, p. 15.
[47] C. Wilson, *The problem of perturbation analytically treated: Euler, Clairaut, d'Alembert*, in: *Planetary Astronomy from the Renaissance to the rise of astrophysics, The General History of Astronomy* 2B, edited by René Taton and Curtis Wilson, Cambridge University Press, 1995, p. 103.

CHAPTER 7

On the Existence of Closed Geodesics

NALINI ANANTHARAMAN

7.1. Introduction

In 1905, Henri Poincaré published in the *Transactions of the American Mathematical Society* a text entitled "Sur les lignes géodésiques des surfaces convexes" ["On geodesic lines of convex surfaces"] **[28]**, in which he studied the geodesics of surfaces with positive curvature. The article is essentially dedicated to the search for closed geodesics and the study of their dynamical and topological properties, such as stability, the number of intersections, etc.

Following Poincaré's work, there were naturally many questions to be asked about the conditions for the existence of one, or even infinitely many, closed geodesics on a compact Riemannian manifold. It must be emphasized that Poincaré himself did not seem to consider the study of geodesics as an end in itself, but as a paradigm for the study of Hamiltonian flows more generally **[31]**; or again, as a simplified frame in which the techniques developed ten or so years earlier in the *Méthodes nouvelles de la mécanique céleste* [*New methods in celestial mechanics*] could be established and deepened, for the study of the three-body problem. It was for this reason, he explained, that he chose to study the case of positively curved surfaces. The geodesic flow on convex surfaces and the three-body problem can in fact both be understood as "perturbations" of a completely integrable Hamiltonian system: the geodesic flow on the sphere in the first case, and in the second, the system formed of two satellites with negligible mass orbiting a massive body. In the 1905 article, Poincaré therefore again took up certain techniques introduced in the *Méthodes nouvelles*, taking advantage of the topological and dynamical aspects characteristic of geodesic flows on surfaces. In this framework, said Poincaré, one encounters the *"principal difficulty"* of the three-body problem, while finding oneself *"free of all secondary difficulties"*. The problem of geodesic lines is *"the simplest of all dynamical problems; first there are but two degrees of freedom, and also ... there is nothing comparable with the difficulty one encounters in the dynamical problems at points where the velocity is zero."* The phase space is compact, because the speed of geodesics is constant, and moreover the problem of collisions, which is a major difficulty in celestial mechanics, does not present itself. Poincaré emphasized, however, that in positive curvature *"the problem is much more difficult than the one that has been solved by M. Hadamard"* in his 1898 memoir on negatively curved surfaces **[18]**: he announced that he would only give *"partial results, especially related to closed geodesics."*

We will begin by briefly outlining the contents of Poincaré's text; we will then see how the paths that are opened therein led his successors to try to elucidate the question of the existence of infinitely many closed geodesics, in the greatest possible generality. This presentation does not have the ambition to be exhaustive: we will outline, essentially, the dynamical systems approach developed by George Birkhoff, and the theory of Marston Morse concerning the critical points of energy on the space of loops. We will only treat a very small portion of the questions concerning "generic" metrics, and not at all the question of the distribution of lengths of closed geodesics (notions of entropy, Mikhail Gromov's work, etc.). Chapter 10 of the book by Marcel Berger [**3**] presents a vaster panorama of the diverse aspects of the problem.

7.2. *Sur les lignes géodésiques des surfaces convexes* (Poincaré, 1905)

In this article, Poincaré studies the geodesics of "convex surfaces"; more precisely, it deals with analytic surfaces whose Gaussian curvature[1] varies between two strictly positive values. Poincaré assumes implicitly that the surface is diffeomorphic to the sphere S^2 (which is in fact the case, by the so-called Bonnet–Schoenberg–Myers theorem, dating from 1935). He always considers surfaces embedded in $\mathbb{R}^3$, but we may say that this hypothesis does not play a fundamental part in most of the reasoning.

The first section is dedicated to the study of the type of the cut locus.[2] Then Poincaré begins discussing the question of the existence of a closed geodesic. He first shows that at least one closed geodesic persists on every "spheroid"—this is what he calls the surfaces obtained by small perturbations of the usual, rigorously round sphere. Then he extends this to every convex analytic surface, by invoking the "principle of continuity". To conclude, he offers a second proof, based on a variational method.

Let us here clarify a few points of terminology and notation. There are usually two ways of defining geodesics on a Riemannian manifold M. One may consider them to be the parametrized curves that are critical points (in other words, stationary points) of the length function, $l(\gamma) = \int \|\dot{\gamma}(s)\|\, ds$, for variations with fixed endpoints, or on the other hand to be the critical points of the energy function $E(\gamma) = \int \|\dot{\gamma}(s)\|^2\, ds$; the speed $\|\dot{\gamma}\|$ is of course measured in the Riemannian metric. The two definitions give essentially the same curves, up to parametrization: the length of a curve does not depend on its parametrization, and geodesics can

[1]The Gaussian curvature at a point x of a surface embedded in $\mathbb{R}^3$ is the product of its two principal curvatures at x. To define the principal curvatures, one considers an oriented plane generated by the normal at x and a unit vector v tangent to the surface at x: this plane cuts the surface along a curve passing through x. The curvature of this planar curve at the point x is a function of v; its maximum and minimum when v varies are the two principal curvatures. The Gaussian curvature at x is positive if the surface stays on one side of its tangent plane, and negative otherwise. This easily visualized definition has the inconvenience of obscuring the fact that the Gaussian curvature is an intrinsic notion, that is, it depends only on the metric induced on the surface, and not on the choice of embedding into $\mathbb{R}^3$ (Gauss's *theorema egregium*); this allows the extension of the definition to a surface endowed with an arbitrary Riemannian metric, not necessarily induced by an embedding into $\mathbb{R}^3$.

[2]Given a point O on the surface, Poincaré studies the family of caustics formed by the envelope of geodesic rays issuing from O (in other words, the conjugate points to O along a geodesic) and also what he calls the "dividing line", which is the set of points P that can be joined to O by at least two geodesics realizing the shortest path. It is the union of the first caustic and the dividing line that we now call the cut locus of O. A geodesic issuing from O ceases to be minimizing at the moment that it crosses the cut locus.

therefore be arbitrarily parametrized if one is only interested in their lengths. By contrast, the definition of geodesics as the critical points of the energy imposes the condition that geodesics be traversed at a constant speed. Moreover, with this second definition, the geodesics are obtained as solutions of a second order differential equation, which is the Euler–Lagrange equation associated with the variational problem, called in this particular case the *geodesic equation*; a geodesic is completely determined by its initial position and velocity. The solution of the equation defines a flow on the tangent bundle TM, called the geodesic flow. By Legendre duality, one may also consider this flow as a Hamiltonian flow on the cotangent bundle T^*M; the Hamiltonian in question is interpreted as kinetic energy. This approach makes it possible to conceive of geodesics as trajectories of a dynamical system, the parametrization of curves being perceived as "time". This is the point of view that will prevail here—we remark that it immediately places the problem in the general context of Hamiltonian systems, by placing in the background the geometric aspects tied to the definition by length.

We will call a closed geodesic, or periodic geodesic, a curve $\gamma : \mathbb{R} \to M$ that is a geodesic for which there exists $T \neq 0$ such that $\gamma(t+T) = \gamma(t)$ for all times t. We set aside the trivial geodesics in order to occupy ourselves only with those whose (constant) speed is non-zero. We may always choose the parametrization in such a way that the period T equals 1; the closed geodesics are then differentiable maps of $S^1 = \mathbb{R}/\mathbb{Z}$ into M. We will say that a geodesic $\gamma : S^1 \to M$ is primitive if it cannot be written as $\gamma(t) = \gamma_0(mt)$, where γ is another closed geodesic and m is an integer at least equal to 2. In general, two geodesics related in the preceding way, with $m \in \mathbb{Z} \setminus \{1\}$, will not be considered as geometrically distinct. Finally, a geodesic can present transverse self-intersections. To designate curves without self-intersections, we will speak of *simple* closed geodesics.

The surfaces that we consider in this part are all diffeomorphic to the 2-dimensional sphere. They will be, most of the time, called S^2, and endowed with a metric of positive curvature. We call the metric that results from the embedding of S^2 into (Euclidean) $\mathbb{R}^3$ as the set of points at distance 1 from the origin "round", "usual", or "canonical".

7.2.1. Closed geodesics of spheroids. Let us look for closed geodesics on the sphere S^2 endowed with a metric obtained by slightly perturbing the round metric. The usual embedding of S^2 into $\mathbb{R}^3$ gives a parametrization of the cotangent bundle, $p = (\theta, \lambda, i, \omega)$, where i measures the angle of inclination from the great circle generated by p with respect to the horizontal; θ measures the longitude of the "node" formed by this great circle and the equator; λ measures the position of the base point of the vector p along this great circle; and ω measures the norm, for the canonical metric, of the vector p (FIG. 7.1).

The geodesic flow for the usual metric is the flow generated by the Hamiltonian function

$$T_0 = \frac{\omega^2}{2},$$

and the equations of motion are

$$\frac{d\lambda}{dt} = \omega, \quad \frac{d\omega}{dt} = 0, \quad \frac{di}{dt} = 0, \quad \frac{d\theta}{dt} = 0.$$

All geodesics are closed; they are the great circles parametrized at constant speed.

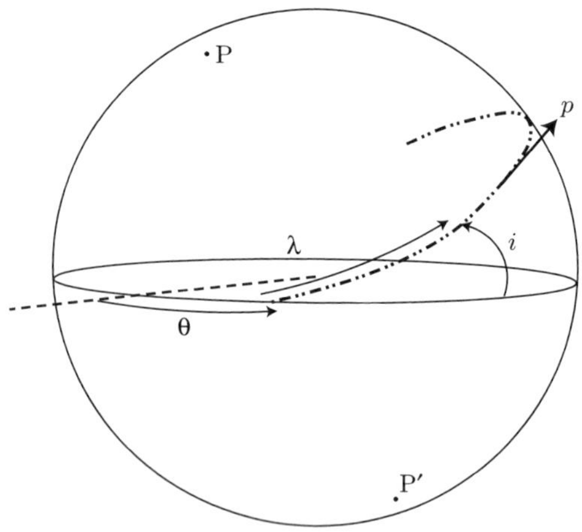

FIGURE 7.1. Definition of the parameters θ, λ, i, ω.

The Hamiltonian is next perturbed as follows:

$$T = T_0 + \mu T_1 \tag{7.1}$$

$$= \frac{\omega^2}{2} + \mu\omega^2 S_1(i, \lambda, \theta), \tag{7.2}$$

in such a way that the new Hamiltonian still corresponds to a metric. We look for closed geodesics of the sphere that persist for small values of μ.

Poincaré takes up again the perturbative method (inspired by Lagrange's method of variation of constants in celestial mechanics) that he developed in the *Méthodes nouvelles* [**29**], and which consists of expanding the equations of the Hamiltonian flow of T by only keeping the order 1 terms of μ. An analysis of the geometric significance of the resulting equation shows that the *simple* closed geodesics for the perturbed metric correspond, at order 1, to critical points of the function $R(i, \theta)$, defined as being the length of the great circle with parameters (i, θ) (FIG. 7.1). This function can be seen as a function on the sphere, by setting $R(P) = R(i, \theta)$ if P is the polar of the circle with parameters (i, θ).

However, for a "generic" function R, in the sense that its critical points are non-degenerate, Poincaré had shown in [**27**] the following relation dealing with the critical points of R: if we denote by m the number of minima, by M the number of maxima, and by c the number of saddle points, we have $m + M - c = 2$ (moreover, this is the number that we now call the Euler–Poincaré characteristic of the sphere). But in reality, the critical points of R come in pairs (see FIG. 7.1), which can be interpreted by remarking that each geodesic can be traveled in either direction. We have therefore: $m = 2m'$, $M = 2M'$, $c = 2c'$, and $m' + M' - c' = 1$; the number of geodesics (each counted only one time) is $m' + M' + c' = 1 + 2c'$, and this number is odd (thus, in particular, it is non-zero).

However, the calculations followed up to this point only represent an asymptotic expansion for the equations of order 1 in μ. In order to truly conclude the existence of a simple closed geodesic for the metric (7.1), Poincaré returned to Chapter 3 of [**29**]. The simplest situation is the one in which the implicit function theorem applies, but Poincaré deals also with the case of more degenerate perturbations, with an eye, as always, on the three-body problem.

We may summarize this section by saying that, for a "generic" perturbation of the standard metric on the sphere, an odd number of simple closed geodesics persists.

7.2.2. The principle of analytic continuity. We now wish to demonstrate the existence of a simple closed geodesic on every analytic surface $\Sigma_1 = (S^2, g_1)$, endowed with a metric g_1 whose Gaussian curvature is contained between two strictly positive real numbers.

We use the fact that two analytic convex surfaces Σ_0, Σ_1 can be joined by an analytic path $(\Sigma_t)_{t\in[0,1]}$ formed of convex surfaces. Thus we connect Σ_1, by a family $\Sigma_t = (S^2, g_t)$, to a surface $\Sigma_0 = (S^2, g_0)$, on which we already know there exist a finite odd number of simple closed geodesics, for example, an ellipsoid whose three axes have distinct lengths, or again one of the "spheroids" of the preceding section. We study the way in which these geodesics evolve with the parameter t, in the hope that one of them persists on Σ_1. Poincaré does not explicitly construct a path connecting Σ_0 to Σ_1, but we can think of several solutions:

- The method of Minkowski sums [**24, 25**], which consists in considering the two convex sets C_0, C_1 whose boundaries are, respectively, the surfaces Σ_0, Σ_1 embedded in $\mathbb{R}^3$, and forming the convex combination $C_t = (1-t)C_0 + tC_1$. Then one takes the boundary C_t as the surface Σ_t.
- If one does not wish to embed the surfaces Σ_t in $\mathbb{R}^3$, one can also use the fact that the metric g_1 is conformally equivalent to the metric g_0 (by the Poincaré–Koebe uniformization theorem), that is, it can be written

$$g_1 = e^{\phi} g_0,$$

where ϕ is a function on S^2. The family of metrics

$$g_t = e^{t\phi} g_0,$$

for $t \in [0,1]$, thus allows us to join g_1 to g_0. The curvature remains positive: in two dimensions, the calculation shows in fact that the Gaussian curvature of Σ_t is at every point a linear combination of those of Σ_0 and Σ_1, with the positive weights $\left((1-t)e^{-t\phi}, te^{(1-t)\phi}\right)$.

From the existence of an odd number of simple closed geodesics on Σ_0, Poincaré deduced the existence of such a geodesic on Σ_1, by the "principle of continuity", whose idea we will now explain.

Searching for closed geodesics on each surface Σ_t reduces to finding the zeroes of an analytic function $G(v,t) = \phi_t^1(v) - v$ on $TS^2 \times [0,1]$, where ϕ_t^1 is the unit time of the geodesic flow on Σ_t. In FIG. 7.2 we have schematically represented the surface $G(v,t) = 0$ in the 5-dimensional space $TS^2 \times [0,1]$. At a given t, each point of the set $G(v,t) = 0$ corresponds to a closed geodesic on Σ_t; there are *a priori* infinitely many branches, corresponding to infinitely many periodic geodesics.

When the parameter t, represented on the horizontal axis of FIG. 7.2, varies from 0 to 1, the periodic solutions disappear or appear in pairs, *"in the same way as real roots of algebraic equations"* [**29**, Chapter 3]: in FIG. 7.2, this is what happens, for example, at the values t_0 and t_2 of the parameter. Moreover, in the course of a single branch, the number of self-intersection points of the closed geodesic remains

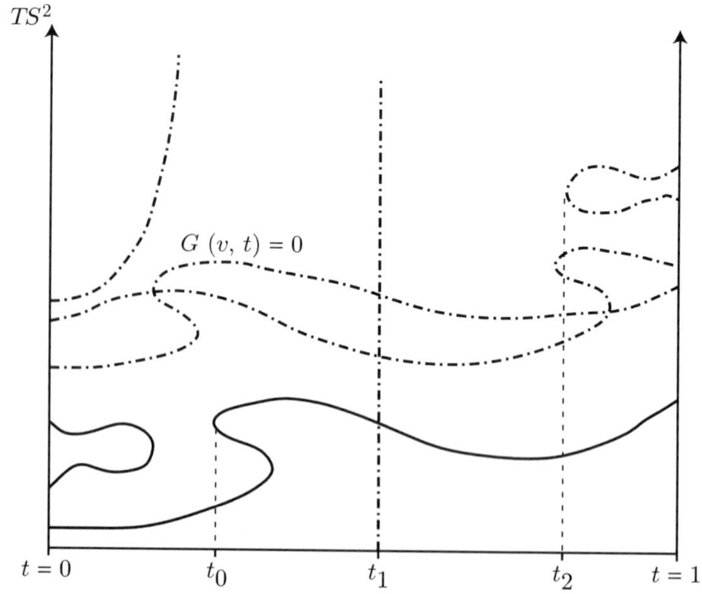

FIGURE 7.2. The principle of analytic continuity.

constant; in fact, during a continuous variation, an intersection can only appear or disappear in one of the ways shown in FIG. 7.3.

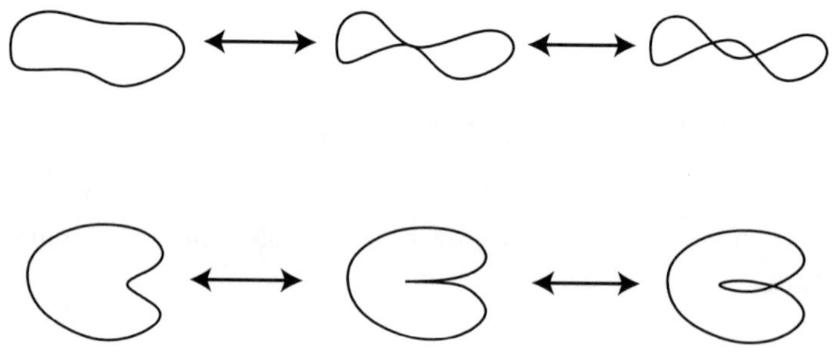

FIGURE 7.3. Appearance/disappearance of self-intersections.

But the first two cases cannot occur because the intermediate situation is impossible for a geodesic. FIG. 7.4 represents a degenerate case that can in fact occur; in this case, the number of intersections is constant, except for isolated values of the parameter, where it is not well defined geometrically because we have a multiple curve, without transverse self-intersection. FIG. 7.2 distinguishes, with solid lines, the branches corresponding to simple geodesics; the others are represented with dash-dotted lines.

This argument shows that only an even number of simple closed geodesics can appear or disappear when t varies from 0 to 1, while there was an odd number of them on Σ_0. There is therefore an odd number of simple geodesics on Σ_1.

This proof calls for several remarks. The first is that Poincaré did not exclude the possibility, which would make his reasoning faulty, of a branch whose projection

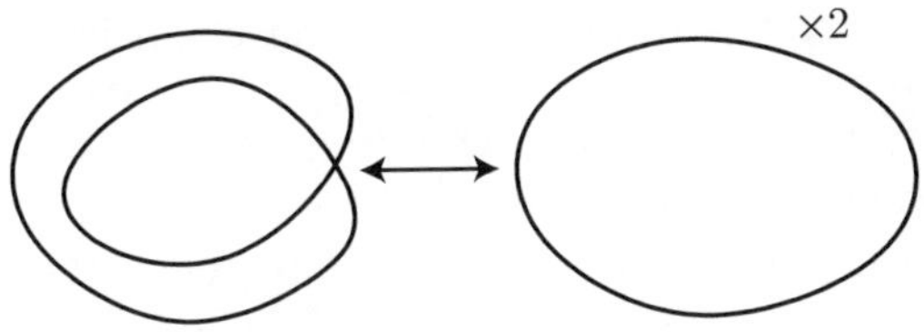

FIGURE 7.4. Degeneration into a multiple curve.

onto TS^2 leaves every compact set (such a branch is shown in FIG. 7.2). This would correspond to the case where the closed geodesics see their lengths tend towards infinity in the course of a single branch. In reality, one can exclude this case by invoking a result of Eugenio Calabi and Jian Guo Cao dating from 1992 [**9**], according to which the length of a simple closed geodesic on a surface of positive curvature is bounded above by $\frac{A}{2}\sqrt{K_{\max}}$, where $K_{\max}$ is the maximum curvature and A is the area of the surface. The hypothesis of positive curvature plays a crucial role here.

Poincaré's proof is commented upon at length in the beginning of Chapter 9 of Marston Morse's book *The calculus of variations in the large* [**26**]. Morse first makes the remark that one must include in the discussion the case where infinitely many closed geodesics appear for a given value of t (for example, if Σ_t turns out to be the round sphere). In FIG. 7.2, this occurs for the value t_1 of the parameter. Poincaré did not comment on this case in [**28**], but he was fully aware of it because he makes mention of it in Section 37 of volume 1 of the *New methods* [**29**]. Morse also remarks that most of the arguments put forward by Poincaré are specific to two dimensions. He proposes, in Chapter 9, to retrace Poincaré's reasoning in arbitrary dimension in order to show the existence of $\frac{m(m+1)}{2}$ geodesics on every Riemannian manifold homeomorphic to S^m, obtained by continuation of the principal ellipses of an ellipsoid with unequal axes. When $\frac{m(m+1)}{2}$ is even, Poincaré's argument does not yield anything. Even in dimension 2, Poincaré only finds one simple closed geodesic, although we now know that there are three (the Lusternik–Schnirelmann theorem [**23**]; see Section 7.5). Finally, the argument of FIG. 7.3, which allows one to say that a principal ellipse cannot disappear except by collapsing with another principal ellipse, is only valid in dimension 2. By trying to introduce the relevant notions for the study of the problem in arbitrary dimension, Morse in reality was tackling the calculation of homology in the space of non-parametrized loops in the sphere, which was not clarified until the 1950s with the introduction of equivariant Morse theory by Raoul Bott (see Section 7.4).

7.2.3. A variational method. In the second part, Poincaré offered another proof of the existence of a simple closed geodesic on any compact surface endowed with a metric of strictly positive curvature. This proof does not require any hypothesis on the analyticity of the surface. The idea is to search for a curve of minimal length among the simple closed curves separating the sphere into two domains D_1, D_2 of total curvature $\int_{D_1} K(x)\,dA(x) = \int_{D_2} K(x)\,dA(x) = 2\pi$, K and A denoting respectively the curvature and the measure of area. The Gauss–Bonnet formula implies that a simple closed geodesic must satisfy this constraint. Therefore, let

$$\kappa = \inf\left\{l(c),\ c \text{ simple closed curve bounding two domains of total curvature } 2\pi\right\},$$

and let c_n be a sequence of curves satisfying the conditions we have just stated and whose lengths approach the infimum, $l(c_n) \to \kappa$. We may extract from the sequence (c_n) a subsequence that converges uniformly to a curve c. If c is simple and differentiable, one shows rather easily that c must be a geodesic.

Concerning the differentiability of c, Poincaré refers the reader *"to the work of M. Hilbert"*. It is, on the other hand, very delicate to determine whether c is simple. One can, in fact, imagine that c is of the type shown in the following figure:

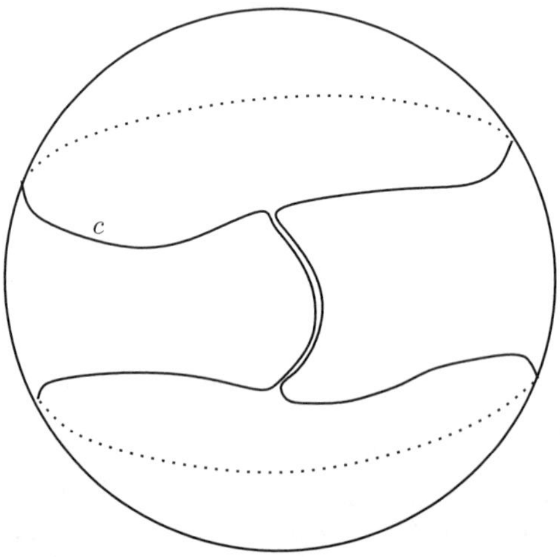

This would signify that the problem that consists in finding, among the domains of total curvature 2π, the one whose boundary has minimal length, in reality admits as a solution a domain whose boundary would have several connected components.

Poincaré ruled out this possibility by a reasoning that appeals to a "physical" interpretation of the problem. After fixing a small constant ε, he invites the reader to imagine a sphere thickened by $\varepsilon K(x)$ at every point x; in the cavity contained between these two walls, an extensible ribbon that slides freely while adhering to the walls by its two boundaries (see FIG. 7.5). In one of the two components bounded by this ribbon, say D_1, an incompressible fluid is located: its volume $\varepsilon \int_{D_1} K(x)\, dA(x)$ is fixed. The system is in equilibrium when the virtual work exerted by the forces on the ribbon is zero:

$$p\,\delta V - T\,\delta U = 0,$$

where p is the fluid pressure, V is its volume, T is the tension of the ribbon, and U is its length. Since the fluid is incompressible, $\delta V = 0$, and equilibrium is therefore attained when $\delta U = 0$, that is, for a minimal length of the ribbon (or, more exactly, a critical length) under the constraint $V = 2\pi\varepsilon$. This problem is therefore equivalent to the preceding one, the ribbon being merely a thickening of the curve c. Poincaré shows, by analyzing the forces being exerted on the ribbon, that at equilibrium the ribbon cannot stick to itself as in FIG. 7.6. He concludes from this that the curve c is simple, and then verifies that it indeed realizes the minimum length.

Poincaré's proof—which does not necessarily satisfy current requirements of mathematical rigor—has been worked out again and again throughout the twentieth century. The first proof to be unanimously accepted is that of Christopher Croke

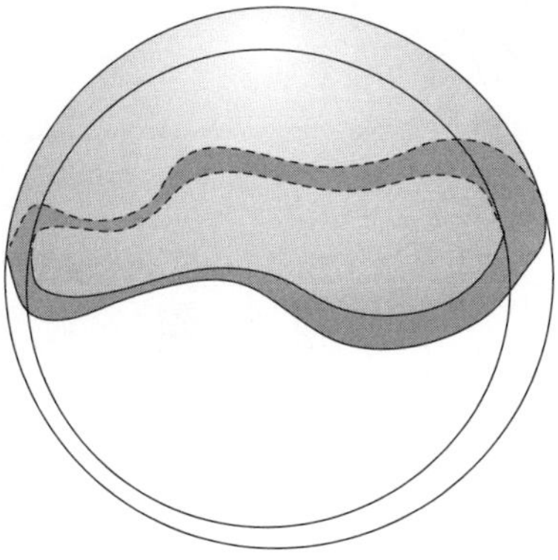

FIGURE 7.5

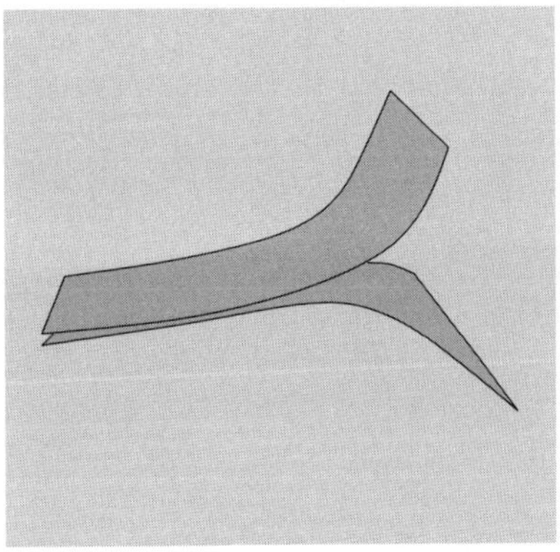

FIGURE 7.6

in 1982 [**10**]; others followed, such as the particularly simple one due to Joel Hass and Frank Morgan [**19**], or the quite recent one of Wilhelm Klingenberg [**21**].

7.3. The work of George D. Birkhoff

7.3.1. Existence of a closed geodesic on (S^n, g). George Birkhoff (1884–1944) proved the existence of closed geodesics on the sphere S^n of dimension n for an arbitrary metric:

THEOREM 7.3.1 ([**5, 6**]). *For any Riemannian metric g on S^n, there exists a closed geodesic.*

Let us choose some Riemannian metric g on S^n. The usual embedding of S^n into $\mathbb{R}^{n+1}$, $S^n = \{(x_0, \dots, x_n) \in \mathbb{R}^{n+1} \mid x_0^2 + x_1^2 + \cdots + x_n^2 = 1\}$, makes it possible to introduce a family of closed curves of the form

$$\{x_2 = \text{constant},\ x_3 = \text{constant},\ \dots,\ x_n = \text{constant}\}, \tag{7.3}$$

some possibly reduced to points, that "cover" the sphere (what Birkhoff calls *"a covering by closed curves"*).

There exists a positive real number ρ, called the radius of injectivity, defined to be the largest real number such that any two points situated at a Riemannian distance less than ρ can be joined by a unique length-minimizing geodesic. On a compact Riemannian manifold, the radius of injectivity is strictly positive. Let us then transform the family (7.3) by replacing each curve with a curve that is piecewise-geodesic for the metric g, by subdividing the curve into subintervals whose endpoints are at a distance less than the radius of injectivity ρ, and by joining these endpoints by a minimizing geodesic segment; see FIG. 7.7. Thus we obtain a new family of closed piecewise-geodesic curves on the sphere, which we will call T_0. Starting from here, Birkhoff proposes to iterate a *"length-reducing operation"* in order to obtain a sequence $(T_k)_{k \geq 0}$ of families of polygonal loops covering the sphere: one passes from T_k to T_{k+1} by joining the midpoints of the segments of each curve constituting T_k by a geodesic segment; see FIG. 7.8. Each new curve is therefore shorter than the previous. When a segment becomes too short, it must be suppressed by concatenating it with its neighbors, in such a way that each segment maintains a length greater than $\rho/2$.

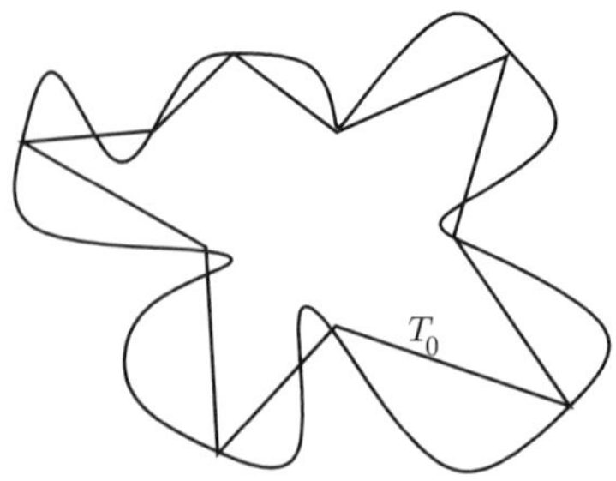

FIGURE 7.7

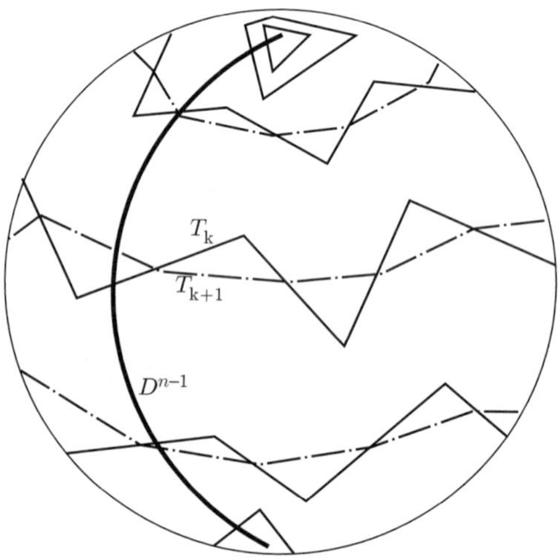

FIGURE 7.8. The process of length reduction.

The initial family (7.3) covered S^n and in fact came from a degree 1 map from S^n to S^n. The degree is preserved by the operation of length reduction, since each family of curves is homotopic to the preceding one. One thus shows that, for every k, the family T_k covers the sphere.[3]

For each integer k, let us now call c_k a curve realizing the maximum length among the curves constituting T_k. The length of c_k, $l(c_k)$, is necessarily larger than the radius of injectivity ρ; otherwise, each curve of T_k could be sent by homotopy, along minimizing geodesic segments, to a point of the submanifold D^{n-1} (FIG. 7.8). We would then have a homotopy equivalence between S^n and a submanifold of

[3]Here I cannot keep myself from opening a parenthesis to cite the explanation provided by Birkhoff [**5**, p. 246], which says a great deal about the interest in having a good mathematical conceptualization: *"For, conceive of the normal covering as a continuous membrane which covers the given surface. Any other covering is obtained by continuous distortion from this particular covering. But this yields merely a distorted membrane which must still cover the surface, so that every point lies on some curve of the distorted covering."* And further on: *"A rigorous proof of the theorem of analysis situs involved is not given here on account of the obvious truth of the theorem. Such a proof can be made by commencing with the case of coverings which vary analytically, in which it is seen that every point is covered an odd number of times, and then passing by a limiting process to the general case."*

dimension $n-1$, which cannot exist because the nth homotopy group $\pi_n(S^n)$ is non-trivial.

We therefore have

$$l(c_{k+1}) \leq l(c_k)$$

and

$$\kappa := \lim_{k\to\infty} l(c_k) \geq \rho.$$

Birkhoff finally showed that c_k converges to a closed geodesic, that is, the angles between the geodesic segments forming c_k tend to 0. For this, it suffices to remark, by a small calculation of local trigonometry, that if c_k somewhere makes an angle δ, we have

$$l(c_{k+1}) \leq l(c_k) - f(\delta)$$

where f is a positive function such that $f(\delta) \sim \delta^2$ when δ tends to 0. If the lengths of the c_k approach the infimum κ, the angles must therefore tend to 0. We obtain in the limit a non-trivial closed geodesic of length $\kappa > 0$, which concludes the proof of the theorem. One speaks of the "minimax technique" because this geodesic was found by taking first a maximum, and then an infimum of the length.

This result can be generalized to an arbitrary compact Riemannian manifold: On any compact Riemannian manifold there exists a closed geodesic (the Lusternik–Fet theorem, 1951). To prove it, one uses the fact that a compact manifold M always has a non-trivial homotopy group $\pi_n(M)$: there exists n such that one can send spheres S^n into M in such a way that their image is not contractible, that is, it cannot be continuously deformed to a point. One can apply to these spheres a minimax process similar to the preceding (the reader may consult Bott's article [**8**] for the complete argument).

7.3.2. Reduction of the study of the geodesic flow on S^2 to the study of a diffeomorphism of the annulus. At this point, it is natural to wonder about the existence of infinitely many closed geodesics. In the case of the sphere of dimension 2, Birkhoff showed, under appropriate hypotheses on the metric, how to construct a "Poincaré section" that allows the reduction of the study of the geodesic flow on S^2 to that of an area-preserving diffeomorphism of the annulus. He thus followed a process introduced by Poincaré for the study of the three-body problem, and answered one of his conjectures by proving the famous "Poincaré–Birkhoff theorem" on the existence of fixed points for certain diffeomorphisms of the annulus.[4] This theorem gives sufficient conditions to prove the existence of infinitely many closed geodesics.

We must first assume the existence of a simple closed geodesic c. Note that the closed geodesic found in the preceding section can *a priori* have self-intersections; it is simple if, for example, the curvature is strictly positive, as Calabi and Cao proved in the article [**9**] already cited.

We must then assume that every geodesic crossing c transversely crosses it again later. This will allow us to define a map T from the open annulus $\mathring{A} = S^1 \times (0,1)$ into itself. This annulus represents the unit tangent vectors to S^2 at a point of c directed to one side of c (toward the top of FIG. 7.9), if one considers that $s \in S^1$ parametrizes the position of the base point of the vector v on c and that $\phi \in (0,1)$ parametrizes the angle made with c. For every $v \in \mathring{A}$, $T(v)$ is defined as the first

[4]EDITOR'S NOTE: See Chapter 6 (p. 133).

return of the geodesic starting from v on $\mathring{A}$ (therefore corresponding to the second intersection of this geodesic with c). The annulus $\mathring{A}$ is what we call a "Poincaré section" of the geodesic flow: in the (three-dimensional) unit tangent bundle, $\mathring{A}$ is a surface that crosses the trajectories transversely. The map T is called the "first return map" or the "Poincaré map" associated to the section $\mathring{A}$. One can show that the map T preserves the area measure $ds \cdot \sin(\pi\phi)\, d\phi$.

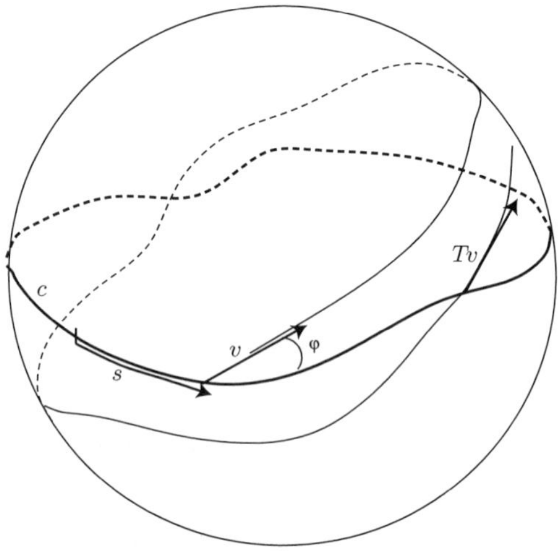

FIGURE 7.9. The diffeomorphism T.

We need one final hypothesis, that T can be continuously extended to the closed annulus $A = S^1 \times [0,1]$. This is the case if there are conjugate points all along c; in this case one can define $T(v)$ for a vector v tangent to c by taking the second conjugate point along the geodesic starting from v.

The three hypotheses stated by Birkhoff are satisfied in the case of positive curvature.

Each periodic point of T in $\mathring{A}$ gives a closed geodesic. Birkhoff answered a question posed by Poincaré by proving the following theorem:

THEOREM 7.3.2 (Poincaré–Birkhoff theorem, 1912 [**4**]). *Let T be an area-preserving homeomorphism of the closed annulus $A = S^1 \times [0,1]$. Assume that T lifts*[5] *to a map $\widetilde{T} : \widetilde{A} = \mathbb{R} \times [0,1] \to \mathbb{R} \times [0,1]$ that translates the points of the boundary component $\mathbb{R} \times \{1\}$ to the left and the points of $\mathbb{R} \times \{0\}$ to the right. Then T has at least two fixed points in $\mathring{A}$.*

This statement generalizes to the study of periodic points of arbitrarily large period. Birkhoff never explicitly wrote down this generalization, but he alludes to it several times in Chapter 6 of his book *Dynamical Systems* [**6**]. The statement that follows is taken from an article by John Franks [**14**].

Let $T : A \to A$ be a homeomorphism that preserves area, the boundary components, and the orientation, and let $\widetilde{T} : \widetilde{A} \to \widetilde{A}$ be a lift of T. The restrictions of

[5] There is a natural projection from $\mathbb{R}$ to $S^1 \simeq \mathbb{R}/\mathbb{Z}$, and therefore a projection $p : \widetilde{A} = \mathbb{R} \times [0,1] \to A = S^1 \times [0,1]$. Given a continuous map $T : A \to A$, it is always possible to lift T to a transformation $\widetilde{T} : \widetilde{A} \to \widetilde{A}$, which means that $T \circ p = p \circ \widetilde{T}$.

T to the boundary components $S^1 \times \{0\}$ and $S^1 \times \{1\}$ define two homeomorphisms of the circle, denoted T_0 and T_1, and the restrictions of $\widetilde{T}$ to the boundary components $\mathbb{R} \times \{0\}$ and $\mathbb{R} \times \{1\}$ define two homeomorphisms of $\mathbb{R}$, $\widetilde{T}_0$ and $\widetilde{T}_1$, which lift respectively T_0 and T_1. We can therefore introduce two "translation numbers"

$$\tau(0) = \lim_{n\to\infty} \frac{\widetilde{T}_0^n(x) - x}{n} \in \mathbb{R} \qquad \text{and} \qquad \tau(1) = \lim_{n\to\infty} \frac{\widetilde{T}_1^n(x) - x}{n} \in \mathbb{R}.$$

These two numbers do not depend on the choice of $x \in \mathbb{R}$, and they measure by what amount points are translated by $\widetilde{T}_0$ and $\widetilde{T}_1$, on average. In the quotient, we obtain the "rotation numbers" $\rho(0) = \tau(0) + \mathbb{Z} \in \mathbb{R}/\mathbb{Z}$ and $\rho(1) = \tau(1) + \mathbb{Z} \in \mathbb{R}/\mathbb{Z}$, which measure by how much T_0 and T_1 turn points on average.

Then, for each rational number $p/q \in (\tau(0), \tau(1))$, there exist at least two periodic orbits for T, with period q and rotation number $p/q + \mathbb{Z}$.

7.3.3. Application to closed geodesics. Let us return to the Poincaré section of the geodesic flow constructed above, $T : A \to A$. The reversibility of the geodesic flow implies that the rotation numbers $\rho(0), \rho(1)$ are opposites in $\mathbb{R}/\mathbb{Z}$. In other words, the translation numbers $\tau(0), \tau(1)$ can be written $\tau(0) = m + r$, $\tau(1) = n - r$, with $m, n \in \mathbb{Z}$. If $m + r \neq n - r$, the Birkhoff theorem permits one to conclude the existence of infinitely many closed geodesics. For the moment the case $m + r = n - r$, which implies that r is a half-integer, remains open; we will return to it in Section 7.5.

We therefore know how to demonstrate the existence of infinitely many closed geodesics on S^2 endowed with a metric that satisfies certain hypotheses, essentially, in the case of positive curvature.

7.4. Using Morse theory

What can we say for manifolds with more complicated topology? The theory of Marston Morse allows us to study the question of the existence of infinitely many closed geodesics by interpreting it as the study of critical points of the energy functional on the space of loops. Let M be a compact Riemannian manifold; we call ΛM the space of loops, that is, maps $\gamma : S^1 \to M$ whose derivative is square integrable. The energy of a loop is by definition $E(\gamma) = \int \|\dot{\gamma}\|^2 \, ds$, and the critical points of this function are precisely the closed geodesics.

Let us remark straightaway that if the fundamental group of M contains infinitely many primitive conjugacy classes (that is, classes that cannot be represented by a power of an element of $\pi_1(M)$), it is easy to find infinitely many closed geodesics. It suffices to look for a curve that minimizes length or energy in each free homotopy class; this method was used since 1898 by Jacques Hadamard, in his memoir on surfaces of negative curvature [**18**]. In this section, we shall describe the completely different situation where the fundamental group is finite. After replacing M with a finite cover, we can even assume that it is trivial, in other words that M is simply connected.[6]

[6]If the fundamental group of M is finite, we can find a simply connected manifold $\widetilde{M}$ with a projection map $p : \widetilde{M} \to M$ such that, for every $x \in M$, there exists a neighborhood V of x whose preimage is a finite disjoint union $p^{-1}(V) = \bigsqcup V_i$; moreover, p realizes a diffeomorphism between each V_i and V. We can endow $\widetilde{M}$ with a metric that makes p a local isometry. Each curve γ drawn in M can be lifted to a curve $\widetilde{\gamma}$ in $\widetilde{M}$, such that $p(\widetilde{\gamma}) = \gamma$. As each point of M has

Recall the principle of Morse inequalities: let f be a C^2 function on a compact manifold X whose critical points are non-degenerate, that is, the second differential of f is a non-degenerate quadratic form. Then the number of critical points of f with index i is greater than or equal to the ith Betti number $b_i(X, K)$, defined as the dimension of the ith homology group $H_i(X, K)$, with coefficients in an arbitrary field K. In other words, a simple piece of information about the topology of X—the non-zero Betti numbers—allows us to detect the existence of critical points of f.

Morse theory applies in this context as long as we take a few precautions. The manifold ΛM is not finite-dimensional, but we can get around this difficulty by restricting ourselves to piecewise geodesic loops. For this, we again use the method of Section 7.3; that is, we divide each loop into small subintervals, and we join the vertices thus obtained by geodesic segments of length less than the radius of injectivity ρ. By considering the polygonal loops with n vertices (for arbitrarily large n), we obtain a submanifold of ΛM, finite-dimensional and compact, that contains all the closed geodesics of length less than $n\rho$. This done, we do not lose any information about the homology of the loop space: Bott showed that, for any integer q, we can choose n such that the space of polygonal loops with n vertices and the space of all loops ΛM have the same homotopy type up through rank q (see [**7, 8**]).

Jean-Pierre Serre had already used Morse inequalities to search for closed geodesics joining two points $x, y \in M$. By introducing the "Serre spectral sequence", he showed, on every compact manifold, that the space of paths joining x and y has infinitely many non-zero Betti numbers. The hypothesis of non-degeneracy of geodesics joining x to y, as critical points of energy, is automatically satisfied for x, y in a set of full measure: it is necessary and sufficient that x and y not be conjugate along a geodesic—in other words, y should not belong to the envelope (in the sense of the envelope of a family of curves) of geodesic rays starting from x. For such points, the Morse inequalities and the result of Serre demonstrate the existence of infinitely many geodesics, of arbitrarily large length, going from x to y. One can then extend this, by continuity, to any two points x and y.

One can again apply this strategy to the search for closed geodesics, and demonstrate the existence of infinitely many non-zero Betti numbers for the space ΛM, and therefore of infinitely many critical points for the energy $E : \Lambda M \to \mathbb{R}$. The problem is, however, that a closed geodesic and its iterates all count as distinct critical points of E, while one would only wish to count them as a single geodesic. It was therefore necessary to understand precisely how the iterates of a single geodesic are accounted for in the calculation of homology; this was made possible by the work of Raoul Bott and Wolfgand Ziller in the 1950s (see [**8**]).

As an additional subtlety, it must be noted that the group $O(2)$ acts on the space of loops by translation and symmetry of the parametrization, while preserving the energy functional E. Neglecting this symmetry of the system can cause the loss of much information. For example, when Morse found $m(m+1)/2$ geodesics on any manifold diffeomorphic to S^m, "continuing" the principal ellipses of an ellipsoid with unequal axes, he needed to take this symmetry into account *a priori*. To solve this problem, one can use the "equivariant Morse theory" introduced by Michael

only finitely many preimages by p, $\widetilde{\gamma}$ will be a closed geodesic if γ is one, and it is equivalent to find infinitely many closed geodesics on $\widetilde{M}$ or on M.

Atiyah and Raoul Bott. However, in the following result, one only needs the usual version of Morse inequalities (see [**8**]).

THEOREM 7.4.1. *Let M be a compact, simply connected Riemannian manifold, having only finitely many closed geodesics, which are moreover assumed to be non-degenerate critical points of the energy. Then the sequence of Betti numbers $(b_i(\Lambda M, K))_{i\in\mathbb{N}}$ is bounded.*

If the sequence of Betti numbers of the space of loops is not bounded, one can conclude the existence of infinitely many closed geodesics, for any "generic" metric on M (sometimes called a "bumpy metric", which means that the critical points of the energy are non-degenerate).

Detlef Gromoll and Wolfgang Meyer succeeded in removing the hypothesis of non-degeneracy:

THEOREM 7.4.2 ([**16, 17**]). *Let M be a compact, simply connected Riemannian manifold, having only finitely many closed geodesics. Then the sequence of Betti numbers $(b_i(\Lambda M, K))_{i\in\mathbb{N}}$ is bounded.*

Therefore, there exist infinitely many closed geodesics for any metric on M as soon as the sequence $(b_i(\Lambda M, K)_{i\in\mathbb{N}})$ is unbounded. This last hypothesis only concerns the topology of M; Micheline Vigué-Poirrier and Dennis Sullivan characterized the manifolds that satisfy it.

THEOREM 7.4.3 ([**32**]). *Let M be a compact, simply connected Riemannian manifold. The following properties are equivalent:*

- *the sequence $(b_i(\Lambda M, \mathbb{R}))_{i\in\mathbb{N}}$ is unbounded;*
- *the cohomology ring $H^*(M, \mathbb{R})$ is not generated by a single element.*

This gives a great number of applications for Theorem 7.4.2, for example, when M is a Cartesian product. On the other hand, spheres are unfortunately not involved... The question of the existence of infinitely many closed geodesics for an arbitrary metric on S^n therefore remains open! Another open case, which does not enter into the previous discussion, is where the fundamental group of M is infinite, but has only finitely many primitive conjugacy classes. The case where $\pi_1(M) = \mathbb{Z}$ was treated by Victor Bangert and Nancy Hingston [**2**].

Nonetheless, for the sphere of dimension 2, the work of Birkhoff was recently completed: we now know how to prove the existence of infinitely many closed geodesics for any metric on the sphere S^2 [**12, 20, 1**].

7.5. Return to the case of the sphere S^2: The work of Franks, Bangert, Hingston

Let us again take up Birkhoff's strategy: it consists in reducing the study of the geodesic flow to that of a diffeomorphism of the closed annulus, $T : A \to A$, by the construction of a Poincaré section. This required certain hypotheses: the existence of a simple closed geodesic having conjugate points, and such that every geodesic that crosses it transversely crosses it again later. The Poincaré–Birkhoff theorem then implied the existence of infinitely many closed geodesics, except perhaps in the case where the rotation number of T along the boundary of A equalled $1/2$.

In this case, however, the rotation number of $T^2 = T \circ T$ is zero, and the theory of homeomorphisms of the circle allows us to conclude that T^2 has a fixed point on the boundary of A. We can then apply to T^2 the following theorem of John Franks, dating from 1992:

THEOREM 7.5.1 ([**12, 14**]). *Let $T : A \to A$ be a diffeomorphism that preserves area, orientation, and the boundary components, having at least one fixed point in A. Then T has infinitely many periodic points in the interior of A.*

This solves the case left in suspense in Section 7.3. In 1993, Nancy Hingston also dealt with this problematic case, by a different approach, similar to that of Bott, and hence to Morse theory [**20**].

What happens now that we cannot put this strategy to use? Suppose that we have found a simple closed geodesic c on our surface, but it does not have conjugate points, or perhaps the first return map is not well defined: in other words, there exists a geodesic crossing c and thereafter remaining "trapped" in one of the two components bounded by c. These situations were elucidated in 1993 by Victor Bangert [**1**], who proved the existence of infinitely many closed geodesics even then.

We have until now assumed the existence of a simple closed geodesic. In positive curvature, we have already mentioned several proofs of the existence of such a geodesic: the variational proof of Poincaré (and its recent avatars, Croke [**10**], Hass and Morgan [**19**], Klingenberg [**21**]), and the minimax method of Birkhoff, which, as Calabi and Cao [**9**] showed, gives a simple geodesic in the case of positive curvature. For an arbitrary metric on S^2, a note of Lusternik and Schnirelmann [**23**] announced in 1929 a proof of the existence of three simple closed geodesics, but their short proof was never accepted as satisfactory. It was not until 1978 that the first complete proof was written down by Werner Ballmann; different proofs were later proposed by Iskander Taĭmanov and Matthew Grayson [**15**]. This latter replaces Birkhoff's process of length reduction by the curvature flow, which has the advantage of not creating double points, but for which the difficulty is to show that it is defined for all time, if the starting curve is well chosen.

Thus ends this history, spread out over nearly a century, for the sphere of dimension 2. It remains entirely to be written for spheres of higher dimension, and more generally for simply connected manifolds whose cohomology has a single generator. The problem also remains open when the fundamental group is infinite, but there are only finitely many primitive conjugacy classes.

While contemplating in a general way the search for infinitely many periodic orbits of a Hamiltonian system, Morse [**26**] compared Birkhoff's approach by dynamical systems to his own—the search for critical points of the energy—by saying that the first is, in most of its applications, a local approach that seeks to find periodic points in the neighborhood of a fixed point given *a priori*, while the second proves the existence of periodic orbits starting from information about the global topology. He perceived that the two approaches coalesce in low dimension, upon which he commented thus:

> *"This is doubtless due to the fact that the theory of the distribution of vectors in a 2-space is essentially equivalent to the theory of critical points of a function of two variables. In fact such a system of vectors, if suitably altered in length, will in general become the gradients of a function. This relation between the fixed point theory and the critical point theory does not persist however in higher spaces."*

And later:

> *"In any event, the further study of the interrelations between the two theories seems likely to be one of the most fascinating subjects for research in the future."*

Bibliography

[1] V. Bangert, *On the existence of closed geodesics on the two-spheres*, Int. J. of Math. **4** (1993), no. 1, pp. 1–10.

[2] V. Bangert, N. Hingston, *Closed geodesics on manifolds with infinite abelian fundamental group*, J. Differential Geom. **19** (1984), no. 2, pp. 277–282.

[3] M. Berger, *A panoramic view of Riemannian geometry*, Springer Verlag, Berlin-Heidelberg 2003.

[4] G. D. Birkhoff, *Proof of Poincaré's geometric theorem*, Trans. Amer. Math. Soc. **14** (1913), no. 1, pp. 14–22.

[5] G. D. Birkhoff, *Dynamical systems with two degrees of freedom*, Trans. Amer. Math. Soc. **18** (1917), no. 2, pp. 199–300.

[6] G. D. Birkhoff, *Dynamical systems*, American Mathematical Society Colloquium Publications, Vol. IX, American Mathematical Society, Providence, R.I., 1966. Republication of the original 1927 edition.

[7] R. Bott, *Nondegenerate critical manifolds*, Ann. of Math. (2) **60** (1954). pp. 248–261.

[8] R. Bott, *Lectures on Morse theory, old and new*, Bull. Amer. Math. Soc. (N.S.) **7** (1982), no. 2, pp. 331–358.

[9] E. Calabi, J. G. Cao, *Simple closed geodesics on convex surfaces*, J. Diff. Geom. **36** (1992), no. 3, pp. 517–549.

[10] C. B. Croke, *Poincaré's problem and the length of the shortest closed geodesic on a convex hypersurface*, J. Differential Geom. **17** (1982), no. 4, pp. 595–634.

[11] J. Franks, *Generalizations of the Poincaré–Birkhoff theorem*, Ann. of Math. (2) **128** (1988), no. 1, pp. 139–151.

[12] J. Franks, *Geodesics on S^2 and periodic points of annulus homeomorphisms*, Invent. Math. **108** (1992), no. 2, pp. 403–418.

[13] J. Franks, *Rotation vectors for surface diffeomorphisms*, Proceedings of the International Congress of Mathematicians, Vols. **1**, **2** (Zurich, 1994), pp. 1179–1186, Birkhäuser, Basel, 1995.

[14] J. Franks, *Rotation numbers and instability sets*, Bull. Amer. Math. Soc. (N.S.) **40** (2003), no. 3, pp. 263–279.

[15] M. A. Grayson, *Shortening embedded curves*, Ann. of Math. (2) **129** (1989), no. 1, pp. 71–111.

[16] D. Gromoll, W. Meyer, *On differentiable functions with isolated critical points*, Topology **8** (1969), pp. 361–369.

[17] D. Gromoll, W. Meyer, *Periodic geodesics on compact Riemannian manifolds*, J. Diff. Geom. **3** (1969), pp. 493–510.

[18] J. Hadamard, *Les surfaces à courbures opposées et leurs lignes géodésiques*, Journal de M. Jordan, 5th series v. IV (1898), p. 27 ff. Œuvres, Vol. II, p. 729.

[19] J. Hass, F. Morgan, *Geodesics and soap bubbles in surfaces*, Math. Z. **223** (1996), no. 2, pp. 185–196.

[20] N. Hingston, *On the growth of the number of closed geodesics on the two-sphere*, Int. Math. Res. Notices **1993**, no. 9, pp. 253–262.

[21] W. P. A. Klingenberg, *Poincaré's closed geodesic on a convex surface*, Trans. Amer. Math. Soc. **356** (2004), no. 6, pp. 2545–2556.

[22] L. A. Lusternik, A. I. Fet, *Variational problems on closed manifolds*, Doklady Akad. Nauk SSSR (N. S.) **81** (1951), pp. 17–18.

[23] L. A. Lusternik, L. G. Schnirelmann, *Sur le problème de trois géodésiques fermées sur les surfaces de genre* 0, C. R. Acad. Sci. Paris **189** (1929), pp. 269–271.

[24] H. Minkowski, *Sur les surfaces convexes fermées*, C. R. Acad. Sci. Paris **132** (1901), pp. 21–24.

[25] H. Minkowski, *Volumen und Oberfläche*, Math. Ann. **57** (1903), pp. 230–276.

[26] M. Morse, *The calculus of variations in the large*, Amer. Math. Soc. (Providence, RI), 1996, reprint of the 1932 original.

[27] H. Poincaré, *Mémoire sur les courbes définies par une équation différentielle*, Journal de mathématiques, 3rd series, **7** (1881), pp. 375–422 and **8** (1882), pp. 251–296.

[28] H. Poincaré, *Sur les lignes géodésiques des surfaces convexes*, Trans. Amer. Math. Soc. **6** (1905), pp. 237–274. Œuvres, Tome VI, pp. 38–85.

[29] H. Poincaré, *Les méthodes nouvelles de la mécanique céleste*, tome I : *Solutions périodiques. Non-existence des intégrales uniformes. Solutions asymptotiques.* Gauthier-Villars (Paris), 1892–1899. Reprintings: Dover Publications Inc. (New York), 1957, and Blanchard (Paris), 1987.

[30] I. A. Taĭmanov, *Closed extremals on two-dimensional manifolds*, Uspekhi Mat. Nauk **47** (1992), no. 2(284), pp. 143–185, p. 223; English transl. in Russian Math. Surveys **47** (1992), no. 2, pp. 163–211.

[31] A. Robadey, *Exploration d'un mode d'écriture de la généralité : l'article de Poincaré sur les lignes géodésiques des surfaces convexes (1905)*, Revue d'histoire des mathématiques **10** (2004), pp. 257–318.

[32] M. Vigué-Poirrier, D. Sullivan, *The homology theory of the closed geodesic problem*, J. Differential Geometry **11** (1976), no. 4, pp. 633–644.

CHAPTER 8

Poincaré's Memoir for the Prize of King Oscar II: Celestial Harmony Entangled in Homoclinic Intersections

FRANÇOIS BÉGUIN

8.1. Is the solar system stable?

A few years ago, a Parisian antiquary, the Galerie J. Kugel, organized an exhibit that I deeply regret having missed.[1] This exhibit assembled objects from diverse periods, all created for the same purpose: to represent the Earth, the stars surrounding it, and their movements. There one could admire a celestial globe from the Roman era, terrestrial globes from the 16th century, armillary spheres, astronomical rings, etc. One of the objects presented was a *mechanical sphere* from the beginning of the 18th century, identical to the one shown in FIG. 8.1.Apart from its undeniable beauty, this sphere possesses a particular historical interest: it is one of the first to have been conceived according to the heliocentric system of Nicolaus Copernicus.[2] Thus, the ball placed on the vertical axis of the mechanical sphere represents the Sun, the ball placed just to the left of the Sun represents the Earth, and the other small balls that one perceives in the engraving represent the other planets known at the time along with the Moon. A clockwork mechanism animates the planets, which turn around the Sun at different speeds, following a choreography that is impeccably regulated, harmonious, and unwavering.

One can let oneself be lulled indefinitely by the waltz of the small metal planets, but at some moment one must awake and ask the crucial question: does this waltz faithfully reflect the motions of the *real* planets in the solar system?

The answer depends on the precision one wants and the scale of time in which one is interested. Over a short period of time (a few years), the trajectories of Mercury, Venus, Mars, Jupiter, and Saturn in our sky coincide quite well with the trajectories predicted by Copernicus' model. But over a longer period,[3] the Copernican trajectories are only poor approximations of the planets' real trajectories: slowly, but inescapably, the real trajectories depart from the Copernican trajectories. Indeed, one must surrender to the facts: even if the heliocentric model of Copernicus constitutes a veritable revolution from a *metaphysical* point of view, it is on the other hand hardly more effective *in practice* than the geocentric model of Ptolemy, once it comes to explaining the trajectories of the planets in our sky.

[1]Fortunately, a glimpse of the magnificent objects that were on display is available on the gallery's website: `http://www.galeriekugel.com/`

[2]More than 150 years after the death of Copernicus!

[3]From generation to generation, astronomers carefully transmitted their notes on the positions of the planets in the sky since ancient times.

FIGURE 8.1. An engraving of the mechanical sphere constructed by Jean Pigeon and Jean-Baptiste Delure and presented to King Louis XIV on November 12, 1706. Taken from *La sphère du monde selon l'hypothèse de Copernic* [*The world's sphere according to the Copernican hypothesis*] by the Abbé de Vallemont, Prosper Marchand, Paris, 1707.

At the beginning of the 17th century, Johannes Kepler proposed a model of the solar system that improves the one given by Copernicus. In Kepler's model, the orbits of the planets around the Sun are not circles or epicycles, but ellipses with the Sun at one focus. The adequacy of this model with observation is much better than that of the models of Ptolemy or Copernicus. Thus, to have a more faithful image of the solar system, one can imagine a mechanical sphere endowed with a mechanism that is a bit more complex than the above, allowing the small metallic planets to describe ellipses rather than circles.

But this image still is not perfect... Above all, imagining more or less sophisticated mechanical spheres will not save us from a more fundamental question: can the motions of the solar system's planets be faithfully represented by a perfectly regular choreography, animated by clockwork motion? Isn't the regularity of the planets' motions a simple illusion? Would these motions seem as regular to us if we were to observe them over several million years? Would the appearance of the solar system be about the same in a few hundred thousand years, or would the slow drift of the planets' real trajectories with respect to the trajectories predicted by Ptolemy, Copernicus, or Kepler have completely upset the planets' configuration? Could this deviation continue until two planets collide? Or a planet is ejected? In one word: is the solar system stable?

While the models of Copernicus or Kepler had as their only goal to *describe* the motions of the planets around the Sun, circa 1685, Isaac Newton invented celestial mechanics by stating two very simple laws that aim to *explain* the planets' motions:

(1) Any two bodies are attracted in direct proportion to their masses and in inverse proportion to the square of the distance between them.
(2) The product of the acceleration of a body and its mass is always equal to the sum of all the forces exerted on this body.

Do these two laws argue in favor of the solar system's stability? At first glance, yes. In fact, from these two laws of Newton, one can *prove* that a planet subjected to the Sun's attraction perpetually describes an ellipse with the Sun at one focus. However, in this case the planet is subject *only* to the Sun's attraction. Yet Newton's laws predict that *all* massive bodies attract each other. Thus, a planet in the solar system is subjected not only to the Sun's attraction, but also to the attraction of the other planets. Of course, the planets are much lighter than the Sun, and therefore they exert a much weaker force of attraction. This is why the Keplerian orbits are, for a short duration, good approximations to the planets' trajectories. But, at every moment, each planet is slightly diverted from its Keplerian orbit by the attraction that the other planets exert. In the course of time, the small deviations accumulate, and nothing tells us that this accumulation will not end up completely changing the solar system's appearance. After all, the most ancient notes we have on the planets' positions dated from the Chaldeans; three millenia is a long time on the human scale, but it is absurdly short before the age of the solar system...

Starting at the end of the 17th century, the problem of the solar system's stability therefore splits into two questions:

(1) Are Newton's laws sufficient to explain the motions of the stars?
(2) Is a system of n bodies, mutually attracting according to Newton's laws, stable?

Over the length of the 18th and 19th centuries, astronomers never ceased collecting results that argue for a positive answer to the first question. In 1750, Alexis Clairaut explained the motion of the Moon using only Newton's laws. A few years later, he calculated with Joseph-Jérôme de Lalande that the attraction exerted by Jupiter and Saturn should, according to Newton's laws, slow the passage of Halley's comet by one year and eight months, give or take a month; the delay would be one year and seven months. About twenty years later, Pierre Simon de Laplace showed that Newton's laws explain the motions of Jupiter and Saturn observed over 2000 years. In 1821, Alexis Bouvard noticed that the motion of Uranus (discovered in 1781) could not be explained by Newton's laws... unless Uranus was subjected to the attraction of an unknown planet. In 1846, the Frenchman Urbain Le Verrier and the Englishman John C. Adams independently determined, following a set of titanic calculations, what position this hypothetical planet should have. The Académie des Sciences hardly paid more attention to Le Verrier's results than the Astronomer Royal George B. Airy gave to those of Adams, but the German astronomer Johann Galle pointed his telescope to the position calculated by Le Verrier... and discovered Neptune![4]

The answers to the second question would be less convincing. In 1785, however, Laplace announced that he had practically solved this question: he showed that the system containing the Sun, Jupiter, and Saturn is stable. But Laplace's result was far from closing the debate. The problem is not so much that Laplace restricted himself to Jupiter and Saturn, but rather that, in his calculations, he made a major approximation. Let us explain.

We can decompose the forces to which Saturn is subjected into an infinite sum of smaller and smaller terms: the principal term is the force of attraction exerted by the Sun, the following term (said to be of *order 1*) is proportional to the ratio between the mass of Saturn and that of the Sun, the next term (called *order 2*) is proportional to the square of this ratio, the next term (called *order 3*) is proportional to the third power of this ratio, etc. In his calculations, Laplace only took the principal term and the order 1 term into account. Over a short period of time (less than a million years), this is not very serious: in fact, since the ratio between the mass of Saturn and that of the Sun is very small, the terms of order greater than 1 are miniscule and hardly perturb the motion at all. But, in the long term, the perturbations due to the terms of order greater than 1 can no longer be neglected. What Laplace had shown is therefore that the Sun-Jupiter-Saturn system is more or less stable over a period on the order of a million years, but his work does not give any information whatsoever about the behavior of Jupiter and Saturn over a longer period.

Around 1850, Le Verrier pushed Laplace's calculations further, this time by taking into account all of the planets. But Le Verrier's calculations rest on the same kind of approximations as those of Laplace. By their very nature, such calculations can only show stability results over a limited period of time. Thus, at the end of the

[4]Le Verrier would not always have such success: a few years later, having noticed that the perihelion of Mercury's orbit turns more quickly than expected (according to the calculations of the perturbations due to the other planets), he sought to explain this advance by the existence of a new planet, Vulcan, whose trajectory he calculated. According to the latest news, the planet Vulcan still has not been discovered (except by the screenwriters of the series *Star Trek*), and the advancement of Mercury's perihelion is due to the space-time distortion in a neighborhood of a massive body (here, the Sun) predicted by the general theory of relativity.

19th century, practically no results were known that had bearing on the behavior of a system of n bodies that attract each other according to Newton's laws *for an arbitrarily long time.*

In 1885, several scientific journals announced that Oscar II, the king of Sweden and Norway, *"desiring to demonstrate his interest in the progress of mathematical science"*, had decided to accord a prize *"to an important discovery in the realm of higher mathematical analysis"*. The prize, which would be conferred on January 21, 1889, on the occasion of the king's 60th birthday, consisted of a gold medal and the sum of two thousand five hundred crowns. The candidates were to present a memoir dealing with one of the four problems proposed by the jury, composed of the Swedish mathematician Gösta Mittag-Leffler (who was at the origin of the competition's organization) and the two leading experts in Analysis of the time: the Frenchman Charles Hermite and the German Karl Weierstrass.

One of the proposed problems concerned the study *"from the algebraic point of view* [...] *of the functions that M. Poincaré has introduced into science under the name of Fuchsian functions"*. This formulation was certainly an invitation for Poincaré to develop the work that had made him famous five years earlier.[5] But he preferred to tackle another of the four proposed problems: the one concerning the study of the behavior *"of a system of arbitrarily many* [bodies] *attracting each other according to Newton's law"*, and in particular the study of the stability of the solar system...

8.2. Poincaré's memoir for the Prize of King Oscar II

In the history of the study of the solar system's stability, and more generally of deterministic systems, Poincaré's memoir for King Oscar's Prize [**16, 17**] constitutes a veritable revolution. And this revolution is above all a change in viewpoint. While previously the search had always been to *calculate* the paths of the planets with the best possible precision, Poincaré attempted to *say something* about the appearance of these paths *without seeking to calculate them.* While previously the differential equations that translated Newton's laws played a primordial role in the calculation of the trajectories, Poincaré most of the time forgot these equations in order to retain only a few of their fundamental properties. Finally, while previously the interest lay only in *the* trajectory that each planet follows, Poincaré was interested *simultaneously* in *all* the trajectories that a planet could have followed if the initial configuration of the system had been different. The genius of Poincaré was to have understood that it is sometimes easier to say something about the behavior of all the trajectories of all systems of a certain type than to calculate a particular trajectory of a particular system.

Poincaré's paper contains such an impressive quantity of results that it is out of the question to explain all of them here. I will simply try to present two of the principal results of the paper: the stability theorem "à la Poisson" and the "theorem of coincidence for asymptotic surfaces".

The restricted three-body problem. The results that will interest us do not concern an arbitrary system of n bodies. They concern a particular system that was introduced for the study of the motions of the Moon, first by Newton[6]

[5]EDITOR'S NOTE: See Chapter 2.

[6]EDITOR'S NOTE: See Chapter 6.

and later by Euler, and has become very classical since then. This system consists of three celestial bodies A, B, C, which attract each other according to Newton's laws, the mass of B being small compared with that of A, and the mass of C being "infinitely small". Because the mass of C is infinitely small, the bodies A and B form an isolated two-body system; the trajectories of A and B are therefore coplanar ellipses. We will moreover place ourselves in the simpler case: the ellipses traversed by A and B are circles, and C moves in the plane of these circles.

Typically, the body A represents the Sun, the body B Jupiter, and the body C the Earth. The system described above models the influence of Jupiter on the Earth's orbit, neglecting the influence of the other planets and the eccentricity of Jupiter's orbit, and assuming that the Earth and Jupiter remain exactly in the plane of the ecliptic. We could equally well assume that the body A represents the Sun, the body B the Earth, and the body C the Moon (as was initially the case for Newton and Euler).

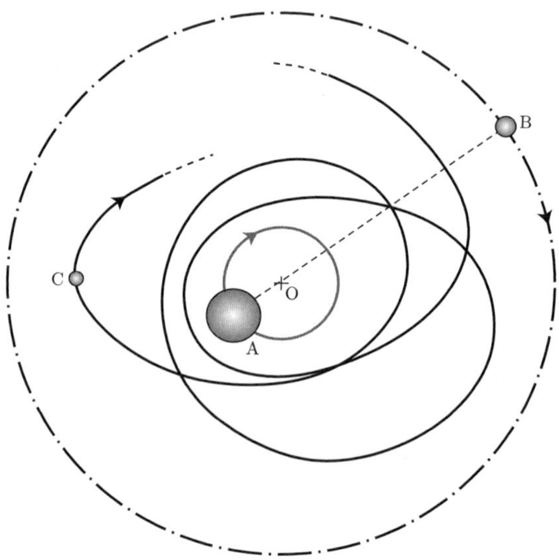

We choose a unit of mass so that the sum of the masses of A and B is 1, we denote by μ the mass of B, and by O the center of gravity of the system consisting of the bodies A and B. We understand perfectly the movement of the bodies A and B (they traverse circles centered at O with constant speed, with respective radii μ and $1-\mu$ in an appropriate unit of length, and, of course, O, A, and B are permanently aligned). The entire problem therefore consists in studying the motion of C. To do so, we put ourselves in a "rotating frame": we use the orthonormal basis, oriented counterclockwise, whose origin is at O and whose first axis is the line passing through A and B (oriented from A to B to fix notation). The state of the system at a given moment is entirely characterized by the coordinates (x, y) of the planet C and the coordinates (x', y') of the velocity vector of C in this basis. The configuration space of the system is therefore four-dimensional: it is the space of quadruples (x, y, x', y').

Newton's laws can be translated into a system of four differential equations[7] that govern the system's evolution. These equations will play practically no part in what follows, however: this is one of the great originalities of Poincaré's viewpoint.

On the other hand, it is important to understand the method of geometric representation that we will adopt. We have defined a space: the *configuration space* of our system. Each possible configuration of the system is represented by a point in this space. Each possible evolution of the system in the course of time will be represented by a curve in the configuration space, parametrized by time; such a curve is called a *trajectory of the system.*[8] Each point of the configuration space is the starting point for a *unique* trajectory of the system (this translates the system's determinism: the knowledge of the initial configuration entirely determines its evolution).[9]

We said in the first section that, before the work of Poincaré, there had been practically no results about the behavior of the solar system over an arbitrarily long period of time. There was, however, one notable exception, a result proved in 1878 by George W. Hill and adapted in 1887 by Karl Bohlin to the particular system described above [**3**]:

THEOREM (Hill, Bohlin). *For the three-body system described above, there exist two values* $E_{\min}$ *and* $E_{\max}$ *such that the trajectories whose energy is contained between* $E_{\min}$ *and* $E_{\max}$ *and for which the initial values of* x *and* y *are not too large never leave a bounded region of space.*[10]

This is already an interesting result on stability: for values of energy contained between $E_{\min}$ and $E_{\max}$ and for initial values of x and y that are "not too large", the body C will never be "ejected to infinity". Moreover, I regret not being able to dwell upon this result, whose proof is very pretty and completely within the spirit of Poincaré's approach: one absolutely does not calculate the trajectories but instead shows that there exists a closed impassable curve which prevents the trajectories from going off to infinity (see [**19**]).

Note that, in many realistic situations, we obtain values for the energy contained between $E_{\min}$ and $E_{\max}$ and initial values for x and y that are "not too large"; for example, once A, B, and C are given masses, speeds, and mutual distances comparable to those of the Sun, the Earth, and the Moon, or to those of the Sun, Jupiter, and the Earth. In what follows in this text, we will consider nothing more than trajectories having an energy contained between $E_{\min}$ and $E_{\max}$ and "not too large" initial values of x and y; therefore the trajectories that appear hereafter never leave a bounded region of the configuration space.

---

[7]They are

$$\begin{cases} dx/dt = x', & dx'/dt = x + 2y' + \partial V/\partial x, \\ dy/dt = y', & dy'/dt = y - 2x' + \partial V/\partial y, \end{cases}$$

where $V(x,y) = \frac{1-\mu}{\sqrt{(x+\mu)^2+y^2}} + \frac{\mu}{\sqrt{(x+\mu-1)^2+y^2}}$.

[8]A trajectory of the system is a curve in the configuration space, and therefore not a trajectory of the body C in physical space. However, the knowledge of a trajectory of the system is equivalent to the knowledge of a trajectory of the body C.

[9]From a mathematical viewpoint, the trajectories of the system are solutions of the above differential equations, and a classical theorem, the Cauchy–Lipschitz theorem, assures us of the local existence and uniqueness of a trajectory passing through a point.

[10]Here the conclusion applies equally well to "physical" space (the plane in which the bodies A, B, and C move) as to the configuration space; the results in these two spaces are equivalent.

A stability theorem "à la Poisson". Poincaré's first result on stability rests on an entirely simple, but fundamental, remark: the three-body system described above is *volume-preserving.*

To explain what this means, I must introduce some notation. Given a region R in the configuration space of our system and a number k, we define another region $R(k)$ in the configuration space as follows: for each point p in the region R, we consider the trajectory of the system passing through p at time 0, and we mark the point where this trajectory is located at the end of k days; the region $R(k)$ is the set of all the points we have marked. Let us reformulate this: the region $R(k)$ is the set of all points in the configuration space where one can find the trajectories that, k days beforehand, were located somewhere in the region R. Now, to say that the *the system is volume-preserving* is to say that, for every region R in the configuration space, and for every k, the volume of the region $R(k)$ equals the volume of the region R. This is a somewhat abstract, but simple, property. One can check that our three-body system is volume-preserving thanks to a calculation of a few lines dealing with the differential equations that govern the system's evolution.[11]

Thanks to this remark, Poincaré was able to prove a result that he called the *stability theorem à la Poisson,*[12] and which today is called the *Poincaré recurrence theorem.*

THEOREM (Poincaré recurrence theorem). *Almost every trajectory*[13] *of the three-body system described above passes arbitrarily close to its initial position infinitely often.*

The expression "almost every trajectory satisfies property (P)" means "the set of initial positions of trajectories that do not satisfy property (P) has zero volume". Thus, to say that "almost every trajectory satisfies property (P)" means that "if we choose *at random* a point in the configuration space, there is zero probability that the trajectory passing through this point at time 0 does not satisfy property (P)".

In particular, the theorem above implies that if we choose at random an initial configuration for our system, the probability that this configuration is such that the body C does not pass close to its initial position infinitely often is zero.

PROOF OF A SIMPLIFIED VERSION OF THE RECURRENCE THEOREM. Let δ be a strictly positive real number (as small as one likes). We will prove the following claim: *Almost every trajectory passes within a distance δ from its initial position.* This simplified version of the Poincaré recurrence theorem is already an interesting result, and allows us to bypass the proof of several technical complications.

Consider a region R in the configuration space. Assume that R has non-zero volume. We can then choose a ball of diameter δ, such that the intersection of the region R with this ball has non-zero volume;[14] we denote this intersection $\widetilde{R}$. By

[11] In fact, this is not a property of the usual volume in the coordinates (x, y, x', y') previously described; a coordinate change is therefore necessary. But once the right coordinate change is found, the calculation is actually elementary: in technical terms, it is sufficient to check that the *divergence* of the system of differential equations is zero.

[12] From the name of the mathematician Siméon Denis Poisson (1781–1840).

[13] Understood: with energy contained between $E_{\min}$ and $E_{\max}$ and for which the initial values of x and y are "not too large".

[14] Such a ball exists. Indeed, if the intersection of the region R with every ball had zero volume, then the region R would have zero volume.

construction, the region $\widetilde{R}$ is contained in the region R and has a non-zero volume, which we denote v. In addition, $\widetilde{R}$ is contained in a ball of diameter δ; two arbitrary points of $\widetilde{R}$ are therefore always situated at a distance less than δ from each other.

Now consider the regions $\widetilde{R}(1), \widetilde{R}(2), \widetilde{R}(3), \dots$ (we recall that the region $\widetilde{R}(k)$ is the set of all the points where those trajectories can be that, k days before, were in the region $\widetilde{R}$). Since the system is volume-preserving, each of these regions has the same volume as the region $\widetilde{R}$, which is v. Recall, moreover, that the trajectories do not leave a bounded region of the configuration space. It follows that the infinite number of regions $\widetilde{R}(1), \widetilde{R}(2), \widetilde{R}(3), \dots$ cannot be pairwise disjoint: indeed, one cannot place infinitely many regions of the same non-zero volume v, pairwise disjoint, into a bounded region of space (you can't put infinitely many oranges into a crate!). Therefore there exist two integers i and j such that the region $\widetilde{R}(i)$ meets the region $\widetilde{R}(j)$. We choose a point in the intersection of these two regions and denote by T the trajectory that passes through this point on day j.

On the one hand, the trajectory T is in the region $\widetilde{R}(j)$ on day j. It was in the region $\widetilde{R}$ on day 0; in other words, its initial position is situated inside $\widetilde{R}$. On the other hand, the trajectory T is in the region $\widetilde{R}(i)$ on day j; it was therefore inside $\widetilde{R}$ on day $j - i$. Thus, we have found a trajectory T whose initial position is situated inside the region $\widetilde{R}$ (thus *a fortiori* inside R), and which passes through $\widetilde{R}$ again (on day $j - i$). Because two points of $\widetilde{R}$ are always less than a distance δ from each other, the trajectory T passes at a distance less than δ from its initial position.

To show the existence of the trajectory T, the only hypothesis we made was that the region R had non-zero volume. Consequently, if no trajectory with initial position in R passed at a distance less than δ from its initial position, then this would imply that the region R has zero volume. In particular, if R is the region consisting of all points p such that the trajectory with initial position p does not pass by again at a distance less than δ from p, then R has zero volume. This means exactly that almost every trajectory passes by at a distance less than δ from its initial position. □

As one can see, the above proof only used two properties of the system being considered: the fact that this system is volume-preserving and the fact that the trajectories stay inside a bounded region. Thus, the Poincaré recurrence theorem is valid for any deterministic system that satisfies these two conditions.[15]

I will here take a few lines to try to explain why I find that this quite simple theorem, which is perfectly typical of Poincaré's method in his memoir, is absolutely admirable.

Throughout the 18th and 19th centuries, the greatest mathematicians sought to show that the solar system is stable. For this, they displayed fabulous intelligence and a colossal power of calculation in order to determine the planets' trajectories as precisely as possible; once the calculations were carried out, there was nothing left but to note that the trajectories thus obtained were periodic, or nearly so. But, because the trajectories were only approximately calculated, one could not thereby gain any information about the solar system except in the relatively near future.

[15]For more on the Poincaré recurrence theorem (and a warning as to what it is not!) see also Chapter 9.

Poincaré "took the high road". He wondered if, without calculating the planets' trajectories, without knowing their initial positions, without even being closely interested in the differential equations that translate Newton's laws, one could nonetheless *say something*. Is there not something very general to be said about the trajectories of a deterministic system? The bet seemed to be completely insane. And still, in one page, Poincaré shows us that, in every volume-preserving deterministic system, if the trajectories stay within a bounded region, then almost every trajectory passes infinitely often close by its initial position. This result applies in particular to the three-body system described above; it is one of the first results on the behavior of this system whose validity is not limited in time.

Of course, there is a price to pay: the information that this theorem provides us is extremely abstract. In fact, we know absolutely nothing about when the trajectories will again pass close to their initial position (for some of them, the solar system may have ceased to exist before the first return), nor what each trajectory does in between two passages. This is why Poincaré's recurrence theorem does not in any way invalidate the titanic calculations of Laplace, Le Verrier, etc. But, in one page, Poincaré opened for us the door to a new world![16]

A theorem on the coincidence of asymptotic surfaces. We are still interested in the three-body system previously described. With the recurrence theorem, Poincaré showed a first stability result for this system: almost every trajectory passes infinitely often close to its original position. This result is doubtless too abstract to have practical consequences. Poincaré therefore continued his study and became interested in more concrete forms of stability: do there exist periodic trajectories? If yes, how do nearby trajectories behave? Can we show that the trajectories "do not go too far away"?

Let us begin by explaining why, even though the configuration space of the system is four-dimensional, we can study these trajectories in three-dimensional spaces.

The system possesses an energy which is greater or less according to the configuration in which it finds itself. By abuse of language, we therefore speak of the energy of a configuration or of a point in the configuration space.[17] For each value E, we can then consider the set of points in the configuration space whose energy equals E. This set is called the *hypersurface of energy E*; one can show that it is a three-dimensional subspace in the interior of the configuration space (except for a few values of E that will not interest us). And, because the system always evolves with constant energy (no external element comes to provide or remove energy and there is no friction), each trajectory of the system stays inside an energy hypersurface. This allows us to study, when we wish, the system restricted to an energy hypersurface.[18] Since the energy hypersurfaces are three-dimensional, this is a practical way to visualize trajectories and make drawings.

---

[16]This world is called *ergodic theory*. It is a field of mathematics that essentially deals with understanding the statistical behavior of "almost all trajectories" of certain systems. Today, hundreds of mathematicians work in this field.

[17]The energy of the point (x, y, x', y') equals $\frac{1}{2}(x'^2 + y'^2) - \frac{1}{2}(x^2 + y^2) - V(x, y)$, but this matters little.

[18]This is also what allows us to speak about the energy of a trajectory, as we have already shamelessly done in the statement of Hill's theorem.

Let us now recall that our system contains a body A of mass $1-\mu$, a body B of mass μ, and a body C of negligible mass; we shall explain the role that the mass μ plays. Indeed, Poincaré used this mass as a parameter. Once μ is zero, the only force to which C is subjected is the attraction of the body A, which stays fixed at the point O. Therefore we understand perfectly the behavior of the system: regardless of its initial position and velocity, the body C traverses an ellipse with one focus at O. Poincaré then seeks to understand how the trajectories are deformed as μ, starting from 0, increases little by little. Most often, he thus obtained results of the type: *When μ is sufficiently small, the trajectories possess such-and-such a characteristic.*

Here is an example of this process. When μ is zero, the system has lots of periodic trajectories. In fact, every choice of initial position and velocity for C gives rise to a periodic (elliptic) trajectory in the absolute reference frame: as long as the ratio of periods of the motions of B and C is rational, C's trajectory is again periodic in the moving reference frame under consideration.[19] Poincaré showed that as μ moves away from 0, certain trajectories remain periodic as long as μ does not become too large. He therefore obtained:

THEOREM (Poincaré). *As long as the mass μ is sufficiently small, the three-body system under consideration possesses periodic trajectories; in addition, the number of periodic trajectories tends to infinity as μ tends to* 0.

Bolstered by this result, Poincaré sought to understand how the neighboring trajectories to a periodic trajectory behave. Fix a periodic trajectory T. What is the behavior of a trajectory T' whose initial position is very close to T? Of course, we would like for such a trajectory T' to remain close to the periodic trajectory T forever (the trajectory T' would then be "periodic up to a small error"). Alas, the situation is not always so idyllic. On the contrary, it can happen that we find, arbitrarily close to the periodic trajectory T, points from which trajectories depart that thereafter distance themselves exponentially fast from T; in this case, Poincaré says that the periodic trajectory T is *unstable*.[20] Poincaré remarked that at least half of the periodic trajectories whose existence he had proved are unstable; we cannot therefore save ourselves the trouble of studying these trajectories.

Let us consider, therefore, an unstable periodic trajectory T. Poincaré defines two types of trajectories having particularly interesting behavior:

- The trajectories whose behavior consists of approaching the periodic trajectory closer and closer; these trajectories are called *future asymptotes of* T.
- The trajectories that have the same behavior as above when the flow of time is reversed; in other words, the trajectories that, when we turn time towards the past, approach the periodic trajectory closer and closer; these trajectories are called *past asymptotes of* T. These are the trajectories that in the past were very close to the periodic trajectory T, then distanced themselves from it little by little.

We will denote by S_T^s the set of initial positions of all trajectories asymptotic to T in the future, and by S_T^u the set of initial positions of trajectories asymptotic

[19] Moreover, if the motions of B and C are circular, then the motion of C in the moving reference frame is periodic, regardless of the ratio of absolute periods.

[20] Today, we would say *hyperbolic* instead, but *unstable* is more evocative.

to T in the past.[21] Poincaré proved that S_T^s and S_T^u are surfaces (which he called the *asymptotic surfaces* associated to T). Then, he stated a second stability result:

"THEOREM" ON THE COINCIDENCE OF ASYMPTOTIC SURFACES. *As long as the mass μ is sufficiently small, the asymptotic surface S_T^s and the asymptotic surface S_T^u associated to the unstable periodic orbit T coincide.*

Even if it does not seem immediately apparent, this statement implies that, in a certain sense, the system is stable; I will try to explain why.

First of all, recall that the points of the surface S_T^u are the initial positions of trajectories that were close to the periodic trajectory T in the past and have moved away from it. As for the points of the surface S_T^s, they are the initial positions of trajectories that, in the future, will move closer to the periodic trajectory T. To assert that the surfaces S_T^s and S_T^u coincide is therefore to assert that the trajectories that were close to the periodic trajectory T in the past and afterwards moved away from it will end by "returning to the fold" by moving closer to T once again. In other words (with a bit of caricature): the theorem on the coincidence of asymptotic surfaces asserts that the trajectories that had practically periodic motion in the past, but whose motion was afterwards disturbed, will end by "returning to the right path" and find their initial practically periodic motion.

In fact, the "theorem on the coincidence of asymptotic surfaces" equally teaches us something about the behavior of *all* the other trajectories of the system. Let us place ourselves in an energy hypersurface, which we denote H (recall that this is a 3-dimensional space). Consider an unstable periodic trajectory T located in H. With a bit of reflection, we can see that the asymptotic surfaces S_T^s and S_T^u are contained in H. Now, the crucial point is the following: the coincidence of the asymptotic surfaces S_T^s and S_T^u implies that $S_T^s = S_T^u$ is a closed surface[22] (alas, I cannot explain why here). The trajectories of the system cannot traverse this closed surface $S_T^s = S_T^u$, which therefore acts as a "trajectory barrier". For the moment, we have only considered a single unstable periodic trajectory, but, as long as μ is small, there are many such trajectories. Let us suppose that there were many of them in the energy hypersurface H. Each of these unstable periodic trajectories would then give us a closed surface in H that acts as a trajectory barrier. And each trajectory would find itself confined to a thin region of H located between two surfaces that, for it, constitute insurmountable barriers. This seems to summarize perfectly the epigraph Poincaré chose for his memoir: *Nunquam praescriptos transibunt sidera fines.*[23]

Before describing the way in which Poincaré proposed to prove the theorem on the coincidence of asymptotic surfaces, it is important to explain how he reduced the study of the asymptotic surfaces S_T^s and S_T^u to the study of two curves drawn on a surface.

We place ourselves in an energy hypersurface. Inside this 3-dimensional space, Poincaré found a surface Σ and a number k such that each trajectory of the system

[21]The letters s and u indicate "stable" and "unstable" in the language of our English-speaking friends. In contemporary terminology, we indeed talk about the *stable* and *unstable surfaces* of a periodic orbit. In my humble opinion, we would have done better to keep Poincaré's terminology...

[22]More precisely, in technical language, it is an immersed torus (there is a circle of double points).

[23] *"Never will the stars cross their prescribed limits."* Nonetheless, we note that Poincaré never, to my knowledge, mentions the implications of the theorem on the coincidence of asymptotic surfaces in terms of confining trajectories between barriers.

crosses this surface Σ at least one every k years. Such a surface is called a *section.* Poincaré thus defined a map f from the surface Σ to itself: for each point p of Σ, the image of p by f is the point of Σ where the trajectory starting from p crosses Σ again for the first time (see FIG. 8.2). We say that f is the *first return map* of the trajectories on the surface Σ.

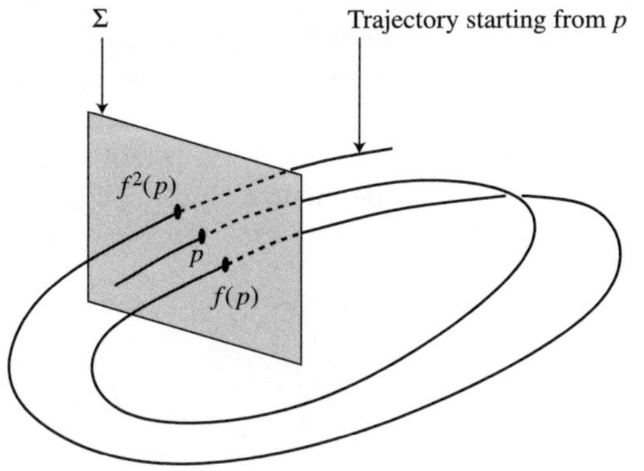

FIGURE 8.2. The first return map f on the section Σ.

For every point p of the section Σ, the trajectory passing through p intersects Σ successively at p, then at $f(p)$, then at $f^2(p) = f(f(p))$, then at $f^3(p) = f(f^2(p))$, etc. The study of the trajectory passing through p thus reduces essentially to the study of the sequence of points $p, f(p), f^2(p), f^3(p), \ldots$ For example, the trajectory passing through p is periodic if and only if the sequence of points $p, f(p), f^2(p), f^3(p), \ldots$ is periodic.

Most properties of the system can be read off f. Thus, the fact that the system is volume-preserving (in the 4-dimensional configuration space) is equivalent to the fact that *the map f is area-preserving.* This means that, for any closed curve Γ drawn on the section Σ, the area enclosed by Γ equals the area enclosed by the curve $f(\Gamma)$.

Likewise, the study of asymptotic surfaces reduces to the study of curves drawn on the section Σ. Indeed, consider an unstable periodic trajectory T, and let p denote one of the points of intersection of the trajectory T with the section Σ. Because the trajectory T is periodic, the sequence of points $p, f(p), f^2(p), f^3(p), \ldots$ is also periodic; in other words, there exists an integer i such that $f^i(p) = p$. Replacing the first return map f by the ith return map f^i, if necessary, we can then assume that $f(p) = p$. The asymptotic surfaces S^s_T and S^u_T of the trajectory T intersect the section Σ along two curves C^s_T and C^u_T that pass through p. To prove the theorem on the coincidence of asymptotic surfaces, it suffices to show that the curves C^s_T and C^u_T coincide.

Let us now turn to the proof that Poincaré gave for the theorem on the coincidence of asymptotic surfaces. This proof is far from being transparently clear, but I will nevertheless try to explain the principle by simplifying it.[24]

Recall that we want to show that the curves C^s_T and C^u_T defined above coincide. Poincaré begins by showing that, as long as μ is small, the curves C^s_T and C^u_T pass

[24]I do not pretend here to reconstitute the arguments of Poincaré (that would be much too long, and much too technical); I hope nevertheless not to betray the spirit of those arguments.

very close to each other.[25] To accomplish this, he *expands the equation of the curve* C_T^s *in powers of* $\sqrt{\mu}$; that is, he writes the equation of the curve C_T^s as a sum of infinitely many terms—an *order 0* term that does not depend on $\sqrt{\mu}$, an *order 1* term proportional to $\sqrt{\mu}$, an *order 2* term proportional to the square of $\sqrt{\mu}$, etc. He does the same for the equation of the curve C_T^u. The sum of the order 0 term and the order 1 term in the equation of C_T^s corresponds to the equation of a curve C that traces a double loop passing near to p; the sum of the order 0 term and the order 1 term in the equation of C_T^u corresponds to the equation of the same curve C, but traversed in the opposite direction. Moreover, the terms whose orders are greater than 2 in the equations are all negligible when the mass μ is small. We deduce that, as long as μ is small, the curves C_T^s and C_T^u both remain very close to the curve C. This implies that the curve C_T^s necessarily passes very close to the curve C_T^u (see FIG. 8.3).

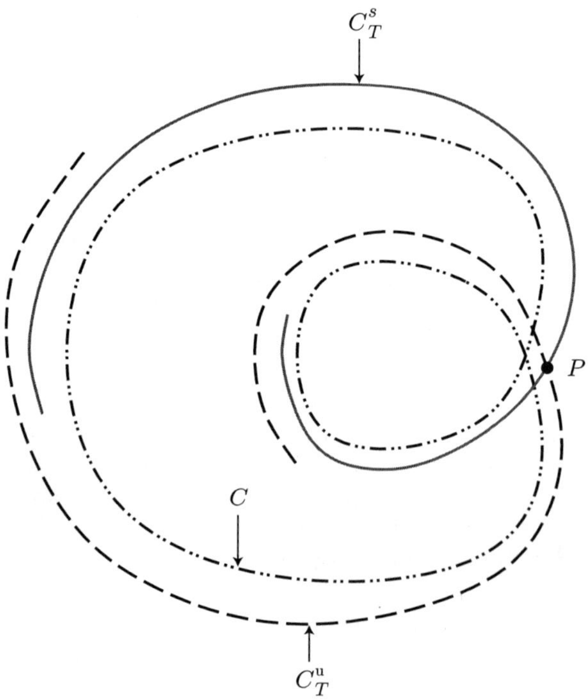

FIGURE 8.3

Poincaré thus proposes two different methods to end the proof. The first method is analytic. It consists simply in proving that each term in the equation for C_T^s coincides with the term of the same order in the equation for C_T^u. Poincaré explains how to calculate the terms one after another, but does not push his investigations far enough to decide if this is sufficient to end the proof. He scarcely lingers on this point; without any doubt he prefers the second method.

[25]A technical precision: in this sentence, it is understood that one can "follow the periodic orbit T as μ varies". More precisely, there is a continuous family $(T_\mu)_{\mu\in[0,\varepsilon]}$ of periodic orbits parametrized by μ, and, each time I write T, p, C_T^s, etc., it must be read as T_μ, p_μ, $C_{T_\mu}^s$, etc.

This second method is geometric. Since the curves C_T^s and C_T^u pass near each other, we can find a point a on C_T^s and a point b on C_T^u close to each other. Since the curves C_T^s and C_T^u are invariant under the map f, the points $f(a)$ and $f(b)$ are located respectively on the curves C_T^s and C_T^u. We then choose a small arc α that joins the point a to the point b. We denote by $[a, f(a)]^s$ the arc of the curve C_T^s joining a to $f(a)$, and by $[f(b), b]^u$ the arc of the curve C_T^u joining $f(b)$ to b. Poincaré shows that one can begin in such a way that the arc α is disjoint from its image $f(\alpha)$, and so that the arcs α and $f(\alpha)$ are disjoint from the arcs $[a, f(a)]^s$ and $[f(b), b]^u$. He asserts that this excludes all geometric situations except three: the situation where the curves C_T^s and C_T^u coincide, and the two situations represented in FIG. 8.4.

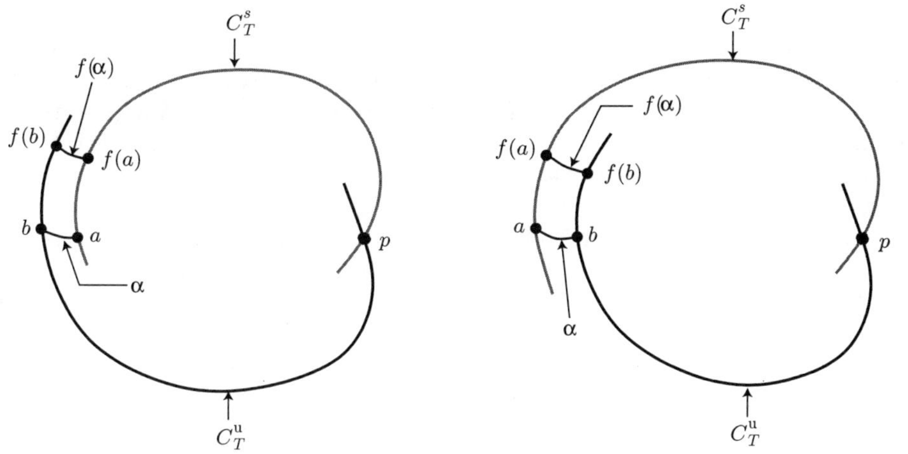

FIGURE 8.4

It therefore remains to exclude these last two situations. To do so, Poincaré considers the closed curve Γ obtained by following the curve C_T^s from p to a, then the arc α from a to b, then the curve C_T^u from b to p. The image $f(\Gamma)$ of this curve is obtained by following the curve C_T^s from p to $f(a)$, then the image of the arc $f(\alpha)$ from $f(a)$ to $f(b)$, then the curve C_T^u from $f(b)$ to p. Since the map f is area-preserving, the area enclosed by the curve Γ must equal the area enclosed by $f(\Gamma)$. In the situation represented on the left of the figure above, the area enclosed by the curve Γ is strictly smaller that the area enclosed by the curve $f(\Gamma)$ (the difference between the two is the area of the quadrilateral $(a, f(a), f(b), b)$); this situation is therefore impossible. In the situation represented on the right, the area enclosed by the curve Γ is strictly greater than the area enclosed by the curve $f(\Gamma)$; this situation is therefore equally impossible. Thus we have excluded all the situations except the one where the curves C_T^s and C_T^u coincide, which concludes the proof.

8.3. The discovery of homoclinic intersections

About a dozen memoirs competed for the Prize of King Oscar II. However, as soon as Mittag-Leffler, Weierstrass, and Hermite began studying the manuscripts they had received, it quickly became apparent that one of the memoirs, entitled *Sur le problème des trois corps et les équations de la dynamique*, was distinctly superior to the others. This memoir impressed them as much for the results it contained as for the completely new light it shed on the old problem to which it was devoted: the stability of the solar system. They would write that it is "the deep and original work of a mathematical genius whose place is marked among the great geometers of the century." And, even though the anonymity of the candidates would be respected until the end, there is no doubt that Mittag-Leffler, Hermite, and Weierstrass from the start guessed the name of this "mathematical genius."

On January 21, 1889, Mittag-Leffler wrote to Poincaré to announce to him that he had been awarded the prize, and without further delay entrusted Poincaré's manuscript to a young mathematician of Stockholm, Lars Edvard Phragmén, to prepare it for the printer.

In July, Phragmén had finished his work. He nonetheless brought to Mittag-Leffler's attention five "obscure points" in Poincaré's memoir, which seemed to him worth enlightening. Mittag-Leffler transmitted the list of these "obscure points" to Poincaré, while indicating that in his opinion, only the fifth required real attention. This last point concerns certain quantities that Poincaré expanded in powers of $\sqrt{\mu}$ (for example, the equations of the curves C_T^s and C_T^u in the proof of the theorem on the coincidence of asymptotic curves). One can always carry out such an expansion, but in order to extract certain information from it, the expansion must be *convergent.* In his memoir, Poincaré hinted that the expansions he used were convergent, but often he did not take the trouble to justify this. And Phragmén thought that a few supplementary explanations on this matter would be useful to the readers.

Poincaré would quickly realize that Phragmén had put his finger on an important problem. Let us remember that, in his memoir, Poincaré envisaged two possible proofs for the theorem on the coincidence of asymptotic surfaces: an analytic proof (which he did not bring to fruition), and a geometric proof. Phragmén's remark implies that there is a "hole" in the "proofs" devised by Poincaré: in order for these to work, it would have to be shown that the expansions in terms of powers of $\sqrt{\mu}$ that he considered are convergent. Taking up his calculations again, Poincaré perceived that he could not hope to "plug this hole": in general, the expansions in terms of powers of $\sqrt{\mu}$ of the equations of C_T^s and C_T^u *are not convergent* [**1, 17**]. Fortunately, he thought he could get around this problem in the geometric proof... until he discovered another important error in this proof!

Let us again consider the geometric "proof" of the theorem on the coincidence of asymptotic surfaces, and try to uncover the error. Recall that we consider an unstable periodic trajectory T, the asymptotic surfaces S_T^s and S_T^u associated to T, and the curves C_T^s and C_T^u obtained as the intersections of the surfaces S_T^s and S_T^u with the section Σ. We must prove that these curves coincide. Recall that the first step consisted in showing that the curves C_T^s and C_T^u pass very close to each other. This first step *a priori* uses an expansion in powers of $\sqrt{\mu}$, but the problem created by the divergence of this expansion is not insurmountable; this is not where the true error is located. We consider a point a on the curve C_T^s and a point b on the curve C_T^u near each other, as well as an arc α joining these two points. The principal argument of the proof then consists in remarking that neither of the geometric situations represented by FIG. 8.4 is compatible with the fact that the first return map f is area-preserving (the area enclosed by the curve $\Gamma = [p,a]^s \cup \alpha \cup [b,p]^u$ must equal the area enclosed by the image curve $f(\Gamma) = [p, f(a)]^s \cup f(\alpha) \cup [f(b), p]^u)$. This argument is incontestable.

Indeed, the error is the following: Poincaré tacitly assumed that the curves C_T^s and C_T^u could not cross each other. Doing so, he was not on the lookout for, and hence forgot, the situation represented in FIG. 8.5, where the arcs $[a, f(a)]^s$ and $[b, f(b)]^u$ intersect. Yet, no argument allows one to eliminate this situation (and, in any case, not the argument of area preservation invoked above: in FIG. 8.5, the area enclosed by the curve $\Gamma = [p, a]^s \cup \alpha \cup [b, p]^u$ equals the area enclosed by the curve $f(\Gamma) = [p, f(a)]^s \cup f(\alpha) \cup [f(b), p]^u)$. Poincaré soon had to accept the facts: there is no reason at all for the asymptotic surfaces S_T^s and S_T^u to coincide.

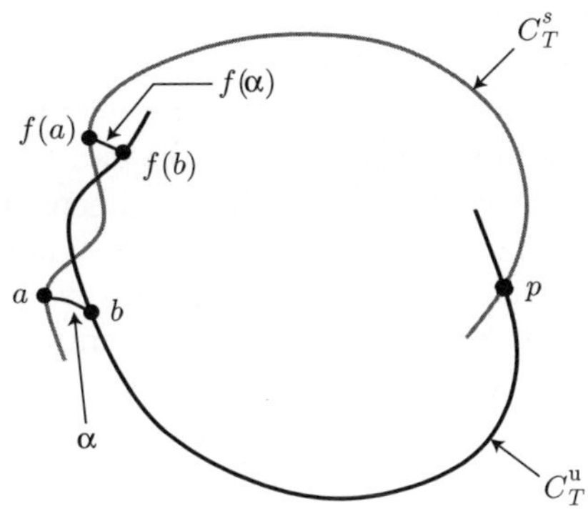

FIGURE 8.5

The origin of Poincaré's error is quite clear. In a previous memoir [**15**], Poincaré had studied in detail the behavior of the trajectories of systems whose configuration space is 2-dimensional. At that time, he introduced certain notions (*unstable fixed points*, *asymptotic curve*, etc.) whose generalizations in higher dimensions gave rise, in the memoir for the Prize of King Oscar II, to the notions of *unstable periodic trajectory*, *asymptotic surface*, etc. Yet, when the configuration space is 2-dimensional, it is obvious that the asymptotic curves of an unstable fixed point cannot intersect, unless they coincide. Poincaré simply forgot that this fact is specific to 2-dimensional configuration spaces, and does not at all generalize to higher dimensions.

Thus the "theorem" on the coincidence of asymptotic surfaces crumbled. It is not true that the asymptotic surfaces S_T^s and S_T^u of an unstable periodic trajectory T always coincide. By using the intersections of the asymptotic surfaces, Poincaré would certainly prove (see Section 8.5) that, for any unstable periodic trajectory T, there exist infinitely many trajectories that, after moving away from T, end by moving afresh toward this same trajectory. But this is a very weak stability result by comparison with the theorem that was initially announced. Recall that one of the interests of this theorem was that it provided closed surfaces that acted as barriers between which the trajectories were confined. Since nothing indicated the existence of such barriers anymore, nothing *a priori* prevented certain trajectories from going all over the configuration space. And the original epigraph of the memoir, *Nunquam praescriptos transibunt sidera fines*, no longer seems to really reflect the content of this situation.

In the letter he addressed to Mittag-Leffler in order to reveal the existence of an error in his memoir, Poincaré did not at all hide his disappointment [**18**]:

> *"It is not true that the asymptotic surfaces are closed, at least not in the sense in which I first understood it. What is true is that, if I consider the two parts of this surface (which yesterday I still believed to be connected with each other), they intersect along infinitely many asymptotic trajectory curves.* [...] *I will not conceal from you the chagrin this discovery caused me. Most importantly, I do not know if you will judge the results that remain, which is to say the existence of periodic solutions, that of asymptotic solutions, the theory of characteristic exponents, the non-existence of uniform integrals, and the divergence of the series of M. Linstedt*[26] *still deserve the great recompense that you had wanted to grant me."*

There is not a shadow of a doubt that, even marred by this error, Poincaré's memoir largely deserved the prize. But the memoir had to be revised... And there was a practical problem: the memoir had already been printed! The first copies had just been sent out![27] There thus followed a somewhat comical episode where Mittag-Leffler, who feared that the existence of an error in the winning memoir would damage his reputation, or provoke the king's anger, sought to reacquire the different copies of the memoir in the most discreet way possible. In a letter, he explained to Poincaré how he went about it, here stalling the mail distribution, there invoking such-and-such a pretext for the recipient to send back the memoir before even glancing at it, rejoicing that certain mathematicians with whom he had delicate relations, such as Kronecker, had not yet received copies, etc. He would go so far as to hide the existence of the error from Weierstrass, which the latter took very poorly...

For his part, Poincaré amended his memoir, of course removing the false theorem, and lightly modifying the Introduction in order to place the emphasis on the other results [**1, 17**]. And he took responsibility for the costs of printing the new memoir, which came to 3585 crowns... about 1000 crowns more than the prize he had received as the winner of the contest!

Let us return to more mathematical considerations. The geometric arguments that Poincaré elaborated to prove the theorem on the coincidence of asymptotic surfaces, even though they did not form a complete proof, nonetheless teach us something: for every unstable periodic trajectory T, either the curves C_T^s and C_T^u coincide, or they intersect. Equivalently, either the asymptotic surfaces S_T^s and S_T^u coincide, or they intersect. In the second case, we say that the system has a *homoclinic intersection.* And since everything indicates that systems admitting homoclinic intersections exist, it becomes necessary to study them...

8.4. "A kind of lattice, tissue, or web of infinitely tight stitches"

As paradoxical as it may seem, Poincaré's mistake would be extremely fruitful. Indeed, by exploring the consequences of the existence of homoclinic intersections,

---

[26] Poincaré has forgotten his theory of integral invariants and his Poisson-type stability theorem.

[27] It is true that Poincaré delayed in answering Mittag-Leffler: Mittag-Leffler's letter to Poincaré, indicating the obscure points revealed by Phragmén, is dated July 16; Poincaré's response is dated December 1.

Poincaré would make a discovery of considerable mathematical, physical, and even philosophical importance: the simplest deterministic systems can present horribly complex long-term behavior.

Beginning his exploration, Poincaré perceived first of all that a homoclinic intersection never appears alone [**19**]:

PROPOSITION. *Consider an unstable periodic trajectory T and an intersection point p of this trajectory with the section Σ. Denote by C_T^s and C_T^u the curves passing through p, obtained as the intersections of the asymptotic surfaces S_T^s and S_T^u with the section Σ.*

Assume that the curves C_T^s and C_T^u intersect elsewhere than at p. Then the curves C_T^s and C_T^u intersect infinitely often. In addition, if q and q' are two arbitrary points of intersection of C_T^s and C_T^u, then C_T^u intersects C_T^s infinitely often along the arc of C_T^s that joins q to q', and C_T^s intersects C_T^u infinitely often along the arc of C_T^u that joins q to q'. Finally, the intersection points of the curves C_T^s and C_T^u belong to infinitely many different trajectories.

This proposition is essentially a consequence of the recurrence theorem. Here is a quick proof (whose reading may be entirely omitted):

NEARLY COMPLETE PROOF OF THE ABOVE PROPOSITION. We denote by f the first return map on the section Σ. Replacing f by one of its iterates, if necessary, we can assume that $f(p) = p$.

If the curves C_T^s and C_T^u intersect at a point $q \neq p$, then the invariance of the curves C_T^s and C_T^u under f implies that the successive returns $f(q), f^2(q), f^3(q)$, etc., of q on the section Σ are all points of intersection of the curves C_T^s and C_T^u. Thus, the curves C_T^s and C_T^u intersect at infinitely many points.

Now consider two points of intersection q and q' of the curves C_T^s and C_T^u. Let I^s denote the arc of the curve C_T^s that joins q to q', and I^u the arc of the curve C_T^u that joins q to q'. The union of the arcs I^s and I^u bounds a region R of the section Σ. The key argument is then the following: the recurrence theorem tells us that there exist trajectories whose initial position is inside R and which pass inside of R infinitely many times. From this we deduce that there exist infinitely many integers n (the return times of the trajectories mentioned above) such that the region $f^n(R)$ encounters the region R. The conservation of area implies that the region $f^n(R)$ cannot be entirely contained in the region R, nor conversely. When the region $f^n(R)$ meets the region R, the boundary of the region $f^n(R)$ must therefore meet the boundary of R. We then note that, for any n, the boundary of the region $f^n(R)$ is contained in the union of the curves C_T^s and C_T^u (since the boundary of the region R is contained in the union of these two curves and since these curves are invariant under f). For each integer n such that the boundary of the region R intersects the boundary of the region $f^n(R)$, there are therefore two possibilities: either the curve C_T^s intersects the arc I^u of the curve C_T^u that joins q to q', or the curve C_T^u intersects the arc I^s of the curve C_T^u that joins q to q'. It follows that the curve C_T^s must intersect the arc of C_T^u that joins q to q' infinitely often, or that the curve C_T^u must intersect the arc of C_T^s that joins q to q' infinitely often. By working a bit more, we can replace the "or" of the previous sentence by an "and", then show that the points of intersection that have been found belong to infinitely many different orbits. □

At first, the above proposition appeared to Poincaré to be a result on stability. Indeed, it implies that for every unstable periodic trajectory T, the asymptotic surfaces S_T^s and S_T^u intersect along infinitely many trajectories; this means that there exist infinitely many trajectories that are asymptotic in the past and asymptotic in the future to the trajectory T. In other words, among the trajectories that were close to the periodic trajectory T in the past and then moved away from it, there are infinitely many that end by coming back toward T in the future.

But then Poincaré tried to imagine the consequences of this proposition on the appearance of the curves C_T^s and C_T^u. Each time that the curve C_T^u crosses the curve C_T^s in two points q and q', it then must return to cross this curve *infinitely many times* between q and q'. The same if the roles of the curves C_T^s and C_T^u are exchanged. And all this happens without either of the two curves C_T^s or C_T^u ever crossing itself. Try to draw two such curves; you will better understand Poincaré's dread [**19**]:

> *"Let one seek to represent the figure formed by these two curves and their infinite number of intersections [...], these intersections form a kind of lattice, or tissue, or web of infinitely tight stitches. Each of these curves must never cross itself, but it must fold upon itself in a very complex manner in order to cross infinitely often over all of the stitches in the network. One will be struck by the complexity of this figure, which I will not even attempt to draw. Nothing is more appropriate to give us an idea of the complication of the three-body problem and in general of all problems in dynamics where there is no uniform integral."*

At this point, it appeared clear to Poincaré that the proposition stated above is completely the opposite of a result on stability. The curves C_T^s and C_T^u are certainly nothing more than auxiliary mathematical objects, introduced for the purpose of studying the system in which we are interested.[28] But it is easy to understand that, from the moment that the curves C_T^s and C_T^u fold onto themselves in a very complex manner, certain trajectories of the system cannot have a simple appearance. Recall, in fact, that if r is a point of the curve C_T^s, then the successive points of intersection $r, f(r), f^2(r), \ldots$ of the trajectory starting from r with the section Σ are located on the curve C_T^s, closer and closer[29] to the point p. Thus, if the curve C_T^s makes complicated detours to go from the point r to the point p, then the sequence of points $r, f(r), f^2(r), \ldots$ will have to follow these meanderings. And the trajectory passing through r will also be forced to follow a very complex path in order to pass successively through the points $r, f(r), f^2(r), \ldots$ Not to speak of the way in which the different trajectories will become enmeshed...

Let us summarize. By discovering the mistake in his memoir, Poincaré understood that a deterministic system *has every likelihood of presenting homoclinic intersections.* He then came to realize that any system that presents a homoclinic intersection automatically has extremely complex trajectories.

Poincaré did not stop there, and he continued to reflect on the implications of homoclinic intersections. He then realized that a system that has a homoclinic

[28] Recall that C_T^s and C_T^u are not trajectories.

[29] The idea here is not proximity "as the crow flies", but proximity *by following the curve* C_T^s.

intersection is *sensitive to initial conditions*: a tiny change in the initial position of a trajectory of the system can completely alter the long term behavior of this trajectory.

In order to explain this, let us recall that a trajectory has radically different behaviors according to whether it crosses the section Σ at points located on the curve C_T^s, on the curve C_T^u, at the intersection of these two curves, or outside of these two curves.[30] The fact that the curves C_T^s and C_T^u form a *network of infinitely tight stitches* therefore means that we have an inextricable mix of trajectories having radically different behaviors. More precisely: arbitrarily close to a point located on the curve C_T^s, in general there exist points located on the curve C_T^u, points located at the intersection of the curves C_T^s and C_T^u, and of course points that are not located on either of these two curves. It therefore suffices to have a tiny change in the initial position of a trajectory for the trajectory, instead of crossing the section Σ at a point of the curve C_T^s, to cross it for example at a point of the curve C_T^u... which will completely change the long-term behavior of this trajectory.

The existence of systems that are sensitive to initial conditions obliges us to completely rethink the connections between the determinism of a system and the possibility of predicting the system's evolution. Poincaré understood this perfectly well, and explained it with admirable clarity, a few years later, in *Science and Hypothesis* [**20**]:

> *"If we knew exactly the laws of nature and the arrangement of the universe at the initial instant, we would be able to predict exactly the arrangement of this same universe at a later instant. But, even if the natural laws held no more secrets from us, we could only know the initial arrangement approximately. If that allows us to predict the later arrangement with the same approximation, then that is all that we need to do* [...] *but it is not always thus. It can happen that very small differences in the initial conditions thence engender very large ones in the final phenomena; a small error at the beginning produces an enormous error at the end. Forecasting becomes impossible* [...]*"*

Thus, although at the beginning he was looking to prove results on the stability of the solar system, Poincaré arrived at the discovery that, on the contrary, deterministic systems, even the simplest ones, in general have horribly complex behavior, and are extremely sensitive to the slightest perturbation of initial conditions—in short, a behavior that no one would ever describe as "stable"!

Today, it is easy to ask a computer to calculate the trajectories of the three-body system considered by Poincaré. The result of such a calculation is reliable as long as one is only interested in the behavior of the trajectories over a relatively

[30]In the first case, we know that the trajectory will move closer and closer to the periodic trajectory T in the future, but we also know that it does not approach T closer and closer as we "move toward the past". In the second case, we know that the trajectory was close to the periodic trajectory T in the past, but that it will not move closer and closer to T in the future. In the third case, we know that the trajectory was very close to the periodic trajectory in the past, then moved away from it, but will, in the future, again move close to this periodic trajectory. Finally, in the last case, we know that the trajectory does not move closer and closer to T in either the past or the future.

short period. FIG. 8.6 shows an example of the trajectories of the bodies A, B, and C thus calculated.[31]

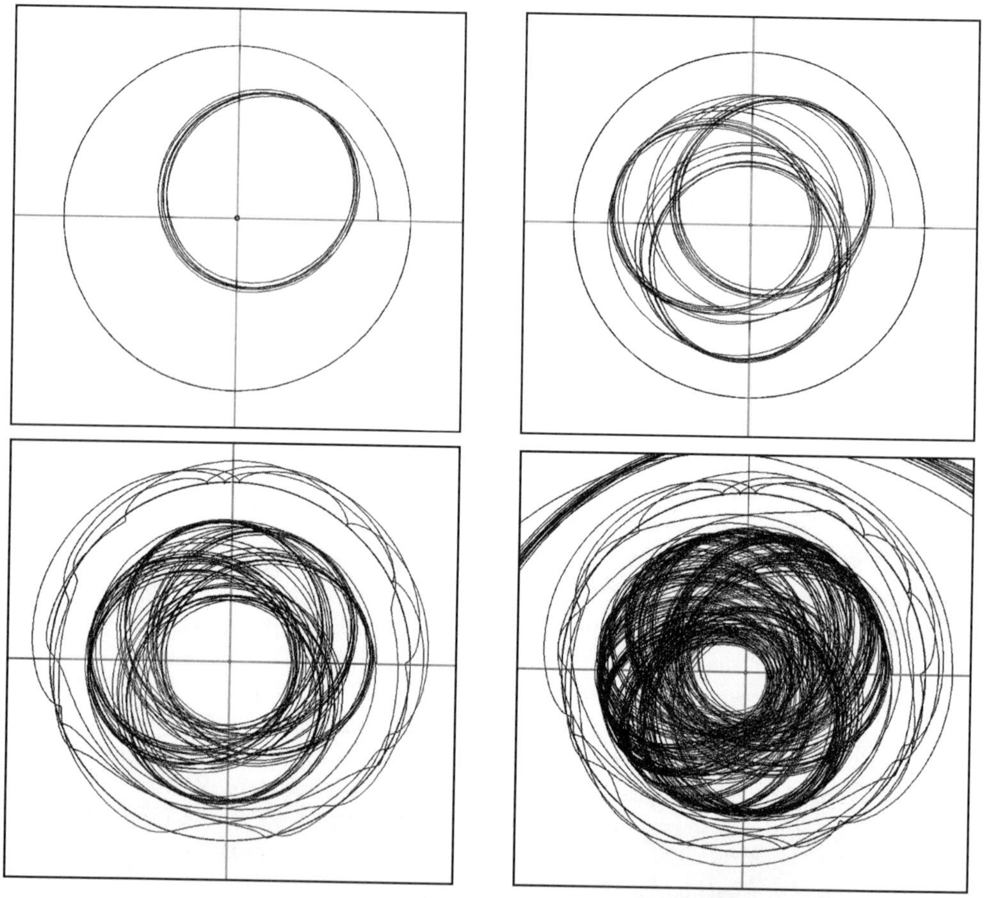

FIGURE 8.6. An example of the behavior of a three-body system.

In this example, the mass of B is 100 times less than that of A.[32] The body A describes the tiny circle in the middle of the figure, and the body B describes a circle whose radius is 99 times greater. At the beginning of the calculation (top left figure), we see that the trajectory of the body C is very close to a Keplerian ellipse. But the attraction that B exerts on C lightly perturbs this Keplerian trajectory, and the ellipse does not close up. As we observe the trajectory of C over a long period of time (top right figure), we see that it is far from being periodic: at every moment, the trajectory of the body C is close to a Keplerian ellipse, but little by little, the major axis of this ellipse turns around A, so much that the trajectory of C begins to "doodle" an entire annulus around A. If we continue to observe the trajectory of C (lower left figure), we perceive the kind of catastrophe that

[31]The figures were realized thanks to the java applet *Kepler3C*, available on the web at the following address: `http://www.physique.usherbrooke.ca/~dsenech/uploads/docs/phq110/simul/Kepler3C.htm`.

[32]TRANSLATOR'S NOTE: Recall that the mass of C is assumed to be negligible (see page 166).

can occur: at a given moment, C has passed too close to B, and has become a satellite of B. If we wait a little longer (lower right figure), we see another possible catastrophe: the body C has undergone a strong acceleration that has ejected it to a (nearly elliptic) trajectory with a large diameter that exceeds the figure.[33]

Another interesting experiment is to calculate the trajectories of the bodies A, B, and C for an initial figure very close to the one used for the figures above. For example, it suffices to move the initial positions of C and A by less than one per cent (without changing B's initial position or the initial velocities of the three bodies) for C not to be "ejected" onto an orbit that leaves the screen. This is a superb illustration of sensitivity to initial conditions!

We note that the behavior of the trajectories calculated above obviously proves nothing about the stability of the solar system, or even of a system composed of the Sun, Jupiter, and Earth. In fact, the force of attraction that Jupiter exerts on Earth is about 4000 times less than the one that planet B exerts on planet C in the figures above; this is why the trajectory of the Earth will not be "upset" as quickly as the trajectory of the body C. But we now understand that Poincaré's discoveries go far beyond the framework of the solar system.

8.5. Have we untangled the web of homoclinic intersections?

With homoclinic intersections, Poincaré discovered that very natural deterministic systems, governed by simple differential equations, can have very complex behavior that is sensitive to initial conditions. But he had almost nothing to say on the behavior of these systems (other than that their complexity seems inextricable, which is rather meager...). I will try to share with you in brief how mathematicians of the 20th century have, at least in certain cases, "untangled the web of homoclinic intersections".

Behavior of orbits in the presence of homoclinic intersections. After Poincaré's death in 1912, the discovery of homoclinic intersections fell into almost general neglect. Indeed, during the first half of the 20th century, only a few mathematicians pursued the way opened by Poincaré and studied geometrically the qualitative properties of solutions to deterministic systems. The most famous was the American George D. Birkhoff.

Poincaré had shown "that a homoclinic intersection never appears alone": the existence of a single intersection between the curves C_T^s and C_T^u associated to an unstable periodic trajectory T implies the existence of a "web of infinitely tight stitches". Birkhoff went farther, and showed that a system that displays a homoclinic intersection has infinitely many periodic trajectories of arbitrarily large periods [**2**]. In addition, Birkhoff was not only interested in periodic trajectories; he also showed, for example, the existence of trajectories that follow very closely one periodic trajectory for a very long time, then little by little move away from it, next moving close to another periodic trajectory and following this second periodic trajectory for a very long time, then little by little move away from this second periodic trajectory, in order to move close to a third periodic trajectory, etc.

[33]At this stage, it is difficult to give assurance that the numerical errors are always negligible. But the ejection of a body with very small mass (here, the body C) onto a trajectory with large diameter after passing near a much more massive body (here, the body B) is absolutely possible. It is even a technique used to send space probes where we wish.

To summarize Birkhoff's work on homoclinic intersections in a few words, we may say that they give a precise meaning to the sentence, "The presence of homoclinic intersections forces the existence of all kinds of extremely complicated trajectories."

We can provide entertaining illustrations of Birkhoff's discoveries by considering very simple physical systems. Consider, first of all, a simple pendulum: a rod that can swing within a vertical plane, around an axis located at one of its endpoints. In the absence of friction, the behavior of such a system is among the simplest imaginable: regardless of the initial position and velocity, the rod has a periodic motion (if its initial velocity is weak, it swings; if its velocity is greater, it completes full turns around the axis). Now consider a double pendulum: at the free end of our first rod, we attach a second rod (in general lighter than the first) with the help of a joint (see FIG. 8.7). Intuitively, we expect that the motions of such a

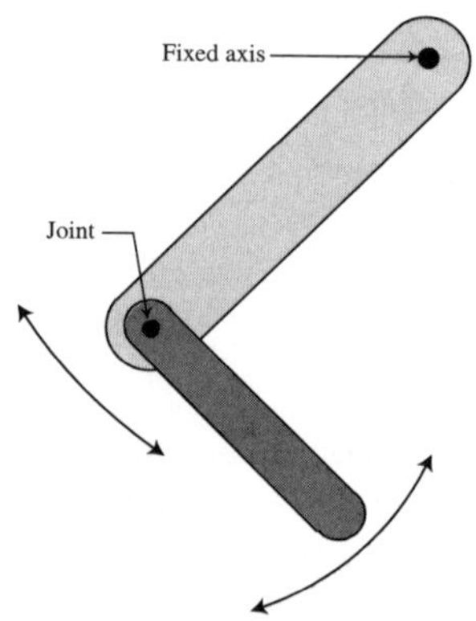

FIGURE 8.7

system would hardly be complicated (similar to the motion of the Moon which, to first approximation, turns around the Earth, itself turning around the Sun). What a surprise when we launch the double pendulum: the articulated rods describe a completely disordered motion—the secondary rod starts turning in one direction, then in the other, the changes in direction of rotation appearing perfectly random! In addition, if we launch the double pendulum with only slightly different initial positions and velocities, the motions described in the two cases have nothing in common between them. Indeed, if we draw, with the help of a computer, the asymptotic surfaces of certain periodic trajectories of the double pendulum, we see homoclinic intersections appear (provided that certain conditions on the ratios of the masses and lengths of the rods are satisfied). If these homoclinic intersections really exist (i.e., if they are not artifacts of numeric approximation), then, by using techniques developed by Birkhoff, we obtain the following entertaining result:

> *For any sequence of integers $a_1, a_2, a_3, \ldots$, there exists a way to launch the double pendulum such that the second rod will carry out at least a_1 turns in the clockwise direction, then make a half-turn, carry out at least a_2 turns in the counterclockwise direction, then make a half-turn, carry out at least a_3 turns in the clockwise direction, etc.*[34]

[34]This statement was inspired by my reading of a very lovely article [**8**], where John Hubbard proves a similar result for a physical system that differs only slightly: the forced pendulum.

The Smale horseshoe. Homoclinic intersections would reappear on the mathematical scene at the start of the 1960s. The architect of this rediscovery was the American mathematician Stephen Smale.[35] At this time, Smale hoped to generalize the recent work of Mauricio Peixoto dealing with the systems whose configuration space is 2-dimensional [**13, 14**]; he had conjectured that a system governed by differential equations can always be perturbed into a system whose trajectories have very simple behavior (in particular, the new system will have only a finite number of periodic trajectories; see [**22**]). Of course, the work of Poincaré and Birkhoff showed that this conjecture is grossly mistaken... but it seems that Smale was completely unfamiliar with this work... and after a certain time, he himself constructed a counterexample to his conjecture.[36]

Smale's counterexample [**23, 24**] is remarkably simple. The idea is to construct a map[37] f on the plane $\mathbb{R}^2$, whose action on a certain square S consists of stretching S vertically and bending it into the shape of a horseshoe, in such a way that $f(S)$ crosses S vertically once from bottom to top, then a second time from top to bottom, as shown in FIG. 8.8, on the left:

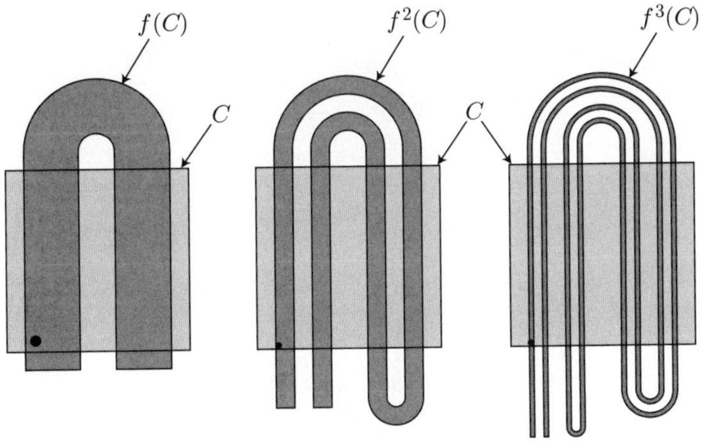

FIGURE 8.8

In addition, the map f must contract horizontal segments and expand vertical segments when restricted to the square S.[38]

Thanks to a classical procedure, we can construct a system governed by differential equations such that the return map of the trajectories of this system on a certain section is exactly the map f constructed by Smale.

[35] This was not the first time that Smale explored a world whose door had been opened by Poincaré. A few years beforehand, he received the Fields Medal for his proof of the Poincaré conjecture for $n \geq 5$: Every compact manifold of dimension $n \geq 5$ that is homotopically equivalent to the sphere $\mathbb{S}^n$ is homeomorphic to $\mathbb{S}^n$. [TRANSLATOR'S NOTE: See Chapter 12 for the case $n = 3$.]

[36] To tell the story in brief, Smale says that he came to the idea of this counterexample not by staying hidden in his office for long days, but by reflecting on the behavior of deterministic systems on the beach of Copacabana (read [**23**]).

[37] More precisely, in technical terms: a C^∞-diffeomorphism.

[38] More precisely, if p is a point of the square S such that $f(p)$ is again in S, then the image by f of the horizontal segment centered at p of length d is a horizontal segment centered at $f(p)$ with length less than $d/2$, and the image by f of the vertical segment centered at p of length d is a vertical segment centered at $f(p)$ with length greater than $2d$.

It is easy to show that the map f implies that there exists a point p, located in the lower left of the square S, such that $f(p) = p$. The immense interest of Smale's example is that the asymptotic curves C^s and C^u associated to this point p can be easily represented. In fact, the image of the square S by f is obtained by a very simple geometric procedure: we stretch vertically, and we bend into the form of a horseshoe. To obtain the successive images $f(S), f^2(S), f^3(S), \ldots$ of the square S by the iterates of f, it suffices to apply this procedure several times in sequence: stretch, fold, stretch, fold, etc. (see FIG. 8.8). As n increases, the set $f^n(S)$ is a rectangle becoming thinner and thinner, longer and longer, and more and more folded onto itself (it crosses the square S from one end to the other exactly 2^n times). And it is not difficult to show that, as n tends to infinity, the set $f^n(S)$ tends to (the accumulation set of) the asymptotic curve C^u of the point p. Similarly, the asymptotic curve C^s of the point p can be obtained as the limit of the sequence of the successive images of the square S by the negative iterates of f. Thus, it is easy to draw finite arcs (but as long as one wishes) of the asymptotic curves C^s and C^u of the point p. As we see in FIG. 8.9, these two curves intersect: the point p has a homoclinic intersection.

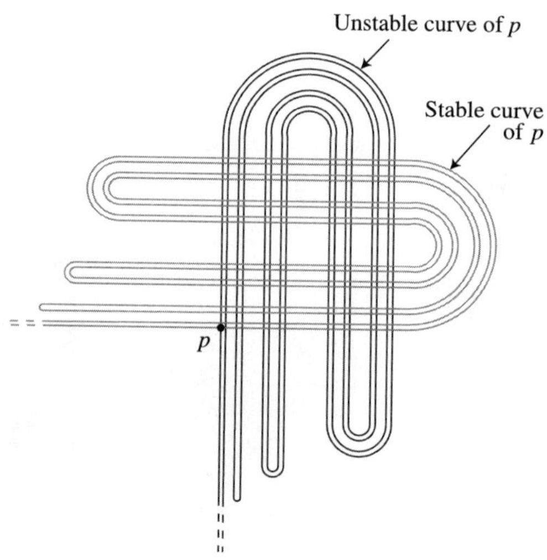

FIGURE 8.9

It is moreover not hard to see that the *homoclinic class* of the point p—that is, the accumulation set of intersection points of the asymptotic curves of p—equals the set of points of the square S whose orbit never leaves the square S.

With this example, Smale achieved a huge step towards understanding homoclinic intersections. Remember that Poincaré, in fact, was not able to represent the asymptotic curves of a point that has a homoclinic intersection. He knew only, thanks to the use of an extremely abstract result (his recurrence theorem), that these two curves must fold on themselves indefinitely and intersect on a *web of infinitely tight stitches.* With the horseshoe map f, we finally understand how, in the presence of a homoclinic intersection, the asymptotic curves C^s and C^u fold on themselves, intersect, ...

For pleasure's sake, I include here two figures (see FIG. 8.10) representing the asymptotic curves of another famous map: the Hénon map. This map is defined by an extremely simple formula:

$$(x, y) \mapsto (x^2 - y, x).$$

What contrast between the simplicity of this formula and the complexity of the curves in FIG. 8.10! What a beautiful illustration of Poincaré's discovery: even the simplest systems can have extremely complex behavior!

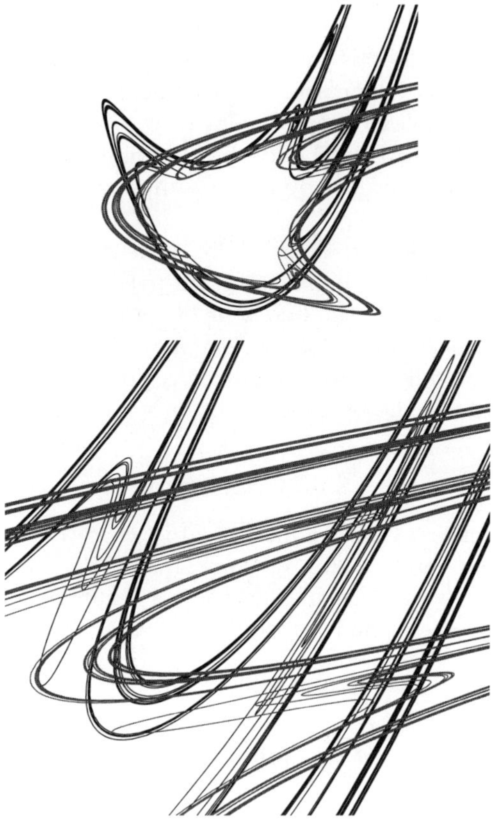

FIGURE 8.10. Asymptotic curves for the Hénon map $(x, y) \mapsto (x^2 - y, x)$.

The notion of hyperbolicity. Another of Smale's fundamental contributions to the understanding of homoclinic intersections is the introduction of the notion of *hyperbolicity*.

Given a map f of a surface S, we say that a subset K of S is *hyperbolic* if, when restricted to K, the map f preserves two directions, contracting in one of these directions and expanding in the other.[39] Typically, if f is the Smale horseshoe map,

[39]More precisely and more generally, in technical terms: Let f be a diffeomorphism of a compact manifold S. A globally f-invariant compact subset K is *hyperbolic* if we can write the tangent bundle of S over K as the sum of two continuous, supplementary subbundles $T_K S = E^s \oplus E^u$, such that the differential of f leaves each of these two subbundles invariant, uniformly contracting the norm of the vectors of E^s and uniformly expanding the norm of the vectors of E^u.

the set K of the points whose orbits never leave the square S is a hyperbolic set[40] (see [**25**]).

The behavior of a map when restricted to a hyperbolic homoclinic class can be extremely complicated (as Smale's example shows). Nonetheless, Rufus Bowen showed at the end of the 1960s (see [**5**]) that the behavior of any map when restricted to a hyperbolic homoclinic class can be characterized by a very simple object: a finite graph (or, equivalently, a finite grid in which certain squares have been marked).[41]

In a certain sense, the loop is therefore closed: Poincaré discovered a very simple geometric phenomenon (the existence of homoclinic intersections) that implies unbelievably rich and complex behavior; conversely, Bowen showed that, in the case where the homoclinic class is hyperbolic, the system's behavior, albeit unbelievably complex, can be entirely described by a very simple object.

Non-hyperbolic systems. Alas, there exist systems that have homoclinic intersections, but whose homoclinic classes are not hyperbolic. Much recent work provides partial descriptions of the behavior of certain of these systems (typically, this work shows that, for certain systems, one may calculate the proportion of time that a trajectory statistically spends in a given region), but we have almost nothing truly general to say about the behavior of these systems[42]...

8.6. Conclusion

By way of conclusion, I would like to cite two recent results that show the importance of the discoveries made by Poincaré a little over a century ago, while he sought to correct the error in his memoir: a result of Jacques Laskar on the stability of the solar system, and a result of Sylvain Crovisier, who established a connection between the complexity of the behavior of a system and the existence of homoclinic intersections.

Stability or instability of the solar system. We begin with Laskar's result. Remember that by discovering the error in his memoir, Poincaré perceived that a system consisting of a star and at least two planets that orbit around this star *has every chance of having homoclinic intersections.* For certain initial configurations, the behavior of such a system will be *unstable*: the trajectories of the planets will be

[40]Recall that this set K is nothing other than the homoclinic class of the fixed point p, that is, the set of accumulation points of the intersection points of the asymptotic curves of p.

[41]More precisely, in technical terms: if G is a finite graph, we denote by Σ_G the set of bi-infinite paths in the graph G, that is, the set of sequences $(a_i)_{i\in\mathbb{Z}}$ of vertices of G such that there exists an edge of G that goes from vertex a_i to vertex a_{i+1} for every i. We can endow Σ_G with a metric topology: the ball of radius $1/2^n$ around a path $(a_i)_{i\in\mathbb{Z}}$ is the set of paths $(b_i)_{i\in\mathbb{Z}}$ such that $b_i = a_i$ for $-n \leq i \leq n$. We then denote by σ_G the shift on Σ_G, that is, the homeomorphism of Σ_G that associates to a path $(a_i)_{i\in\mathbb{Z}}$ the same path shifted one notch to the left: $\sigma_G((a_i)_{i\in\mathbb{Z}}) = (b_i)_{i\in\mathbb{Z}}$, where $b_i = a_{i+1}$ for all i. Bowen showed that, for every diffeomorphism f of a compact manifold and every homoclinic class K of f, if K is hyperbolic, then there exists a finite graph G such that the restriction of f to K is (topologically) conjugate to the shift σ_G on Σ_G. For a proof of this result, see for example [**6**] or [**21**].

[42]We know that the dynamics of a system that has a homoclinic intersection "is at least as complicated as the Smale horseshoe diffeomorphism", and that is just about everything we can say that is perfectly general!

so complex that they will not appear to obey any rules,[43] and a tiny perturbation of the system (a comet's passage, for example) could radically alter the appearance of the planets' trajectories.

However, this does not prove that the solar system is unstable. In fact, even for a system with homoclinic intersections, not all initial configurations give rise to complicated trajectories or those that are sensitive to perturbations. In 1954, Andrei Kolmogorov, Vladimir Arnold, and Jürgen Moser proved a theorem that asserts that, for a system consisting of n planets with small masses that orbit around a star, there are many initial configurations that give rise to quasi-periodic motions.

By simplifying a little, we can summarize all of this as follows: Poincaré showed that there is a risk that the solar system is unstable, and Kolmogorov, Arnold, and Moser showed that there is a good chance that it is stable.

In 1989, Laskar was able to calculate numerically (while controlling the numeric errors) the evolution of the various planets' positions in the solar system during 200 million years. The conclusion of his work is the following [**10**]: *The motion of the inside planets of the solar system (Mercury, Venus, Earth, Mars) is sensitive to initial conditions.* More precisely, an uncertainty about the position of a planet at a given moment induces an uncertainty 10 billion times greater about the planets' positions 100 million years later. Thus, in order to know the position of a planet in the solar system after 100 million years with a precision of one million kilometers, we must know the current positions of all the planets with a precision of 10 centimeters (which is of course impossible). It is then tempting to calculate the trajectories of the planets resulting from several nearly identical initial configurations. This is what Laskar did [**11**]: he then observed the following behaviors: Jupiter, Saturn, Uranus, and Neptune never strayed far from their current trajectories; Earth, Mars, and Venus can move a great distance from their current trajectories, but stay sufficiently far from each other to not risk colliding; on the other hand, Mercury can move so far from its current trajectory to risk colliding with Venus (or, during a very close approach to Venus, to undergo an acceleration so that it is ejected far away); finally, Pluto's trajectory is even more unstable![44]

I believe that this shows that the abstract mathematical phenomena discovered by Poincaré a little more than a century ago (the existence of intersections between two curves C_T^s and C_T^u that have no physical existence) have highly concrete implications!

> *"Even though the idea of a chaotic solar system may be surprising, and may even upset our conception of the world built on several centuries of stability,"* Laskar remarks, *"it is rather the contrary that has been*

[43] Caution: here we are dealing with the system's *long term* behavior. Thus, the trajectories will appear almost periodic if we only observe them for a few hundred thousand years. It is only by observing them for a very long time that we will clearly see that their motion is not at all periodic.

[44] Here is another observation, as interesting as it is surprising, made by Laskar, Joutel, and Robutel [**12**]: If the Moon did not exist, then the inclination of the Earth's axis of rotation with respect to the ecliptic could pass from 60 to 0 degrees very quickly (which would offer the penguins free vacations in the tropics... on the condition that it did not prevent the existence of all forms of life on Earth!).

extraordinary, to the extent that since Poincaré we know that most dynamical systems are not integrable and are therefore subject to this kind of behavior."

Homoclinic intersections *versus* simple behavior. Now here is Crovisier's result:[45]

THEOREM (Crovisier). *Consider a system that is governed by differential equations and admits a closed section. Then:*

- *either the system has very simple behavior (in particular, it only has finitely many periodic trajectories, and each non-periodic trajectory is asymptotic in the past and the future to a periodic trajectory),*
- *or the system is close to a system that has homoclinic intersections.*[46]

More than a century ago, Poincaré discovered homoclinic intersections, and proved a statement of the type: *The presence of a homoclinic intersection forces the system to have very complex behavior.* Crovisier's above result constitutes a kind of converse to these statements, since it asserts that *if a system has complex behavior, then there are homoclinic intersections in the vicinity.* This is a supplementary proof, if such was necessary, of the importance of Poincaré's discovery.

Bibliography

[1] J. Barrow-Green, *Poincaré and the three-body problem*, Amer. Math. Soc. (Providence, RI), 1997.

[2] G. D. Birkhoff, *Dynamical systems with two degrees of freedom*, Trans. Amer. Math. Soc. **18** (1917), pp. 199–300.

[3] K. Bohlin, *Über die Bedeutung des Princips der lebendigen Kraft für die Frage von der Stabilität dynamischer Systeme*, Acta Math. **10** (1887), pp. 109–130.

[4] U. Bottazzini, *Henri Poincaré, philosophe et mathématicien*, Les génies de la Science, vol. **4**, Belin/Pour la Science (Paris), 2000.

[5] R. Bowen, *Markov partitions for Axiom A diffeomorphisms*, Amer. J. Math. **92** (1970), pp. 725–747.

[6] R. Bowen, *On axiom A diffeomorphisms*, CBMS Regional Conference Series in Mathematics **35**, Amer. Math. Soc., 1978.

[7] S. Crovisier, Exposition at the *International Congress on Dynamical Systems*, Angra dos Reis, August 2005.

[8] J. H. Hubbard, *The Forced Damped Pendulum: Chaos, Complication and Control*, Amer. Math. Monthly **8** (1999), pp. 741–758.

[9] J. Laskar, *La stabilité du système solaire*, Chaos et déterminisme, under the direction of A. Dahan-Dalmenico, J.-L. Chabert and K. Chemla, Seuil (Paris), 1992.

[10] J. Laskar, *A numerical experiment on the chaotic behaviour of the Solar System*, Nature **338** (1989), pp. 237–238.

[45] This result was conjectured by Jacob Palis a dozen or so years ago. It was proved in dimension 2 by Enrique Pujals and Martin Sambarino in 2000, and in dimension 3 by Christian Bonatti, Shaobo Gan, and Lan Wen in 2005, a few weeks before Crovisier announced the proof in complete generality.

[46] Here is a more precise statement (in technical terms). Let f be a C^1-diffeomorphism of a compact manifold. Then:

- either f only has finitely many periodic trajectories, and each non-periodic trajectory is asymptotic in the past and the future to a periodic trajectory,
- or there exists a sequence of diffeomorphisms that approach f in the C^1-topology and have homoclinic intersections.

[11] J. Laskar, *The chaotic motion of the Solar System. A numerical estimate of the size of chaotic zones*, Icarus **88** (1990), pp. 266–291.
[12] J. Laskar, F. Joutel and P. Robutel, *Stabilization of the Earth's obliquity by the Moon*, Nature **361** (1993), pp. 615–617.
[13] M. M. Peixoto, *Structural stability on two-dimensional manifolds*, Topology **1** (1962), pp. 101–120.
[14] M. M. Peixoto, *On classification of flows on 2-manifolds*, Proc. Symp. Dynam. Syst., Salvador (1973), pp. 389–419.
[15] H. Poincaré, *Mémoire sur les courbes définies par des équations différentielles*, Œuvres d'Henri Poincaré, vol. 1: *Équations différentielles*, Gauthier-Villars (Paris), 1951; facsimile: éditions Jacques Gabay (Sceaux), 2004.
[16] H. Poincaré, *Sur le problème des trois corps et les équations de la dynamique* (uncorrected version). Not published. The differences between this version and the published version are described in [**1**].
[17] H. Poincaré, *Sur le problème des trois corps et les équations de la dynamique*, Memoir crowned with the Prize of H. M. Oscar II of Sweden: Acta Math. **13** (1890), pp. 1–270. Œuvres d'Henri Poincaré, vol. 7 : *Masses fluides en rotation. Principes de Mécanique analytique. Problème des trois corps*, Gauthier-Villars (Paris), 1952 ; facsimile: éditions Jacques Gabay (Sceaux), 2005.
[18] H. Poincaré, *Correspondance.* Available on the website at Archives H. Poincaré, at the Université de Nancy: `http://www.univ-nancy2.fr/poincare/chp/hpcochron.html`.
[19] H. Poincaré, *Méthodes nouvelles de la mécanique céleste*, vol. 3. Reprintings: Dover Publications Inc. (New York), 1957, and Blanchard (Paris), 1987.
[20] H. Poincaré, *La Science et l'Hypothèse.* Flammarion (Paris), 1902.
[21] M. Shub, *Stabilité globale des systèmes dynamiques*, Astérisque **56**, Soc. Math, France, 1978.
[22] S. Smale, *On dynamical systems*, Boletin de la Sociedad Mathematics Mexicana (1960), pp. 195–198.
[23] S. Smale, *Chaos: Finding a Horseshoe on the Beaches of Rio.* Available at `http://math.berkeley.edu/~smale/biblio/biblio.html`.
[24] S. Smale, *A structurally stable differentiable homeomorphism with an infinite number of periodic points*, Report on the Symposium on Non Linear Oscillations, Kiev Mathematics Institute (1963), pp. 365–366.
[25] S. Smale, *Differentiable dynamical systems*, Bull. of the Amer. Math. Soc. **73** (1967), pp. 747–817.

CHAPTER 9

Variations on Poincaré's Recurrence Theorem

ÉTIENNE GHYS

9.1. Introduction

HENRI POINCARÉ is the founder of the theory of dynamical systems.[1] Confronted with the impossibility of explicitly solving the differential equations governing the trajectories of planets (as soon as their number is greater than or equal to two), he developed an ambitious program that sought to qualitatively describe the motion of celestial bodies. In his famous 1890 memoir, *"Sur le problème des trois corps et les équations de la dynamique"*[2] [**14**], he proved an extremely surprising theorem. We shall give a precise statement later on, but it is preferable to cite directly the analysis H. Poincaré made of his own work [**14**]:

> *"I was not able to solve rigorously and completely the problem of the solar system's stability, taking this phrase in a strict mathematical sense. The use of integral invariants did, however, permit me to attain some partial results, which apply especially to the so-called restricted problem where the two principal bodies travel in circular orbits, while the disturbed body has negligible mass. In this case, if one leaves aside certain exceptional trajectories, whose realization is infinitely unlikely, one may show that* **the system will return infinitely often as close as one likes to its initial position.** *This is the case that I have called Poisson stability."*

Two popular articles published in *Scientific American* and *Science & Vie* describe this theorem by an example where one *sees* an astounding recurrence [**3, 5**]. Starting with a reproduction of a photograph of H. Poincaré, they apply to it a certain transformation and iterate the process. By the third iteration, not much remains of the great man's visage, but miraculously, after 241 iterations, Henri Poincaré returns without even missing a single hair of his beard! We have reproduced the result at the end of the chapter (p. 205).

We will try to explain here why this example, even if it is striking, does not illustrate Poincaré's theorem *in any case!* This phenomenon is in fact the result of a series of small "miracles" of a number-theoretic nature which we will analyze. We hope that on the way this will offer a better understanding of the *true* theorem of Poincaré and of its limitations. In passing, we shall encounter several open questions that some readers will perhaps have the pleasure of attacking.

This chapter is a lightly modified version of an article that appeared in the first volume of the *Journal de Maths des élèves de l'ENS Lyon* (1994).

[1] Among others...

[2] Honored with the Prize of H. M. King Oscar II of Sweden. See Chapter 8.

Acknowledgements. I am happy to thank Christophe Bavard, Patrick Iglesias, Bruno Sévennec, and my daughter Élise for their precious help in the preparation of this text.

9.2. Poincaré's recurrence theorem

We will in fact content ourselves with a very special case of this theorem which will suffice for our needs. Let C be the square $[0,1] \times [0,1]$ (which will soon be a computer screen...). By gluing the opposite sides of C, we obtain a torus T. In other words, T is obtained from C by identifying, for each t of $[0,1]$, the points $(0,t)$ and $(1,t)$ as well as $(t,0)$ and $(t,1)$. We could also consider T as the quotient of the plane $\mathbb{R}^2$ by the lattice $\mathbb{Z}^2$ of integer points, that is, where one identifies the points (x,y) and (x_0,y_0) if their difference has integer coordinates. We can naturally make a metric space from T by defining the distance between two points to be the minimum of the distances between the various points of $\mathbb{R}^2$ that represent them.

Let $F : \mathbb{R}^2 \to \mathbb{R}^2$ be an area-preserving homeomorphism, that is, such that for any domain $\Omega \subset \mathbb{R}^2$, the area of Ω is equal to that of $F(\Omega)$. If F is a diffeomorphism, this reduces to saying that the determinant of the Jacobian matrix of F equals ± 1 at every point. We assume in addition that F passes in the quotient to a homeomorphism of T. In other words, F and its inverse send points of $\mathbb{R}^2$ that differ by an element of the lattice $\mathbb{Z}^2$ to points that also differ by an element of this lattice. We obtain a homeomorphism f of the torus T that is area-preserving in a more or less obvious sense. If k is a positive integer, we will denote by f^k the composition $f \circ \cdots \circ f$ of the homeomorphism f with itself k times. If a is a point of T, its *orbit* under f is the sequence $(a_k)_{k \geq 0}$ defined by $a_k = f^k(a)$. To study the dynamics of f means to describe the asymptotic behavior of these orbits as the *time* k tends to infinity. We say that a point a is *recurrent* if it is an accumulation point of its orbit.

The following theorem is one version of Poincaré's recurrence theorem[3] (1890):

THEOREM. *The set of recurrent points of an area-preserving homeomorphism of the torus is dense in the torus.*

PROOF. For n a positive integer, we will say that a point a is $1/n$-recurrent if there exists a strictly positive integer k such that the distance (in T) between a and $f^k(a)$ is strictly less than $1/n$. The set of $1/n$-recurrent points is an open set R_n in T, and the intersection of all the R_n is the set of recurrent points of T. According to Baire's theorem, it therefore suffices to show that each R_n is a dense open set. The idea of the proof is very simple. Let Ω be an open ball in T of radius less than $1/(2n)$, and consider the sequence of open sets $f^k(\Omega)$ with $k \geq 0$. All have the same area, and the finiteness of the total area of T implies that they cannot be pairwise disjoint. Therefore there exist two integers k_1 and k_2 with $0 \leq k_1 < k_2$ such that $f^{k_1}(\Omega)$ and $f^{k_2}(\Omega)$ are not disjoint. If we set $k = k_2 - k_1$, we therefore have $k > 0$ and $\Omega \cap f^k(\Omega)$ is not empty. But this means precisely that each of these balls Ω contains at least one $1/n$-recurrent point, and therefore that R_n is dense in T. This establishes the theorem. □

Some remarks are necessary.

[3]For a version and a proof close to the originals, with the historical context, see Chapter 8.

- We have used Baire's theorem, which post-dates that of Poincaré by ten years or so! We suggest that the reader examine the articles of H. Poincaré to discern therein the original statements of the theorem [**11, 12, 13, 14**].
- It would be easy to show that the set of points that are not recurrent has zero Lebesgue measure (see for example [**9**]). H. Poincaré did not have the general theory of measure, but he had the clear ideas! He writes in [**11**]:

  > *"We may say that the* [non-recurrent trajectories] *are the exception and that the* [recurrent trajectories] *are the rule in the same way that rational numbers are the exception and incommensurable numbers are the rule. I will prove, in fact, that the probability for the initial conditions of the movement to be those corresponding to an instable* [non-recurrent] *solution, that this probability, I say, is zero. This word does not by itself have any meaning: in my Mémoire I give it a precise meaning."*

- The fact that f is a homeomorphism of the torus is not at all important, of course. In general, one uses a space X which can be the phase space of a mechanical system, and a homeomorphism f of X that sends each initial position to the position one second later, for example. Classical mechanics teaches us that in many cases X is a *symplectic manifold*, but from this we will only retain that there exists a notion of volume on X and that f preserves this volume (Liouville's theorem). Poincaré's theorem generalizes to this new context as long as the crucial (but quite general) hypothesis is satisfied: the finiteness of the total volume of X. To know more about this, consult for example [**1**].

9.3. The example of *Scientific American* and *Science & Vie*

Let Φ be the linear map of $\mathbb{R}^2$ with matrix

$$\begin{pmatrix} 0 & 1 \\ 1 & 1 \end{pmatrix}.$$

This is an area-preserving diffeomorphism because the determinant of this matrix is -1. On the other hand, the fact that the coefficients of Φ and its inverse are integers shows that Φ bijectively sends the lattice of integer points onto itself. Thus, by passing to the quotient, Φ defines an area-preserving diffeomorphism of the torus T, which we will also denote Φ, without any great danger of confusion. This example is of great importance in the theory of dynamical systems, despite the great simplicity of its definition. In a certain sense, it is much more than an example.

THEOREM (Anosov 1967). *Let Ψ be a diffeomorphism of T close to Φ (in the C^1 topology*[4]*). Then there exists a homeomorphism h of the torus T such that*

$$\Psi = h \circ \Phi \circ h^{-1}.$$

One may find a proof in [**2**]. This means that the diffeomorphisms close to Φ are *the same* as Φ, once the coordinates have been changed by a homeomorphism. If we understand the structure of Φ, we therefore understand in the same stroke

[4]Two diffeomorphisms f_1 and f_2 of the torus are C^1 close if they arise from two diffeomorphisms F_1 and F_2 of $\mathbb{R}^2$ such that $F_1 - F_2$ and its first partial derivatives are small.

the structure of an open subset of the group of diffeomorphisms of T. This is the famous *structural stability* that has had so many developments these last forty years. Moreover, the reader will have probably guessed that the result is much more general than we stated here; in particular, it remains true if we replace Φ by another 2×2 matrix with integer coefficients and determinant ± 1, as long as it does not have an eigenvalue of modulus 1.

Let us apply Poincaré's recurrence theorem to Φ: almost all the points of T are recurrent. In the articles of *Scientific American* and *Science & Vie* (*cf.* p. 205), one considers a reproduction of Poincaré's face in the square $[0,1] \times [0,1]$ that serves as the basis for constructing[5] the torus T, and then one iterates this image by Φ.

After 241 iterations, the miracle occurs: Poincaré has returned! Was this predicted by the recurrence theorem?

9.4. Why is the example surprising?

It is of course the speed of Poincaré's return that is surprising. The recurrence theorem does not specify the likely length of time to return, but a simple heuristic argument allows us to estimate it. If Ω is a non-empty open subset of T, we have seen that the essential argument of the proof is that the open sets $\Phi^k(\Omega)$ cannot all be disjoint. In fact, one can only place at most $\text{area}(T)/\text{area}(\Omega)$ open subsets of T whose area equals $\text{area}(\Omega)$. One therefore has the right to expect that the return time is on the order of $\text{area}(T)/\text{area}(\Omega)$. This is in fact a theorem, under a technical hypothesis, called ergodicity, which is satisfied in our case:[6]

THEOREM (Kac 1947). *Let f be an ergodic, area-preserving homeomorphism of the torus T. Let Ω be a non-empty open subset, and for each point a of Ω, let us denote by $u(a)$ the smallest non-zero integer k such that $f^k(a)$ belongs to Ω. Then the average value of u on Ω is given by:*

$$\frac{1}{\text{area}(\Omega)} \iint_{\Omega} u(a)\, da = \frac{\text{area}(T)}{\text{area}(\Omega)}.$$

Let us apply this theorem to the image of H. Poincaré's face. Of course, we must discretize the image and replace it with a finite number of points (the pixels) whose orbits we consider. We therefore assume that we have replaced the square by a finite set of N^2 elements, namely, the points $(i/N, j/N)$, where $1 \leq i \leq N$ and $1 \leq j \leq N$. Each pixel corresponds in fact to a small square that covers a $1/N^2$ proportion of the large square $[0,1] \times [0,1]$. According to the previous theorem, we can therefore hope that a pixel will return to its place after a number of iterations on the order of N^2. If, to fix ideas, we take a value of N equal to 1000, the return time of a pixel will be about a million. This value is remarkably larger than 241!

That is not all! If each pixel comes back after about a million iterations, the return time probably varies from one pixel to another. The return time of the complete image is therefore the LCM of a million integers which are all about on the order of a million. The image should therefore take a very, very long time before coming back... To illustrate this phenomenon, we will cite several combinatorial

[5] In the example, the coordinate origin is in the center of the screen: Poincaré's nose is a fixed point.

[6] We say that a homeomorphism of the torus is ergodic if every invariant Borel set has zero Lebesgue measure or has a complement with zero Lebesgue measure. The brave reader may establish by himself that Φ is ergodic, and thence the theorem (in case ideas run out, the reader may consult [**2, 10**]).

results concerning the permutations of a finite set E_M with M elements (where we will later think that $M = N^2$). There are of course $M!$ permutations of such a set, and it is well known that each permutation σ is a product of disjoint cycles. If a is a point of E_M, the period of a under the action of σ is the length of the cycle containing a. The period of σ, that is, its order in the symmetric group, is the LCM of the lengths of the cycles that compose it. One can seek an estimate of the arithmetic averages of these quantities over the set of $M!$ permutations of E_M. Here are the results that may be found in [**4**] (in which one can also locate many references and complementary results).

THEOREM. *As M approaches infinity:*

(1) *the average value of the logarithm of the order of a permutation of a set of M elements is equivalent to $\frac{1}{2}\ln^2 M$;*
(2) *the average value of the length of the longest cycle of a permutation of a set of M elements is equivalent to $(0.62432965\ldots)M$;*
(3) *the average value of the number of cycles of a permutation of a set of M elements is equivalent to $\ln M$.*

By the time M equals a million, the preceding theorem gives such a large value for the average order of a permutation that one loses all hope of seeing H. Poincaré return!

But it gets worse! As soon as we discretize a homeomorphism f onto a finite set E_M with M elements, we obtain a map f_M from E_M to itself that has no reason to be a bijection. Two pixels a and b can be different but have images under f sufficiently close to be identified in E_M. Thus, the computer in fact iterates a not necessarily bijective map from a finite set to itself.

Let us introduce several elementary notions related to the structure of a map f from a finite set E into E. For each positive integer k, let $E(k)$ be the image $f^k(E)$. This defines a decreasing sequence of parts of E, which is therefore stationary. We denote by R the intersection of the $E(k)$; it is an f-invariant part, and the restriction of f to R is a bijection. We will say that R is the *recurrent set* of E. The complement of R is the *wandering set* of E. A point a of E is in the recurrent set or the wandering set according to whether the orbit $f^k(a)$ passes again by a or not. We also define the *recurrence degree* of f as the ratio between the size of the recurrent set and that of E. One also finds the following theorem in [**4**].

THEOREM. *As M approaches infinity:*

(1) *the average recurrence degree among the M^M maps from a set with M elements to itself is equivalent to $\sqrt{\pi/(2M)}$;*
(2) *the average number of cycles of the restriction to the recurrent set is equivalent to $\ln M$.*

In other words, a *random* map on a set with a million elements has on average a recurrent set containing scarcely more than a thousand elements partitioned into about ten cycles. The vast majority of points never return to their place as we iterate the map. Yet another reason to no longer believe in the return of the visage of H. Poincaré.

9.5. Why the example is very special

First miracle. *The finite set* E_{N^2} *formed of the* N^2 *points* $(i/N, j/N)$ *with* $1 \leq i \leq N$ *and* $1 \leq j \leq N$*, over which we discretize, is left invariant by the transformation* Φ.

This is clear because the linear map Φ preserves the lattice $\mathbb{Z}^2$ as well as all of the lattices $\frac{1}{N}\mathbb{Z}^2$. The discretization of Φ is therefore a bijection of E_{N^2}, and we have already pointed out that this is a very special property. If we conjugate Φ by just about any homeomorphism of the torus, the new homeomorphism thus obtained no longer preserves the set E_{N^2}, and the discretization is no longer a bijection.

Second miracle. *The permutation induced by* Φ *on the set* E_{N^2} *is not at all arbitrary, and its order is much smaller than the order which we might expect (about* 10^{41} *for* $N^2 = 10^6$*).*

Before justifying this assertion, let us introduce a piece of notation. If $\mathcal{A}$ is a commutative ring with unity, we denote by $\mathrm{GL}(2, \mathcal{A})$ the group of 2×2 matrices with coefficients in $\mathcal{A}$ and invertible determinant. This group acts naturally on $\mathcal{A} \times \mathcal{A}$ by "linear" maps. We can in particular consider $\mathrm{GL}(2, \mathbb{Z})$ and $\mathrm{GL}(2, \mathbb{Z}/N\mathbb{Z})$. Reducing modulo N gives a homeomorphism

$$\rho_N : \mathrm{GL}(2, \mathbb{Z}) \to \mathrm{GL}(2, \mathbb{Z}/N\mathbb{Z}).$$

The set of discretization E_{N^2} can of course be identified with $\mathbb{Z}/N\mathbb{Z} \times \mathbb{Z}/N\mathbb{Z}$ and the action of Φ on this set is simply that of the element $\rho_N(\Phi)$ on $\mathbb{Z}/N\mathbb{Z} \times \mathbb{Z}/N\mathbb{Z}$ (we identify the diffeomorphism Φ and the 2×2 matrix that defines it). The order of the restriction of Φ to E_{N^2} is therefore the order of the element $\rho_N(\Phi)$ in the group $\mathrm{GL}(2, \mathbb{Z}/N\mathbb{Z})$. Thus we obtain an estimate, rough but effective, of this order: it is bounded above by the cardinality of this finite group, which is obviously smaller than N^4. We still do not understand why 241 is so small, but we now have an upper bound on the order of 10^{12}, which is more "reasonable" than 10^{41}.

Third miracle. *The order of the elements of* $\mathrm{GL}(2, \mathbb{Z}/N\mathbb{Z})$ *is in fact much smaller than the preceding bound* N^4.

To simplify, we will work in fact in the subgroup $\mathrm{SL}(2, \mathbb{Z}/N\mathbb{Z})$ of matrices of determinant 1. We remark that the matrix Φ is not in this subgroup, but its square of course has determinant 1. The following theorem is due to Dyson and Falk [**6**].

THEOREM. *The order of an element of* $\mathrm{SL}(2, \mathbb{Z}/N\mathbb{Z})$ *is less than or equal to* $3N$.

The proof that follows is perhaps a bit technical, but the nervous reader may simply ignore it with no harm for understanding what follows.

PROOF. Let A be an element of the group $\mathrm{SL}(2, \mathbb{Z}/N\mathbb{Z})$. We will distinguish several cases.

First case. Let us first assume that N is a prime number (which we will of course denote p). The two eigenvalues of A are therefore either in $\mathbb{Z}/p\mathbb{Z} = \mathbb{F}_p$ or in a field[7] $\mathbb{F}_{p^2}$ which is a quadratic extension of $\mathbb{F}_p$.

---

[7]In case of problems, see [**15**].

(1) Suppose that A is diagonalizable over the finite field $\mathbb{F}_p$ with p elements. Because the two eigenvalues of A are inverses, the order of the element A in $\mathrm{SL}(2, \mathbb{F}_p)$ is therefore equal to the order of one of the eigenvalues in the multiplicative group of non-zero elements of $\mathbb{F}_p$. This is a divisor of $p-1$ and therefore less than $3p$.

(2) Now suppose that the eigenvalues of A are not in $\mathbb{F}_p$ but in $\mathbb{F}_{p^2}$. The two eigenvalues λ_1 and λ_2 of A are then exchanged by the Frobenius automorphism:

$$x \in \mathbb{F}_{p^2} \mapsto x^p \in \mathbb{F}_{p^2}.$$

We therefore have $\lambda_2 = \lambda_1^p$ and $\lambda_1^{(p+1)} = 1$, because the determinant of A equals 1. It follows that the order of A is a divisor of $p+1$, therefore in particular less than $3p$.

(3) If the matrix A is not diagonalizable in either $\mathbb{F}_p$ or $\mathbb{F}_{p^2}$, then it is the case that the two eigenvalues are equal and therefore have one of the values ± 1.
   - If the matrix A is unipotent—that is, if its two eigenvalues are equal to 1—then it is conjugate to a matrix of the form:

$$\begin{pmatrix} 1 & v \\ 0 & 1 \end{pmatrix}$$

     with v in $\mathbb{F}_p$. The order of such a matrix is obviously 1 or p and therefore less than $3p$.
   - If the two eigenvalues of A equal -1, then $-A$ is unipotent; the order of A is therefore a divisor of $2p$ and hence less than $3p$.

Second case. Next assume that N is a power p^n of a prime number; we will prove the theorem by induction on n. The case $n = 1$ has just been established above, and we therefore assume the theorem to be proved up to n. By reducing modulo p^n, we have a homomorphism

$$\rho : \mathrm{SL}(2, \mathbb{Z}/p^{n+1}\mathbb{Z}) \to \mathrm{SL}(2, \mathbb{Z}/p^n\mathbb{Z}).$$

We know by induction that the order of an element of $\mathrm{SL}(2, \mathbb{Z}/p^n\mathbb{Z})$ is less than $3p^n$, and it suffices to show that the order of an element of the kernel of ρ is 1 or p. But this is elementary because this kernel consists of matrices of the form $\mathrm{Id} + p^n B$ so that their pth power is the identity in $\mathrm{SL}(2, \mathbb{Z}/p^{n+1}\mathbb{Z})$.

Third case. It remains to show that if the theorem is true for the relatively prime integers $p_1^{n_1}, \ldots, p_\ell^{n_\ell}$, then it is equally true for the product $N = p_1^{n_1} \cdots p_\ell^{n_\ell}$. By considering the reductions modulo $p_1^{n_1}, \ldots, p_\ell^{n_\ell}$, we define a homomorphism of $\mathrm{SL}(2, \mathbb{Z}/N\mathbb{Z})$ onto the product of the $\mathrm{SL}(2, \mathbb{Z}/p_i^{n_i}\mathbb{Z})$, and the *Chinese remainder theorem* guarantees that it is an isomorphism. The order of an element of $\mathrm{SL}(2, \mathbb{Z}/N\mathbb{Z})$ is therefore the LCM of the orders of the various components of its image in the $\mathrm{SL}(2, \mathbb{Z}/p_i^{n_i}\mathbb{Z})$. The second case in fact showed us that the order of an element of $\mathrm{SL}(2, \mathbb{Z}/p^n\mathbb{Z})$ has one of the following forms:

(1) αp^β where α divides $p-1$ and $0 \leq \beta \leq n-1$;
(2) αp^β where α divides $p+1$ and $0 \leq \beta \leq n-1$;
(3) p^β where $\beta \leq n$;
(4) $2p^\beta$ where $\beta \leq n$.

If p is an odd number, $p+1$ is even! If $p \neq 2$, this order is therefore less than or equal to p^n or is twice an integer less than or equal to p^n. It follows easily that if N

is odd—that is, if none of the p_i are 2—the LCM of the various orders modulo $p_i^{n_i}$ is less than or equal to $2N$. If $N = 2^n$, the order is less than or equal to $3 \cdot 2^{n-1}$, that is $3N/2$.

In the general case, N is the product of a power of 2 and an odd number so that the order modulo N is bounded above by $((3/2) \times 2)N = 3N$, and the theorem is proved. □

Let us note that the bound we obtained is optimal. We can find, for example, an element of $\mathrm{SL}(2, \mathbb{Z}/10\mathbb{Z})$ that has order 3 modulo 2 and order $10 = 2 \times 5$ modulo 5 and therefore has order $30 = 3 \times 10$ modulo 10.

REMARK. Fix a matrix A in $\mathrm{SL}(2, \mathbb{Z})$ and a prime number p. It is not difficult to convince oneself that there exist two integers α and β such that, for any *sufficiently large* n, the order of the projection of A in $\mathrm{SL}(2, \mathbb{Z}/p^n\mathbb{Z})$ is exactly $\alpha \cdot p^{n-\beta}$. Thus, as n tends to infinity, the order of A modulo p^n is comparable to p^n. For "most" integers N (not necessarily powers of a prime number), we can therefore think that the order of A modulo N has the same order of magnitude as N. We leave to the reader the care of giving meaning to this assertion and then proving it if it turns out to be exact! In any case, if $N = 1000$, we now have an upper bound of 3000 for the order of an element of $\mathrm{SL}(2, \mathbb{Z}/N\mathbb{Z})$, and we have seen that in many cases, this bound is still too large, so that our 241 finally becomes reasonable!

Fourth miracle. *There exists a sequence of integers $(\varphi_k)_{k \geq 0}$ tending to infinity (exponentially fast) as k tends to infinity, such that if we discretize the screen into $\varphi_k \times \varphi_k$ points, then H. Poincaré's return occurs after* **exactly** *$2k$ iterations...*

Of course, to choose the discretization as a function of the desired return time is cheating a bit. But this shows even better the illusory character of these recurrences. The quality of the discretizations is excellent because we will see that we have, for example, for a return after 30, 34, 40, 106, 238, 240, 242 or 246 iterations:

$$\begin{aligned}
\varphi_{15} &= 1364,\\
\varphi_{17} &= 3571,\\
\varphi_{20} &= 15127,\\
\varphi_{53} &= 119218851371,\\
\varphi_{119} &= 7405070366464951264563599,\\
\varphi_{120} &= 5358359254990966640871840,\\
\varphi_{121} &= 19386725908489881939795601,\\
\varphi_{123} &= 50755107359004694554823204.
\end{aligned}$$

Thus, by using a (good) image consisting of 3571×3571 pixels, Poincaré returns after just 34 iterations! But who would have the idea of using exactly 3571 pixels![8]

On the other hand, if we choose an integer n not of the form φ_k, we know that the return will certainly be quick (in fewer than $3n$ steps), but perhaps not quick enough.

[8]Exercise: When would Poincaré return with a 3570×3570 image? with 3572×3572?

To justify these assertions, we introduce two Fibonacci sequences a_k and b_k, which are defined by[9]

$$a_0 = 0, \qquad a_1 = 1, \qquad a_{k+2} = a_{k+1} + a_k,$$
$$b_0 = 2, \qquad b_1 = 1, \qquad b_{k+2} = b_{k+1} + b_k$$

for $k \geq 0$, and we define the sequence φ_k by:

$$\varphi_{2k} = a_{2k}, \qquad \varphi_{2k+1} = b_{2k+1}.$$

Of course, ϕ_k tends to infinity exponentially fast as k tends to infinity.

THEOREM. *The* $2k$*th power of the matrix* Φ *is congruent to the identity modulo* φ_k.

PROOF. According to the Hamilton–Cayley theorem (in the case of 2×2 matrices!), we have:

$$\Phi^2 = \Phi + \mathrm{Id}.$$

It follows that:

$$\Phi^{-2} = -\Phi^{-1} + \mathrm{Id}.$$

Consequently, for every integer k, we have:

$$\Phi^{k+2} = \Phi^{k+1} + \Phi^k,$$
$$\Phi^{-(k+2)} = -\Phi^{-(k+1)} + \Phi^{-k}.$$

For each k, we set:

$$A_k = \Phi^k - (-1)^k \Phi^{-k}, \qquad B_k = \Phi^k + (-1)^k \Phi^{-k}.$$

We therefore have, for every integer k:

$$A_{k+2} = A_{k+1} + A_k, \qquad B_{k+2} = B_{k+1} + B_k.$$

Moreover, we easily calculate:

$$A_0 = 0, \qquad A_1 = \Phi + \Phi^{-1},$$
$$B_0 = 2 \cdot \mathrm{Id}, \qquad B_1 = \Phi - \Phi^{-1} = \mathrm{Id}.$$

We therefore obtain by recursion the following expression for A_k and B_k:

$$A_k = a_k \cdot (\Phi + \Phi^{-1}), \qquad B_k = b_k \cdot \mathrm{Id}.$$

Finally

$$\Phi^{4k} - \mathrm{Id} = (\Phi^{2k} - \Phi^{-2k})\Phi^{2k} = a_{2k} \cdot (\Phi + \Phi^{-1})\Phi^{2k},$$
$$\Phi^{4k+2} - \mathrm{Id} = (\Phi^{2k+1} - \Phi^{-2k-1})\Phi^{2k+1} = b_{2k+1} \cdot \Phi^{2k+1}.$$

We have thus shown that $\Phi^{2k} - \mathrm{Id}$ is divisible by φ_k. □

REMARK. The odd powers of Φ do not present this phenomenon of high recurrence that we encountered for the even powers. The matrix Φ, having determinant -1, reverses orientation.[10] It is therefore necessary to raise Φ to an even power to obtain an oriented return! More precisely, if the integer matrix Φ^{2k+1} equals Id modulo some integer l, then by comparing determinants we find that l equals ± 1 or ± 2. It is for this reason that we think that the recurrence presented in *Scientific*

[9] TRANSLATOR'S NOTE: The numbers b_k are often also called *Lucas numbers*.

[10] In the initial photograph, one can see the left ear of the Master, although in its transformation by Φ it is the right ear that one distinguishes, even if it is deformed.

American and *Science & Vie* does not occur after 241 iterations, and that there was probably an error of a unit in counting the iterations (the first figure being the order 0 iteration). *The recurrence surely occurs at the* 240*th iteration!*

It remains to understand the value 240 even if it is now clear that this value is necessarily artificial. One possible explanation comes from the following observation. Once we decompose the integers φ_k into prime factors (thanks to Maple), we remark that the prime numbers that appear are almost all extremely large, with the notable exception of φ_{120}. Taking a few examples chosen above, we have for example:

$$\begin{aligned}
\varphi_{53} &= 119218851371,\\
\varphi_{119} &= 29 \times 239 \times 10711 \times 3571 \times 27932732439809,\\
\varphi_{120} &= 2^5 \times 3^2 \times 5 \times 7 \times 11 \times 23 \times 31 \times 41 \times 61 \times 241 \times 2161 \times 2521 \times 20641,\\
\varphi_{121} &= 199 \times 97420733208491869044199,\\
\varphi_{123} &= 2^2 \times 4767481 \times 370248451 \times 7188487771.
\end{aligned}$$

Computer manufacturers have a natural tendency to choose the size of their screens among integers that are products of small integers such as 2, 3, or 5, and it seems that in the end it is this property of φ_{120} that lies at the origin of the return after 240 iterations... Moreover, this is no surprise. In fact, the polynomial $X^{2k}-1$ is the product of cyclotomic polynomials indexed by the divisors of $2k$ so that if $2k$ has many divisors (like 120), then $A^{2k} - \mathrm{Id}$ equally has many divisors.

To summarize, the return after 241 *iterations happens in fact after* 240 *iterations, and the reason for this return has little to do with the recurrence theorem but rather with the fact that* 120 *is an integer with lots of divisors!*

We can even try to guess the number n of pixels used in the articles of *Scientific American* and *Science & Vie*. The idea is to find a divisor of φ_{120} that is probably divisible only by 2, 3, and 5. Would it be 160, 180, or 1440?

Exercise 1. If we carefully examine the reproduction included at the end of this chapter, we see a bizarre phenomenon: after 48 iterations five copies of H. Poincaré's face appear, shifted with respect to each other. We leave to the reader the pleasure of finding an explanation. We will content ourselves with remarking that

$$240 = 5 \times 48,$$
$$\varphi_{24} = 46368 = 2^5 \times 3^2 \times 7 \times 23$$

(thus 5 divides φ_{120} but not φ_{24}). The discriminant of the polynomial $X^2 - X - 1$ is 5, so that the reduction of Φ modulo 5 has a double eigenvalue (and 5 is the only prime number having this property).

Exercise 2. How could we fabricate an image of Poincaré such that after 241 iterations the face of Newton would appear!?

9.6. A question

The space of all area-preserving homeomorphisms of the torus can be endowed (for example) with the topology of uniform convergence. We say that a property of a homeomorphism is generic if the set of homeomorphisms with this property is a residual set in the sense of Baire (i.e., a countable intersection of dense open sets).

The problem we would like to pose to the reader is to find out to what extent a computer is able to take into account the generic dynamics of a homeomorphism. Let us make the question more precise. If N is a positive integer, we again denote by E_{N^2} the set of points of the torus with coordinates $(i/N, j/N)$, where $1 \leq i \leq N$ and $1 \leq j \leq N$. Let f be an area-preserving homeomorphism of the torus. Let $f_N : E_{N^2} \to E_{N^2}$ be the order N discretization, that is, the map that sends a point of E_{N^2} to the point in E_{N^2} that is the closest to its image under f (generically unique). Let $r_N(f)$ be the recurrence degree of f_N.

Problem. *Is it possible to describe the asymptotic behavior of the sequence $r_N(f)$ as N tends to infinity for a generic area-preserving homeomorphism of the torus?*

Of course, the discretization of a generic homeomorphism is not an arbitrary map of a finite set with N^2 elements into itself: the continuity of f forces a kind of weak continuity for f_N on the finite set E_{N^2}. It is therefore not clear that the estimate of $\sqrt{\pi/2}/N$ for the recurrence degree that we described above still holds for $r_N(f)$. If this estimate (or another one, not quite as good) held for a generic homeomorphism all the same, one could think about the use of computing in the theory of dynamical systems...

One may find an interesting discussion of this type of question for *non-invertible* dynamical systems (as well as a complementary bibliography) in [**7**]. The length of the cycles of discretizations of generic diffeomorphisms of the circle is discussed in [**8**].

9.7. Mixing

To end this article, we would like to cite a result that carries the *coup de grâce* to the hope of detecting a recurrence in the figures (see [**10**] for more information). Even though almost all the points are recurrent, the figures have on the other hand a tendency to mix:

THEOREM. *Let Ω_1 and Ω_2 be two Borel sets of the torus T. Then the area of the intersection $\Phi^k(\Omega_1) \cap \Omega_2$ tends to the product of the areas of Ω_1 and Ω_2 as the integer k tends to infinity.*

PROOF. Let u_1 and u_2 be the indicator functions of Ω_1 and Ω_2 respectively. The idea is to show that

$$\lim_{k\to\infty} \iint_T u_1 \circ f^{-k}(x,y) \cdot u_2(x,y)\,dx\,dy = \left(\iint_T u_1(x,y)\,dx\,dy\right)\left(\iint_T u_2(x,y)\,dx\,dy\right).$$

A fortiori, it suffices to prove this for all pairs (u_1, u_2) of square-integrable functions on T. Expanding u_1 and u_2 into Fourier series reduces to the case where u_1 and u_2 are of the form

$$u_1(x,y) = \exp 2i\pi(m_1 x + n_1 y), \qquad u_2(x,y) = \exp 2i\pi(m_2 x + n_2 y),$$

where x and y denote elements of $\mathbb{R}/\mathbb{Z}$ and m_1, m_2, n_1, n_2 are integers. Once we evaluate $u_1 \circ \Phi^{-k}$, we find

$$u_1 \circ \Phi^{-k}(x,y) = \exp 2i\pi(m(k)x + n(k)y)$$

with

$$\begin{pmatrix} m(k) \\ n(k) \end{pmatrix} = {}^t\Phi^{-k} \begin{pmatrix} m_1 \\ n_1 \end{pmatrix}.$$

If (m_1, n_1) or (m_2, n_2) equals $(0, 0)$, then the product of the integrals of $u_1 \circ \Phi^{-k}$ and u_2 is of course constant, and the convergence of this product is not too difficult to show...

If (m_1, n_1) and (m_2, n_2) are different from $(0, 0)$, one can easily check that, for sufficiently large k, $(m(k), n(k)) \neq (m_2, n_2)$ so that $u_1 \circ \Phi^{-k}$ and u_2 are orthogonal for the Hermitian product in the Hilbert space of square-integrable functions on T, and, of course, each of these two functions has zero integral. Here again, convergence is easy to show... This establishes the theorem. □

The theorem means that if we iterate a small domain Ω_1 by Φ, it will scatter throughout the torus in a uniform way: asymptotically, the proportion of $\Phi^k(\Omega_1)$ in Ω_2 is the same as that of Ω_1 in T. Everything blurs and fades...

Avec le temps...
Avec le temps, va, tout s'en va
On oublie le visage et l'on oublie la voix
Le cœur quand ça bat plus, c'est pas la peine d'aller
Chercher plus loin, faut laisser faire et c'est très bien
Avec le temps...
Avec le temps, va, tout s'en va
L'autre qu'on adorait, qu'on cherchait sous la pluie
L'autre qu'on devinait au détour d'un regard
Entre les mots, entre les lignes et sous le fard
D'un serment maquillé qui s'en va faire sa nuit
Avec le temps tout s'évanouit.

[TRANSLATION:

With time...
With time, you know, everything goes away
You forget the face and you forget the voice
When the heart no longer beats, it's not worth going
To look farther, you must let it be and it's all right
With time...
With time, you know, everything goes away
The one that you adored, that you sought in the rain
The one that you could discern with only a glance
Between the words, between the lines and under the make-up
Of a disguised oath that is leaving to fall asleep
With time everything fades away.]

—LÉO FERRÉ

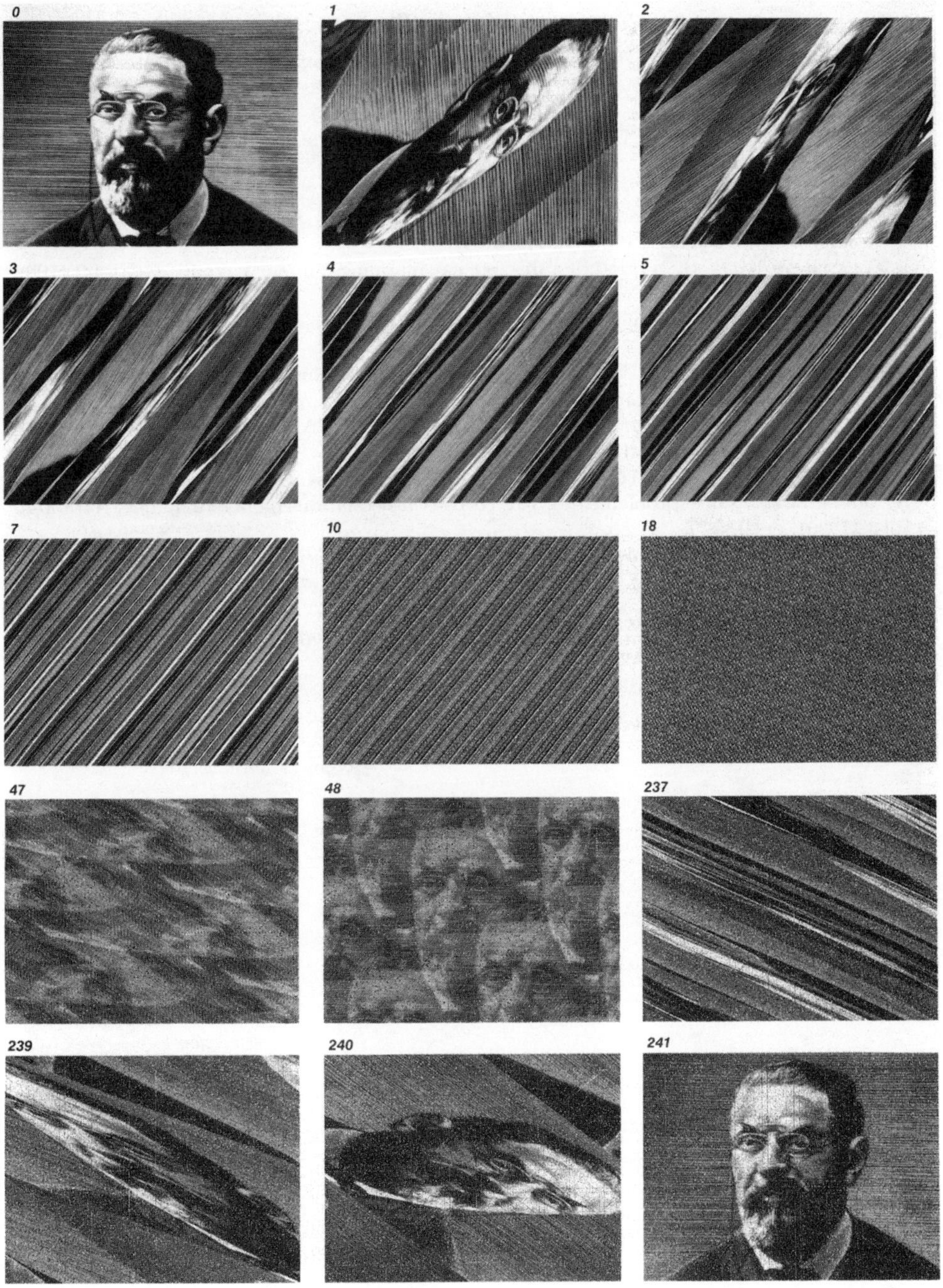

FIGURE 9.1. Image reproduced in the journal *Scientific American* in December 1986 [**5**].

Bibliography

[1] ARNOLD, V.: *Méthodes mathématiques de la mécanique classique.* Mir, Moscow, 1976.

[2] ARNOLD, V.: *Chapitres supplémentaires de la théorie des équations différentielles ordinaires.* Mir, Moscow, 1980.

[3] BERGER, M.: Maths 89: l'école française 3ème du monde. *Science & Vie*, March 1989, pp. 274–283.

[4] BOLLOBÁS, B.: *Random Graphs.* Academic Press, London, 1985.

[5] CRUTCHFIELD, J. P., FARMER, J. D., PACKARD, N. H., & SHAW, R. S.: Chaos. *Scientific American* 255, December 1986, pp. 46–57.

[6] DYSON F. & FALK, H.: Period of a discrete cat mapping. *Amer. Math. Monthly* **99** (1992), no. 7, pp. 603–614.

[7] LANFORD, O.: Informal remarks on the orbit structure of discrete approximations to chaotic maps. *Experiment. Math.* **7** (1998), no. 4, pp. 317–324.

[8] MIERNOWSKI, T.: Discrétisations des homéomorphismes du cercle, *Ergod. Th. & Dynam. Sys.* **26** (2006), no. 6, pp. 1867–1903.

[9] OXTOBY, J.-C.: *Measure and Category.* Graduate Texts in Mathematics 2, Springer Verlag, New York, Heidelberg, Berlin, 1971.

[10] PETERSEN, K.: *Ergodic theory.* Cambridge University Press, Cambridge, 1983.

[11] POINCARÉ, H.: Sur le problème des trois corps et les équations de la dynamique. *Acta Math.* **13** (1890), pp. 1–270.

[12] POINCARÉ, H.: Sur le problème des trois corps. *Bulletin Astronomique* **8** (1891), pp. 12–24.

[13] POINCARÉ, H.: *Les méthodes nouvelles de la mécanique céleste,* vol. 3, Gauthier-Villars, Paris, 1899.

[14] POINCARÉ, H.: Analyse des travaux scientifiques, *Œuvres* of Henri Poincaré, vol. 7, Gauthier-Villars, Paris, 1952.

[15] SERRE, J.-P.: *Cours d'arithmétique.* PUF, Paris, 1970.

CHAPTER 10

Low-Dimensional Chaos and Asymptotic Time Behavior in the Mechanics of Fluids

GUIDO BOFFETTA, GUGLIELMO LACORATA, AND ANGELO VULPIANI

10.1. Introduction

The contribution of H. Poincaré in Celestial Mechanics has an extremely rare, possibly even unique, character in Physics and Mathematics. While, usually, scientific books become obsolete in a few years (or several decades), the *Méthodes Nouvelles de la Mécanique Céleste* [**36**] more than a century later still remains a source of inspiration not only for beginners but also for confirmed scientists in many branches of the Theory of Dynamical Systems. Many of the methods, concepts and ideas introduced there by Poincaré are still useful tools even in the fields apparently distant from Celestial Mechanics.

We discuss here Chaotic Advection in laminar incompressible flows and long-time diffusive behavior. The basic mechanism for the chaotic behavior generated by homoclinic intersections[1] in quasi-integrable Hamiltonian systems, as well as the multiscale expansion, turn out to be still the main ingredients of this issue. The multiscale approach for partial differential equations, which will be explained below, can be seen as a spatio-temporal extension of the methods introduced by Lindstedt, and improved by Poincaré, for the treatment of the secular terms in Celestial Mechanics.

10.2. Lagrangian chaos

A problem of great interest concerns the study of the spatial and temporal structure of the so-called passive fields, indicating by this term quantities passively driven by the flow (i.e. without perturbing the dynamics of the fluid), such as the temperature under certain conditions [**31**]. The equation for the evolution of a passive scalar field $\theta(\mathbf{x}, t)$, advected (i.e. drifted) by a given velocity field $\mathbf{v}(\mathbf{x}, t)$, is

$$\partial_t \theta + \nabla \cdot (\mathbf{v}\, \theta) = D\, \nabla^2 \theta, \tag{10.1}$$

where D is the molecular diffusion coefficient.

The problem (10.1) can be studied through two different approaches. Either one deals, at any time t and any point $\mathbf{x}$ of the space domain covered by the fluid, with the field $\theta(\mathbf{x}, t)$, or one deals with the trajectory of each fluid particle. The two approaches are usually designated as "Eulerian" and "Lagrangian", although both of them are due to Euler [**23**]. The two points of view are in principle equivalent.

[1] See Chapter 8.

The motion of a fluid particle is determined by the differential equation

$$\frac{d\mathbf{x}}{dt} = \mathbf{v}(\mathbf{x}, t), \tag{10.2}$$

which also describes the motion of test particles, for example a powder embedded in the fluid, provided that the particles are neutral and small enough not to perturb the velocity field, although large enough not to perform a Brownian motion. Particles of this type are commonly used for flow visualization in fluid mechanics experiments; see [**40**]. Let us remark that the precise equation for the motion of a material particle in a fluid can be rather complicated [**30**].

It is now well established that even in a regular (i.e. laminar) velocity field the motion of fluid particles may be very irregular (i.e. turbulent) [**19, 2**]. In this case, initially nearby trajectories diverge exponentially and one speaks of *Lagrangian chaos*. In general, a chaotic behavior can arise in two-dimensional flows only for time dependent velocity fields, while it can be present even for stationary velocity fields in three dimensions [**1, 15**]. Let us observe that for the dynamical system (10.2) the phase space coincides with the physical space.

In the following we will consider only incompressible flows,

$$\nabla \cdot \mathbf{v} = 0, \tag{10.3}$$

for which the dynamical system (10.2) is conservative (the volume is preserved under the time evolution). If $D = 0$, it is easy to realize that (10.1) is equivalent to (10.2). Indeed, we can write

$$\theta(\mathbf{x}, t) = \theta_o(T^{-t}\mathbf{x}), \tag{10.4}$$

where $\theta_o(\mathbf{x}) = \theta(\mathbf{x}, t = 0)$ and T is the formal evolution operator of (10.2),

$$\mathbf{x}(t) = T^t\mathbf{x}(0), \tag{10.5}$$

hence $\theta(\mathbf{x}(t), t) = \theta_o(\mathbf{x}(0))$ and, if $d\mathbf{x}/dt = \mathbf{v}(\mathbf{x}, t)$ (condition (10.2)),

$$\partial_t\theta + (\mathbf{v} \cdot \nabla)\theta = \frac{d}{dt}\,\theta = 0.$$

But $\nabla \cdot (\mathbf{v}\,\theta) = (\nabla \cdot \mathbf{v})\theta + (\mathbf{v} \cdot \nabla)\theta$ and $\nabla \cdot \mathbf{v} = 0$ give $(\mathbf{v} \cdot \nabla)\theta = \nabla \cdot (\mathbf{v}\,\theta)$, thus (10.2) implies (10.1) (and conversely).

Taking into account the molecular diffusion D, (10.1) is nothing but the Fokker–Planck equation associated to the Langevin equation [**13**]

$$\frac{d\mathbf{x}}{dt} = \mathbf{v}(\mathbf{x}, t) + \eta(t), \tag{10.6}$$

where η is a three-dimensional Gaussian process with zero mean and variance

$$\langle \eta_i(t)\, \eta_j(t') \rangle = 2D\delta_{ij}\, \delta(t - t'). \tag{10.7}$$

In two dimensions, the constraint (10.3) is formally satisfied assuming

$$v_1 = \frac{\partial \psi}{\partial x_2}, \qquad v_2 = -\frac{\partial \psi}{\partial x_1}. \tag{10.8}$$

The function $\psi(\mathbf{x}, t)$, which is assumed to be sufficiently smooth, is called the *stream function*. Inserting (10.8) into (10.2), the evolution equations become

$$\frac{dx_1}{dt} = \frac{\partial \psi}{\partial x_2}, \qquad \frac{dx_2}{dt} = -\frac{\partial \psi}{\partial x_1}. \tag{10.9}$$

Formally (10.9) is a Hamiltonian system with the Hamiltonian given by the stream function ψ, and the canonical variables are the spatial coordinates x_1 and x_2: $(q, p) = (x_1, x_2)$.

10.2.1. Examples of Lagrangian chaos. As a first example we consider a $3D$ stationary velocity field, the so-called ABC flow (Arnold, Beltrami, Childress):

$$\mathbf{v} = (A \sin z + C \cos y,\ B \sin x + A \cos z,\ C \sin y + B \cos x)\,, \tag{10.10}$$

where A, B and C are non-zero real parameters. Because of the incompressibility condition (10.3), the evolution $\mathbf{x}(0) \to \mathbf{x}(t)$ defines a conservative dynamics (i.e. phase space volumes are preserved).

Arnold [**4, 5**] argued that (10.10) is a good candidate for chaotic motion; namely the trajectories of the dynamical system (10.2) may exhibit sensitivity to initial conditions. Let us briefly repeat his elegant argument. For a steady state solution of the $3D$ Euler equation one has:

$$\begin{aligned} \nabla \cdot \mathbf{v} &= 0, \\ \mathbf{v} \times (\nabla \times \mathbf{v}) &= \nabla \alpha, \\ \alpha &= \frac{P}{\rho} + \frac{\mathbf{v}^2}{2}, \end{aligned} \tag{10.11}$$

where P is the pressure and ρ the density. As a consequence of the Bernoulli theorem [**24**], $\alpha(\mathbf{x})$ is constant along a streamline—that is, a Lagrangian trajectory $\mathbf{x}(t)$. One can easily verify that chaotic motion can appear only if $\alpha(\mathbf{x})$ is constant (i.e. $\nabla \alpha(\mathbf{x}) = 0$) in a volume of space; otherwise the trajectory would be confined on a $2D$ surface $\alpha(\mathbf{x}) = \text{const}$, where the motion must be regular as a consequence of general arguments [**35**]. In order to satisfy such a constraint, from (10.11) one has the Beltrami condition:

$$\nabla \times \mathbf{v} = \gamma(\mathbf{x})\, \mathbf{v}. \tag{10.12}$$

Thus chaotic motion can appear if we have (10.12) with $\gamma(\mathbf{x}) = \text{const}$ in a volume of space: in this case, nothing more, it seems, compels the trajectory to stay on a surface. Arnold [**4**] concludes: "It is probable that [such] flows have trajectories with complicated topology. Such complications occur in celestial mechanics." (As we see, the work of Poincaré on celestial mechanics acts as a guide here.) It is easy to verify that the field $\mathbf{v}$ given by (10.10) satisfies (10.12) with $\gamma(\mathbf{x}) = \text{const}$ (this example is given by Arnold himself in [**4**]), and, indeed, numerical experiments by Hénon [**19**] provided evidence that Lagrangian motion under velocity (10.10) is chaotic for typical values of the parameters A, B, and C (see an example in FIG 10.1).

Hénon wrote [**19**]: "In conclusion, one is led to this curious result that the particular case [$\alpha = \text{const}$, $\gamma = \text{const}$] seems to have a topology much more complicated than the general case."

In a two-dimensional incompressible stationary flow the motion of fluid particles is given by a time independent one-dimensional Hamiltonian system (i.e. with two degrees of freedom) and, since trajectories follow iso-ψ lines, it is impossible to have chaos. However, for explicit time dependent stream function ψ, the system (10.9) can exhibit chaotic motion [**35**]. Indeed, generically in one-dimensional Hamiltonian systems, a periodic perturbation gives origin to a so-called "stochastic layer" around the separatrices (i.e. the orbits of infinite period connecting hyperbolic fixed points)

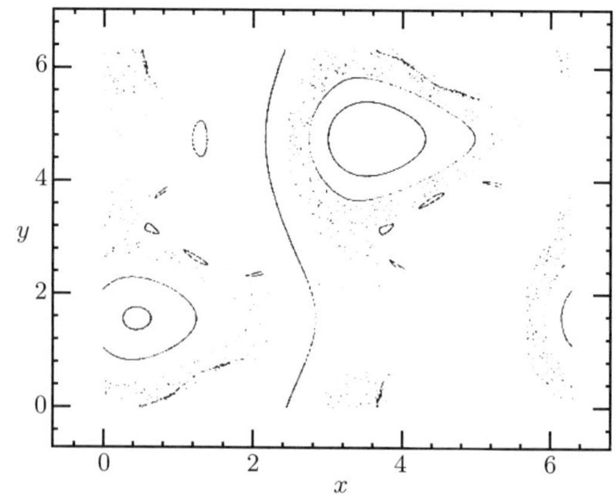

FIGURE 10.1. Intersections with the Poincaré section, plane $z = 0$, of eight trajectories of the ABC flow with parameters $A = 2.0$, $B = 1.70$, $C = 1.50$.

where the motion is chaotic, as a consequence of folding and crossing of the stable and unstable manifolds in domains centered at the hyperbolic fixed points: this is the celebrated mechanism of the homoclinic intersection, formerly discovered by Poincaré in his work on Celestial Mechanics.[2]

In the particular case of time periodic velocity fields, $\mathbf{v}(\mathbf{x}, t + T) = \mathbf{v}(\mathbf{x}, t)$, the trajectory of (10.2) can be studied in terms of discrete dynamical systems: the position $\mathbf{x}(t+T)$ is determined in terms of $\mathbf{x}(t)$, the *Poincaré map* $\mathbf{x}(t) \to \mathbf{x}(t+T)$ will not depend explicitly on t, and thus (10.2) can be written in the form

$$\mathbf{x}(n + 1) = \mathbf{F}[\mathbf{x}(n)], \tag{10.13}$$

where now the time is measured in units of the period T. Because of incompressibility, the map (10.13) is conservative and the Lebesgue measure $d\mathbf{x}$ is invariant:

$$\big|\det A[\mathbf{x}]\big| = 1, \qquad \text{where} \qquad A_{ij}[\mathbf{x}] = \frac{\partial F_i[\mathbf{x}]}{\partial x_j}. \tag{10.14}$$

The explicit form of $\mathbf{F}$ for a general $2D$ or $3D$ flow is usually very difficult to find. However, in some simple cases, this can be deduced on the basis of physical features [**3, 12**].

10.3. Eulerian properties and Lagrangian chaos

In principle, the evolution of the velocity field $\mathbf{v}$ is described by partial differential equations, e.g. Navier–Stokes or Boussinesq equations. However, often in weakly turbulent situations, a good approximation of the flow can be obtained by using a Galerkin approach, i.e. writing the velocity field in terms of suitable functions (usually a Fourier series), and reducing the Eulerian problem, which is, in principle, described by a partial differential equation, to a (small) system of a number F of ordinary differential equations [**9, 27**]. (We shall present an example

[2]See Chapter 8.

below, p. 212.) The motion of a fluid particle is then described by the $(d+F)$-dimensional dynamical system

$$\frac{d\mathbf{Q}}{dt} = \mathbf{f}(\mathbf{Q},t) \qquad \text{with } \mathbf{Q},\ \mathbf{f}(\mathbf{Q},t) \in \mathbb{R}^F, \tag{10.15}$$

$$\frac{d\mathbf{x}}{dt} = \mathbf{v}(\mathbf{x},\mathbf{Q}) \qquad \text{with } \mathbf{x},\ \mathbf{v}(\mathbf{x},\mathbf{Q}) \in \mathbb{R}^d, \tag{10.16}$$

where d is the space dimension and $\mathbf{Q} = (Q_1, \dots, Q_F)$ are the F variables (typically normal modes) which give a representation of the velocity field $\mathbf{v}$. Note that the equations (10.15) describe the Eulerian evolution and do not depend on the Lagrangian part (10.16), so that (10.15) can be solved independently of (10.16). In order to characterize the degree of chaos, three different Lyapunov exponents can be defined [**17**]:

a) λ_{E} for the Eulerian part (10.15);
b) λ_{L} for the Lagrangian part (10.16), where the time evolution of the velocity field (i.e. $\mathbf{Q}$(t)) is assumed to be known;
c) λ_{T} for the total system of the $d+F$ equations.

These Lyapunov exponents [**15**] are defined as:

$$\lambda_{\mathrm{E,L,T}} = \lim_{t\to\infty} \frac{1}{t} \ln \frac{|\mathbf{z}(t)^{(\mathrm{E,L,T})}|}{|\mathbf{z}(0)^{(\mathrm{E,L,T})}|} \tag{10.17}$$

(the limit in (10.17) exists almost everywhere), where the evolution of the three tangent vectors $\mathbf{z}^{(\mathrm{E})}$, $\mathbf{z}^{(\mathrm{L})}$, $\mathbf{z}^{(\mathrm{T})}$ is given by the linearized stability equations for the Eulerian part, for the Lagrangian part and for the total system, respectively:

$$\frac{dz_i^{(\mathrm{E})}}{dt} = \sum_{j=1}^{F} \left.\frac{\partial f_i}{\partial Q_j}\right|_{\mathbf{Q}(t)} z_j{}^{(\mathrm{E})}, \qquad \mathbf{z}(t)^{(\mathrm{E})} \in \mathbb{R}^F, \tag{10.18}$$

$$\frac{dz_i^{(\mathrm{L})}}{dt} = \sum_{j=1}^{d} \left.\frac{\partial v_i}{\partial x_j}\right|_{\mathbf{x}(t)} z_j{}^{(\mathrm{L})}, \qquad \mathbf{z}(t)^{(\mathrm{L})} \in \mathbb{R}^d, \tag{10.19}$$

$$\frac{dz_i^{(\mathrm{T})}}{dt} = \sum_{j=1}^{d+F} \left.\frac{\partial G_i}{\partial y_j}\right|_{\mathbf{y}(t)} z_j{}^{(\mathrm{T})}, \qquad \mathbf{z}(t)^{(\mathrm{T})} \in \mathbb{R}^{F+d}, \tag{10.20}$$

with $\mathbf{y} = (Q_1, \dots, Q_F, x_1, \dots, x_d)$ and $\mathbf{G} = (f_1, \dots, f_F, v_1, \dots, v_d)$. The meaning of these Lyapunov exponents is evident:

a) λ_{E} is the mean exponential rate of the increase of the uncertainty in the knowledge of the velocity field (which is represented by $\mathbf{Q}$ and is independent of the Lagrangian motion);
b) λ_{L} estimates the rate at which the distance $\delta x(t)$ between two fluid particles initially close increases with time when the velocity field is given, i.e. a particle pair in the same Eulerian realization (i.e. the same $\mathbf{Q}(t)$);
c) λ_{T} is the rate of growth of the distance between initially close particle pairs when the velocity field is not known with infinite precision.

Thus two different dynamical systems can be defined: one for the evolution of the velocity field in the F-dimensional Fourier space; the other for the evolution of the Lagrangian trajectories in the d-dimensional physical space. Let us notice that, since the dynamical system (10.16) is conservative, one may have the coexistence of non-communicating regions and the Lyapunov exponent may depend on the initial conditions (see below). We stress, at this point, that there is no general relation between λ_E and λ_L. One could expect that in the presence of a chaotic velocity field the particle motion has to be chaotic. However, the inequality $\lambda_L \geq \lambda_E$—even if generic—sometimes does not hold, e.g. in some systems like the Lorenz model [**17**] and in generic $2D$ flows when the Lagrangian motion occurs around well-defined vortex structures [**6**] as discussed in the following. On the contrary, one has [**15**]

$$\lambda_T = \max(\lambda_E, \lambda_L). \tag{10.21}$$

10.3.1. Lagrangian chaos in two-dimensional flows. Let us now consider the two-dimensional Navier–Stokes equations with periodic boundary conditions at low Reynolds numbers, for which we can expand the stream function ψ in Fourier series (in a reference frame where the spatial mean of $\mathbf{v}$ vanishes) and take into account only the first F terms [**9, 27**],

$$\psi = -i\sum_{j=1}^{F} k_j^{-1} Q_j e^{i\mathbf{k}_j\mathbf{x}} + \text{c.c.}, \tag{10.22}$$

where c.c. indicates the complex conjugate term and $\mathbf{Q} = (Q_1, \ldots, Q_F)$ are the Fourier coefficients. Inserting (10.22) into the Navier–Stokes equations and by an appropriate time rescaling, we obtain the system of F ordinary differential equations

$$\frac{dQ_j}{dt} = -k_j^2 Q_j + \sum_{l,m} A_{jlm} Q_l Q_m + f_j, \tag{10.23}$$

in which f_j represents an external forcing. Let us briefly discuss the transition to chaos of the above Eulerian equations. Franceschini and coworkers have numerically studied this truncated model with $F = 5$ and $F = 7$ (see [**9, 27**]). The forcing was restricted to the 3rd mode [**27**]: $f_j = \text{Re}\,\delta_{j,3}$, where Re is the Reynolds number. For $F = 5$ and $\text{Re} < \text{Re}_1 = 22.85\ldots$ there are four stable stationary solutions, say $\widehat{\mathbf{Q}}$, and $\lambda_E < 0$. At $\text{Re} = \text{Re}_1$, these solutions become unstable, via a Hopf bifurcation [**29**], i.e. four stable periodic orbits appear, and therefore $\lambda_E = 0$. For $\text{Re}_1 < \text{Re} < \text{Re}_2 = 28.41\ldots$ one thus finds the stable limit cycles:

$$\mathbf{Q}(t) = \widehat{\mathbf{Q}} + (\text{Re} - \text{Re}_1)^{1/2}\delta\mathbf{Q}(t) + O(\text{Re} - \text{Re}_1), \tag{10.24}$$

where $\delta\mathbf{Q}(t)$ is periodic with period

$$T(\text{Re}) = T_0 + O(\text{Re} - \text{Re}_1), \qquad T_0 = 0.7328\ldots \tag{10.25}$$

At $\text{Re} = \text{Re}_2$, these limit cycles lose stability and there is a period doubling cascade towards Eulerian chaos.

Let us now discuss the Lagrangian behavior of a fluid particle for the above flow. For $\text{Re} < \text{Re}_1$, the stream function is asymptotically stationary, $\psi(\mathbf{x}, t) \to \widehat{\psi}(\mathbf{x})$, and the corresponding one-dimensional Hamiltonian is time independent; therefore Lagrangian trajectories are regular. For $\text{Re} = \text{Re}_1 + \varepsilon$ the asymptotic stream function becomes time dependent

$$\psi(\mathbf{x}, t) = \widehat{\psi}(\mathbf{x}) + \sqrt{\varepsilon}\,\delta\psi(\mathbf{x}, t) + O(\varepsilon), \tag{10.26}$$

where $\widehat{\psi}(\mathbf{x})$ is given by $\widehat{\mathbf{Q}}$ and $\delta\psi$ is periodic in $\mathbf{x}$ and in t (with time period T). The region of phase space, here the real two-dimensional space, adjacent to a separatrix is very sensitive to perturbations, even of very weak intensity. FIG. 10.2 shows the structure of the separatrices (orbits of infinite period connecting hyperbolic fixed points) at $\mathrm{Re} = \mathrm{Re}_1 - 0.05$. One can observe the presence of two kinds of separatrices: the isolated "eights" labeled by A and the connected periodic separatrices, labeled by B.

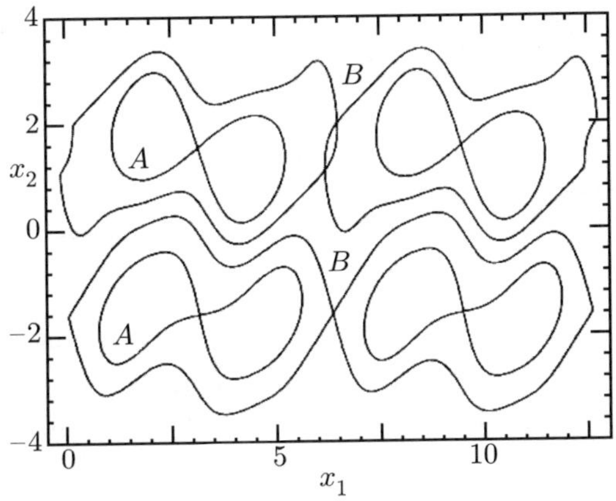

FIGURE 10.2. Structure of the separatrices in the 5-mode model (10.22) with $\mathrm{Re} = \mathrm{Re}_1 - 0.05$.

Generically, in one-dimensional Hamiltonian systems, a periodic perturbation gives origin to chaotic motion in a domain containing the hyperbolic fixed points [**14, 35**]. A method due to Melnikov [**35**] allows one to prove the existence of the chaotic layer in the phase space. However such a technique is not straightforward (one has to compute explicitly an integral which can involve complicated functions). On the other hand we have a rather strong numerical evidence for the existence of the chaotic regions; see FIG. 10.3.

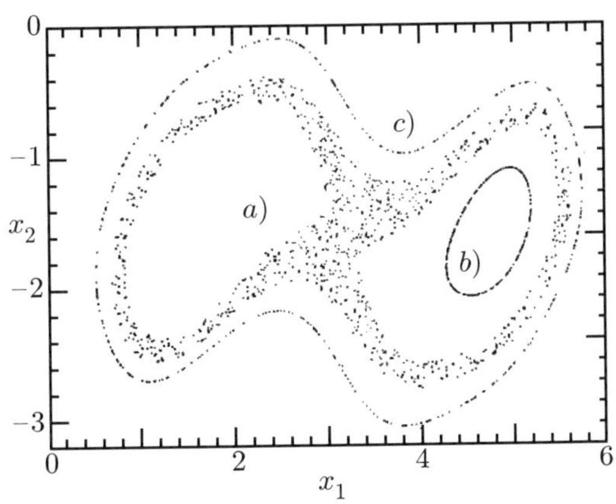

FIGURE 10.3. Poincaré map for three trajectories of the 5-mode model with $\mathrm{Re} = \mathrm{Re}_1 + 0.05$. The initial conditions are selected close to a separatrix, case a) ($x_1(0) = 3.2$, $x_2(0) = -1.6$), or far from the separatrices, cases b) ($x_1(0) = 4.3$, $x_2(0) = -2.0$) and c) ($x_1(0) = 4.267$, $x_2(0) = -3.009$).

Let us explain. Chaotic and regular motion for small $\varepsilon = \mathrm{Re} - \mathrm{Re}_1$ can be studied by the Poincaré map

$$\mathbf{x}(nT) \to \mathbf{x}(nT + T), \tag{10.27}$$

where (at a given ε) the period $T(\varepsilon)$ is computed numerically. The sizes of the stochastic layers rapidly increase with ε. At $\varepsilon = \varepsilon_c \approx 0.7$ they overlap, according to the resonance overlap mechanism introduced by Chirikov [**14**], and it is practically impossible to distinguish between regular and chaotic zones. At $\varepsilon > \varepsilon_c$ there is always diffusive motion.

We stress that this scenario for the onset of Lagrangian chaos in two-dimensional incompressible fluids is generic and does not depend on the particular truncated model. In fact, it is only related to the appearance of stochastic layers under the effect of small time-dependent perturbations in one-dimensional integrable Hamiltonian systems. As a consequence of general features of one-dimensional Hamiltonian systems we expect that a stationary stream function becomes time periodic through a Hopf bifurcation as occurs for all known truncated models of Navier–Stokes equations.

We have seen that there is no simple relation between Eulerian and Lagrangian behaviors, i.e. there is no simple relation between λ_L and λ_E. In the following, we shall discuss two important points:

(i) What are the effects on the Lagrangian chaos of the transition to Eulerian chaos, i.e. from $\lambda_{\mathrm{E}} = 0$ to $\lambda_{\mathrm{E}} > 0$?
(ii) Does a chaotic velocity field ($\lambda_{\mathrm{E}} > 0$) always imply an erratic motion of fluid particles?

The first point can be studied again within the $F = 5$ modes model (10.23). Increasing Re, the limit cycles bifurcate to new double period orbits followed by a period doubling transition to chaos and a strange attractor appears at $\mathrm{Re}_c \approx 28.73$, where λ_{E} becomes positive. These transitions have no signature on Lagrangian behavior, as shown in FIG. 10.4, i.e. the onset of Eulerian chaos has no influence on Lagrangian properties.

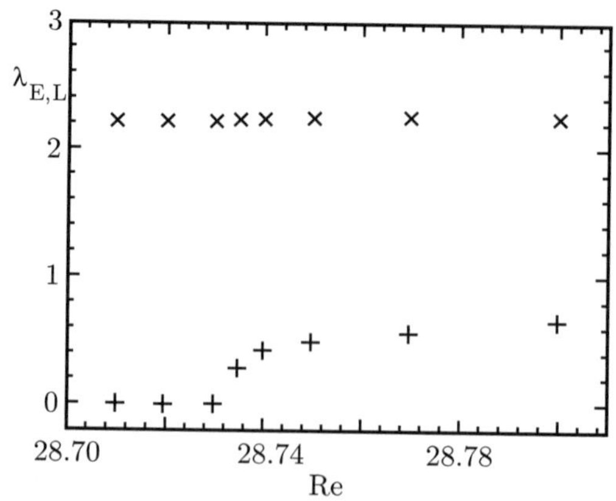

FIGURE 10.4. Lyapunov exponents λ_{E} (+) and λ_{L} (×) as functions of Re around Re_c, for the 5-mode model.

This feature should be valid in most situations, since it is natural to expect that in generic cases there is a strong separation of the characteristic times for Eulerian and Lagrangian behaviors.

The second point—the conjecture that a chaotic velocity field always implies chaotic motion of particles—looks very reasonable. Indeed, it appears to hold in many systems [**15**]. Nevertheless, one can find a class of systems where this is false: e.g. the equations (10.15), (10.16) may exhibit Eulerian chaoticity $\lambda_{\mathrm{E}} > 0$, even if $\lambda_{\mathrm{L}} = 0$ (see [**6**]).

10.3.2. Lagrangian chaos in point-vortex systems. Let us consider again the two-dimensional Euler equation. We assume that the vorticity $\omega = \nabla \times \mathbf{v}$, namely its component $\omega(\mathbf{r}, t)$ perpendicular to the plane of the flow, at the initial time $t = 0$ is localized on N point-vortices, i.e.:

$$\omega(\mathbf{r}, 0) = \sum_{i=1}^{N} \Gamma_i \, \delta(\mathbf{r} - \mathbf{r}_i(0)), \tag{10.28}$$

where Γ_i is the circulation of the ith vortex. As a consequence of Kelvin's circulation theorem, which states that in an ideal incompressible fluid, in the absence of forcings, the circulation around a closed material curve moving with the fluid is constant, one has that, at any time, $\omega(\mathbf{r}, t)$ must still remain localized on N point-vortices:

$$\omega(\mathbf{r}, t) = \sum_{i=1}^{N} \Gamma_i \delta(\mathbf{r} - \mathbf{r}_i(t)), \tag{10.29}$$

where $\mathbf{r}_i = (x_i, y_i)$ evolves according to

$$\Gamma_i \frac{dx_i}{dt} = \frac{\partial H}{\partial y_i}, \tag{10.30}$$

$$\Gamma_i \frac{dy_i}{dt} = -\frac{\partial H}{\partial x_i}, \tag{10.31}$$

where

$$H = -\frac{1}{4\pi} \sum_{i \neq j} \Gamma_i \Gamma_j \ln r_{ij} \tag{10.32}$$

and $r_{ij}^2 = (x_i - x_j)^2 + (y_i - y_j)^2$.

Therefore the motion of N point-vortices is described by a Hamiltonian system whose canonical coordinates are $(x_i, \Gamma_i y_i)$. Because of the presence of global conserved quantities, namely $\sum_i \Gamma_i x_i = \text{const}$, $\sum_i \Gamma_i y_i = \text{const}$ (due to the translation invariance), $\sum_i \Gamma_i (x_i^2 + y_i^2) = \text{const}$ (in view of the rotation invariance) and $H = \text{const}$, a system of three vortices is integrable (and thus there is no exponential divergence of nearby trajectories in the phase space) [**1**]. Indeed, the phase space is $2N$-dimensional, and the four constraints above confines the motion on a $(2N-4)$-dimensional surface: for $N = 3$, this is a 2-dimensional surface, and the motion cannot be chaotic. For $N \geq 4$, numerical studies show that, apart from non-generic initial conditions and/or values of the parameters Γ_i, the system is chaotic [**1**].

The motion of a passively advected particle located in $(x(t), y(t))$ in the velocity field defined by (10.30)–(10.31) is given by

$$\frac{dx}{dt} = -\sum_i \frac{\Gamma_i}{2\pi}\frac{y - y_i}{R_i^2}, \tag{10.33}$$

$$\frac{dy}{dt} = \sum_i \frac{\Gamma_i}{2\pi}\frac{x - x_i}{R_i^2}, \tag{10.34}$$

where $R_i^2 = (x - x_i)^2 + (y - y_i)^2$.

It is interesting to discuss the analogy between point-vortex systems and Celestial Mechanics. We notice that the problem of passive particles advected by N vortices is formally equivalent to the case of $N + 1$ vortices with $\Gamma_{N+1} = 0$; in particular, the case $N = 2$ is integrable. In Celestial Mechanics, in the restricted 3-body problem, the satellite has a negligible mass, while the Sun and the planet follow Keplerian orbits. At first sight, one could think that the problem of a passive particle (i.e. a vortex with negligible circulation) advected by *two* vortices is rather similar to the restricted 3-body problem, and that the system (10.30), (10.31), (10.33), and (10.34) with $N \geq 3$ corresponds, in Celestial Mechanics, to the case of Sun, two or more planets, and a satellite. But there is a difference: in Celestial Mechanics we have a *force* field (with positions and velocities as variables), whereas here we have a *velocity* field (with positions as variables). In the problem of advection of a passive particle by N vortices (thus $N + 1$ "bodies"), the phase space is $2(N+1)$-dimensional and there are 4 conserved quantities (see above), hence $2N-2$ independent variables, and a chaotic behavior may appear if $N+1 \geq 4$; while in the N-body problem of Celestial Mechanics, the phase space is $4N$-dimensional (four coordinates x, y, $\dot{x}$, $\dot{y}$ for each body) and there are 6 conserved quantities (energy, the two momentum components, angular momentum, the two components of the center of mass), hence $4N - 6$ independent variables; thus a chaotic behavior may appear as soon as $N \geq 3$. Therefore, a better analogue of the restricted *three*-body problem is advection of a passive particle by *three* vortices (for a discussion of this analogy, see [**33**]).

Let us first consider the motion of passive particles advected by a three-vortices (thus integrable) system in which $\lambda_E = 0$. In this case, the stream function for the passive particles is periodic in time and the expectation is that they may display chaotic behavior. The typical trajectories of passive particles which have initially been placed respectively in close proximity of a vortex center or in the background field between the vortices display a very different behavior. The particle seeded close to the vortex center displays a regular motion around the vortex and thus $\lambda_L = 0$; by contrast, the particle in the background field undergoes an irregular and aperiodic trajectory, and λ_L is positive.

We now discuss a case where the Eulerian flow is chaotic, with $N = 4$ point-vortices. Let us consider again the trajectory of a passive particle deposited in proximity of a vortex center. As before, the particle rotates around the moving vortex. But the vortex motion is chaotic; consequently, the particle position is unpredictable on long times as is the vortex position. Nevertheless, the Lagrangian Lyapunov exponent for this trajectory is zero (i.e. two initially close particles around the vortex remain close), even if the Eulerian Lyapunov exponent is positive; see FIG. 10.5.

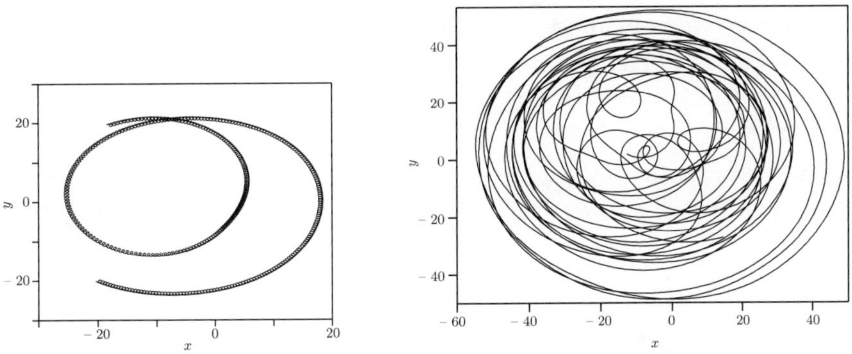

FIGURE 10.5. Particle trajectories in the four-vortex system. Eulerian dynamics in this case is chaotic. The left panel shows a regular Lagrangian trajectory, while the right panel shows a chaotic Lagrangian trajectory. The different behavior of the two particles is due to different initial conditions: particles initially close to one vortex (left panel) and particles initially distant from the vortices (right panel).

This result indicates once more that there is no strict link between Eulerian and Lagrangian chaoticity.

One may wonder whether a much more complex Eulerian flow, such as $2D$ turbulence, may give the same scenario for particle advection: i.e. regular trajectories close to the vortices and chaotic behavior between the vortices. It has been shown that this is indeed the case [**6**] and that the chaotic nature of the trajectories of advected particles is not strictly related to the complex time evolution of the turbulent flow.

We have seen that there is no general link between Lagrangian and Eulerian chaos. In the typical situation Lagrangian chaos may appear also for regular velocity fields. However, it is also possible to have the opposite situation, with $\lambda_L = 0$ in the presence of Eulerian chaos, as in the example of Lagrangian motion inside vortex structures. As an important consequence of this discussion we remark that it is not possible to separate Lagrangian and Eulerian properties in a measured trajectory, e.g. a buoy in the oceanic currents [**34**]. Indeed, using the standard methods for data analysis [**18**], from Lagrangian trajectories one extracts the total Lyapunov exponent λ_T and not λ_L or λ_E.

10.4. Asymptotic time behavior: Transport and diffusion

Let us now discuss the asymptotic behavior, i.e. at long times and large spatial scales, of the Lagrangian trajectories. The simplest model of diffusion is the Brownian motion, i.e. the erratic movement of particles suspended in a liquid, observed by the botanist Robert Brown as early as in 1827. After the fundamental work of Einstein [**16**] and Smoluchowski [**37**] (see also Langevin [**25**]), Brownian motion became the prototypical example of a stochastic process.

From a discrete random walk model, i.e. a walker that moves randomly making jumps of $\pm\Delta x$ every time step Δt, it is easy to obtain the diffusion equation:

$$\frac{\partial p(x,t)}{\partial t} = D\frac{\partial^2 p(x,t)}{\partial x^2}, \tag{10.35}$$

where one takes the limit $\Delta x, \Delta t \to 0$ in such a way that $(\Delta x)^2/\Delta t \to 2D$ (the factor 2 is purely conventional). The solution to (10.35) is readily obtained as

$$p(x,t) = \frac{1}{\sqrt{4\pi Dt}} \exp\left(-\frac{x^2}{4Dt}\right), \tag{10.36}$$

with $\langle x^2 \rangle = 2Dt$. Diffusion equation (10.35) is written here for the probability density function $p(x,t)$ of observing a marked particle (a tracer) in a position x at time t. The same equation can have another interpretation, in which $p(x,t) = \theta(x,t)$ represents the concentration of a scalar quantity (marked fluid, temperature, pollutant) as function of time. The only difference is, of course, in the normalization.

It is rather clear that time decorrelation is the key ingredient for normal diffusion. In the random walk model, it is related to randomness: the jumps are independent random variables and this assures the applicability of the central limit theorem. But we can have a finite time correlation and thus diffusion also without randomness. A popular example is the following symplectic system (standard map [**14**]):

$$\begin{aligned} J(t+1) &= J(t) + K\sin\theta(t), \\ \theta(t+1) &= \theta(t) + J(t+1). \end{aligned} \tag{10.37}$$

This map is a paradigmatic model which shows the generic behavior of chaotic symplectic systems. In particular one has large-scale chaotic behavior for $K > K_c \simeq 0.9716$ and, as a consequence of deterministic chaos, $J(t)$ has diffusive behavior. For long times, $J(t)$ is large and thus the angle $\theta(t)$ rotates rapidly. In this limit, we can assume that $\theta(t)$ decorrelates at each step and thus write

$$J(t)^2 = K^2\left(\sum_{t'=1}^{t} \sin\theta(t')\right)^2 \simeq K^2\langle \sin^2\theta\rangle t = 2Dt. \tag{10.38}$$

The diffusion coefficient D, in the random phase approximation, i.e. assuming that $\sin\theta(t)$ is not correlated with $\sin\theta(t')$ for $t \neq t'$, is obtained from the above expression as $D_{RPA} = K^2/4$. As a consequence of non-trivial dynamical effects (long memory due to the presence of accelerator modes) the random phase approximation is rather rough. Nevertheless, at least at a qualitative level, this can grasp the basic ingredients of the diffusion process.

The two examples discussed above are in completely different classes: stochastic for the random walk and deterministic for the standard map (10.37). Despite this difference in the microscopic dynamics, both lead to a macroscopic diffusion equation and Gaussian distribution. This suggests how the diffusion equation (10.35) is of general applicability, and, therefore, a diffusive behavior at macroscopic scale says nothing on the underlying microscopic dynamics.

Let us now consider the more complex situation of dispersion in a non-steady fluid with velocity field $\mathbf{v}(\mathbf{x},t)$. For simplicity we consider incompressible flow (i.e. $\nabla\cdot\mathbf{v} = 0$), which can be laminar or turbulent. In the presence of $\mathbf{v}(\mathbf{x},t)$, the diffusion equation (10.35) becomes the *advection-diffusion* equation for the concentration $\theta(\mathbf{x},t)$ (10.1). This equation is linear in θ but nevertheless it can display very interesting and non-trivial properties even in the presence of simple velocity fields, as a consequence of Lagrangian chaos.

Let us start with some general considerations. We know that in physical systems the molecular diffusivity is typically very small. Thus in (10.1) the advection term dominates over diffusion. This is quantified by the Péclet number, which is the ratio of the typical value of the advection term to the diffusive term,

$$\mathrm{Pe} = \frac{v_0 l_0}{D}, \tag{10.39}$$

where v_0 is the typical velocity at the typical scale of the flow l_0. We denote by $\tau_0 \simeq l_0/v_0$ the typical correlation time of the velocity.

The central quantity in the following discussion is the *eddy diffusivity*. The idea is rather simple and dates back to the classical work of Taylor [**39**]. To illustrate this concept, let us consider a Lagrangian description of dispersion in which the trajectory of a tracer $\mathbf{x}(t)$ is given by (10.2). Being interested in the limit $\mathrm{Pe} \to \infty$, in the following we will neglect, just for simplicity, the molecular diffusivity D, which is generally much less than the effective dynamical diffusion coefficient.

Starting from the origin, $\mathbf{x}(0) = 0$, and assuming that $\langle \mathbf{v} \rangle = 0$ we have $\langle \mathbf{x}(t) \rangle = 0$ forever. The mean square displacement, on the other hand, grows according to

$$\frac{d}{dt}\left\langle \frac{1}{2}\mathbf{x}(t)^2 \right\rangle = \langle \mathbf{x}(t) \cdot \mathbf{v}_L(t) \rangle = \int_0^t \langle \mathbf{v}_L(s) \cdot \mathbf{v}_L(t) \rangle \, ds, \tag{10.40}$$

where we have introduced, for simplicity of notation, the Lagrangian velocity $\mathbf{v}_L(t) = \mathbf{v}(\mathbf{x}(t), t) + \eta(t)$; see (10.6). Let us determine the Lagrangian correlation time τ_L from

$$\int_0^\infty \langle \mathbf{v}_L(s) \cdot \mathbf{v}_L(0) \rangle ds = \langle \mathbf{v}_L(0)^2 \rangle \tau_L \tag{10.41}$$

and assume that the integral converges so that τ_L is finite. From (10.40), for $t \gg \tau_L$ we get

$$\langle \mathbf{x}(t)^2 \rangle = 2\tau_L \langle \mathbf{v}_L^2 \rangle t, \tag{10.42}$$

i.e. the diffusive behavior with diffusion coefficient (eddy diffusivity) $D^E = \tau_L \langle \mathbf{v}_L^2 \rangle$. This simple derivation shows, once more, that diffusion in general is to be expected in the presence of a finite correlation time τ_L.

Coming back to the advection-diffusion equation (10.1), the above argument means that, for $t \gg \tau_L$, we expect that the evolution of the concentration, for scales larger than l_0, can be described by an effective diffusion equation, i.e.

$$\frac{\partial \langle \theta \rangle}{\partial t} = D_{ij}^E \frac{\partial^2 \langle \theta \rangle}{\partial x_i \partial x_j}. \tag{10.43}$$

Here, the effective diffusivity D^E is a tensor. The coefficients D_{ij}^E are related to the statistical properties of the trajectories:

$$D_{ij}^E = \lim_{t \to \infty} \frac{1}{2t} \langle x_i(t) x_j(t) \rangle. \tag{10.44}$$

The original idea of a hydrodynamic analysis to derive (10.43) from (10.1) is due to Maxwell. However, only in recent years the multiscale techniques allowed for a rigorous derivation of Fick's equation (10.43), as well as its range of validity, and a systematic method for the computation of the effective diffusion coefficients

D^E_{ij} [**8, 28**]. The idea behind this method has close connection with Lindstedt–Poincaré's perturbation theory, in Celestial Mechanics, where the asymptotic effects of the non-linearities, in an ordinary differential equation, are treated with the introduction of (slow) frequencies which depend on the initial conditions and the non-linear terms [**32**]. Since in the diffusion process one is interested in large scales and long times, let us assume the existence of some (small) parameter ε: one can thus introduce, in addition to the natural (fast) variables $\mathbf{x}$ and t, slow variables $\mathbf{X} = \varepsilon\mathbf{x}$ and $T = \varepsilon^2 t$. Treating the two sets of variables (slow and fast) as independent, and expanding θ in a perturbation series

$$\theta(\mathbf{x}, t; \mathbf{X}, T) = \theta^{(0)}(\mathbf{X}, T) + \varepsilon\theta^{(1)}(\mathbf{x}, t; \mathbf{X}, T) + \cdots \tag{10.45}$$

one has a set of equations for $\theta^{(0)}, \theta^{(1)}, \ldots$ Using the linear structure of the problem and imposing the solvability conditions [**8, 28**] (this is the analogue of the suppression of the secular terms in the multiscale approach for the ordinary differential equations [**7**]) one has that $\theta^{(0)}$ obeys Fick's equation (10.43).

The computation of the eddy diffusivity for a given Eulerian flow is not an easy task. However it is possible to show that

$$D^E_{ij} = D\delta_{ij} - \frac{1}{2}[\langle v_i w_j\rangle + \langle v_j w_i\rangle], \tag{10.46}$$

where the vector $\mathbf{w}$ satisfies the equation

$$\partial_t \mathbf{w} + (\mathbf{v}\cdot\nabla)\mathbf{w} = D\nabla^2\mathbf{w} - \mathbf{v}. \tag{10.47}$$

Let us stress that with the multiscale approach one has that the asymptotic behavior of the transport equation (10.1) is well approximated by Fick's law (10.43) where the eddy diffusivity tensor D^E depends on the advection field $\mathbf{v}$. This is rather close, both from a technical and conceptual point of view, to the Lindstedt–Poincaré method in Mechanics where the asymptotic effect of the non-linear terms is to induce a renormalization of the frequencies, which depend on the initial conditions and the non-linear terms; see [**36**, Vol. II, Chap. 9] and [**32**]. The explicit computation can be performed only in the case of simple flows [**28, 8**]. On the other hand, in the general case, it is relatively easy [**8**] to give some bounds, the simplest one being $\mathrm{tr}(D^E) \geq D \cdot d$, with d equal to the space dimension, i.e. the presence of an incompressible velocity field enhances large-scale transport.

Diffusion equation (10.43) is the typical long-time behavior in a generic flow. However there exists, also, the possibility of the so-called *anomalous diffusion*, i.e. when the spread of particles does not grow linearly in time, but obeys a power law:

$$\langle x^2(t)\rangle \sim t^{2\nu} \tag{10.48}$$

with $\nu \neq 1/2$. The case of $\nu > 1/2$ (formally $D^E = \infty$) is called *superdiffusion*; *subdiffusion*, i.e. $\nu < 1/2$ (formally $D^E = 0$), is possible only for compressible velocity fields.

Superdiffusion arises when the Taylor argument above for deriving (10.42) fails and formally $D^E \to \infty$. This can be due to one of the following mechanisms:

a) the divergence of $\langle \mathbf{v}_L^2\rangle$ (which is the case of *Lévy flights*), or
b) the lack of decorrelation, i.e. $\langle \mathbf{v}_L(t)\mathbf{v}_L(0)\rangle \sim t^{-\beta}$, with $\beta \leq 1$, and thus $\tau_L \to \infty$ (*Lévy walks*).

The second case is more physical, and it is related to the existence of strong correlations in the dynamics, even at long times and large scales.

Avellaneda and Majda [**28**] obtained a very general important result for the validity of equation (10.43) in an incompressible velocity field. If the molecular diffusion coefficient D is non-zero and the infrared contribution to $\mathbf{v}$ is weak enough, namely

$$\int d\mathbf{k} \frac{\langle|\widehat{\mathbf{v}}(\mathbf{k})|^2\rangle}{k^2} < \infty, \tag{10.49}$$

where $\langle\cdot\rangle$ indicates the time average and $\widehat{\mathbf{v}}$ is the Fourier transform of $\mathbf{v}$, then one has standard diffusion, i.e. the D^E_{ij} are finite. Therefore there exist two possible origins for superdiffusion:

I) $D > 0$ and, in order to violate (10.49), a velocity field with very long spatial correlations;

II) $D = 0$ and strong correlation between $\mathbf{v}[\mathbf{x}(t)]$ and $\mathbf{v}[\mathbf{x}(t+\tau)]$ at large τ.

An example of anomalous diffusion, according to the mechanism I, is a $2D$ random shear, i.e. $\mathbf{v} = (v(y), 0)$ where $v(y)$ is a random function obtained with a spatial Brownian process. For such a case it is possible to show that $\langle x^2(t)\rangle \sim t^{3/2}$ and, in addition, the probability density function is non-Gaussian [**28, 22, 11**]:

$$p(x,t) \sim t^{-3/4} \exp\left(-\frac{c|x^4|}{t^3}\right). \tag{10.50}$$

Let us now briefly discuss anomalous diffusion due to the mechanism II: consider a simple model simulating Rayleigh–Bénard convection:

$$\Psi(x,y,t) = \Psi_0 \sin(x + B\sin(\omega t))\sin(y). \tag{10.51}$$

The even oscillatory instability is accounted for the term $B\sin(\omega t)$ representing the lateral oscillations of the rolls. In spite of its apparent simplicity, this model is able to capture the essential features of the convection problem [**38**]. For a generic value of ω one has standard diffusion, and the only role of the advection term is to enhance D^E. For special values of ω (resonance with the characteristic frequency of the passive particle motion in the steady case) one has, in the limit $D \to 0$, superdiffusion, i.e. $\langle x^2(t)\rangle \sim t^{2\nu}$ with $\nu > 1/2$.

A similar behavior, i.e. standard diffusion for generic values of the control parameters and superdiffusion for specific values, has been observed for the standard map. In this system, at certain values of K, the accelerator modes induce a superdiffusion with a non-Gaussian shape of the probability density function [**11, 21**].

Therefore, on the basis of the above results (and similar results found for other symplectic systems [**26**]), one can say that the mechanism II is non-generic. Introducing a small perturbation $O(\varepsilon)$ in the evolution law, superdiffusion disappears, i.e. it is present only for a transient time up to $t_* \sim \varepsilon^{-\alpha}$. At longer times standard diffusion takes place with a rather large diffusion coefficient: we can say that, even if the true (i.e. asymptotic) anomalous diffusion is very rare, its ghost is still visible [**11**].

10.5. Conclusions

The recent advances in the understanding of the chaotic behavior and the (asymptotic) diffusive properties of Lagrangian trajectories, advected by laminar velocity fields, stem from ideas and techniques borrowed from Poincaré's contributions to Celestial Mechanics (mainly the homoclinic intersection mechanism and multiscale expansion). This bears the indelible mark of the genius of Henri Poincaré and shows how many areas of Mathematics and Physics have been affected by his works.

We acknowledge the fruitful and friendly collaboration of many colleagues who shared with us their knowledge and research, in particular P. Castiglione, A. Celani, M. Cencini, M. Falcioni, A. Mazzino, P. Muratore-Ginanneschi, S. Musacchio, and M. Vergassola.

Bibliography

[1] H. Aref, *Integrable, and turbulent vortex motion in two-dimensional flows*, Ann. Rev. Fluid Mech. **15** (1983), p. 345.

[2] H. Aref, *Stirring by chaotic advection*, J. Fluid Mech. **143** (1984), p. 1.

[3] H. Aref and S. Balachandar, *Chaotic advection in a Stokes flow*, Phys. Fluids **29** (1986), p. 3515.

[4] V. I. Arnold, *Sur la topologie des écoulements stationnaires des fluides parfaits*, C. R. Acad. Sci. Paris A **261** (1965), p. 17.

[5] V. I. Arnold and B. A. Khesin, *Topological methods in hydrodynamics*, Applied Math. Sciences **125**, Springer-Verlag, Berlin (1998).

[6] A. Babiano, G. Boffetta, A. Provenzale, and A. Vulpiani, *Chaotic advection in point vortex models and two-dimensional turbulence*, Phys. Fluids A **6** (1994), p. 2465.

[7] C. Bender and S. A. Orszag, *Advanced mathematical methods for scientists and engineers*, McGraw-Hill, New York (1978).

[8] L. Biferale, A. Crisanti, M. Vergassola, and A. Vulpiani, *Eddy diffusivities in scalar transport*, Phys. Fluids **7** (1995), p. 2725.

[9] C. Boldrighini and V. Franceschini, *A five-mode truncation of the plane Navier-Stokes equations*, Commun. Math. Phys. **64** (1979), p. 159.

[10] J.-P. Bouchaud and A. Georges, *Anomalous diffusion in disordered media: Statistical mechanisms, models and physical applications*, Phys. Rep. **195** (1990), p. 127.

[11] P. Castiglione, A. Mazzino, P. Muratore-Ginanneschi, and A. Vulpiani, *On the strong anomalous diffusion*, Physica D **134** (1999), p. 75.

[12] J. Chaiken, C. K. Chu, M. Tabor, and Q. M. Tan, *Lagrangian turbulence in Stokes flow*, Phys. Fluids **30** (1987), p. 687.

[13] S. Chandrasekhar, *Stochastic problems in Physics and Astronomy*, Rev. Mod. Phys. **15** (1943), p. 1.

[14] B. V. Chirikov, *A universal instability of many dimensional oscillator systems*, Phys. Rep. **52** (1979), p. 263.

[15] A. Crisanti, M. Falcioni, G. Paladin, and A. Vulpiani, *Lagrangian chaos: Transport, mixing and diffusion in fluids*, Riv. Nuovo Cim. **14** (1991), p. 1.

[16] A. Einstein, *On the movement of small particles suspended in a stationary liquid demanded by the molecular kinetic theory of heat*, Ann. Phys. (Leipzig) **17** (1905), p. 549.

[17] M. Falcioni, G. Paladin and A. Vulpiani, *Regular and chaotic motion of fluid-particles in two-dimensional fluids*, J. Phys. A: Math. Gen. **21** (1988), p. 3451.

[18] P. Grassberger and I. Procaccia, *Estimation of the Kolmogorov entropy from a chaotic signal*, Phys. Rev. A **28** (1983), p. 2591.

[19] M. Hénon, *Sur la topologie des lignes de courant dans un cas particulier*, C. R. Acad. Sci. Paris A **262** (1966), p. 312.

[20] P. J. Holmes, J. L. Lumley and G. Berkooz, *Turbulence, coherent structures, dynamical systems and symmetry*, Cambridge University Press, Cambridge (1996).

[21] Y. H. Ichikawa, T. Kamimura, and T. Hatori, *Stochastic diffusion in the standard map*, Physica D **29** (1987), p. 247.

[22] M. B. Isichenko, *Percolation, statistical topography, and transport in random media*, Rev. Mod. Phys. **64** (1992), p. 961.

[23] H. Lamb, *Hydrodynamics*, New York Dover Publ., New York (1945).

[24] L. D. Landau and L. Lifshitz, *Fluid Mechanics*, New York Pergamon Press, New York (1987).

[25] P. Langevin, *Sur la théorie du mouvement brownien*, Comptes Rendus **146** (1908), p. 530.

[26] P. Lebœuf, *Normal and anomalous diffusion in a deterministic area-preserving map*, Physica D **116** (1998), p. 8.

[27] J. Lee, *Triad-angle locking in low-order models of the 2D Navier-Stokes equations*, Physica D **24** (1987), p. 54.

[28] A. J. Majda and P. R. Kramer, *Simplified models for turbulent diffusion: Theory, numerical modeling and physical phenomena*, Phys. Rep. **314** (1999), p. 237.

[29] J. E. Marsden and M. McCracken, *The Hopf bifurcation and its applications*, MIT Press, Cambridge, Mass. (1975).

[30] M. R. Maxey and J. J. Riley, *Equation of motion for a small rigid sphere in a non-uniform flow*, Phys. Fluids **26** (1983), p. 883.

[31] H. K. Moffatt, *Transport effects associated with turbulence, with particular attention to the influence of helicity*, Rep. Prog. Phys. **46** (1983), p. 621.

[32] J. A. Murdock, *Perturbations: theory and methods*, J. Wiley, New York (1991).

[33] Z. Neufeld and T. Tél, *The vortex dynamics analogue of the restricted three-body problem: Advection in the field of three identical point vortices*, J. Phys. A: Math. Gen. **30** (1997), pp. 2263–2280.

[34] A. R. Osborne, A. D. Kirwan, A. Provenzale, and L. Bergamasco, *A search for chaotic behaviour in large and mesoscale motions in the Pacific Ocean*, Physica D **23** (1986), p. 75.

[35] E. Ott, *Chaos in dynamical systems*, Cambridge University Press, Cambridge (1993).

[36] H. Poincaré, *Les méthodes nouvelles de la Mécanique Céleste*, Gauthier-Villars, Paris (1892, 1893, 1899).

[37] M. Smoluchowski, Œuvres, Acad. Pol. Sci. Lettres (Cracovie), Librairie Polytechnique Ch. Béranger, Paris (1924).

[38] T. H. Solomon and J. P. Gollub, *Passive transport in steady Rayleigh-Bénard convection*, Phys. Fluids **31** (1988), p. 1372.

[39] G. I. Taylor, *Diffusion by continuous movements*, Proc. London Math. Soc. **2** (1921), p. 196.

[40] D. J. Tritton, *Physical fluid dynamics*, Oxford Science Publ., Oxford (1988).

CHAPTER 11

The Concept of "Residue" after Poincaré: Cutting across All of Mathematics

ALAIN YGER

11.1. Introduction

In a memoir published in the *Acta Mathematica* in 1887 and entitled *Sur les résidus des intégrales doubles* [**38**, *On the residues of double integrals*], Henri Poincaré wrote:

> *"Cauchy receives the glory for having founded the theory of integrals taken between imaginary limits; this theory has, in a manner of speaking, doubled the power of mathematical analysis and has been the point of departure of all the work that has followed, in every country in which the exact sciences are cultivated, and particularly in Germany and France. It seemed that nothing but a small step remained to extend this theory to double integrals and that one could expect of this extension such great conquests as in the consideration of simple integrals. Therein lay what could tempt the geometers' ambition, yet after forty years we are hardly any further along than on day one. Most attempts that have been made were nothing more than failures or half-successes."* [**38**, p. 321]

One first difficulty is that by passing from simple integrals to double integrals, we pass from the plane $\mathbb{C} \cong \mathbb{R}^2$ to the four-dimensional space $\mathbb{C}^2 \cong \mathbb{R}^4$:

> *"A double integral must be extended to all the points of a surface. We will therefore have a* surface of integration *in the same way that we have, in the theory of simple integrals, a path of integration. In addition, the set of singular points will form a surface."*

For example, if we integrate a rational function P/Q, the singular points are the points where $Q = 0$, a condition which, by separating the real and imaginary parts, translates into two real equations that define a 2-dimensional surface (in general) in $\mathbb{R}^4$.

> *"It will then be necessary that the surface of integration and the singular surfaces have no points in common."* [**38**, p. 325]

We must therefore begin by finding convenient *"modes of representation"* [**38**, no. 1, pp. 324–327]. Of course, we can speak of "points", "curves", "surfaces" (of dimension 2), and "hypersurfaces" (of dimension 3) in "hyperspace" $\mathbb{R}^4$, but Poincaré claimed, *"This language of hypergeometry is still repugnant to many great minds"*, and he proposed to avoid it by the following *"tricks"*. He (analytically) embedded in *"hyperspace"* $\mathbb{R}^4$ an auxiliary 3-dimensional space, or more generally an open subset (say, V_3) of such a space: $V_3 \xrightarrow{\varphi} \mathbb{R}^4$, and he considered the (2-dimensional) surfaces in $\mathbb{R}^4$ obtained as images $\varphi(S)$ of surfaces S in V_3. The integration of a function over $\varphi(S)$ reduces to an integration over S; if the set of singular points is defined by a polynomial (or more generally holomorphic) equation $Q = 0$, hence by two real equations $Q_1 = 0$, $Q_2 = 0$, the equations $Q_1 \circ \varphi = 0$, $Q_2 \circ \varphi = 0$ define inside of V_3 a curve that Poincaré called a *"singular curve"*. With this mode of representation, the situation is reduced to surfaces and curves *"in ordinary space and we are freed from hypergeometry."*

Next, we must find an analogue of *Cauchy's theorem*, which says that, in $\mathbb{C} \cong \mathbb{R}^2$, the integral of a holomorphic function does not change value when we continuously deform the path of integration while fixing the endpoints, as long as the path does not meet a singular point. Or (what is the same) that the integral of a holomorphic function over a simple closed curve (i.e., one without self-intersections) that does not pass through a singular point or enclose any singular points is zero. This is valid, in fact, for any integral of a regular differential form $\omega = P(x,y)\,dx + Q(x,y)\,dy$ that satisfies the *integrability conditions* $\partial P/\partial y - \partial Q/\partial x = 0$. This result is intimately connected to the so-called "Green–Riemann" formula (which seems in reality to be due to Cauchy[1]), according to which

$$\int_\gamma (P\,dx + Q\,dy) = \iint_K \left(\frac{\partial Q}{\partial x} - \frac{\partial P}{\partial y}\right) dx\,dy,$$

where γ is a simple closed curve, traversed in the counterclockwise direction, and K is the bounded part of the plane defined by γ. In the case of double integrals, Poincaré asked for

> *"the conditions under which the integral does not depend on the surface of integration, but only on the curve that is the boundary of this surface, in the same way that simple integrals, when applied to exact differentials, do not depend on the path of integration, but only on the endpoints of this path."*

He found these *"integrability conditions"* [**38**, no. 2, pp. 328–337], not only for double integrals, but for all multiple integrals. If $\omega = \sum f_{ij}\,dx_i \wedge dx_j$, where $\wedge$ denotes the *exterior product*[2] and $f_{ji} = -f_{ij}$, then Poincaré's condition is that, for any set of three indices $\{i,j,k\}$,

$$\frac{\partial}{\partial x_k} f_{ij} + \frac{\partial}{\partial x_i} f_{jk} + \frac{\partial}{\partial x_j} f_{ki} = 0$$

(which makes $n(n-1)(n-2)/6$ distinct conditions if the space is n-dimensional). For a triple integral, with $\omega = \sum f_{ijk}\,dx_i \wedge dx_j \wedge dx_k$ (the f_{ijk} being anti-symmetric, i.e., they change sign when two indices are exchanged), Poincaré's condition is that, for any $\{i,j,k,l\}$,

$$\frac{\partial}{\partial x_l} f_{ijk} - \frac{\partial}{\partial x_i} f_{jkl} + \frac{\partial}{\partial x_j} f_{kli} - \frac{\partial}{\partial x_k} f_{lij} = 0.$$

---

[1]Cf. Œuvres de Cauchy, series 1, vol. 10, pp. 70–74.

[2]Which has the property that $dx_i \wedge dx_j = -dx_j \wedge dx_i$ and allows one to account for the orientation of the surface. Poincaré did not use this notation $dx_i \wedge dx_j$, but he turned to a (Jacobian) determinant $\frac{D(x_i,x_j)}{D(u,v)}\,du\,dv$ in the integral, which amounts to the same thing.

The general rule, Poincaré said, is that for forms ω of even degree p, all of the signs are $+$, and for odd p the signs alternate. Later, Élie Cartan introduced the notation $d\omega$ (the "exterior derivative" of ω),

$$\omega = \sum_{i_1,\dots,i_k} f_{i_1,\dots,i_k}\, dx_{i_1} \wedge \cdots \wedge dx_{i_k}$$

$$\implies \quad d\omega = \sum_{i_0,i_1,\dots,i_k} \frac{\partial f_{i_1,\dots,i_k}}{\partial x_{i_0}}\, dx_{i_0} \wedge dx_{i_1} \wedge \cdots \wedge dx_{i_k}$$

which allows the Poincaré conditions to be written in the condensed form $d\omega = 0$ (which we express by saying that ω is *closed*).[3] This condition is automatically satisfied for holomorphic functions in an open set of $\mathbb{C}^n$, as in the classical case ($n = 1$). A fundamental (and immediate) property of the exterior derivative is $d^2 = 0$, which Élie Cartan called the "Poincaré property". This property simply translates the fact that an *exact* differential form (i.e., one that can be written as $\omega = d\alpha$) is closed.

Poincaré's reasoning leads to a generalization of the "Green–Riemann" formula, but it only appears implicitly in his 1887 memoir; it would be made explicit by Volterra in 1889 (inspired by Poincaré's memoir), then (in an even simpler form) by Poincaré himself in 1899, in Volume 3 of his *Méthodes nouvelles de la mécanique céleste*: this is the famous generalized "Stokes formula", which, in modern notation, takes the extremely simple and elegant form:

$$\int_{\flat\Omega} \omega = \int_{\Omega} d\omega,$$

where Ω is a $(p+1)$-dimensional domain of integration, $\flat\Omega$ is its p-dimensional boundary,[4] ω is a differential form of degree p (a p-form), and $d\omega$ is its exterior derivative, of degree $p+1$. This formula[5] provides a kind of duality between the operators $\flat$ (boundary) and d (exterior differentiation). It also implies that if Ω has no boundary, i.e., $\flat\Omega = 0$ (in which case we say that Ω is a *cycle*), then $\int_\Omega d\omega = 0$. In other words, *the integral of an exact form over a cycle is zero:* this is the analogue of Cauchy's theorem. Dually, *the integral of a closed form on a boundary is zero:* $\int_{\flat\Omega} \omega = 0$ if $d\omega = 0$.

Let us return to Poincaré's 1887 memoir. Having obtained his "integrability conditions", Poincaré passed to singular differential forms [**38**, no. 3: *"Théorème fondamental"*; no. 4: *"Résidus des fonctions rationnelles"*]. It is well known that if, for example, one integrates a differential form $\omega = f(z)\,dz$ over a closed path c in the plane $\mathbb{C} \cong \mathbb{R}^2$, the value of this integral will be zero if c does not enclose any of the singular points of f, but otherwise it will not, in general, be zero: its value will depend on the singular points that c encloses, and more precisely on the *residues* associated to these points. The *"fundamental theorem"* generalizes this fact to double integrals (and, potentially, to p-fold integrals, for any p). Poincaré ends

[3] The rule $dx_i \wedge dx_j = -dx_j \wedge dx_i$ (along with its corollary $dx_i \wedge dx_i = 0$) automatically incorporates the rule of signs, given by Poincaré, when one regroups the terms in $d\omega$.

[4] The notation ∂ is used more often than $\flat$, but in complex analysis ∂ represents the differential operator $\sum_j \frac{\partial}{\partial \zeta_j} d\zeta_j$.

[5] Which encompasses all the classical formulas known under the names of "Green–Riemann", "Stokes", "Ostrogradski", "gradient formula", etc.

Section 3 with these words (which refer to the mode of representation explained above):

> *"We may summarize the preceding by saying that the integral taken along a closed surface S only depends on the singular curves contained in the interior of this surface."*

This extremely concise sentence signals the birth not so much of a concept (that of the "residue") as of a new vision, one that would, half a century later, lead Georges de Rham [**12**] and then Jean Leray [**30**] to think in terms of homology/cohomology duality (see Section 11.2 below) about the problem of integrating a singular differential form, a problem that before Poincaré was ill-posed because it was attacked frontally.

Two brilliant ideas appear in germinal form in this famous memoir of Henri Poincaré. To express them in a convenient way, first we introduce the necessary language, fixed subsequently by Poincaré and his successors (in particular, de Rham). Let X be an analytic manifold and $S = \{f = 0\} \subset X$ a *smooth* hypersurface (i.e., $f = 0 \implies df \neq 0$); let ω be a regular differential p-form in $X \setminus S$ that is *closed* ($d\omega = 0$). In general, the integral of ω over a p-cycle Γ in $X \setminus S$ (i.e., $\flat\Gamma = 0$, $\dim \Gamma = p$) will be non-zero. But, according to the "Stokes" formula, if we add to ω an exact form $d\alpha$, the integral will not change (because $\int_\Gamma d\alpha = 0$). We say that a p-form ω' belongs to the *cohomology class* of ω if $\omega' - \omega$ is an exact form. We can therefore say that the integral $\int_\Gamma \omega$ only depends on the cohomology class of ω: if we want to study its integrals, it suffices to consider a single p-form in each cohomology class. We write $H^p(X \setminus S)$ for the set of cohomology classes of closed p-forms on $X \setminus S$. This is naturally an additive group: if $\dot\varphi, \dot\psi$ denote the cohomology classes of two closed p-forms φ, ψ, then we write $\dot\varphi + \dot\psi$ for the class of $\varphi + \psi$.

The "Stokes" formula also teaches us that if Γ is a boundary ($\Gamma = \flat\Omega$), then $\int_\Gamma \omega = 0$ (since $d\omega = 0$). Formally, we may say that if the "difference" between two p-cycles Γ and Γ' is a boundary, then

$$\int_\Gamma \omega = \int_{\Gamma'} \omega.$$

But what is the meaning of an equality such as $\Gamma' - \Gamma = \flat\Omega$? Let us take an example: in $\mathbb{C}^* = \mathbb{C} \setminus \{0\}$, consider the circles $\Gamma = C(0, r)$, $\Gamma' = C(0, r')$ ($r' > r$), oriented counterclockwise. We denote by $-\Gamma, -\Gamma'$ the same circles with the orientation reversed. Then $\Gamma' - \Gamma$ is the oriented boundary of the annulus $r \leq |z| \leq r'$; for any closed 1-form ω,

$$\int_{\Gamma' - \Gamma} \omega = \int_{\Gamma'} \omega - \int_\Gamma \omega,$$

which is the integral of ω over the oriented boundary of the annulus. It is also natural to use $n\Gamma$ to denote the circle Γ traversed n times ($n \in \mathbb{N}$), $-n\Gamma$ if it is traversed n times in the opposite direction, and to set

$$\int_{n\Gamma} \omega = n \int_\Gamma \omega \qquad (n \in \mathbb{Z}).$$

More generally, we can create (formal) linear combinations of p-cycles (this again was Poincaré's idea: *Analysis situs*, 1895), with the property that for every closed differential p-form ω on $X \setminus S$,

$$\int_{\lambda_1\Gamma_1+\lambda_2\Gamma_2} \omega = \lambda_1 \int_{\Gamma_1} \omega + \lambda_2 \int_{\Gamma_2} \omega.$$

Here we take complex coefficients ($\lambda_1, \lambda_2 \in \mathbb{C}$). We extend the boundary operator $\flat$ by linearity:

$$\flat(\lambda_1\Gamma_1 + \lambda_2\Gamma_2) = \lambda_1\flat\Gamma_1 + \lambda_2\flat\Gamma_2.$$

A linear combination of p-boundaries is again called a p-boundary. Poincaré (1895) would say that two p-cycles Γ and Γ' are *homologous* (one also says that they belong to the same *homology class*) if their difference is a boundary. The "Stokes" formula tells us that the integral $\int_\Gamma \omega$ only depends on the homology class of Γ: if we want to study these integrals, it therefore suffices to consider a single p-cycle in each homology class. We denote by $H_p(X \setminus S, \mathbb{C})$ the set of homology classes of p-cycles on $X \setminus S$. It is naturally a vector space over $\mathbb{C}$. In fact, Poincaré defined homology (in 1895), and it was de Rham, later, who made explicit the dual notion of cohomology (whence the name), which Poincaré used implicitly without giving it a name.

The first brilliant idea that we find germinating in Poincaré's 1887 memoir is the following. Let us again take a closed regular p-form ω in $X \setminus S$ (X being an analytic manifold, $S = \{f = 0\} \subset X$ a smooth hypersurface). We assume that ω has simple poles along S (which means that the form $f\omega$ is regular in a neighborhood of S). Then we know how to associate to ω a closed differential $(p-1)$-form on S, its *residue* $\mathrm{Res}_S[\omega]$, which in fact only depends on the cohomology class of ω in $X \setminus S$, and we thus obtain a *residue* morphism $\mathrm{Res} : H^p(X \setminus S) \to H^{p-1}(S)$; the ideas developed by Poincaré also carry the germ of the notion of a "dual" morphism δ (one speaks of the *coboundary* morphism), from the homology group $H_{p-1}(S, \mathbb{C})$ to the homology group $H_p(X \setminus S, \mathbb{C})$, which allows us to generalize the Cauchy formula in this context, in the form

$$\int_{\delta(\gamma)} \omega = 2i\pi \int_\gamma \mathrm{Res}_S[\omega] \tag{11.1}$$

for any $(p-1)$-cycle γ in S. In Section 11.2, we will detail the principle of constructing the coboundary morphism (following Jean Leray, inspired by the path sketched by Poincaré in [**38**]). This reduces the calculation of a multiple integral along a $(p-1)$-cycle (traced on the *polar locus* S of ω) of the residue form $\mathrm{Res}_S[\omega]$. We thus extend Cauchy's result to the setting of multiple integrals, according to which

$$\int_c \frac{h(z)}{z}\, dz = 2i\pi \,\mathrm{Ind}(c, 0)\, \mathrm{Res}_{\{0\}} \left[\frac{h(\zeta)}{\zeta}\, d\zeta\right]$$

for every loop c in an open set U of $\mathbb{C}^*$ and for every holomorphic function on U, where $\mathrm{Ind}(c, 0)$ denotes the index of c with respect to the origin (i.e., the number of turns, counted positively in the counterclockwise direction, and negatively in the opposite direction). In this setting, the image of the cycle $\{0\}$ under δ is of course the class of $t \mapsto e^{2i\pi t}$!

We can then pursue the process further and pass to iterated residues on subvarieties of codimension > 1, as sketched by F. Didon [**15**] in 1873 (starting from the Cauchy residue) in the special case of integrals of the form

$$\frac{1}{(2\pi i)^n} \int_{c_1 \times \cdots \times c_n} \frac{df_1}{f_1} \wedge \cdots \wedge \frac{df_n}{f_n}$$

(where the f_i are holomorphic or meromorphic functions of $z_1, \ldots, z_n$ and where for each i, c_i is a simple closed curve in the plane of the variable z_i), with the goal of estimating the number of solutions to a system of algebraic or analytic equations $f_1 = \cdots = f_n = 0$ in a product of disks $D_1 \times \cdots \times D_n$.

After being taken up in the work of Émile Picard [**36**] in 1897, then much later (once the concept of differential form was solidly established and the duality between homology and cohomology was well understood), in those of G. de Rham [**12, 13**] (1932–1936) and André Weil [**43**] (1947), all of these ideas would find their culmination in the memoir of Jean Leray [**30**] (1959); they would thereafter be enriched by the "current-theoretic" viewpoint (see Section 11.3 below), which tightly links the analytic and geometric viewpoints.

The second innovative idea that I shall lift from the 1887 memoir concerns the notion of *regularity* of a meromorphic differential form in the neighborhood of an analytic hypersurface, a notion that precisely translates the nullity of what would become in our modern language the *residual current*: In the one-variable case, a function h that is meromorphic in a punctured neighborhood of 0 is holomorphic at the origin if and only if $\mathrm{Res}_0[gh\,d\zeta] = 0$ for every germ g of a holomorphic function at 0. This notion of regularity, attached (in arbitrary dimension) to the residue morphism introduced by Poincaré, would be led to play a very important role in the modern development of the theory of residues (as much from the algebraic angle as the analytic). One finds it in the developments centered around Abel's theorem (*the trace of a rational form is rational*: see Section 11.5 below) and especially its converse (*if the trace of a differential form along a family of germs of hypersurfaces is rational, then these germs can be joined into an algebraic hypersurface*). It is moreover important to emphasize that in an 1885 note to the *Comptes Rendus de l'Académie des Sciences* [**37**], Poincaré himself sketched an approach to these problems (in the spirit that we have just indicated), an approach that P. Griffiths would bring to light in a rich memoir [**19**] that appeared in 1976, thus profiting from the massive development that the method proposed by Poincaré in [**37**] experienced throughout the twentieth century.

Poincaré's ideas (as he developed them, for example, in [**39**]), also led to the introduction of another class of remarkable formulas, those which make it possible to express elementary solutions of hyperbolic equations with constant coefficients in terms of periods of abelian integrals. The work of I. Fredholm (1900), and later of I. G. Petrovskiĭ (1945) and Jean Leray (1953) is drawn in the line of this legacy. The fundamental Ehrenpreis–Palamodov principle[6] (1960) admits, with residue theory in its early phase, an effective formulation, at least in the context of complete intersection[7] (a favorable condition for realizing the duality principle: see below).

[6]This is a generalization of the famous Euler principle—according to which every solution of a differential equation in the form $P(d/dt)\,x(t) = 0$ (P polynomial) can be expressed as a sum of exponential-polynomial solutions $e^{\alpha_j t}Q_j(t)$ (α_j roots of P, Q_j polynomials)—to *systems of partial differential equations* with constant coefficients. According the Ehrenpreis–Palamodov principle, the solutions of a system $P_i(\partial/\partial t_1, \ldots, \partial/\partial t_n)\,x(t_1, \ldots, t_n) = 0$ are sums of terms of the form $\int_{P^{-1}(0)} Q_k(\alpha; t)\, e^{\langle \alpha, t\rangle}\, d\mu_k(\alpha)$, where $\alpha = (\alpha_1, \ldots, \alpha_n)$, $t = (t_1, \ldots, t_n)$, $\langle \alpha, t\rangle = \alpha_1 t_1 + \cdots + \alpha_n t_n$, the Q_k are polynomials, the μ_k are measures carried by $P^{-1}(0)$, and $Q_k(\alpha; t)e^{\langle \alpha; t\rangle}$ is a solution to the system for every α within the support of μ_k.

[7]Meaning that each of the analytic sets involved can be defined by a number of equations equal to its codimension.

We will attempt in this brief exposition to unearth (in the various directions we have just indicated) the development of the concept brought to light by Poincaré, its increasing ubiquity in diverse mathematical fields, as well as its interface with, for example, modern physics.

11.2. The Poincaré residue from the cohomological angle

Using an effective lemma on the division of differential forms (adapted by J. Leray from G. de Rham), we can better outline Poincaré's approach to the notion of residue. Let X be an analytic manifold of dimension n, U an open subset of X, and S a smooth hypersurface in U defined locally by $S = \{f = 0\}$. Let k be an arbitrary strictly positive integer and p be an integer between 1 and $2n$. Then every closed regular p-form φ in $U \setminus S$ that is *semimeromorphic and presents a pole of order k along S* (which means that $f^k\varphi$ is a C^∞-form in U) can be divided by df in the following manner:

$$\varphi = \frac{df}{f^k} \wedge \alpha + \frac{\beta}{f^{k-1}},$$

where α and β are regular forms in U.

The case where $k = 1$ is particularly interesting, because then the restriction of α to a smooth hypersurface S is a closed form, which, in the case where φ is a $(p, 0)$-form[8] in $U \setminus S$, is precisely the Poincaré residue. This lemma on the division of forms, once iterated, makes possible what is called *lowering the order of the pole*: given a closed p-form in $U \setminus S$, if its cohomology class in $H^p(U \setminus S)$ has a semimeromorphic representative, then this class also has a semimeromorphic representative with at most a simple pole along S. Of course, the possibility of lowering the order in this manner comes at the cost of using a partition of unity[9] and departs from the framework Poincaré had in mind[10] in 1887. Nonetheless, it makes it possible to associate to each cohomology p-class $\dot{\varphi}$ in $H^p(U \setminus S)$ that admits a semimeromorphic representative with poles along S, a cohomology class $\text{Res}[\dot{\varphi}]$ in $H^{p-1}(S)$, which is the class represented by the residue form of φ_1, where φ_1 is a representative of the class $\dot{\varphi}$ having at most a simple pole along S.

In fact, the hypothesis that the cohomology class $\dot{\varphi}$ in $H^p(U \setminus S)$ contains a semimeromorphic representative turns out to be unnecessary. The merit of the theory developed by Jean Leray in 1959 [**30**] (in the wake of Poincaré and G. de Rham) is that, by playing with the homology/cohomology duality conceived by Poincaré, it makes obvious the fact that the residue morphism is well defined from $H^p(U \setminus S)$ to $H^{p-1}(S)$, and that there is an associated *coboundary morphism* δ from $H_{p-1}(S, \mathbb{C})$ to $H_p(U \setminus S, \mathbb{C})$, the pairing of the two morphisms (residue and coboundary) being

---

[8]We say that a differential $(p+q)$-form is of bidegree (p, q), or that it is a (p, q)-form, if it is homogeneous of degree p in $d\zeta_1, \dots, d\zeta_n$ and of degree q in $d\bar{\zeta}_1, \dots, d\bar{\zeta}_n$, i.e., if it can be written as

$$\sum_{i_1,\dots,i_p,j_1,\dots,j_q} f_{i_1,\dots,i_p,j_1,\dots,j_q}(\zeta_1, \dots, \zeta_n, \bar{\zeta}_1, \dots, \bar{\zeta}_n)\, d\zeta_{i_1} \wedge \dots \wedge d\zeta_{i_p} \wedge d\bar{\zeta}_{j_1} \wedge \dots \wedge d\bar{\zeta}_{j_q}.$$

[9]I.e., a decomposition $1 = \sum_i \varphi_i$ with the functions φ_i regular (for example, C^∞) and having support in a family of given open sets. This allows the decomposition of *every* regular function f into $\sum_i \varphi_i f$ and enables us to pass certain properties from "local" to "global" by "gluing" (see for example [**42**]).

[10]His method for managing the case of poles of higher order arose rather from a method of derivation with respect to auxiliary parameters, which we will discuss again in Section 11.5.

realized precisely by the *residue formula* (11.1). The *coboundary morphism* is defined as follows. Let $\dot{\tau}$ be an element of the homology group $H_{p-1}(S, \mathbb{C})$. A p-cycle γ in $U \setminus S$ is a representative of $\delta(\dot{\tau})$ *if and only if* there exists a representative τ of $\dot{\tau}$ and a $(p+1)$-simplex σ in U, retractable onto S, such that its support intersects S along the support of τ transversely and γ is homologous to the boundary of σ in $U \setminus S$. This construction merely reproduces the well-known situation in dimension 1, where $\tau = \{0\} = S$ and where σ is a disk in the complex plane centered at 0 (therefore retractable onto this point): a cycle γ in $U \setminus \{0\}$ is a representative of $\delta(\{0\})$ if it is homologous to the boundary of such a disk in $U \setminus \{0\}$, thus conforming to the classical residue formula in one variable.

This mechanism can be iterated (as Poincaré already understood). Indeed, let $S_1, \ldots, S_m$ $(1 \leq m \leq n)$ be m hypersurfaces in U, in transverse position. We have the chain of inclusions

$$\begin{gathered} S_1 \cap \cdots \cap S_m \subset S_1 \cap \cdots \cap S_{m-1}, \\ (S_1 \cap \cdots \cap S_{m-1}) \setminus S_m \subset (S_1 \cap \cdots \cap S_{m-2}) \setminus S_m, \\ (S_1 \cap \cdots \cap S_{m-2}) \setminus (S_{m-1} \cup S_m) \subset (S_1 \cap \cdots \cap S_{m-3}) \setminus (S_{m-1} \cup S_m), \\ \vdots \\ S_1 \setminus (S_2 \cup \cdots \cup S_m) \subset U \setminus (S_2 \cup \cdots \cup S_m). \end{gathered}$$

Iterating the coboundary morphisms then allows us to construct an *iterated coboundary morphism* (denoted $\delta^{S_1,\ldots,S_m}$) that transports the homology on the smooth variety $S_1 \cap \cdots \cap S_m$ into the homology on $U \setminus (S_1 \cup \cdots \cup S_m)$. This morphism is paired with an *iterated residue morphism* (denoted $\mathrm{Res}_{S_1,\ldots,S_m}$) that transports the cohomology on $U \setminus (S_1 \cup \cdots \cup S_m)$ into the cohomology on $S_1 \cap \cdots \cap S_m$; the two morphisms are paired via the *iterated residue formula*

$$\int_{\delta^{S_1,\ldots,S_m}(\gamma)} \varphi = (2i\pi)^m \int_\gamma \mathrm{Res}_{S_1,\ldots,S_m} [\varphi] \,. \tag{11.2}$$

Within the global framework of rational forms envisioned by Poincaré (where the flexibility of analytic beings, notably partitions of unity, can no longer be exploited), the idea of lifting the order of a pole has also continued along its path since the work of Leray, and then later that of P. Griffiths [**18**]. By introducing methods based on "mixed Hodge theory", P. Deligne and A. Dimca [**11**] were able to extend to the singular case some of the lowering results that were unreachable, except in the case where S is smooth, in the theory inaugurated by Poincaré and followed by G. de Rham and J. Leray [**18**]. For example, a rational p-form on $\mathbb{P}^n(\mathbb{C})$ of denominator Q^k (where Q is a homogeneous polynomial defining a possibly singular hypersurface) admits in its cohomology class (in $H^p(\mathbb{P}^n(\mathbb{C}) \setminus \{Q = 0\})$) a rational representative of denominator Q^l with $l \leq \min(k, n)$. To my way of thinking, this is a deep result that, for many reasons—such as the desire to lower the order of a pole or the use of the complex of forms with logarithmic poles (see Section 11.4) and its two filtrations (the "mixed Hodge" filtration, and another according to the order of the pole)—is truly part of Henri Poincaré's legacy.

We cannot here develop all the cohomological aspects of the theory begun by Poincaré and enriched throughout the 20th century at the crossroads of geometry and the nascent pluricomplex analysis; F. Norguet's article [**34**] (which contributed an active part) presents the state of affairs, with which we invite the reader to make acquaintance.

11.3. The Poincaré residue revisited in the light of the theory of currents

Within the context of modern pluricomplex analysis, the definition (today both classical and essential) of the Poincaré residue is the following. We consider a meromorphic differential n-form

$$\varphi = \frac{h(\zeta)\, d\zeta_1 \wedge \cdots \wedge d\zeta_n}{f(\zeta)}$$

on an open subset U of $\mathbb{C}^n$, in which h and f are holomorphic functions, the complex hypersurface $V(f) := \{f = 0\}$ being defined in reduced form (i.e., f is a product of pairwise distinct irreducible factors). Here the polar locus of ω is the hypersurface $V(f)$; the set of singular points of $V(f)$ is

$$\operatorname{sing} V(f) = \{z : f(z) = 0,\ df(z) = 0\},$$

and the regular part of $V(f)$ is the smooth ($(n-1)$-dimensional) surface

$$S = \operatorname{reg} V(f) = V(f) \setminus \operatorname{sing} V(f).$$

The Poincaré residue of φ along S is the $(n-1)$-form defined on each open set

$$U_j := \left\{\zeta \in V(f) : \frac{\partial f}{\partial \zeta_j}(\zeta) \neq 0\right\}, \qquad j = 1, \ldots, n,$$

by

$$\operatorname{Res}_S[\varphi] = \left[(-1)^{n-j}\, \frac{h(\zeta)\, d\zeta_1 \wedge \cdots \wedge d\zeta_{j-1} \wedge d\zeta_{j+1} \wedge \cdots \wedge d\zeta_n}{\partial f(\zeta)/\partial \zeta_j}\right]\Bigg|_{U_j},$$

the definitions being "glued together" on the intersections of the various U_j. The idea here is exactly the construction proposed by Poincaré in [**38**, no. 4] with the goal of being able to exploit the Cauchy formula (11.1).

The Poincaré residue of the n-form φ is a closed regular form on the smooth hypersurface $S = \operatorname{reg} V(f)$, and one may naturally wonder if $\operatorname{Res}_S[\varphi]$ can be extended to all of $V(f)$. Poincaré certainly did not get rid of the idea (whose inspiration is unabashedly analytic) according to which *the inessential singularities of a holomorphic function of one variable are precisely those that can be integrated in the framework of distribution theory* (in the sense of Laurent Schwartz [**42**])—more generally, any meromorphic function of several complex variables can be regarded as a distribution on the open set where it is defined, *including the polar locus.* Today the duality between differential forms and currents[11] (for the moment in the open set $U \setminus \operatorname{sing} V(f)$) makes it possible to associate to the Poincaré residue an $(n, 1)$-current in $U \setminus \operatorname{sing} V(f)$, which we also denote $\operatorname{Res}_S[\varphi]$, whose action on a

[11] The notion of a current (due to de Rham and reinterpreted by Schwartz: see [**42, 29**]) generalizes that of a differential form, in the sense that the coefficients of the basis forms $d\zeta_{i_1} \wedge \cdots \wedge d\zeta_{i_p} \wedge d\bar{\zeta}_{j_1} \wedge \cdots \wedge d\bar{\zeta}_{j_q}$ are distributions and no longer functions. We say that a current is of *bidegree* (p, q) (or that it is a (p, q)*-current*) if it is homogeneous of degree p in $d\zeta_1, \ldots, d\zeta_n$ and of degree q in $d\bar{\zeta}_1, \ldots, d\bar{\zeta}_n$.

test $(0, n-1)$-form ψ (of class C^∞ and having compact support in $U \setminus \operatorname{sing} V(f)$) can be expressed as

$$\langle \operatorname{Res}_S[\varphi]\,,\,\psi\rangle := \int_S \operatorname{Res}_S[\varphi] \wedge \psi,$$

the integration being unambiguously defined here because S is smooth and ψ has compact support in $U \setminus \operatorname{sing} V(f)$. Moreover, do we not again find in the notion the principle of *Poincaré duality*:

$$H^p(\cdot) \times H^{n-p}(\cdot) \to \mathbb{C} : (\dot{\varphi}, \dot{\psi}) \mapsto \int_X \varphi \wedge \psi$$

(where $\dot{\varphi}, \dot{\psi}$ denote, as before, the cohomology classes of φ, ψ)? Even better, based on the model of constructing distributions of the "Principal Value" type, we know how to extend this $(n, 1)$-current $\operatorname{Res}_S[\varphi]$ to a current on U (called the *standard extension*, from the fact of its construction), which we again denote $\operatorname{Res}_S[\varphi]$ and whose action on a test $(0, n-1)$-form (of class C^∞ and having compact support in all of U, this time) can be expressed as

$$\langle \operatorname{Res}_S[\varphi]\,,\,\psi\rangle := \lim_{\varepsilon\to 0} \int_{S\cap\{\|df(\zeta)\|\geq\varepsilon\}} \operatorname{Res}_S[\varphi] \wedge \psi\,.$$

In fact, the current $\operatorname{Res}_S[\varphi]$ thus constructed belongs to the class of currents introduced in 1978 by N. Coleff and M. Herrera [**7**]: of course, these authors profited from the *theorem of resolution of singularities* that had just been established by H. Hironaka [**23**], but at the same time they gained inspiration from the method inaugurated by Poincaré and followed by G. de Rham. One can show, in fact, that our current $\operatorname{Res}_S[\varphi]$ coincides with the $\overline{\partial}$-closed[12] $(n, 1)$-current $\overline{\partial}(1/f) \wedge (h(\zeta)\, d\zeta_1 \wedge \cdots \wedge d\zeta_n)$, whose action on a test $(0, n-1)$-form ψ in U can be expressed as

$$\langle \overline{\partial}(1/f) \wedge (h\, d\zeta_1 \wedge \cdots \wedge d\zeta_n)\,,\,\psi\rangle = \lim_{\varepsilon\to 0} \frac{1}{2i\pi} \int_{|f|=\varepsilon} \varphi \wedge \psi\,.$$

Apart from these two major properties (being the *standard extension* and $\overline{\partial}$-closed), the current thus defined by Coleff–Herrera is *annihilated by all antiholomorphic functions vanishing on* $V(f)$*:* intuitively, one can say that it treats antiholomorphic objects as "neutral" objects, which it assimilates as it would constants. We thus see that this current is primarily an algebraic object, even though our construction was of an analytic nature. For a current of bidegree (n, p) supported on an analytic set V of pure codimension equal to p (here $V = V(f)$ and $p = 1$), these three properties characterize belonging to the Coleff–Herrera class CH_V. We mention in passing that this class is today known to play an essential part in the decomposition

$$\operatorname{Ker}_{\overline{\partial}} \mathcal{C}_V^{n,p} = \mathrm{CH}_V \oplus \overline{\partial}(\mathcal{C}_V^{n,p-1})\,,$$

where $\mathcal{C}_V^{n,p}$ denotes the sheaf of currents of bidegree (n, p) supported on the analytic set V (cf. [**14**]).

Thus, *the Poincaré residue of the form* φ*, considered initially as a closed regular form on* $\operatorname{reg} V(f)$*, and then reinterpreted as an* $(n, 1)$*-current in* $U \setminus \operatorname{sing} V(f)$*, has a standard extension to an* $(n, 1)$*-current in the Coleff–Herrera class on all of* U.

---

[12]That is, annihilated by the operator $\overline{\partial} = \sum_i \frac{\partial}{\partial\bar{\zeta}_i}\, d\bar{\zeta}_i$ (which transforms a form or a current of bidegree (p, q) into a form or a current of bidegree $(p, q+1)$). This is certainly the case for the form given here, because $\overline{\partial}^2 = 0$ and $\overline{\partial} h = 0$ (h is holomorphic).

What explains the importance of the idea brought to light by Poincaré is that one can prove that *every* $(n-1,0)$-form that is defined and regular on the smooth hypersurface $S = \operatorname{reg} V(f)$ and induces by the standard extension a $\overline{\partial}$-closed $(n,1)$-current in U (following the principle described above beginning with the form $\operatorname{Res}_S[\varphi]$) is indeed the Poincaré residue α/g of a meromorphic $(n,0)$-form ω in U, with poles on $V(f)$, that can be written as

$$\omega = \frac{1}{g}\left(\frac{df}{f} \wedge \alpha + \beta\right), \tag{11.3}$$

where α and β are respectively holomorphic $(n-1,0)$- and (n,n)-forms on U, and g is a holomorphic function on U that, together with f, defines a complete intersection in U. We also note that $V(g) \cap V(f)$ is necessarily included in $\operatorname{sing} V(f)$ since the form must be regular on $\operatorname{reg} V(f)$.

What is more natural than to think that the regular $(n-1,0)$-forms on the reduced analytic set $\{f = 0\}$ are the regular $(n-1)$-forms θ on $\operatorname{reg} V(f) = S$ such that the $(n,1)$-current on $U \setminus \operatorname{sing} V(f)$

$$\psi \mapsto \int_S \theta \wedge \psi$$

has a standard extension to a Coleff–Herrera current on U (the regularity on all of $V(f)$ being realized precisely by the fact that this current is $\overline{\partial}$-closed)? *These forms are therefore nothing other than the Poincaré residues along* $\operatorname{reg} V(f)$ *of meromorphic* $(n,0)$*-forms of the form* (11.3), *with* $V(f) \cap V(g) \subset \operatorname{sing} V(f)$.

We now understand better the importance of the ideas introduced by Poincaré, and we notice their legacy contains in its direct lineage such work as that of J. Leray [**30**] (of whom we spoke above, in the section on the cohomological approach) or, more recently, the following result of A. K. Alexandrov and A. Tsikh [**2**] (where I would like to emphasize even more the extent to which the notion of a regular differential form on a singular analytic variety is a crucial one, as much from the analytic angle as from the algebraic one [**3, 19, 28**]):

THEOREM (Alexandrov, Tsikh [**2**]). *Let X be a complex n-dimensional analytic manifold, U an open subset of X, and $f_1, \dots, f_m$ holomorphic functions defining a reduced complete intersection of dimension $n-m$ in U, $V(f_1, \dots, f_m) = \{f_1 = \cdots = f_m = 0\}$. For $k = 0, \dots, n-m$, the regular differential k-forms on $V(f_1, \dots, f_m)$ can be expressed on* $\operatorname{reg} V(f)$ *exactly as the iterated Poincaré residues*[13] *α/g of the meromorphic $(k+m)$-forms that can be written*

$$\varphi = \frac{1}{g}\left(\frac{df_1}{f_1} \wedge \cdots \wedge \frac{df_m}{f_m} \wedge \alpha + \beta\right), \tag{11.4}$$

where g is a holomorphic function on U that together with $(f_1, \dots, f_m)$ defines a complete intersection in U (contained in $\operatorname{sing} V(f_1, \dots, f_m)$*), α is a regular k-form on U, and β is a sum of meromorphic differential k-forms on U, each of whose polar loci is contained in one of the hypersurfaces $S_j := \bigcup_{l \neq j} \{f_l = 0\}$, $j = 1, \dots, m$.*[14]

[13] Relative to $\operatorname{reg} V(f_1) \setminus V(f_2, \dots, f_m)$, etc.

[14] Leray's result dealt with the case where the hypersurfaces $V(f_j)$, $j = 1, \dots, m$, are all smooth and intersect transversely (in this case g, which is in fact locally nothing other than Oka's universal denominator [**35**], can be taken to be equal to 1).

The iterated Poincaré residues relative to $\operatorname{reg} V(f_1, \ldots, f_m)$ therefore appear as regular forms on the set $\operatorname{reg} V(f_1, \ldots, f_m)$ which, once extended in the sense of the theory of currents, no longer has a residue along $\operatorname{sing} V(f_1, \ldots, f_m)$. This interpretation corresponds well with Poincaré's statement cited in the introduction: *"The integral taken along a closed surface* [...] *only depends on the singular curves contained in the interior of this surface"*.

11.4. Logarithmic differential forms; connections with logarithmic poles

Apart from the study of the regularity of a singular variety, mentioned in Section 11.3, the rereading of Poincaré's work we have just done leads us to another trail. We encounter therein, with expressions of type (11.4), the notion of a *logarithmic differential form* (one can equally speak here of a *multilogarithmic differential form*): while φ is a form whose polar locus is contained in $\{f_1 \cdots f_m = 0\}$, the forms $f_j \, \varphi$ and $f_j \, d\varphi$, for each $j = 1, \ldots, m$, are sums of forms, each of whose polar loci is contained in the hypersurface $S_j := \bigcup_{l \neq j} \{f_l = 0\}$.

This notion, which draws its origins from the contributions of Henri Poincaré, reappeared in the 1970s in the work of Pierre Deligne [**10**] and Nicholas Katz [**25**]. In [**10**],[15] P. Deligne, in order to establish a "Riemann–Hilbert correspondence" between monodromy groups and integrable Fuchsian systems of partial differential equations (or flat connections on complex analytic manifolds), was led to introduce the very important notion of a *connection with logarithmic poles.* To give an idea of this notion, let us take an example. Let X be the unit polydisk Δ^n, $S = \{0\} \times \Delta^{n-1}$, and $\mathcal{L}$ a vector bundle over X such that $\mathcal{L}_0 := \mathcal{L}|_{X \setminus S}$ is equipped with an integrable connection ∇ having logarithmic poles along S, which means that the entries of its matrix in any local basis for $\mathcal{L}$ are logarithmic differential forms (as in (11.4)). Then $\pi_1(X \setminus S)$ (the fundamental group of $X \setminus S$: another invention by Poincaré!) is isomorphic to $\mathbb{Z}$, and its generator induces on $\mathcal{L}_0$ a so-called monodromy transformation, which extends to an automorphism σ of $\mathcal{L}$. In a given basis of local sections, the restriction of σ to S has $\exp(-2i\pi \mathrm{Res}_S(\Phi))$ as its matrix, where Φ denotes the matrix of ∇ in the same basis. Here we perceive the importance of the Poincaré residue, through the notion of the *residue of a connection,* in the study of differential equations and systems, in particular for describing the monodromy action.

More generally, S will be a "normal crossing divisor" in an analytic manifold X, and $\mathcal{L}_0$ a vector bundle over $X \setminus S$ equipped with an integrable connection ∇ with logarithmic poles (relative to the divisor S, here in addition having normal crossings, which returns us to the situation where f is a monomial).

For the reasons just mentioned, the theory of logarithmic differential forms was systematically developed by K. Saito [**40**] in 1976. Logarithmic differential forms, which we have just illustrated starting with original ideas of Poincaré, also appear naturally as the coefficients of the "Gauss–Manin connection" associated to a flat family of complex spaces. Such a connection can moreover be considered as an affine connection with logarithmic poles. There are deep, tight relations between

[15] In Section 3 of [**10**], one can also find, within the framework of normal crossing divisors, a presentation (now classical) of the Poincaré residue that is more geometric than the (analytic) presentation we have sketched here.

the properties of logarithmic connections and a certain class of holonomic systems of differential equations with singularities along a hypersurface.

11.5. Poincaré duality and global residues

Another question investigated by Poincaré [**37**] concerned the possibility of generalizing a famous theorem of Abel. Consider a curve Γ of degree d in $\mathbb{P}^2(\mathbb{C})$, intersected by a family of lines $l(a,b) : y = ax + b$. Let $x_\lambda(a,b), y_\lambda(a,b)$ be the coordinates of the d points of intersection (in $\mathbb{P}^2(\mathbb{C})$) of $l(a,b)$ and Γ, and let $\omega = A(x,y)\,dx + B(x,y)\,dy$ be a differential 1-form. Substitute $x_\lambda(a,b)$ and $y_\lambda(a,b)$ for x and y in $A(x,y)\,dx+B(x,y)\,dy$, and add up the contributions of these d points: this sum is called the *trace* of ω relative to Γ (it is a 1-form in a and b). The classical theorem of Abel says that if ω is *rational* (i.e., if A and B are rational functions), then the trace of ω is *rational.* An obvious example is where $\Gamma = \{x^2 + y^2 = 1\}$, $\omega = y^{-1}\,dx$, and $l(a)$ is the familyy $= ax$; then the trace of ω is $-2(1 + a^2)^{-1}\,da$.

The problem Poincaré posed for himself consisted in finding, more generally, a closed-form expression for the trace of a differential form relative to a subvariety defined as a complete intersection of dimension m, intersected by a (continuously parametrized) family of varieties that are themselves complete intersections (of dimension $n - m$). This trace is presented as a complete sum of residues: this is tied to a formula known today under the name of the *Lelong–Poincaré formula,* and it expresses that the current of integration (with bidegree (m,m)) on an analytic set of dimension $n - m$ (defined as a complete intersection) equals the product of the residual current (the "current-theoretic" incarnation of the iterated Poincaré residue) and the Jacobian differential form, which generalizes the classical formula (in one variable):

$$\mathrm{Res}_0\left[\frac{h(\zeta)\,df(\zeta)}{f(\zeta)}\right] = \mathrm{order}_0(f)\,h(0).$$

This problem belongs to a much more general framework. Let X be an n-dimensional complete projective complex analytic variety;[16] let $S_1, \ldots, S_n$ be effective divisors on X defined by global sections. We are asking for closed-form expressions for the complete sums of iterated residues relative to $S_1, \ldots, S_n$, and notably for precise statements on the dependence of these expressions on the perturbation parameters of the defining global sections, once the S_j are perturbed. Some crucial questions, still open, at the heart of the intersecting domains between physics and geometry (such as *mirror symmetry*: see for example [**4**]) can be reinterpreted (in the light of ideas introduced or reintroduced by I. M. Gelfand, M. M. Kapranov, and A. V. Zelevinsky [**16**]) in terms of conjectural formulas that involve such complete sums of residues, and especially their dependence on the parameters defining the perturbed S_j.

A second example of this situation will particularly hold our attention, because in it we find the key ideas Poincaré introduced in [**38**] to consider the calculation of multiple integrals when the differential form under the integral has a multiple pole along the polar locus. Let $f_2, \ldots, f_n$ be Laurent polynomials in n variables whose

---

[16] One might think, for example, of complex projective space $\mathbb{P}^n(\mathbb{C})$, or of a weighted projective space, or even more generally of a toric variety attached to a convex polyhedron in $\mathbb{R}^n$ with non-empty interior and vertices contained in a lattice...

Newton polytopes[17] $\Delta_1, \dots, \Delta_n$ are such that the decomposition $\delta = \delta_1 + \cdots + \delta_n$ of each face δ of $\Delta_1 + \cdots + \Delta_n$ (with δ_j a face of Δ_j) always contains at least one face reduced to a vertex. (In the case $n = 2$, the example where Δ_1 and Δ_2 have no parallel faces illustrates this type of situation.) In this case [**17, 44**], the total iterated residue of a monomial

$$\mathrm{Res}_{V(f_1,\dots,f_n)}\left[\frac{\zeta^\alpha\, d\zeta}{f_1\cdots f_n}\right] := \sum_{\{\zeta\in(\mathbb{C}^*)^n\,;\, f(\zeta)=0\}} \mathrm{Res}_{V(f_1,\dots,f_n),\zeta}\left[\frac{\zeta_1^{\alpha_1}\cdots\zeta_n^{\alpha_n}\, d\zeta_1\wedge\cdots\wedge d\zeta_n}{f_1\cdots f_n}\right]$$

can be presented as a completely explicit Laurent polynomial in the coefficients of the polynomials f_j. If ζ^m is one of the monomials appearing in f_j and a is its coefficient, then, for every $l \in \mathbb{N}$,

$$\left(\frac{\partial}{\partial a}\right)^l \left(\mathrm{Res}_{V(f_1,\dots,f_n)}\left[\frac{\zeta^\alpha\, d\zeta}{f_1\cdots f_n}\right]\right) = (-1)^l\, l!\,\mathrm{Res}_{V(f_1,\dots,f_n)}\left[\frac{\zeta^{\alpha+ml}\, d\zeta}{f_1\cdots f_j^{l+1}\cdots f_n}\right].$$

We note that this idea (differentiating with respect to the parameters to increase the order of the pole) already appears in its germinal form in the calculations contained in no. 4 of Poincaré's memoir [**38**].

Now, let $f_0, \dots, f_n$ be polynomials on $\mathbb{C}^n$ (the affine case) or Laurent polynomials on $(\mathbb{C}^*)^n$ (the "toric" case), inducing on $X = \mathbb{P}^n(\mathbb{C})$ (or on a toric compactification X of $(\mathbb{C}^*)^n$ associated to the sum of the supports of the Laurent polynomials f_j) the divisors defined by global sections $F_0, \dots, F_n$ and not intersecting in X. The complete sum of the iterated residues (in $\mathbb{C}^n$ or $(\mathbb{C}^*)^n$, respectively) of a rational function h/f_0, relative to the system $(f_1, \dots, f_n)$, can again be calculated with the aid of Poincaré duality.[18]

Thus, the test for the regularity of rational differential forms on a complete intersection algebraic subvariety of $\mathbb{P}^n(\mathbb{C})$ can be deduced from the versions (proposed by Carl Jacobi [**24**] or A. Khovanskii [**26**]) of the residue theorem on a compact algebraic variety, versions that lead to interesting geometric applications (in the wake of the Cayley–Bacharach theorems [**19, 27**]). In the same way, the handling of the notion of trace, just as in the explicit inversion procedures of Abel's theorem (if a trace is defined, *a priori* locally, then it comes from global algebraic objects [**18, 22**]), appeals in a crucial way to the expression of sums of iterated residues (as a function of the coefficients involved in the sections at play) as much as to the mechanism for taking the Poincaré residue itself. We emphasize once again that

[17] A Laurent polynomial in n variables is a finite sum of the form $f = \sum a_{\alpha_1,\dots,\alpha_n}\, \zeta_1^{\alpha_1}\cdots\zeta_n^{\alpha_n}$ $(a_{\alpha_1,\dots,\alpha_n} \neq 0)$, which we denote more concisely by $\sum a_\alpha \zeta^\alpha$. The support of such a polynomial is the set of n-tuples of the exponents $\alpha = (\alpha_1, \dots, \alpha_n) \in \mathbb{Z}^n$ that are involved: it is a finite set of points of the lattice $\mathbb{Z}^n$ in $\mathbb{R}^n$. Its convex hull is a convex polytope in $\mathbb{R}^n$, called the *Newton polytope* of f.

[18] For the initiated: one can show (see for example [**9**]) that this complete sum of iterated residues is the image under the trace function $\mathrm{Tr}_X : H^n(X, \Omega_X^n) \to \mathbb{C}$ of the Čech coboundary $[H\Omega/(F_0\cdots F_n)]$ associated to the covering of X by the open sets $\{F_j = 0\}$ $(j = 0, \dots, n)$; here Ω denotes the Euler form on $\mathbb{P}^n(\mathbb{C})$ or on the toric compactification X. (In the projective case, the Euler form is the form obtained by expressing the form $d\zeta_1 \wedge \cdots \wedge d\zeta_n$ in homogeneous coordinates, a procedure that one can imitate in the toric case by using the existence of a ring of homogeneous coordinates $\mathbb{C}[X_0, \dots, X_{n+s}]$, which makes possible a gradation of the Chow $(n-1)$-group $A_{n-1}(X)$ of classes of Cartier divisors modulo rational equivalence [**9**].)

such ideas are in the spirit of Poincaré's 1885 note [**37**] and his 1887 memoir [**38**]. In [**5**], one can find an analysis of the path of the ideas from Abel up to the modern theory of residues such as we have tried (very partially) to describe here from only a few points of view (with Poincaré's own contributions as a basis).

We also note that today one finds (in the situation where all of the divisors involved are affine hyperplanes) the need to make explicit the complete sums of the residues in more analytic settings, namely, the search for "developing" formulas for (multivariable) Mellin–Barnes integrals [**6**] or the study of extending multiple Dirichlet series [**31**]; such results naturally touch on tools of physics, dynamics, or number theory.

11.6. Residues and duality: Conclusion

We have mentioned how the notion of Poincaré residue is found intimately linked to the notion (as algebraic as it is analytic) of *trace*. For example, the *Lelong–Poincaré formula*

$$\langle [V(f)], h\rangle = \left\langle \mathrm{Res}_{V(f_1),\dots,V(f_m)}\left[\frac{df_1\wedge\cdots\wedge df_m}{f_1\cdots f_m}\right], h\right\rangle,$$

where $f_1,\dots,f_m$ define a complete intersection cycle $[V(f)]$ in an analytic n-dimensional manifold, is a trace formula. The same is true of the *Jacobi–Kronecker interpolation formula* (here $m=n$)

$$h(z) = \left\langle \mathrm{Res}_{V(f_1),\dots,V(f_m)}\left[\frac{\Delta(z,\zeta)\,d\zeta_1\wedge\cdots\wedge d\zeta_n}{f_1\cdots f_m}\right], h\right\rangle,$$

which is the "multivariable" version of the Lagrange interpolation formula: Δ denotes a *Bézoutian* resultant. The action of the iterated residue is understood here in the sense of Poincaré duality. The mechanism leading to the construction of the iterated Poincaré residue is, in both cases, very clearly involved, even when one is freed here from the regularity and the transversality of the hypersurfaces $V(f_j)$. The exploitation of these trace formulas, with the goal of effectively realizing duality, is drawn in the program inaugurated by A. Grothendieck in the 1970s and actively followed since (see for example [**21, 32, 41**]). We find these trace formulas again (in the form of substitution within the theory of elimination) in the effective formulation of problems of algebraic geometry (Hilbert's Nullstellensatz, for example) or of analytic geometry (the integral closure of ideals in a local regular ring and the realization of the Briançon–Skoda theorem [**33**]).

The calculation of multivariable residues was thus developed by means of a few key rules, such as the *transformation law* of [**20**] (an analytic version of the algebraic rule given by Wiebe's theorem), all of which are avatars of trace formulas. The technical tool that it produces makes it possible today to extend the calculations began by Poincaré in [**38**] to a more general context, while still maintaining the ideas that presided at their genesis (see [**1**] for numerous examples and references). We thus see that the tool of the Poincaré residue, which may be studied from an algebraic, analytic, or geometric angle, plays a pivotal role on which methods of various disciplines are based, all inspired by the same double legacy: on the one hand, that of Jacobi, Cayley, Macaulay, etc., who, since the 19th century, had explored the "dualizing" character; on the other hand, that of Poincaré himself, who knew how to illuminate it from a new angle, this time more geometric, that of the homology/cohomology duality. We have tried here to present a few facets of this tool, in the light of contemporary mathematical concerns.

Bibliography

[1] L. Aizenberg and A. P. Yuzhakov, *Integral representations and residues in multidimensional complex analysis*, Transl. Amer. Math. Soc. **58**, American Mathematical Society, Providence, 1979.

[2] A. K. Alexandrov and A. Tsikh, *Théorie des résidus de Leray et formes de Barlet sur une intersection complète singulière*, Comptes Rendus de l'Académie des Sciences (Paris), Série I, **333** (2001), pp. 1–6.

[3] D. Barlet, *Le faisceau* $\omega_X^{\bullet}$ *sur un espace analytique* X *de dimension pure*, Lecture Notes in Math. **670**, Springer 1978, pp. 187–204.

[4] V. V. Batyrev and E. N. Materov, *Toric residues and mirror symmetry*, Moscow Math. Journal **2** (2002), no. 3, pp. 435–475.

[5] J. E. Björk, *Residues and $\mathcal{D}$-modules*, in *The legacy of Niels Henrik Abel*, Springer, Berlin, 2004, pp. 605–651.

[6] A. A. Cheshel', M. Passare, and A. Tsikh, *Multiple Mellin–Barnes integrals as periods on Calabi-Yau manifolds with several moduli*, Theoret. and Math. Phys. **109**, no. 3 (1996), pp. 1544–1555.

[7] N. Coleff and M. Herrera, *Les courants résiduels associés à une forme méromorphe*, Lecture Notes in Math. **633**, Springer, 1978.

[8] D. Cox, *The homogeneous ring of a toric variety*, J. Algebraic Geom. **4** (1995), pp. 17–50.

[9] D. Cox, *Toric residues*, Ark. Mat. **34** (1996), pp. 73–96.

[10] P. Deligne, *Equations différentielles à points singuliers-réguliers*, Lecture Notes in Math. **163**, Springer, 1970.

[11] P. Deligne and A. Dimca, *Filtrations de Hodge et par l'ordre du pôle pour les hypersurfaces singulières*, Ann. Scient. Éc. Norm. Sup. **23**, 1990, pp. 645–656.

[12] G. de Rham, *Sur la notion d'homologie et les résidus des intégrales multiples*, Verhandlungen des Internaionalen Mathematike-Kongresse, Zürich (1932), vol. II, p. 195.

[13] G. de Rham, *Relations entre la topologie et la théorie des intégrales multiples*, Ens. Math. **35** (1936), pp. 213–228.

[14] A. Dickenstein and C. Sessa, *Canonical representatives in moderate cohomology*, Inventiones Math. **80** (1985), pp. 417–434.

[15] F. Didon, *Notes sur une formule de calcul intégral*, Ann. Scient. Éc. Norm. Sup., série 2, t. **2** (1873), pp. 31–48.

[16] I. M. Gelfand, M. M. Kapranov, A. V. Zelevinsky, *Discriminants, resultants and multidimensional determinants*, Birkäuser, Boston, 1994.

[17] O. A. Gelfond and A. G. Khovanskii, *Toric geometry and Grothendieck residues*, Moscow Math. Journal **2** (2002), no. 1, pp. 99–112.

[18] P. Griffiths, *On the periods of certain abelian integrals, I*, Ann. of Maths. (2) **90** (1969), pp. 460–495.

[19] P. Griffiths, *Variations on a theorem of Abel*, Inventiones Math. **35** (1976), pp. 321–390.

[20] P. Griffiths and J. Harris, *Principles of algebraic geometry*, Wiley, 1978.

[21] R. Hartshorne, *Residues and duality*, Springer, Berlin, 1966.

[22] G. Henkin and M. Passare, *Abelian differentials on singular varieties and variation on a theorem of Lie-Griffiths*, Inventiones Math. **135** (1999), pp. 297–328.

[23] H. Hironaka, *Resolution of singularities of an algebraic variety over a field of characteristic zero: I, II*, Ann. of Math. (2) **79** (1964), pp. 109–203 and 205–326.

[24] C. G. J. Jacobi, *De relationibus, quae locum habere denent inter puncta intersectionis duarum curvarum vel trium superficierum algebraicarum dati ordinis, simul cum enodatione paradoxi algebraici*, Gesammelte Werke, Band III, pp. 329–354.

[25] N. Katz, *Nilpotent connections and the monodromy theorem: Applications of a result of Turrittin*, Publ. Math. Inst. Hautes Études Sci. **39** (1970), pp. 175–232.

[26] A. Khovanskii, *Newton polyhedra and the Euler–Jacobi formula*, Russian Mathematical Surveys **33** (1978), pp. 237–238.

[27] E. Kunz, *Über den n-dimensionalen Residuensatz*, Jahr. D. M. V. **94** (1992), pp. 170–188.

[28] E. Kunz and R. Waldi, *Regular differential forms*, Contemporary Mathematics **79**, American Mathematical Society, Providence, RI, 1988.

[29] P. Lelong, *Fonctions plurisousharmoniques et formes différentielles positives*, Gordon & Breach, Paris-London-New York / Dunod, Paris, 1968.

[30] J. Leray, *Le calcul différentiel et intégral sur une variété analytique complexe (Problème de Cauchy III)*, Bull. Soc. Math. France **87** (1959), pp. 81–180.

[31] B. Lichtin, *The asymptotics of a lattice point problem associated to a finite number of polynomials, I, II*, Duke Math. Journal **63** (1991), no. 1, pp. 139–192; ibid. **77** (1995), no. 3, pp. 699–751.

[32] J. Lipman, *Dualizing sheaves, differentials and residues on algebraic varieties*, Astérisque **117**, 1984.

[33] J. Lipman and B. Teissier, *Pseudo-rational local rings and a theorem of Briançon-Skoda about integral closures of ideals*, Michigan Math. J. **28** (1981) pp. 97–116.

[34] F. Norguet, *Résidus: de Poincaré à Leray, un siècle de suspense*, in *Géométrie Complexe*, Collection Actualités Scientifiques et Industrielles, vol. **1438**, Hermann, Paris, 1996, pp. 295–319.

[35] K. Oka, *Sur les fonctions analytiques de plusieurs variables*, Iwanami Shoten, Tokyo, 1961.

[36] E. Picard and G. Simart, *Théorie des fonctions algébriques de deux variables indépendantes*, Gauthier-Villars, Paris, 1897.

[37] H. Poincaré, *Sur une généralisation du théorème d'Abel*, Comptes Rendus de l'Académie des Sciences (Paris) **100** (1885), pp. 40–42.

[38] H. Poincaré, *Sur les résidus des intégrales doubles*, Acta Math. **9** (1887), pp. 321–380.

[39] H. Poincaré, *Sur les périodes des intégrales doubles*, Journal de Mathématiques (6-ème série) **2** (1906), pp. 135–189.

[40] K. Saito, *Theory of logarithmic differential forms and logarithmic vector fields*, J. Fac. Sci. Univ. Tokyo (ser. IA) **27** (1980), pp. 265–291.

[41] G. Scheja, U. Störch, *Über Spurfunktionen bei vollständigen Durschschmitten*, J. Reine. Angew. Math. **278/279** (1975), pp. 174–190.

[42] L. Schwartz, *Théorie des distributions*, 5th ed., Hermann, Paris, 1966.

[43] A. Weil, *L'intégrale de Cauchy et les fonctions de plusieurs variables*, Math. Ann. **111** (1935), pp. 178–182.

[44] H. Zhang, *Calculs de résidus toriques*, Comptes Rendus de l'Académie des Sciences (Paris) Série I, **327** (1998), pp. 639–644.

CHAPTER 12

The Proof of the Poincaré Conjecture, according to Perelman

LAURENT BESSIÈRES, GÉRARD BESSON, AND MICHEL BOILEAU

The Poincaré conjecture, *a priori* purely topological, resisted the topologists for 100 years to finally deliver itself to the geometers. The program launched by Richard Hamilton in 1982 and brought to fruition by Grigori Perelman in 2003 rests on the *Ricci curvature flow*, an evolution equation that tends to homogenize the metric.

A shorter version of this text appeared in *Images des Mathématiques* 2006 (Éditions du CNRS).

12.1. Introduction

In his inaugural lecture in 1851, Bernhard Riemann [**16**] conceived of the idea of "n-dimensional multiplicities" (today we say manifolds of dimension n) defined in an intrinsic manner, without using the embedding in Euclidean space $\mathbb{R}^n$.

Since the end of the 19th century, the topology of surfaces (2-dimensional manifolds) has been well understood. Following Riemann, the work of Walther von Dyck, Camille Jordan, Felix Klein, August Ferdinand Moebius, and Ludwig Schläfli made it possible to obtain a complete list of compact surfaces without boundary. Every oriented compact surface without boundary can be described topologically as the boundary of a "pretzel". The number of holes in the pretzel is also the maximum number of disjoint curves one can draw on the surface without separating it: this number, called the *genus*, suffices to classify the surfaces. From a topological point of view, the simplest surface is the sphere $\mathbb{S}^2 \subset \mathbb{R}^3$, which is the boundary of the unit ball: it is the surface of genus 0.

Riemann had envisioned a similar study for the hypersurfaces (or manifolds) of dimension $n \geq 3$. His ideas were developed by Enrico Betti [**3**] in an 1871 memoir. To generalize the notion of the genus of a surface, Betti associated, to an n-dimensional manifold, n *orders of connection* (today called "Betti numbers") by considering the pieces of $(k+1)$-dimensional manifolds bounded by k-dimensional submanifolds. Betti's memoir drew Poincaré's attention to these topological notions, whose importance he sensed in all branches of mathematics.

In his 1885 memoir entitled *Analysis situs*[1] [**15**], Poincaré laid the foundations for what would become algebraic topology. One of his fundamental ideas was to attach to a manifold sophisticated algebraic objects (homology and the fundamental

[1]TRANSLATOR'S NOTE: "The analysis of location", the 19th century term for topology.

group), making possible the translation of topological problems into algebraic problems. These invariants allowed him to give examples of different three-dimensional manifolds that have the same Betti numbers, contrary to the case of surfaces.

Poincaré expressed the fact that, in a manifold, the submanifolds $V_1, V_2, \ldots, V_h$ form the boundary of a submanifold W by the relation $V_1 + V_2 + \cdots + V_h \sim 0$, which he called a "homology". He then defined the manifold's kth Betti number as the largest possible number of k-dimensional submanifolds that are not connected by a homology.

Poincaré stated a duality theorem for the Betti numbers, which would turn out to be fundamental in algebraic topology: Two Betti numbers of a manifold (compact, orientable, and without boundary) equally distant from the extremities are equal. However, in 1898, the thesis of a Danish mathematician, Poul Heegaard [**11**], revealed an error in Poincaré's proof, which led the latter to return to these questions in five supplements to the *Analysis situs*, published between 1899 and 1904 (see [**15**]). In particular, he remarked that his definition of Betti numbers differed from the one given by Betti, and corrected his proof.

Poincaré introduced the notion of the fundamental group in two ways, which are now known to be equivalent: the first as a group of substitutions of the equations defining the manifold, and the second, intrinsic, as a non-commutative version of the homologies between 1-dimensional submanifolds. It is a group that takes into account the closed paths (loops), within the space under consideration, that cannot be retracted onto a point by a continuous deformation (we then say that the loop is *essential*). If no loop is essential, the fundamental group is trivial and the space is called *simply connected*.[2] This is the case for all spheres $\mathbb{S}^n$ of dimension $n \geq 2$

From a topological point of view, the simplest of the manifolds with dimension $n \geq 2$ is the unit sphere $\mathbb{S}^n \subset \mathbb{R}^{n+1}$. In the second supplement to the *Analysis situs*, published in 1900, Poincaré stated without proof that a sphere of dimension $n \geq 2$ is characterized by the vanishing of its mth Betti numbers for $1 \leq m \leq n-1$. It would have to wait until 1904 and the fifth supplement to *Analysis situs* for Poincaré to return to this question and to construct a counterexample to his assertion: a compact 3-dimensional manifold without boundary whose Betti numbers are zero, but whose fundamental group is non-trivial. To construct his example, Poincaré used a method due initially to Heegaard, but to which he gave its modern form. This construction by *Heegaard splitting*, based on the study of curves drawn on a surface, would become fundamental for the study of 3-dimensional manifolds and would give birth to combinatorial group theory. It also contained elements that would later become Morse theory.

At the end of the fifth supplement to the *Analysis situs*, Poincaré poses the following, thereafter famous, question: *"Is it possible for the fundamental group of a 3-dimensional manifold V to reduce to the identity substitution, and yet for V not to be* [the sphere][3] *?"*

After a brief discussion of what should be done to adapt his construction of a homology sphere to the case of a homotopy sphere (that is, a compact 3-dimensional manifold without boundary, and simply connected in the current sense), Poincaré

[2]This is modern terminology. Poincaré used the expression "simply connected" to denote a sphere.

[3]Here Poincaré uses the expression "simply connected".

concludes his memoir with: *"But this question would lead us too far astray."* One hundred years later, we can only appreciate the extreme lucidity of this prophecy.

The hypothesis that a simply connected 3-dimensional manifold (compact, without boundary) can only be the sphere $\mathbb{S}^3$ is known by the name of the *Poincaré Conjecture.* This conjecture turned out to be an extremely difficult problem, more difficult than its analogue in dimension $n \geq 4$.[4] For a century, it garnered the attention and the efforts of eminent mathematicians. It lies at the origin of fundamental results in 3-dimensional topology and finds a natural generalization in the *Geometrization Conjecture*, formulated by William Thurston in the 1970s to describe all 3-dimensional manifolds.

Thurston [**18**] conjectured that eight homogeneous geometries would suffice to describe the elementary blocks that allow one to construct all 3-dimensional manifolds (see also [**17**]). The geometrization conjecture has thus replaced differential geometry at the heart of the study of 3-dimensional manifolds.

At the beginning of the 1980s, Richard Hamilton launched a new program to prove the geometrization conjecture and, in particular, the Poincaré conjecture. His approach is based on the Ricci flow: the idea is to analyze and control the solutions to a differential equation, tied to the curvature (the Ricci flow), on the space of Riemannian metrics of the manifold under consideration. As the flow evolves, the metric becomes homogenized, but the spreading of the curvature is not uniform: in finite time it can accumulate and become infinite (we will say "blow up") at certain points of the manifold. These are phenomena, called singularities of the Ricci flow, that Hamilton did not succeed in controlling.

Recently, Grigori Perelman discovered how to define a monotonic quantity, called *entropy*, for the Ricci flow, which made it possible for him to describe the way in which singularities appear and to classify them. Starting from there, he succeeded in the *tour de force* necessary to bring Hamilton's program to fruition by constructing a solution for Hamilton's equation called *Ricci flow with surgery*, allowing him to get rid of the singularities. We will now describe the proof of the Poincaré conjecture proposed by Perelman.

12.2. The flow associated to Ricci curvature

We are looking for an evolutionary process that produces a distinguished Riemannian metric on a given differential manifold M (see the box "Ricci curvature", below). The hope is that this will be a metric of constant Ricci curvature; such a metric is called an Einstein metric.

1. Ricci curvature

An n-dimensional differentiable manifold is a space locally modeled on the standard Euclidean space $\mathbb{R}^n$: each point has an open neighborhood, called a *chart*, homeomorphic to $\mathbb{R}^n$, and the passage from one chart to another is carried out by a C^∞-diffeomorphism. In dimension 3, the differentiable structure is unique up to diffeomorphism, while $\mathbb{R}^4$ can be endowed with infinitely

[4] EDITOR'S NOTE: The analogue of the Poincaré conjecture for $n \geq 4$ is that an n-dimensional manifold having the same homotopy type as $\mathbb{S}^n$ is homeomorphic to $\mathbb{S}^n$. This was proved by Stephen Smale in 1961 for $n \geq 5$ and by Michael Freedman in 1982 for $n = 4$.

many pairwise non-diffeomorphic differentiable structures. In what follows all manifolds will be of class C^∞ and orientable.

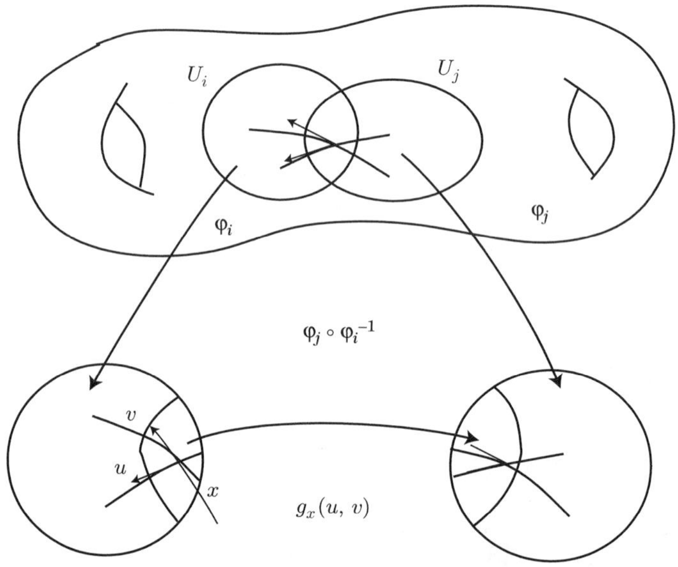

In order to do geometry inside of M (for example, to calculate the length of a curve, distances, volumes, etc.), we need an additional structure, called a **Riemannian metric** and denoted g: this is the specification of a scalar product g_x at each point x of the charts, varying in a C^∞ manner with x and compatible with the changes of charts. The curvatures associated to a metric measure the infinitesimal deviation from the standard metric of $\mathbb{R}^n$. They can be calculated by polynomial expressions in the coefficients of g, ∂g, and $\partial^2 g$.

For example, to each 2-plane P of the tangent space T_xM, we associate the sectional curvature $K(P)$, which can be defined as follows. Let $C(r)$ be a circle centered at x, with radius r, that is obtained by taking the points at distance r along the geodesics starting from x and tangent to P. Then its length satisfies the formula

$$\ell(C(r)) = 2\pi r \left(1 - \frac{K(P)}{6} r^2 + o(r^2)\right),$$

and $K(P)$ measures the defect from the Euclidean circumference. The Ricci curvature (one also speaks of the *Ricci tensor*) is, at each point of M, a symmetric bilinear form on T_xM (not necessarily positive definite). Its value in a direction $v \in T_xM$, denoted $\mathrm{Ric}_g(v, v)_x$, is calculated by taking the sum of the sectional curvatures of the 2-planes generated by v and e_i, where e_i runs through an orthonormal basis of the orthogonal space to v in T_xM. It measures the defect in the area of small spheres. The **scalar curvature** $R(x)$ is a function on M defined at each point x as the trace of the Ricci curvature $\mathrm{Ric}_g(\,\cdot\,,\,\cdot\,)_x$ with respect to the scalar product g_x, that is, the sum of its eigenvalues. It measures the defect of "Euclideanness" in the volumes of small balls.

The flow associated to the Ricci curvature (see box 1) is a differential equation on the (infinite-dimensional) space $\mathcal{M}$ of Riemannian metrics on the manifold M. It would be ideal for this ordinary differential equation to be given by the opposite of the gradient of a function (so that the trajectories converge to the minima). The natural candidate for such a function is the one that physicists call the Hilbert–Einstein functional: this is an integral of curvature (the scalar curvature) whose critical points are the so-called Einstein metrics, that is, the metrics such that $\mathrm{Ric}_g = \lambda g$. Unfortunately, a simple calculation shows that its gradient gives rise to an equation that does not admit solutions, in general. On the other hand, a modification to this sterile equation is appropriate. We call the Ricci flow a family $g(t)$ of Riemannian metrics on M, defined on an interval $[0, T)$, and satisfying the following evolution equation (see the box "Ricci flow", below):

$$\frac{\partial g}{\partial t} = -2\,\mathrm{Ric}_{g(t)}. \tag{12.1}$$

The point of view of differential equations, mentioned above, is very difficult to put into action on the infinite-dimensional space $\mathcal{M}$. We prefer a more efficient approach that consists in writing (12.1) by parametrizing M (by local coordinates). Then (12.1) becomes a parabolic partial differential equation of *reaction-diffusion* type.

2. Example of Ricci flow

On the round sphere, the Ricci flow has the solution $g(t) = (1 - 2\lambda t)g_0$ on the interval $[0, \frac{1}{2\lambda})$ if the initial Ricci curvature is $\mathrm{Ric}_{g_0} = \lambda g_0$ with $\lambda > 0$.

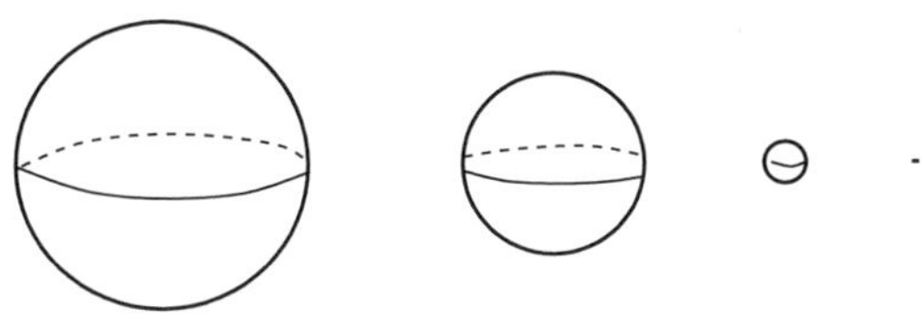

More generally, the flow evolves by homothety if the initial metric is an Einstein metric. We may summarize the evolution by saying that metrics of negative curvature inflate and metrics of positive curvature contract.

12.3. The work of R. Hamilton

Hamilton [**6**] proved the short-term existence of a solution for any smooth initial data (see [**5**] for a simple proof). In addition, the flow can be extended as long as the sectional curvatures (see the box "Ricci curvature", p. 245) remain bounded in absolute value. To control the curvatures, we write their evolution equations and use maximum principles.

12.3.1. Maximum principles over curvatures. We begin with the scalar curvature, which evolves according to the parabolic equation

$$\frac{\partial R}{\partial t} = \Delta R + 2\,|\mathrm{Ric}|^2, \tag{12.2}$$

where all quantities depend on $g(t)$. At a point x that minimizes the scalar curvature of $g(t)$, the Laplacian ΔR is positive or zero, and therefore $\partial R/\partial t \geq 0$. Heuristically, we can think that the minimum of the scalar curvature of $g(t)$ over M, which we denote $R_{\min}(t)$, increases with t. A maximum principle makes it possible to show this rigorously. Better yet, if $R_{\min}(0) > 0$, we can show that $R_{\min}(t)$ tends to $+\infty$ in finite time (if the flow exists). In this case, it is certain that the maximum of the sectional curvatures tends to $+\infty$ in finite time. In order to have more information, we use the evolution equation of the Ricci tensor, which is of the form

$$\frac{\partial \mathrm{Ric}}{\partial t} = \Delta \mathrm{Ric} + Q(\mathrm{Ric}), \tag{12.3}$$

where Q is a quadratic expression. A vectorial maximum principle shows that if $\mathrm{Ric}_{g_0} \geq 0$, then $\mathrm{Ric}_{g(t)} \geq 0$. If in addition $\mathrm{Ric}_{g_0} > 0$, then this inequality holds for all t, and at every point we have the pinching

$$\frac{|\mathrm{Ric} - \frac{R}{3}g|}{R} \leq \frac{\alpha}{R^\beta}, \tag{12.4}$$

where α and β are strictly positive constants. This means that as $R(x,t) \to +\infty$, the relative deviation of $\mathrm{Ric}_{g(t)}$ at x from its average $Rg(t)/3$ tends to 0. With the help of a control on the gradient of the scalar curvature, Hamilton showed that, under the hypothesis of strict positivity on the Ricci curvature, this curvature blows up in finite time everywhere at the same speed. Thus, after renormalizing in order to make the volume constant, $g(t)$ converges to a metric of constant strictly positive sectional curvature, from which we have the following theorem:

THEOREM 12.3.1. *If M is a compact Riemannian manifold without boundary having a metric of strictly positive Ricci curvature, then M can be endowed with a metric of constant strictly positive sectional curvature.*

REMARK. In particular, M is the quotient of the sphere $\mathbb{S}^3$ by a finite group of isometries. Such a manifold is called **spherical**; an example is given by the space of lines in $\mathbb{R}^4$, which is the quotient of the sphere $\mathbb{S}^3$ by the antipodal map (this is projective space, denoted $\mathbb{P}^3(\mathbb{R})$). This theorem is fundamental for the entire theory and the first step toward the proof of the Poincaré conjecture.

The situation is radically different if the Ricci curvature is not strictly positive. The most general result is that, *for every initial set of data, the flow exists on a maximal interval* $[0,T)$, *and if* $T < \infty$, *the maximum of the sectional curvatures at time t tends to* $+\infty$ *as* $t \to T$. In the latter case, we say that T is a *singular* time. In general the curvature blows up only on part of the manifold: we say that the flow encounters a singularity. However, a variant of the preceding results, called the Hamilton–Ivey "pinching" theorem, shows that the negative part of the curvature becomes negligible compared to the scalar curvature. In particular, the scalar curvature controls all of the curvatures. Let us now say a few words about the study of singularities.

12.3.2. The study of singularities: the zoom technique. This technique, classical in analysis, was put into action in this context by Hamilton in [**8**] and [**9**]. A zoom consists of dilating the metric and slowing the passing of time to have a new solution to the flow. We will consider sequences of zooms and attempt to pass to the limit (see the box "Parabolic dilations", p. 249). If we can show the existence of limit flows and classify them, we obtain models for the singularities. The existence of the limit flow was oen of the sticking points of Hamilton's program. This question has been completely solved by Perelman.

In general one considers a sequence of zooms on the points (x_k, t_k) such that $Q_k := R(x_k, t_k) \to +\infty$ and maximizes the scalar curvature on $M \times [0, t_k]$. Then the sequence of parabolic dilations $g_k(t)$ at (x_k, t_k) has bounded curvature over the intervals $[-t_k Q_k, 0]$ converging to $(-\infty, 0]$. A sufficient condition to ensure the convergence of the sequence (or a subsequence) $(M, g_k(t), x_k)$—in a sense that we will not make precise—to a flow $(M_\infty, g_\infty(t), x_\infty)$ is a lower bound on the volume of the unit ball centered at x_k (for the metric $g_k(0)$) by a strictly positive constant independent of k. Perelman's first striking contribution [**12**] was to establish that this lower bound is always satisfied if the curvature's explosion takes place in finite time. By construction, the limit obtained is a flow over $(-\infty, 0]$ of non-zero bounded curvature. In addition, the Hamilton–Ivey pinching theorem makes it possible to show that the sectional curvature is positive or zero on the limit flow.

3. Parabolic dilations

The idea of zooming is formalized by the notion of *parabolic dilation*. Given a Ricci flow $g(t)$ on $M \times [0, T)$, a point x_0, and a time t_0, it is the solution to the flow given by the formula

$$g_0(t) = Q_0 \cdot g\left(t_0 + \frac{t}{Q_0}\right),$$

where $Q_0 = R(x_0, t_0)$. It is defined on $[-t_0 Q_0, (T - t_0)Q_0)$. This is a renormalization such that $R_{g_0}(x_0, 0)$ is equal to 1.

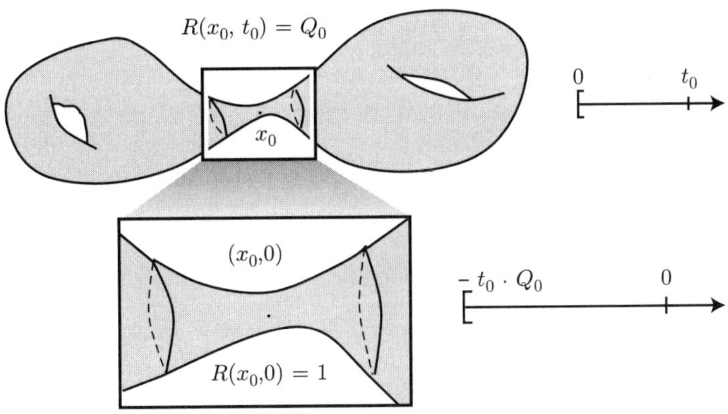

12.4. G. Perelman's flow with surgery

The main result of Perelman's first article [**12**] is the *canonical neighborhood theorem*, which describes the metric $g(t)$ at points of large scalar curvature. If the curvature is large everywhere, we can deduce the classification of M. In this case, we say that the flow becomes extinct. Otherwise the idea (going back to Hamilton [**10**]) is to remove the pieces of M with large curvature by cutting the manifold along spheres $\mathbb{S}^2$ and filling the holes with balls B^3. Of course, this must be done while controlling the topology and the geometry. Then we relaunch the flow on the new, possibly disconnected, manifold, and we iterate the process. Eventually, some connected components may disappear during the surgeries; we also say that they become extinct. Perelman shows in [**13**] that we can follow this flow

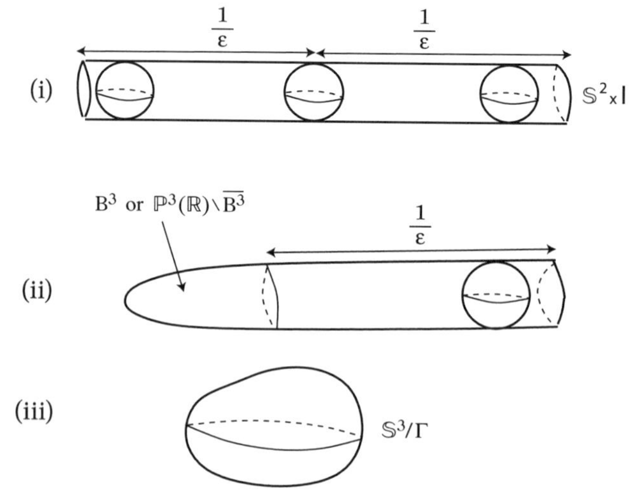

FIGURE 12.1. The canonical neighborhoods.

with surgery indefinitely, for any appropriately normalized initial data. On each finite-time interval, we only carry out finitely many surgeries. If the flow becomes completely extinct in finite time, we can classify all of its connected components and hence the starting manifold. This is what happens in the proof of the Poincaré conjecture. (The long-term classification is harder, and we will not go into it here.)

12.4.1. The canonical neighborhoods. The canonical neighborhood theorem essentially asserts that at points of large scalar curvature of Ricci flow, the geometry is canonical, that is, almost isometric to a finite number of simple models. So as not to include too many parameters, by dilation of the initial metric, we can assume that the flow lives on $[0,1]$, at least, and require that the unit balls of the initial metric be almost Euclidean: we say that the initial data are *normalized.* Then:

THEOREM 12.4.1. *For every sufficiently small $\varepsilon > 0$, there exists a universal constant $r = r(\varepsilon) > 0$ with the following property. Let $(M, g(t))$ be a Ricci flow of normalized initial data, $x \in M$, and $t \geq 1$ such that $R(x,t) \geq r^{-2}$. Then x has a neighborhood that is ε-almost isometric, after a dilation of factor $\sqrt{R(x,t)}$, to one of the following models (see* FIG. 12.1*):*

(i) *A cylinder $\mathbb{S}^2 \times (-1/\varepsilon, 1/\varepsilon)$, with the canonical product metric, of scalar curvature 1. We call this neighborhood an ε-neck.*
(ii) *A ball B^3 or the complement of a ball in projective space, that is, $\mathbb{P}^3(\mathbb{R}) \setminus \overline{B^3}$, endowed with a metric of strictly positive curvature that is close, outside of a compact subset, to a spherical cylinder as above. We call such a neighborhood an ε-cap.*
(iii) *A closed manifold of strictly positive sectional curvature.*

We shall say that $g(t)$ satisfies the canonical neighborhood hypothesis at scale r.

By two ε-almost isometric manifolds, we mean two diffeomorphic manifolds whose Riemannian metrics are ε-close, as well as their derivatives of order $\leq 1/\varepsilon$. In particular, the curvatures on these neighborhoods are comparable to the scalar curvature $R(x,t)$. The size of the neighborhoods corresponding to (i) and (ii) is comparable to $R(x,t)^{-1/2} \times 2/\varepsilon$. In addition, the spatial and temporal oscillations of the scalar curvature are controlled by universal constants.

REMARK. In the last case, by connectivity, all of M is contained in the neighborhood, and by Theorem 12.3.1, M is diffeomorphic to a spherical manifold.

12.4.2. Description of the first singular time. In this section, we assume that $\varepsilon > 0$ and a scale $r > 0$ are fixed for which $g(t)$ satisfies the canonical neighborhood hypothesis. We describe the metric $g(t)$ as $t \to T$ where $T < \infty$ is a singular time. Let Ω denote the set of points where the scalar curvature remains bounded, that is,

$$\Omega = \{x \in M; R(x, \cdot\,) \leq c(x) < +\infty\}.$$

By hypothesis, there exists $x \in M$ such that $R(x,t) \to +\infty$, so Ω is strictly smaller than M.

First case: Ω is empty, and the flow becomes extinct. Then we can show that M is a spherical manifold, or a product $\mathbb{S}^2 \times \mathbb{S}^1$, or a connected sum of two projective spaces, denoted $\mathbb{P}^3(\mathbb{R}) \# \mathbb{P}^3(\mathbb{R})$. In fact, if the curvature blows up everywhere, we can find a time t_0 near the singular time T such that $(M, g(t_0))$ is covered by a finite number of canonical neighborhoods. If there is a neighborhood of type (iii), M is diffeomorphic to a spherical manifold. Otherwise, we place ε-necks end-to-end until they close to form an $\mathbb{S}^2 \times \mathbb{S}^1$, or we fill them in by ε-caps to obtain $\mathbb{S}^3$, $\mathbb{P}^3(\mathbb{R})$, or $\mathbb{P}^3(\mathbb{R}) \# \mathbb{P}^3(\mathbb{R})$.

REMARK. In the case where M is simply connected, we obtain that it is diffeomorphic to $\mathbb{S}^3$.

Second case: Ω is non-empty. The controls on the curvature make it possible to show that Ω is an open set on which the metic $g(t)$ converges to a regular metric $g(T)$. Essentially, $g(T)$ satisfies the canonical neighborhood hypothesis by passing to the limit. In order to understand the structure of Ω, we choose a scale of curvature $\rho < r$, and we define the set

$$\Omega_\rho = \{x \in \Omega; R(x,T) \leq \rho^{-2}\}.$$

The set $\Omega \setminus \Omega_\rho$ is covered by ε-necks and ε-caps. An examination of the different combinations shows that every point $x \in \Omega \setminus \Omega_\rho$ is in one of the following sets:

(i) *An ε-tube*, that is, a cylinder $\mathbb{S}^2 \times I$ that is a union of finitely many ε-necks and whose boundary is in Ω_ρ.

(ii) *An ε-horn*, that is, a union of infinitely many ε-necks, diffeomorphic to $\mathbb{S}^2 \times \mathbb{R}^+$. The end $\mathbb{S}^2 \times \{0\}$ is in Ω_ρ, while at the other end, the scalar curvature tends to $+\infty$.

(iii) A union of finitely many ε-necks closed by an ε-cap and connected by its boundary to Ω_ρ.

(iv) Connected components of Ω disjoint from Ω_ρ: *double horns* diffeomorphic to $\mathbb{S}^2 \times \mathbb{R}$ or to $\mathbb{P}^3(\mathbb{R}) \setminus \overline{B^3}$, the union of infinitely many ε-necks, and *ε-capped horns* diffeomorphic to $\mathbb{R}^3$, the union of infinitely many ε-necks closed by an ε-cap (see FIG. 12.2).

REMARK. If Ω_ρ is empty, we show, as in the case where Ω is empty, that M is diffeomorphic to $\mathbb{S}^2 \times \mathbb{S}^1$, $\mathbb{P}^3(\mathbb{R}) \# \mathbb{P}^3(\mathbb{R})$, or a spherical manifold. We again say that the flow becomes extinct, even if the curvature does not blow up everywhere.

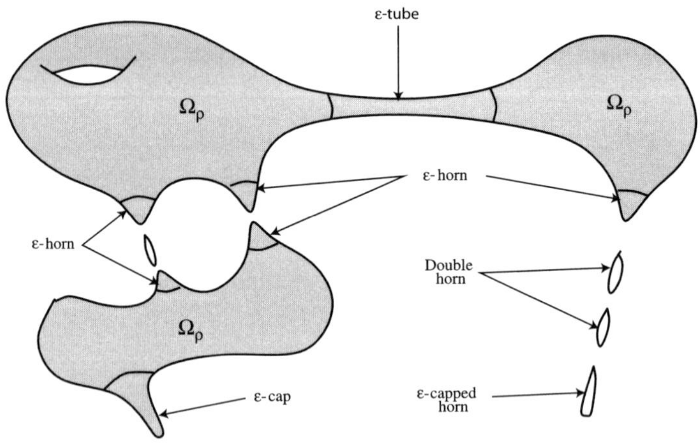

FIGURE 12.2. The structure of Ω.

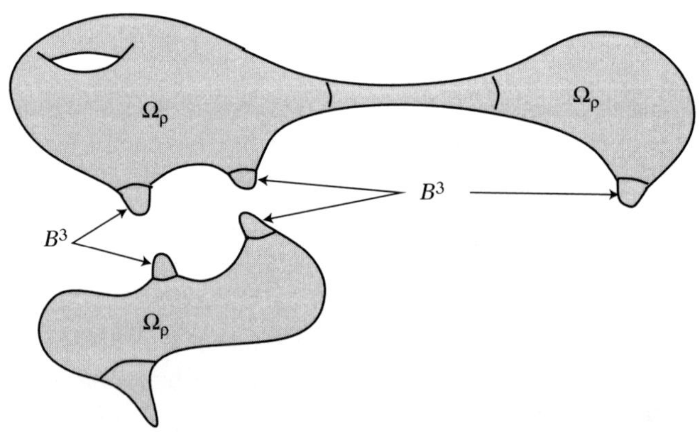

FIGURE 12.3. The manifold M_1 obtained by surgery.

Surgery. We operate on this set Ω using the surgery described as follows:

(i) we eliminate all the connected components of Ω disjoint from Ω_ρ;
(ii) we truncate the horns (connected to Ω_ρ) and fill them in by an ε-cap diffeomorphic to a ball (see FIG. 12.3).

Thus we obtain a new differentiable manifold, possibly disconnected, which we denote M_1. At every time $t < T$ near T, $(M \setminus \Omega_\rho, g(t))$ is covered by canonical neighborhoods. We thus verify that M is the sum of the different connected components of M_1, and possibly also of a finite number of handles $\mathbb{S}^2 \times \mathbb{S}^1$ and projective spaces $\mathbb{P}^3(\mathbb{R})$.

This surgery must be carried out in a metric way, that is, while precisely controlling the gluings. For this, we choose to truncate the horns at the middle of a δ-neck, for a parameter $0 < \delta \ll \varepsilon$. We thus define a surgery with parameter (r, δ), where r is the scale of curvature (depending on ε) starting at which we have canonical neighborhoods, and the parameter ρ is fixed by setting $\rho = \delta r$. We thus endow M_1 with a well-chosen Riemannian metric $g_1(T)$, which becomes the initial data of the equation (12.1), and we relaunch the flow simultaneously on the connected components of M_1.

REMARK. If the set Ω_ρ is empty, the surgery above has meaning. But in this case $M_1 = \varnothing$ and the flow becomes extinct.

12.4.3. Flow with surgery. Perelman's *tour de force* in [**13**] is to have succeeded in iterating the construction above indefinitely. We fix an $\varepsilon > 0$.

DEFINITION. Let $r(t)$, $\delta(t)$ be strictly positive decreasing functions on $[0, +\infty)$. We call the following set of data a *flow with surgery*:

(i) a strictly increasing discrete sequence $(t_k)_{0 \leq k \leq N \leq \infty}$ in $[0, +\infty)$,

and for each integer k,

(ii) a compact manifold M_k, which may be disconnected or empty,
(iii) a Ricci flow $g_k(t)$ on $M_k \times [t_k, t_{k+1})$, singular at t_{k+1}, that satisfies the canonical neighborhood hypothesis at the scale $r(t)$,

such that $(M_{k+1}, g_{k+1}(t_{k+1}))$ is derived from $(M_k, g_k(t))$ by a surgery with parameters (r, δ) at time t_{k+1}.

We say that a Riemannian manifold (M, g_0) is normalized if the sectional curvatures are bounded in absolute value by 1 and if the volume of every unit ball is at least half of the Euclidean volume. Perelman proved that there exist universal strictly decreasing functions $(r(t), \delta(t))$ such that the flow with surgery exists on $[0, \infty)$ for all normalized initial data (M, g_0).

In particular, there are only finitely many surgeries on each finite interval. If M_k is the manifold obtained after the kth singular time, by taking into account the components that become extinct, M can be obtained as a connected sum of the connected components of M_k with a certain number of copies of $\mathbb{S}^2 \times \mathbb{S}^1$ and finite quotients of $\mathbb{S}^3$. If M_k is empty, the flow becomes extinct and M is diffeomorphic to a connected sum of a finite number of $\mathbb{S}^2 \times \mathbb{S}^1$ and finite quotients of $\mathbb{S}^3$. In particular, if M is simply connected, it is diffeomorphic to $\mathbb{S}^3$.

12.5. The Poincaré conjecture

The existence of a flow with surgery having been established, the proof of the Poincaré conjecture consists in showing that it becomes extinct in finite time on a homotopy sphere, that is, a simply connected manifold. According to the preceding, it is then diffeomorphic to $\mathbb{S}^3$. We now give more details.

We give ourselves a compact, simply connected manifold M_0, which we assume to be irreducible (see the box "Irreducibility", below). We endow it with a normalized metric g_0. For this initial data, we construct a flow with surgery $(M_k, g_k(t))$ defined on $[0, \infty)$. We know that each M_k (as long as it is not empty) contains a component M_k^1 diffeomorphic to M_0, and the other components are spheres. We can therefore consider the restriction of the flow with surgery to just this component. To show that it becomes extinct in finite time, we sketch the argument of T. Colding and W. Minicozzi [4], which is technically simpler than that of Perelman [**14**]. It consists in showing that the width of $(M_0, g(t))$ reaches 0 in finite time.

4. Irreducibility

We say that an orientable manifold M is irreducible if every embedded sphere $\mathbb{S}^2 \subset M$ bounds a ball B^3. This implies that M is a connected sum of two manifolds, one of which is diffeomorphic to M and the other to $\mathbb{S}^3$. Kneser's theorem asserts that every orientable manifold is the connected sum of a finite number of irreducible manifolds and copies of $\mathbb{S}^2 \times \mathbb{S}^1$. In particular, if M is simply connected, it is the connected sum of finitely many irreducible simply connected manifolds.

The width of $(M_0, g(t))$ is defined by the minimax of the energy of the spheres $\mathbb{S}^2$ that sweep out M_0. This is a strictly positive geometric quantity as long as the loop defining the sweeping is essential in the space $\mathcal{H}$ of maps from $\mathbb{S}^2$ into $(M_0, g(t))$ that are continuous and have bounded energy. The existence of an essential loop in $\mathcal{H}$ is a consequence of the simple connectivity of M_0. We then fix once and for all the homotopy class β of an essential loop in $\mathcal{H}$. We define the width $W([\beta], g(t))$ of the Riemannian manifold $(M_0, g(t))$ by

$$W([\beta], g(t)) = \inf_{\gamma\in[\beta]} \sup_{s\in[0,1]} E(\gamma(s)),$$

where

$$E(f) = \int_{\mathbb{S}^2} |df|^2_{g(t)} \, \mathrm{dvol}_{\mathbb{S}^2}$$

is the energy of the map $f : \mathbb{S}^2 \to (M_0, g(t))$.

The proof of extinction in finite time then rests on the following two facts:

(i) On the smooth portions of the flow, the width $W([\beta], g(t))$ decreases sufficiently fast along the flow, according to the following inequality of Colding and Minicozzi [**4**]:

$$\frac{dW([\beta], g(t))}{dt} \leq -4\pi + \frac{3}{4(t+C)} W([\beta], g(t))$$

(where C is a constant that can be calculated). This ensures extinction in finite time if the flow remains smooth, since the width reaches 0 in finite time and, in addition, must be strictly positive.

(ii) If t_{k+1} is a singular time for the flow $G_k(t)$ on M_0, we have

$$\lim_{t\to t_{k+1}^-} W([\beta], g_k(t)) \geq W([\beta], g_{k+1}(t_{k+1})).$$

This is a consequence of the existence of a $(1+\xi(t))$-Lipschitz diffeomorphism between $(M_0, g_k(t))$ and $(M_0, g_{k+1}(t_{k+1}))$, with $\xi(t) \to 0$ as $t \to t_{k+1}$.

Bibliography

[1] L. Bessières, *Conjecture de Poincaré : la preuve de R. Hamilton et G. Perelman*, Gazette des Mathématiciens **106** (2005), pp. 7–35.

[2] G. Besson, *Preuve de la conjecture de Poincaré en déformant la métrique par la courbure de Ricci, d'après G. Perelman*, à paraître dans Astérisque. Modified version from the séminaire Bourbaki n0. 947 (2004–2005), 57th year.

[3] E. Betti, *Sopra gli spazi di un numero qualunque di dimensioni*, Ann. Mat. Pura Appl. (2) **4** (1871), pp. 140–158.

[4] T. Colding and W. Minicozzi, *Estimates for the extinction time for the Ricci flow on certain three-manifolds and a question of Perelman*, J. Amer. Math. Soc. **18**:3 (2005), pp. 561–569.
[5] D. DeTurck, *Deforming metrics in the direction of their Ricci tensor*, J. Differential Geom. **18** (1983), pp. 157–162.
[6] R. Hamilton, *Three-manifolds with positive Ricci curvature*, J. Differential Geom. **17**:2 (1982), pp. 255–306.
[7] R. Hamilton, *Four-manifolds with positive curvature operator*, J. Differential Geom. **24** (1986), pp. 153–179.
[8] R. Hamilton, *The formations of the singularities of the Ricci flow*: pp. 7–136 of *Surveys in Differential Geometry*, vol. II, International press, Cambridge MA, 1995.
[9] R. Hamilton, *A compactness property for solutions of the Ricci flow*, Amer. Jour. Math. **117** (1995), pp. 545–572.
[10] R. Hamilton, *Four-manifolds with positive isotropic curvature*, Comm. Anal. Geom. **5** (1997), pp. 1–92.
[11] P. Heegaard, *Sur l'Analysis situs*, Bull. Soc. Math. France **44** (1916), pp. 161–242.
[12] G. Perelman, *The entropy formula for the Ricci flow and its geometric applications.* `arXiv:math.DG/0211159`.
[13] G. Perelman, *Ricci flow with surgery on three-manifolds.* `arXiv:math.DG/0303109`.
[14] G. Perelman, *Finite extinction time for the solutions to the Ricci flow on certain three-manifolds.* `arXiv:math.DG/0307245`.
[15] Henri Poincaré, *Œuvres*, tome VI, Gauthier-Villars (Paris), 1953.
[16] B. Riemann, *Œuvres mathématiques*, translated by L. Laugel, Gauthier-Villars (Paris), 1898.
[17] P. Scott, *The geometries of* 3*-manifolds*, Bull. London Math. Soc. **15**:5 (1983), pp. 401–487.
[18] W. P. Thurston. *Three-dimensional manifolds, Kleinian groups and hyperbolic geometry*, Bull. Amer. Math. Soc. **6** (1982), pp. 357–381.

CHAPTER 13

Henri Poincaré and the Partial Differential Equations of Mathematical Physics

JEAN MAWHIN

13.1. Introduction

After being lecturer in analysis at the Université de Caen (1879), assistant professor in analysis (1881) and lecturer in physical and experimental mechanics (1885) at the Faculté des Sciences de l'Université de Paris, Henri Poincaré was given the title of professor of mathematical physics and probability there in 1886. At this time, the young and brilliant mathematician was known for his work on the number theory of quadratic, cubic, and quaternary forms; automorphic functions; entire functions; abelian functions; the qualitative theory of non-linear differential equations; and celestial mechanics. The context of his surprising nomination to the chair of mathematical physics and probability is described in [**2**]. Although there is no doubt that Poincaré had thought about questions of mathematical physics beforehand, and although he may have desired this chair, all of his work on the partial differential equations of mathematical physics came after 1886. It is also worth mentioning the publication of his numerous lecture notes on this subject, which often contain original contributions or presentations.

Poincaré's most important results pertaining to the partial differential equations of mathematical physics are contained in three long memoirs [**23, 25, 27**] published between 1890 and 1896. Their contents were generally announced by one or several notes in the *Comptes Rendus de l'Académie des Sciences.* We will analyze the main ideas and results as well as the impact of each of these articles and a few other important contributions. We will not hesitate to provide details for certain proofs, which are replaced by others in most expositions today. They remain exemplary for their simplicity of means, doubled by a great inventiveness. This is in particular the case for the proof of the Poincaré inequality and the proof of the existence of eigenvalues for the Dirichlet problem. We will concentrate primarily on the (elliptic) stationary problems for linear equations, briefly mentioning, in the final section, the no less important contributions Poincaré made to evolution problems. The case of non-linear partial differential equations will be the subject of another work.

13.2. The equations of mathematical physics

The introduction to the memoir [**23**] is interesting from many viewpoints. First of all, it insists upon the unifying role of mathematical physics:

> *"When we consider the various problems of the Integral Calculus that arise naturally when we want to deepen the most different parts of*

Physics, it is impossible not to be struck by the analogies that all of these problems share. Whether dealing with electrostatics or electrodynamics, the propagation of heat, optics, elasticity, hydrodynamics, we are always led to differential equations of the same family, and the boundary conditions, although different, are not, however, without a few similarities [...] We must therefore expect to find that they have common properties. Unfortunately, the first property common to all of these problems is their extreme difficulty. Not only are we most often unable to solve them completely, but it is only at the expense of the greatest efforts that we can rigorously prove the possibility."

Poincaré wondered about the necessity of proving the existence of a solution.

"Most physicists do not think it worth considering. Since experiment does not allow one to doubt, for example, the possibility of electrical equilibrium, neither can one doubt, it appears, the possibility of equations that express this equilibrium. We do not know how to be content with this defeat; the analysis must be sufficient unto itself, and moreover, a similar reasoning, if perhaps applicable to problems that one encounters directly in physics, would not apply in the same way to a host of simpler problems, which themselves arise as soon as one seeks to solve the first ones."

Curiously, Poincaré does not mention the problem of the mathematical model's adequacy to the reality it hopes to describe. If experiment shows the existence of a unique equilibrium, a necessary condition for the validity of the mathematical model is that it have one and only one solution.

Poincaré then discusses the necessity of *rigorously* proving the existence of a solution:

"The differential equations that are obeyed by physical phenomena have often only been established by reasoning that has little rigor. They are regarded merely as approximations; the experimental results, to which the consequences of the theory are to be compared, are themselves approximate. In these conditions, rigor is little prized, and it often seems that there is no cause to search for it if one must pay with too much effort. But then [...] when will we have the right to say that such a proof, insufficient for Analysis, is rigorous enough for Physics? The limit is quite hard to draw. I will, however, attempt to do so."

Here at last is Poincaré's plea for rigor:

"Every time that I can, I seek absolute rigor, and this for two reasons: in the first place, it is always hard for a geometer to approach a problem without completely solving it; in the second place, the equations that I study are susceptible, not only to physical applications, but even to analytic applications."

Poincaré cites the example of Riemann's study of abelian functions by means of the Dirichlet principle. And he concludes prophetically:

"And who can tell us if the other problems of Mathematical Physics will not one day [...] be called to play a considerable role in Analysis?"

13.3. The Dirichlet problem and the sweeping method

Given a domain D in $\mathbb{R}^3$ bounded by a surface S, and a continuous function Φ on S, the *Dirichlet problem* consists in finding a function V such that

$$\Delta V := \frac{\partial^2 V}{\partial x^2} + \frac{\partial^2 V}{\partial y^2} + \frac{\partial^2 V}{\partial z^2} = 0 \quad \text{in} \quad D, \qquad V = \Phi \quad \text{on} \quad S. \tag{13.1}$$

If $\nabla U := (\partial U/\partial x, \partial U/\partial y, \partial U/\partial z)$ and $\|\cdot\|$ denotes the Euclidean norm in $\mathbb{R}^3$, it is easy to show that every function V that minimizes the integral

$$\mathcal{I}(U) := \int_D \|\nabla U\|^2 \tag{13.2}$$

over the set of functions U that are of class C^2 on D, continuous on $D \cup S$, and equal to Φ on S, is a solution of (13.1). The argument, used previously by Dirichlet and Thomson (Lord Kelvin), that deduces the existence of a minimum from the positivity of $\mathcal{I}(U)$, was called the *Dirichlet principle* by Riemann. Weierstrass justly criticized the rigor of the Dirichlet principle, and so mathematicians at the end of the 19th century, in particular C. Neumann and Robin, looked for other proofs of the existence of a solution to the Dirichlet problem. Poincaré comments upon them as follows:

> *"They are at the same time methods of proof intended to establish the possibility of the problem, and methods of calculation intended to solve them effectively. As methods of proof, they are rather complicated, but they complement each other in such a way as to apply to all cases and to satisfy even the most severe judges on the subject of rigor. As methods of calculation, they are worthless; even the simplest among them, those of Neumann or Robin, leads to inextricable calculations at the second approximation."*

In [**23**], Poincaré introduces a new approach that solves the Dirichlet problem for all bounded domains D with sufficiently regular boundary S. With his usual modesty and foresight, Poincaré immediately judges the impact of his method:

> *"What would have been most interesting would be to replace the current methods of calculation by other, less defective ones. I was not able to do so; I have limited myself to looking for a method of proof that is simpler than the ones that have been proposed to date and applicable to all cases."*

The outline of his method was announced in a note to the *Comptes Rendus* of January 3, 1887 [**20**]. Poincaré first remarks that the work of Murphy makes it possible to reduce the problem to the case where D is simply connected, and by virtue of results of Green, it suffices to solve the problem of the electrostatic distribution on the surface of a conductor S bounding D, which is to find a function V on $\mathbb{R}^3 \setminus D$ such that

$$\begin{aligned} &\Delta V = 0 \quad \text{in} \quad \mathbb{R}^3 \setminus D, \\ &V = 1 \quad \text{on} \quad S, \qquad V \to 0 \quad \text{as} \quad r := \sqrt{x^2+y^2+z^2} \to \infty. \end{aligned} \tag{13.3}$$

In his memoir, Poincaré introduced the *sweeping method* to solve the latter problem. The same method is applied directly to the Dirichlet problem (13.1) in his work [**29**]. We describe it within this framework, more familiar to mathematicians.

The *sweeping of a sphere S* consists in replacing the mass (or the charge) inside S by a distribution of an equivalent quantity of mass (or of charge) on the surface

S. It is easy to show that a unit point mass M can thus be replaced by a unit mass distributed over the boundary with a surface density inversely proportional to the cube of the distance from M. The result extends effortlessly to a distribution of an arbitrary mass inside of S. The sweeping process does not introduce negative masses, does not change the potential outside of S, and reduces the potential inside of S. In fact, if U is the interior potential of the surface distribution equivalent to a particle for which the exterior potential is $1/r$, the Green function is $G = \frac{1}{r} - U$, the difference between the particle's interior potential and the interior potential of the equivalent surface potential. As the Green function is always strictly positive, $U < 1/r$ at each interior point. The result extends to the case of an arbitrary distribution of mass.

Let us assume first that we are given a (strictly) positive Φ on S, such that $0 < m < \Phi(x) < M$ for appropriate constants m and M. If S is the boundary of a closed ball B containing D in its interior, then we can find a sequence of polynomials (P_n) such that

$$0 < P_n(x) < \varepsilon_n \quad (n \neq 1), \qquad \sum_{n=1}^{\infty} \varepsilon_n < +\infty, \qquad U = \sum_{n=1}^{\infty} P_n = \Phi \quad \text{on} \quad S.$$

Assume that, for each n, there exists a function V_n that is harmonic on D and equals P_n on S. Then, if m_n and M_n respectively denote the minimum and the maximum of P_n on S, we have, by the maximum principle,

$$0 < m_n \leq V_n(x) \leq M_n < \varepsilon_n \quad (x \in D \cup S,\ n \geq 1).$$

Consequently the series $\sum_{n=1}^{\infty} V_n$ converges absolutely and uniformly on $D \cup S$, and, by Harnack's theorem, its sum V is harmonic on D. The positivity constraint of Φ on S is easily lifted, since if $|\Phi(x)| < C$ for all $x \in S$, then $\Phi = C - (C - \Phi)$, and if W is a function that is harmonic on D and equals $C - \Phi$ on S, the condition of Harnack's theorem is satisfied, and the function $V = C - W$ is harmonic on D and equals Φ on S. *It therefore suffices to solve the Dirichlet problem for initial data on S that is the restriction of a polynomial P.*

In order to do this, Poincaré begins by constructing a sequence of balls (B_n) contained in D and such that $D = \bigcup_{n=1}^{\infty} B_n$. He then assumes that $\Delta P < 0$ on the ball B, with boundary Σ and containing $D \cup S$ in its interior. Let $\Delta P = -4\pi\sigma$, and let $W_0 = \int_B \frac{\sigma}{r}\, dv$ be such that W_0, the volume potential for a distribution of positive masses, is positive and satisfies the equation

$$\Delta W_0 = -4\pi\sigma = \Delta P$$

in Σ. Considering σ as an attracting mass density, Poincaré sweeps the balls B_n in the order

$$B_1, B_2;\ B_1, B_2, B_3;\ B_1, B_2, B_3, B_4;\ \ldots$$

in such a way that each ball is swept infinitely often. If W_n denotes the attracting mass potential after the nth operation, it is obvious that $W_n > 0$ and that $W_n \leq W_{n-1}$. As a consequence, the sequence (W_n) converges pointwise, say to W, and we have $0 < W < W_0$ in D and $W = W_0$ outside of D. If the ball B_k has been swept during the operations numbered $\alpha_1, \alpha_2, \alpha_3, \ldots$, the functions W_{α_j} are harmonic in B_k, and the sequence (W_{α_j}) converges to W. By virtue of Harnack's theorem, W is harmonic in B_k, and hence in $D = \bigcup_{k \geq 1} B_k$. We have thus obtained a function W that is *harmonic on D*.

Poincaré then shows that W *is continuous at every point Q of S at which S has a tangent plane and two principal radii of curvature.* If S_0 is a sufficiently small sphere contained in $\mathbb{R}^3 \setminus D$ and tangent to S at Q, it is possible to construct a surface distribution of mass over S_0 whose potential coincides with the values of W_0 on S_0. If U is the potential of this distribution of mass, U is harmonic outside of S_0, equal to W_0 on S_0, and tends to zero at infinity. Thus, $W - U$ could have a maximum outside of S_0 but not a minimum, and it vanishes at Q and at infinity. Therefore, at every $q \in D$, we have $U \leq W \leq W_0$, and, as q approaches S, U tends to W_0, and it is the same with W.

The function $V = W - W_0 + P$ is harmonic on D since $\Delta W = 0$ and $\Delta W_0 = \Delta P$. It equals P on S since $W = W_0$ on S. The Dirichlet problem is therefore solved if $\Delta P < 0$ and if S has the specified regularity at each point or even, as Poincaré shows, with the exception of finitely many conical points. The restriction $\Delta P < 0$ can easily be lifted: every polynomial can be written as the difference $P_2 - P_1$ of two polynomials such that $\Delta P_1 < 0$ and $\Delta P_2 < 0$. If V_j is the harmonic function on D that equals P_j on S $(j = 1, 2)$, then $V = V_1 - V_2$ is the sought-after function. We can therefore conclude with Poincaré that

> *"The Dirichlet principle is established for every conductor whose surface is such that the tangent plane at each point is well defined except at a limited number of ordinary conical points."*

The conditions on the boundary would be generalized by Lebesgue and by Wiener.

The current version of Poincaré's sweeping method consists in introducing the class $\mathcal{S}(D)$ of *subharmonic* functions $v \in C(D)$ satisfying the inequality

$$v(x) \leq \frac{1}{\mu(B(x,R))} \int_{B(x,R)} v$$

over all balls $B(x, R)$ with compact support in D. Functions satisfying the reverse inequality form the class $\mathcal{T}(D)$ of *superharmonic* functions. To solve the Dirichlet problem (13.1), we begin by continuously extending, using Tietze's theorem, Φ to $\mathbb{R}^n$ with $\sup_{\mathbb{R}^n} |\Phi| = \sup_{\partial D} |\Phi|$. Using the Poisson integral, we associate to each open ball B with compact closure in D the linear map $P_B : C(\overline{D}) \to C(\overline{D})$ defined by $(P_B v)(x) = v(x)$ for all $x \in \overline{D} \setminus B$ and $(P_B v)(x) = h(x)$ for all $x \in \overline{B}$, where h is the harmonic function on B that equals v on ∂B. The map P_B is a projection, and its norm equals 1 by virtue of the maximum principle. In addition, P_B leaves the spaces $\mathcal{S}(D)$ and $\mathcal{T}(D)$ invariant, and $v \leq P_B(v)$ for $v \in \mathcal{S}(D)$ and $P_B w \leq w$ for all $w \in \mathcal{T}(D)$. By then choosing, as Poincaré did, a sequence of open balls B_i with compact closure in D, such that $D = \bigcup_{i=1}^{\infty} B_i$, and by setting $P_i = P_{B_i}$, we form the sequence of indices i_k defined by $12123123412345\ldots$, and we define the sequence $u_k \in C(\overline{D})$, such that $u_k = \Phi$ on ∂D, by

$$u_0 := \Phi, \qquad u_k := P_{i_k} u_{k-1} \quad (k = 1, 2, \ldots).$$

One then proves that the sequence u_k converges uniformly on compact subsets of D to a function V, harmonic on D. By defining, as Lebesgue did, a *barrier* for D at $x_0 \in \partial D$ by a function b that is continuous on $\overline{D}$ and superharmonic on D such that $b(x_0) = 0$ and $b(x) > 0$ for $x \in \overline{D} \setminus \{x_0\}$, and by calling a point $x_0 \in \partial D$ *regular* if a barrier exists at x_0, one proves that $V(x_0) = \Phi(x_0)$ at every regular point.

It is superfluous to stress the consequences and the developments of the sweeping method in potential theory and in the study of elliptic equations, in the hands of mathematicians such as Zaremba, Lebesgue, Perron, Wiener, F. Riesz, Frostman, de la Vallée Poussin, H. Cartan, Brelot and many others [**4**]. This method is central to a recent elementary proof of the continuity of minimizers of the Dirichlet integral when the given data are continuous along a boundary formed of regular points [**12**].

13.4. Eigenvalues and eigenfunctions of the Fourier problem: A heuristic approach

The separation of the temporal and spatial variables in studying the *Fourier problem for the heat equation* leads to the search for positive numbers k and functions U such that

$$\Delta U + kU = 0 \quad \text{in} \quad D \subset \mathbb{R}^3, \qquad \frac{\partial U}{\partial n} + hU = 0 \quad \text{on} \quad S = \partial D, \qquad \int_D U^2 = 1,$$

where $D \subset \mathbb{R}^3$ is a bounded domain having boundary S, $\partial U/\partial n$ is the exterior normal derivative, and h is a positive constant. For particular domains D, the problem can be explicitly solved, but, for an arbitrary bounded D, according to Poincaré,

> "*the first point is to establish the existence of these functions U. This has not yet been done, to my knowledge, in the general case, and this is what I will seek to do.*"

The first approach to this problem, also contained in [**23**], was announced by two notes in the *Comptes Rendus* of July 20, 1887 [**21**], and December 17, 1888 [**22**]. Poincaré knew that it was not rigorous, but he was surely unaware that it was not new, since he does not mention the work [**45**] by H. Weber published in 1869, which proposed a wholly similar approach in dimension 2. The common idea of these papers is an extension of the Dirichlet principle. The expression

$$B(F) := h \int_S F^2 \, d\sigma + \int_D \|\nabla F\|^2 \tag{13.4}$$

is positive and cannot vanish on the set Σ_1 of C^1-functions F such that

$$A(F) := \int_D F^2 = 1. \tag{13.5}$$

If $B(U_1)$ is the minimum of $B(F)$ on Σ_1, the method of Lagrange multipliers leads to the existence of a number k_1 such that

$$\Delta U_1 + k_1 U_1 = 0 \quad \text{in} \quad D \subset \mathbb{R}^3, \qquad \frac{\partial U_1}{\partial n} + hU_1 = 0 \quad \text{on} \quad S. \tag{13.6}$$

The existence of a *fundamental function* U_1, in Poincaré's terminology, is thus "proved", and Poincaré remarks that

$$k_1 \leq \frac{h \int_S F^2 \, d\sigma + \int_D \|\nabla F\|^2}{\int_D F^2} \tag{13.7}$$

for all non-zero functions F.

If $B(U_2)$ is the minimum of $B(F)$ on the set $\Sigma_2 \subset \Sigma_1$ of functions F that satisfy

$$\int_D F^2 = 1, \qquad \int_D F U_1 = 0, \tag{13.8}$$

then there exist Lagrange multipliers k_2 and λ such that

$$\Delta U_2 + k_2 U_2 + \lambda U_1 = 0 \quad \text{in} \quad D \subset \mathbb{R}^3, \qquad \frac{\partial U_2}{\partial n} + hU_2 = 0 \quad \text{on} \quad S.$$

Taking into account the conditions

$$\int_D U_1^2 = 1, \qquad \int_D U_1 U_2 = 0,$$

and using Green's formulas, we see easily that we must have $\lambda = 0$, and U_2, which satisfies the conditions

$$\Delta U_2 + k_2 U_2 = 0 \quad \text{on} \quad D \subset \mathbb{R}^3, \qquad \frac{\partial U_2}{\partial n} + hU_2 = 0 \quad \text{on} \quad S, \tag{13.9}$$

is a second fundamental solution. Now by imposing on F the conditions

$$\int_D F^2 = 1, \qquad \int_D FU_1 = 0, \qquad \int_D FU_2 = 0, \tag{13.10}$$

we obtain in the same way the existence of a number k_3 and a third fundamental function U_3, and, continuing in this way, the existence of infinitely many fundamental functions. Poincaré was aware of the non-rigorous character of this reasoning, because he wrote:

> *"This proof is completely analogous to the one used by Riemann to establish the Dirichlet principle and which the analysts have since replaced by more rigorous reasoning."*

Poincaré, with his sweeping method, is one of them, but he added:

> *"I believe nonetheless that we can here be content with this."*

Poincaré went further than Weber by proposing a method for obtaining an *upper bound* on k_n. If $F_1, \ldots, F_n$ are given, and if we take a linear combination $F = \sum_{j=1}^n \alpha_j F_j$, then $B(F)$ and $A(F)$ are positive definite quadratic forms in $\alpha = (\alpha_1, \ldots, \alpha_n)$, and Poincaré showed, by quite simple algebraic reasoning, that the roots $\lambda_1 \leq \cdots \leq \lambda_n$ of the characteristic equation

$$|B(F) - \lambda A(F)| = 0$$

satisfy the inequalities

$$k_j \leq \lambda_j \quad (1 \leq j \leq n). \tag{13.11}$$

As remarked by Pólya [**42**], if we take $F_j = U_j$ $(1 \leq j \leq n)$, then

$$k_j = \lambda_j \quad (1 \leq j \leq n),$$

and if $S_n = \left\{\sum_{j=1}^n \alpha_j F_j : \alpha \in \mathbb{R}^n\right\}$, then

$$\lambda_n = \max_{F \in S_n \setminus \{0\}} \frac{B(F)}{A(F)}.$$

As a consequence, Poincaré's reasoning *implicitly* contains the *minimax characterization of* k_n:

$$k_n = \min_{S_n} \max_{F \in S_n} \frac{B(F)}{A(F)}, \tag{13.12}$$

made explicit for the first time, for a spectral problem in finite dimensions, by E. Fisher [**7**] in 1905, without any reference to the previous work, and by an algebraic method, independent of the calculus of variations. It is unnecessary to stress the importance and the development of the minimax methods for the study of eigenvalues, in the hands of Weyl, Courant, Weinstein, Pólya, Schiffer and many others [**9, 46**].

Once again using Green's formula, Poincaré showed the growth, with respect to h, of the quantities k_j. By construction they also grow with j, and therefore to show that $k_j \to +\infty$ as $j \to +\infty$, it suffices to treat the case where $h = 0$, that is, what today is called the *Neumann problem*. In order to do this, Poincaré looked for a *lower bound* for k_n. By decomposing D into p parts D_i $(1 \leq i \leq p)$, calling $U_{i,j}$ and $k_{i,j}$ respectively the fundamental functions and the characteristic numbers associated to the Neumann conditions on D_i, and setting $V = \sum_{l=1}^{n} \alpha_l U_l$, where the U_l are the first n characteristic functions on D and the α_l are indeterminate coefficients, we easily see that

$$\frac{\int_D \|\nabla V\|^2}{\int_D V^2} = \frac{\sum_{l=1}^{n} k_l \alpha_l^2}{\sum_{l=1}^{n} \alpha_l^2} \leq k_n. \tag{13.13}$$

If λ_i $(1 \leq i \leq p)$ are positive integers such that $\sum_{i=1}^{p} \lambda_i = n - 1$, Poincaré chose the coefficients α_l in such a way that

$$\int_{D_i} V U_{i,j} = 0, \qquad j = 1, \ldots, \lambda_i, \quad i = 1, \ldots, p,$$

and from this easily concluded that

$$\int_D \|\nabla V\|^2 = \sum_{i=1}^{p} \int_{D_i} \|\nabla V\|^2 \geq \min_{1 \leq i \leq p} \{k_{i,\lambda_i+1}\} \left(\int_D V^2 \right),$$

and hence, using (13.13),

$$k_n \geq \min_{1 \leq i \leq p} \{k_{i,\lambda_i+1}\}.$$

In particular, if D is a polyhedron bounded by faces parallel to the coordinate planes, we can decompose it into $n-1$ rectangular parallelepipeds whose longest side tends to zero as $n \to \infty$. In this case,

$$k_n \geq \min_{1 \leq i \leq n-1} \{k_{i,2}\}.$$

Also, if the longest sides of these parallelepipeds are all less than or equal to a_n, we have, by separation of variables,

$$k_{i,2} \geq \frac{\pi^2}{a_n^2} \qquad (i = 1, \ldots, n-1),$$

and hence

$$k_n \geq \frac{\pi^2}{a_n^2}.$$

This leads to the conclusion that $k_n \to +\infty$ as $n \to \infty$. We can approximate D by similar polyhedra, and therefore the result must remain valid for D. But Poincaré was not content with this reasoning, and he posed himself the problem of finding a lower bound for k_2 when D is a bounded convex open set. This is the origin of the famous *Poincaré inequality*, which is the subject of the next section.

13.5. The Poincaré inequality

Let $D \subset \mathbb{R}^3$ be a bounded convex open set. Poincaré first remarked that, if $\mu(D)$ denotes the volume of D,

$$\int_{D\times D} [V(x) - V(x')]^2 \, dx \, dx' = 2\mu(D) \int_D V^2 - 2\left(\int_D V\right)^2.$$

Hence,

$$\begin{aligned} k_2 &= \min_{V\neq 0:\, \int_D V=0} \frac{\int_D \|\nabla V\|^2}{\int_D V^2} \\ &= 2\mu(D) \min_{V\neq 0:\, \int_D V=0} \frac{\int_D \|\nabla V\|^2}{\int_{D\times D}[V(x) - V(x')]^2 \, dx \, dx'}. \end{aligned}$$

Since the second minimized expression does not change if we replace V by $V + C$, where C is an arbitrary constant, we have

$$k_2 = 2\mu(D) \min_{V\neq 0} \frac{\int_D \|\nabla V\|^2}{\int_{D\times D}[V(x) - V(x')]^2 \, dx \, dx'}. \tag{13.14}$$

Starting with a complicated argument—based on a spherical change of variables for the two variables with a common moving origin—to which we will return later, and a reduction to a problem in the one-dimensional calculus of variations, Poincaré was led to the inequality

$$k_2 \geq \frac{6K_0\mu(D)}{\pi d^5}, \tag{13.15}$$

where K_0 is a certain numeric constant and d is the diameter of D.

This first formulation (13.15) of the *Poincaré inequality* allowed him to extend his result that $k_n \to +\infty$ as $n \to \infty$ to the case of a "general" domain D. If we decompose D into $n-1$ convex pieces D_j $(1 \leq j \leq n-1)$ with diameters d_j such that, for each D_j, $\mu(D_j)/d_j^5 \geq \alpha$ for some $\alpha > 0$, we easily obtain

$$k_n \geq \frac{6K_0}{\pi}\alpha. \tag{13.16}$$

The quantity α (which is of the order of d_j^{-2}) may be taken to be arbitrarily large by taking n sufficiently large and by appropriately choosing the pieces D_j. Therefore, for such a domain D, $k_n \to +\infty$ as $n \to \infty$.

Poincaré returned to inequality (13.15) in the third section ("Preliminary lemma") of his second memoir [**25**], dedicated to the equations of mathematical physics. There, he substantially simplified his argument and obtained an expression more explicit than (13.15). In order to reduce the technical weight, we will explain the case of a bounded convex set D in $\mathbb{R}^2$, for which Poincaré was content to give the result. To estimate the right-hand side of (13.14), Poincaré introduced, as in the preceding memoir, the change of variables

$$x = \xi + \rho\cos\varphi, \qquad y = \rho\sin\varphi, \qquad x' = \xi + \rho'\cos\varphi, \qquad y' = \rho'\sin\varphi. \tag{13.17}$$

Geometrically, ξ is the first coordinate of the intersection of the line xx' with the first coordinate axis, φ is the angle of the segment joining $(\xi, 0)$ to x with this axis, and ρ (respectively, ρ') is the location of x (respectively, of x') on the line xx' with $(\xi, 0)$ as the origin. When the direction $\varphi \in [0, 2\pi]$ is fixed, ξ varies between two extreme values $\xi_0(\varphi) < \xi_1(\varphi)$, outside of which the lines in the direction φ no longer intersect D, and, when the line $d_{\varphi,\xi}$ with direction φ passing through ξ is given, ρ and ρ' will vary between the values $\rho_0 < \rho_1$ corresponding to the two points of intersection of $d_{\varphi,\xi}$ with ∂D. The Jacobian of the transformation (13.17) is given by $(\rho-\rho')\sin\varphi$. To simplify notation, we set $T(\xi,\varphi,\sigma) = (\xi+\sigma\cos\varphi, \sigma\sin\varphi)$. The Cauchy inequality implies that

$$\begin{aligned}\left[\frac{\partial}{\partial\rho}(V\circ T)\right]^2 &= \left[\left(\frac{\partial V}{\partial x}\circ T\right)\cos\varphi + \left(\frac{\partial V}{\partial y}\circ T\right)\sin\varphi\right]^2 \\ &\leq \left(\frac{\partial V}{\partial x}\circ T\right)^2 + \left(\frac{\partial V}{\partial y}\circ T\right)^2 = \|\nabla V\circ T\|^2.\end{aligned}\tag{13.18}$$

Moreover, if $\mu(D)$ denotes the area of D, we have

$$\begin{aligned}\mu(D)\int_D \|\nabla V\|^2 &= \int_{D\times D}\|\nabla V(x)\|^2\,dx\,dx' \\ &= \int_0^{2\pi}\sin\varphi\,d\varphi\int_{\xi_0(\varphi)}^{\xi_1(\varphi)}d\xi\int_{\rho_0(\xi,\varphi)}^{\rho_1(\xi,\varphi)}\|\nabla V\circ T\|^2\,d\rho\int_{\rho_0(\xi,\varphi)}^{\rho_1(\xi,\varphi)}|\rho-\rho'|\,d\rho'.\end{aligned}\tag{13.19}$$

But, writing ρ_j in place of $\rho_j(\xi,\varphi)$ $(j=0,1)$ for simplicity, we have by (13.18)

$$\begin{aligned}&\int_{\rho_0}^{\rho_1}\|\nabla V\circ T\|^2\,d\rho\int_{\rho_0}^{\rho_1}|\rho-\rho'|\,d\rho' \\ &\quad\geq \int_{\rho_0}^{\rho_1}\left[\frac{\partial}{\partial\rho}(V\circ T)\right]^2 d\rho\int_{\rho_0}^{\rho_1}|\rho-\rho'|\,d\rho' \\ &\quad= \int_{\rho_0}^{\rho_1}\left[\frac{\partial}{\partial\rho}(V\circ T)\right]^2\left[\frac{(\rho-\rho_0)^2+(\rho_1-\rho)^2}{2}\right]d\rho \\ &\quad\geq \frac{(\rho_1-\rho_0)^2}{4}\int_{\rho_0}^{\rho_1}\left[\frac{\partial}{\partial\rho}(V\circ T)\right]^2 d\rho.\end{aligned}\tag{13.20}$$

On the other hand, if we simplify notation by denoting $T' = T(\xi,\varphi,\sigma')$,

$$\begin{aligned}&\int_{D\times D}[V(x)-V(x')]^2\,dx\,dx' \\ &= \int_0^{2\pi}\sin\varphi\,d\varphi\int_{\xi_0(\varphi)}^{\xi_1(\varphi)}d\xi\int_{\rho_0(\xi,\varphi)}^{\rho_1(\xi,\varphi)}d\rho\int_{\rho_0(\xi,\varphi)}^{\rho_1(\xi,\varphi)}[V\circ T - V\circ T']^2|\rho-\rho'|\,d\rho'.\end{aligned}\tag{13.21}$$

By the symmetry of ρ and ρ',

$$\begin{aligned}&\int_{\rho_0}^{\rho_1}d\rho\int_{\rho_0}^{\rho_1}[V\circ T - V\circ T']^2\,|\rho-\rho'|\,d\rho' \\ &\quad= 2\int_{\rho_0}^{\rho_1}d\rho\int_{\rho}^{\rho_1}[V\circ T - V\circ T']^2\,(\rho'-\rho)\,d\rho'.\end{aligned}\tag{13.22}$$

But, if $\rho' > \rho$, then by the Schwarz inequality,

$$[V\circ T(\xi,\varphi,\sigma') - V\circ T(\xi,\varphi,\sigma)]^2 \leq (\rho'-\rho)\int_\rho^{\rho'}\left[\frac{\partial}{\partial\sigma}[V\circ T(\xi,\varphi,\sigma)]\right]^2 d\sigma,$$

and hence, by using (13.22), we get

$$\begin{aligned}
&\int_{\rho_0}^{\rho_1} d\rho \int_{\rho_0}^{\rho_1} [V\circ T - V\circ T']^2 \, |\rho-\rho'| \, d\rho' \\
&\le 2\int_{\rho_0}^{\rho_1} d\rho \int_{\rho}^{\rho_1} (\rho'-\rho)^2 \, d\rho' \int_{\rho}^{\rho'} \left[\frac{\partial}{\partial\sigma}[V\circ T(\xi,\varphi,\sigma)]\right]^2 d\sigma \\
&= 2\int_{\rho_0}^{\rho_1} \left[\frac{\partial}{\partial\sigma}[V\circ T(\xi,\varphi,\sigma)]\right]^2 \left[\frac{(\rho_1-\rho_0)^4-(\rho_1-\sigma)^4-(\sigma-\rho_0)^4}{12}\right] d\sigma \\
&\le \frac{7}{48}(\rho_1-\rho_0)^4 \int_{\rho_0}^{\rho_1} \left[\frac{\partial}{\partial\sigma}[V\circ T(\xi,\varphi,\sigma)]\right]^2 d\sigma.
\end{aligned} \tag{13.23}$$

By comparing (13.20) and (13.23), we obtain

$$\begin{aligned}
&\int_{\rho_0}^{\rho_1} \|\nabla V\circ T\|^2 \, d\rho \int_{\rho_0}^{\rho_1} |\rho-\rho'| \, d\rho' \\
&\qquad \ge \frac{12}{7(\rho_1-\rho_2)^2} \int_{\rho_0}^{\rho_1} d\rho \int_{\rho_0}^{\rho_1} [V\circ T - V\circ T']^2 \, |\rho-\rho'| \, d\rho',
\end{aligned}$$

which immediately gives, by using (13.19) and (13.21), the *Poincaré inequality in the plane*:

$$\int_D \|\nabla V\|^2 \ge \frac{24}{7d^2}\int_D V^2 \tag{13.24}$$

for every continuously differentiable function V such that $\int_D V = 0$.

The calculation we have just sketched extends to the case where D is a convex open subset of $\mathbb{R}^3$. It suffices to replace the transformation (13.17) by

$$\begin{aligned}
x &= \xi + \rho\cos\varphi\sin\theta, & y &= \eta + \rho\sin\varphi\sin\theta, & z &= \rho\cos\theta, \\
x' &= \xi + \rho'\cos\varphi\sin\theta, & y' &= \eta + \rho'\sin\varphi\sin\theta, & z' &= \rho'\cos\theta.
\end{aligned}$$

We then obtain the *Poincaré inequality in space*

$$\int_D \|\nabla V\|^2 \ge \frac{16}{9d^2}\int_D V^2 \tag{13.25}$$

for every continuously differentiable function V such that $\int_D V = 0$. As Poincaré wrote:

> *"In the memoir for the* American Journal *I had found a different limit. The one that I adopt here is more manageable."*

Most modern works that deal with the Poincaré inequality are content with proving, by a compactness argument, the *existence* of a constant $c > 0$ such that

$$\int_D \|\nabla V\|^2 \ge c\int_D V^2 \tag{13.26}$$

when $\int_D V = 0$, without giving any information about the dependance of c on D. A recent work [**3**]—which corrects the previous work [**17**], whose argument is invalid for $n = 2$—shows that the best constant c for a convex $D \subset \mathbb{R}^n$ is equal to π^2/d^2 for any n. Moreover, the inequality (13.26) for all functions *that vanish on* ∂D, often called the Poincaré inequality in the literature, does not appear in the work of the French scholar.

Poincaré used the inequalities (13.24)–(13.25) to show that, given p functions $\varphi_1, \ldots, \varphi_p$ and $V = \sum_{i=1}^{p} \alpha_i \varphi_i$, one can choose the α_i in such a way that the quotient $\int_D \|\nabla V\|^2 / \int_D V^2$ is larger than a number given in advance. First of all, *if we can decompose D into $p-1$ convex open sets D_i with diameters less than d, we can choose the α_i in such a way that*

$$\frac{\int_D \|\nabla V\|^2\, dx}{\int_D V^2} \geq \frac{16}{9d^2}.$$

In fact, by determining the α_i in such a way that $\int_{D_i} V = 0$ $(1 \leq i \leq p-1)$, we can apply inequality (13.25) to D_i to deduce that

$$\frac{\int_{D_i} \|\nabla V\|^2\, dx}{\int_{D_i} V^2} \geq \frac{16}{9d^2} \qquad (1 \leq i \leq p-1),$$

and hence

$$\int_D \|\nabla V\|^2 = \sum_{i=1}^{p-1} \int_{D_i} \|\nabla V\|^2 \geq \frac{16}{9d^2} \int_D V^2.$$

Now if D *is convex* and contained in a cube of side length Λ, we divide this cube into q^3 equal cubes of side length Λ/q. The intersection of one of these small cubes with D is convex and has diameter less than $\Lambda\sqrt{3}/q$. If these intersections number $p-1$, the preceding reasoning shows that we can choose the α_i in such a way that

$$\int_D \|\nabla V\|^2 \geq \frac{16q^2}{27\Lambda^2} \int_D V^2.$$

We can do so *a fortiori* if p is larger than the number of intersections plus 1, and it suffices to take $p \geq q^3 + 1$. Thus, *for convex D, we can choose the α_i in such a way that*

$$\int_D \|\nabla V\|^2 \geq L_p \int_D V^2 \tag{13.27}$$

with $L_p = 16q^2/(27\Lambda^2)$ and q^3 the largest perfect cube less than or equal to $p-1$. If D can be decomposed into m appropriate convex sets, we again obtain the inequality (13.27) for an appropriate choice of the α_i, with q^3 the largest perfect cube less than or equal to $(p-1)/m$. In dimension 2, we have the same inequality, with $L_p = 12q^2/(7\Lambda)$, q^2 being the largest perfect square less than or equal to $(p-1)/m$. As $p \to \infty$, $L_p \simeq p^{2/3}$ if $n = 3$, and $L_p \simeq p$ if $n = 2$.

13.6. Eigenvalues and eigenfunctions of the Dirichlet problem: A rigorous approach

One of the principal contributions of the 1894 memoir [**25**] is the first rigorous proof of the existence of eigenvalues and eigenfunctions of the Dirichlet problem in an arbitrary bounded domain. The existence of the first eigenvalue had been proved in 1885 by Schwarz [**44**], and the existence of the second in 1893 by Picard [**19**]. The result was announced in a note in the *Comptes Rendus* of February 26, 1894 [**24**]. Given a bounded open set $D \subset \mathbb{R}^3$, f continuous on S, and $\xi \in \mathbb{R}$, Poincaré began by studying the Dirichlet problem

$$\Delta v + \xi v + f = 0 \quad \text{in} \quad D, \qquad v = 0 \quad \text{on} \quad \partial D. \tag{13.28}$$

To do so, and following an idea of Schwarz, he looked for the solution in the form $v = \sum_{k=0}^{\infty} \xi^k v_k$, which leads formally to the sequence of Dirichlet problems for the Laplacian

$$\begin{aligned} &\Delta v_0 + f = 0 \quad \text{in} \quad D, \qquad v_0 = 0 \quad \text{on} \quad \partial D, \\ &\Delta v_k + v_{k-1} = 0 \quad \text{in} \quad D, \qquad v_k = 0 \quad \text{on} \quad \partial D \qquad (k = 1, 2, \dots), \end{aligned} \tag{13.29}$$

which have a unique solution by virtue of the results on his sweeping method. For $m, n = 0, 1, 2, \dots$, Poincaré introduced the "Schwarz integrals"

$$W_{m,n} := \int_D v_m v_n, \qquad V_{m,n} := \int_D \nabla v_m \cdot \nabla v_n, \tag{13.30}$$

and recalled the results of that German mathematician,

$$\begin{aligned} &W_{m,n} = V_{m+1,n}, \qquad W_{m,n} = W_{m-1,n+1} = \cdots = W_{0,m+n} =: W_{m+n}, \\ &W_{n,n} = W_{2n} = \int_D v_n^2, \qquad W_{2n-1} = W_{n-1,n} = V_{n,n} = \int_D \|\nabla v_n\|^2, \end{aligned} \tag{13.31}$$

obtained by integration by parts. In addition, upon multiplying the two sides of $\Delta v_q = -v_{q-1}$ (resp. $\Delta v_{q+1} = -v_q$) by v_{q+1}, integrating by parts and using the Schwarz inequality, we have

$$W_{2q}^2 \leq W_{2q+1} W_{2q-1}, \qquad W_{2q+1}^2 \leq W_{2q} W_{2q+2},$$

which implies that the sequence $(W_{n+1}/W_n)_n$ is increasing. Moreover, if G is the Green function of the Laplacian,

$$v_n(x) = \int_D G(x, x')\, v_{n-1}(x')\, dx',$$

we conclude by the Schwarz inequality that, for all $x \in \overline{D}$,

$$|v_n(x)|^2 \leq \int_D G^2(x, x')\, dx' \int_D \|v_{n-1}\|^2 \leq Q^2 W_{2n-2}, \tag{13.32}$$

with $Q = d/(4\pi)$ if $n = 2$ and $Q = d^2/(4\pi)$ if $n = 3$. This leads to

$$W_{2n} = \int_D v_n^2 \leq Q^2 \mu(D) W_{2n-2},$$

and $Q\sqrt{\mu(D)}$ as an upper bound for the sequence $(W_{n+1}/W_n)_n$. As a consequence, if we set

$$\lim_{n\to\infty} \frac{W_{n+1}}{W_n} = \frac{1}{R},$$

we have $R \geq 1/(Q\sqrt{\mu(D)})$, $W_{2n-2} \leq (1/R)^{2n-2} W_0$, and hence, by (13.32),

$$\limsup_{n\to\infty} \left[\max_{x\in\overline{D}} |v_n(x)|\right]^{1/n} \leq \lim_{n\to\infty} \left[(Q W_0^{1/2})^{\frac{1}{n}} \left(\frac{1}{R}\right)^{1-\frac{1}{n}}\right] = \frac{1}{R},$$

which leads to the convergence (uniform in x) of the series $\sum_{n=0}^{\infty} \xi^n v_n(x)$ for $|\xi| < R$. We easily conclude that *the problem* (13.28) *has a solution for all* $|\xi| < R$, *and in particular for all* $|\xi| < 1/(Q\sqrt{\mu(D)})$. Poincaré denotes this solution $v = [f, \xi]$.

Now let φ_j $(1 \leq j \leq p)$ be p given functions, and $w_j = [\varphi_j, \xi]$ $(1 \leq j \leq p)$. If

$$\varphi = \sum_{j=1}^{p} \alpha_j \varphi_j, \qquad v = \sum_{j=1}^{p} \alpha_j w_j,$$

then, by linearity,

$$v = [\varphi, \xi] = \sum_{n=0}^{\infty} \xi^n v_n.$$

If, for this problem, $W_{n+1}/W_n \to 1/R$, *we can choose the* α_j *in such a way that* $R \geq L_p$, *where* L_p *is given by* (13.27). In fact, for fixed n, we can choose the α_j in such a way that $V_{n,n}/W_{n,n} \geq L_p$, that is, $W_{2n}/W_{2n-1} \leq 1/L_p$, which leads to

$$\frac{W_2}{W_1} \leq \frac{W_3}{W_2} \leq \cdots \leq \frac{W_{2n}}{W_{2n-1}} \leq \frac{1}{L_p}. \tag{13.33}$$

Let δ_n be the set of the $(\alpha_1, \ldots, \alpha_p)$ for which (13.33) is satisfied. Then $\delta_{n+1} \subset \delta_n$, and, if we choose $(\alpha_1, \ldots, \alpha_p) \in \bigcap_{n=1}^{\infty} \delta_n$, we will have $W_{n+1}/W_n \leq 1/L_p$, independent of n, that is, $R \geq L_p$.

Having acquired this fact, let us return to the problem (13.28), and let

$$v = [f, \xi] = \sum_{n=0}^{\infty} \xi^n v_n, \qquad u_j = [v_{j-2}, \xi] \quad (2 \leq j \leq p).$$

If $u_j = \sum_{n=0}^{\infty} \xi^n u_{j,n}$, we have, with the Dirichlet condition of vanishing on the boundary,

$$\Delta u_{j,0} + v_{j-2} = 0, \qquad \Delta u_{j,1} + u_{j,0} = 0, \quad \ldots, \qquad \Delta u_{j,n} + u_{j,n-1} = 0, \quad \ldots$$

and hence, by the uniqueness of the Dirichlet problem,

$$u_{j,0} = v_{j-1}, \qquad u_{j,1} = v_j, \quad \ldots, \qquad u_{j,n} = v_{j+n-1}, \quad \ldots,$$

that is,

$$u_j = \sum_{n=0}^{\infty} \xi^n v_{j+n-1} \qquad (2 \leq j \leq p). \tag{13.34}$$

Now let $w = \alpha_1 v + \sum_{j=2}^{p} \alpha_j u_j$. By linearity,

$$w = \sum_{n=0}^{\infty} \xi^n w_n = \left[\alpha_1 v + \sum_{j=2}^{p} \alpha_j u_j, \xi\right] = \sum_{n=0}^{\infty} \xi^n \left(\alpha_1 v_n + \sum_{j=2}^{p} \alpha_j v_{j+n-1}\right). \tag{13.35}$$

By the result of the previous paragraph, we can choose the α_j in such a way that the series (13.35) has a radius of convergence greater than or equal to L_p. Since

$$\xi u_2 = \sum_{n=0}^{\infty} \xi^{n+1} v_{n+1} = v - v_0,$$

$$\xi u_j = \sum_{n=0}^{\infty} \xi^{n+1} v_{j+n-1} = u_{j-1} - v_{j-2} \qquad (3 \leq j \leq p),$$

we have

$$
\begin{aligned}
\alpha_1 v + \sum_{j=2}^{p} \alpha_j u_j &= w, \\
v - \xi u_2 &= v_0, \\
u_2 - \xi u_2 &= v_1, \\
&\vdots \\
u_{p-1} - \xi u_p &= v_{p-2}.
\end{aligned} \tag{13.36}
$$

Cramer's rule gives $v = P(x,y,z,\xi)/D(\xi)$, with

$$
D(\xi) = \begin{vmatrix}
\alpha_1 & \alpha_2 & \alpha_3 & \cdots & \alpha_{p-1} & \alpha_p \\
1 & -\xi & 0 & \cdots & 0 & 0 \\
0 & 1 & -\xi & \cdots & 0 & 0 \\
0 & 0 & 1 & \ddots & \vdots & \vdots \\
\vdots & \vdots & \vdots & \ddots & -\xi & 0 \\
0 & 0 & 0 & \cdots & 1 & -\xi
\end{vmatrix} = \sum_{j=0}^{p-1} (-1)^j \alpha_{p-j} \xi^j,
$$

and

$$
P(x,y,z,\xi) = \begin{vmatrix}
w(x,y,z,\xi) & \alpha_2 & \alpha_3 & \cdots & \alpha_{p-1} & \alpha_p \\
v_0(x,y,z) & -\xi & 0 & \cdots & 0 & 0 \\
v_1(x,y,z) & 1 & -\xi & \cdots & 0 & 0 \\
v_2(x,y,z) & 0 & 1 & \ddots & \vdots & \vdots \\
\vdots & \vdots & \vdots & \ddots & -\xi & 0 \\
v_{p-2}(x,y,z) & 0 & 0 & \cdots & 1 & -\xi
\end{vmatrix}.
$$

Since $v_0, v_1, \ldots, v_{p-2}$ do not depend on ξ, and w is holomorphic in ξ for $|\xi| < L_p$, the function P/D is meromorphic in ξ for $|\xi| < L_p$. Since we can take p as large as we like, *v is meromorphic in ξ on the entire complex plane.* It is not hard to see that $P(\,\cdot\,,\xi)$ vanishes on the boundary of D, has first and second order partial derivatives with respect to x, y, z, and satisfies the partial differential equation

$$
\Delta P = \Delta(vD) = D\Delta v = -\xi D v + Df = -\xi P + Df. \tag{13.37}
$$

The poles of modulus less than L_p of the meromorphic function v are the zeroes of the polynomials D. If $\xi = k$ is a simple zero of D and $P_k = P(\,\cdot\,,k)$, then (13.37) leads to

$$
\Delta P_k + k P_k = 0,
$$

which shows that k is a *"characteristic number"* associated to the *"harmonic function"* P_k; that is, in current terminology, P_k is an eigenfunction of $-\Delta$ with Dirichlet conditions on ∂D, associated to the eigenvalue k. Poincaré justified his terminology, which would not be retained, as follows:

> *"The various simple sounds that a membrane can emit are characterized by equations of the form $\Delta u + ku = 0$, the function being subjected to vanishing on the boundary. We know that these simple sounds have received the name of harmonics."*

Poincaré then showed that *v only has simple poles,* and hence, up to a constant factor, *P_k is the residue of v at k.* By relatively elementary considerations, he also showed that v *cannot be holomorphic in the entire plane,* which ensures that harmonic functions exist, that there can be no more that $p-1$ of them that are linearly independent and have characteristic number less than L_p, and therefore that *only finitely many linearly independent harmonic functions can correspond to a characteristic number,* and finally that *all of the characteristic numbers are real and positive.* Poincaré *normalized* the harmonic functions by the condition $\int_D P^2 = 1$.

Poincaré's proof was quickly forgotten, following the development, by Fredholm and Hilbert, of the *spectral theory of linear integral equations,* which was nonetheless inspired by Poincaré's proof. One can find information about the development of spectral theory during the 20th century in [**6**].

The next chapter of the memoir [**25**] studies the *maximum principle* for the problems

$$\Delta v = f \quad \text{on} \quad D, \qquad \frac{\partial v}{\partial n} + hv = 0 \quad \text{on} \quad \partial D, \tag{13.38}$$

$$\Delta v = -f \quad \text{on} \quad D, \qquad \frac{\partial v}{\partial n} + hv = \varphi \quad \text{on} \quad \partial D, \tag{13.39}$$

when $f > 0$, $h > 0$, and $\varphi > 0$. Poincaré's most original contribution is the introduction of the idea of a *weak solution* to the problem (13.39). Poincaré remarked that if v is a solution to (13.39) and if u is any twice continuously differentiable function, then the condition

$$\int_D uf\, dx + \int_D v\Delta u\, dx + \int_{\partial D} v\varphi\, d\sigma = \int_{\partial D} v\left(hu + \frac{\partial u}{\partial n}\right) d\sigma \tag{13.40}$$

—which, for simplicity, Poincaré called the *modified condition*—is satisfied. Conversely, if the condition (13.40) is satisfied for all u, the equations (13.39) will also be, provided that v and $\partial v/\partial n$ are continuous functions. Poincaré thus showed that if v is merely continuous and satisfies (13.40) for all u, the maximum principles developed within the classical framework remain valid. After recalling the idea of Neumann's method for the Dirichlet problem, to which we will return later, Poincaré applied it to the mixed problem

$$\Delta v + f = 0 \quad \text{in} \quad D, \qquad \frac{\partial v}{\partial n} + hv = 0 \quad \text{on} \quad \partial D, \tag{13.41}$$

when D is a bounded convex open set. He did not succeed in proving that the solution v obtained by this method of successive approximations has a normal derivative on ∂D and that, *a fortiori*, it satisfies the boundary condition. But he proved that it satisfies the condition (13.40) (with $\varphi = 0$), once again anticipating the notion of a weak solution:

> *"We are therefore uncertain that the expression $\frac{\partial v}{\partial n}$ has a meaning and even less certain that the boundary condition $\frac{\partial v}{\partial n} + hu = 0$ is fulfilled. On the other hand, we can assert that we will have the* modified condition. *It is obviously equivalent [...] from the physical point of view."*

Finally, the rest of the memoir is an attempt to extend his proof of the existence of eigenvalues for the problem

$$\Delta v + \xi v + f = 0 \quad \text{in} \quad \Omega, \qquad \frac{\partial v}{\partial n} + hv = 0 \quad \text{on} \quad \partial D, \tag{13.42}$$

which runs into the same difficulty concerning the regularity of the coefficients v_n on the boundary. Poincaré wrote:

> *"One can see that I have not been able to achieve results as satisfactory in the general case as in the case of $h = \infty$; one sees how many gaps still remain. I will not force myself any longer to fill them. [...] I believed, however, that these results, incomplete as they are, were not stripped of interest, and I decided to publish them. I would be happy if this publication could provoke new research on the subject."*

No doubt that Poincaré's wish was fully realized. The memoir ends with a few partial results on expanding harmonic functions into series.

13.7. The Neumann method for the Dirichlet problem on an arbitrary domain

In 1878, Carl Neumann [**15**] justified a formal method, introduced by Beer around 1860, to prove the existence of a solution to the Dirichlet problem

$$\Delta V = 0 \quad \text{in} \quad D, \qquad V = \Phi \quad \text{on} \quad \partial D, \tag{13.43}$$

when Φ is continuous and D is bounded and convex, with ∂D sufficiently regular. This method consists in searching for V in the form of a *double layer potential* of unknown density ρ, that is, in the form of the surface integral

$$V(x) = \int_{\partial D} \rho(s) \frac{\partial}{\partial n} \frac{1}{\|x - s\|} \, d\sigma. \tag{13.44}$$

We know that V is harmonic outside of ∂D and that, if $y \in \partial D$, the limits

$$V^i(y) = \lim_{x \to y;\, x \in D} V(x), \qquad V^e(y) = \lim_{x \to y;\, x \notin \overline{D}} V(x)$$

exist and satisfy the relations

$$V^i(y) - V^e(y) = -4\pi\rho(y).$$

We set, for $y \in \partial D$,

$$V(y) = \frac{1}{2}[V^i(y) + V^e(y)].$$

Neumann proposed to determine V such that one has, on ∂D, the relation

$$V^i - V^e = \lambda(V^i + V^e) + 2\Phi, \tag{13.45}$$

with λ being a real parameter. If the problem is solved for $\lambda = -1$, we will have $V^i = \Phi$ on ∂D, and V will be a solution to (13.43). By looking for V in the form of a series $V = \sum_{n=0}^{\infty} \lambda^n V_n$, which gives $V^i = \sum_{n=0}^{\infty} \lambda^n V_n^i$, $V^e = \sum_{n=0}^{\infty} \lambda^n V_n^e$, and, on ∂D, $V_n = \frac{1}{2}(V_n^i + V_n^e)$, by setting the coefficients of like powers of λ in (13.45) to be equal, we find that

$$\begin{aligned} V_0^i - V_0^e &= 2\Phi, \\ V_1^i - V_1^e &= V_0^i + V_0^e = 2V_0, \\ V_2^i - V_2^3 &= V_1^i + V_1^e = 2V_2, \\ &\vdots \end{aligned} \tag{13.46}$$

which leads to taking for V_0 the potential of a double layer with density $-\Phi/2\pi$, for V_1 the potential of a double layer with density $-V_0/2\pi$, for V_2 the potential of a double layer with density $-V_1/2\pi$, etc. In other words, the V_n are determined by the recurrence relations, for $x \in \partial D$,

$$(13.47)\qquad \begin{aligned} V_0 &= -\int_{\partial D} \frac{\Phi(s)}{2\pi} \frac{\partial}{\partial n} \frac{1}{\|x-s\|}\, d\sigma, \\ V_n &= -\int_{\partial D} \frac{V_{n-1}(s)}{2\pi} \frac{\partial}{\partial n} \frac{1}{\|x-s\|}\, d\sigma \qquad (n = 1, 2, \dots). \end{aligned}$$

When D is convex, Neumann proved the existence of a constant C such that the series $\sum_{n=0}^{\infty} \lambda^n (V_n - C)$ converges for $|\lambda| \leq 1$. The Dirichlet problem (13.43) can therefore be solved in this case. Poincaré's 1896 memoir [**27**], announced by the note in the *Comptes Rendus* [**26**] of February 18, 1895, proves the convergence of the Neumann method when D *is simply connected and at every point has a tangent plane and a radius of curvature.* We will not linger on the details of this highly technical proof, which is nearly forgotten today. The historical importance of the work [**27**] is that, along with the memoir [**25**], it directly inspired Fredholm to develop his theory of *linear integral equations.* In fact, the relations (13.47) give

$$\sum_{n=1}^{\infty} \lambda^n V_n(x) = -\lambda \int_{\partial D} \frac{1}{2\pi} \sum_{n=1}^{\infty} \lambda^{n-1} V_{n-1}(s) \frac{\partial}{\partial n} \frac{1}{\|x-s\|}\, d\sigma,$$

that is,

$$V(x) + \lambda \int_{\partial D} \frac{1}{2\pi} \frac{\partial}{\partial n} \frac{1}{\|x-s\|} V(s)\, d\sigma + \int_{\partial D} \frac{1}{2\pi} \frac{\partial}{\partial n} \frac{1}{\|x-s\|} \Phi(s)\, d\sigma = 0.$$

In other words, V satisfies what would be called a *Fredholm integral equation of the second kind* with kernel $\frac{1}{2\pi} \frac{\partial}{\partial n} \frac{1}{\|x-s\|}$. The article in which Fredholm introduces his first study of an integral equation of the second kind is entitled *Sur une nouvelle méthode pour la résolution du problème de Dirichlet* [*On a new method for solving the Dirichlet problem*] [**8**], and begins as follows:

> *"In his deep research* (Acta Mathematatica, vol. 20) *on the convergence of the well-known Neumann method in potential theory, M. Poincaré considered the Dirichlet problem as a particular case of another problem, which he called the Neumann problem.* [...] *Neumann solved this problem by expanding the unknown function in powers of the parameter* λ*. But it results from the research of M. Poincaré that* V *is a meromorphic function. Thus it is clear that the Neumann series cannot converge for all values of* λ*. But because we know that a meromorphic function can always be expressed as the quotient of two entire functions, it seemed natural to me to look for these entire functions directly."*

Poincaré immediately grasped the importance of Fredholm's work on integral equations, and gave several generalizations [**32, 33, 40**] and applications of it, in particular the theory of tides [**38**] and the diffraction of Hertzian waves[1] [**34, 39**]. We may observe that half of the lectures given by Poincaré during his visit to Göttingen in April 1909, at Hilbert's invitation, are devoted to the Fredholm integral equations and their applications [**35, 36, 37**].

[1] EDITOR'S NOTE: See Chapter 18.

13.8. The heat equation and the telegraph equation

In the third section, entitled *"Lois de refroidissement"* ["Cooling laws"], of his memoir [**23**], Poincaré studies the *non-stationary Fourier problem*, which consists in finding $V = V(t,x)$ such that $V(0,x) = V_0(x)$ $(x \in D)$ and

$$\frac{\partial V}{\partial t} = a^2 \Delta V \quad \text{in} \quad (0,\infty) \times D, \qquad \frac{\partial V}{\partial n} + hV = 0 \quad \text{in} \quad (0,\infty) \times \partial D. \tag{13.48}$$

If U_n are fundamental functions of the stationary problem and k_n are the corresponding characteristic numbers, we must be able to expand V_0 as follows:

$$V_0 = \sum_{n=1}^{\infty} A_n U_n, \tag{13.49}$$

and from this to conclude that, for all $t > 0$,

$$V(t,x) = \sum_{n=1}^{\infty} A_n e^{-a^2 k_n t} U_n(x). \tag{13.50}$$

Remarking that he is *"still not able to prove generally"* the possibility of the expansion, Poincaré shows that the *average error*

$$S_0 := \int_D \left[V_0 - \sum_{p=0}^{n} A_p U_p \right]^2$$

is minimized when $A_p = \int_D V_0 U_p =: J_p^0$ $(1 \leq p \leq n)$. For this choice of the coefficients A_p, Poincaré shows that if $J_p(t) = J_p^0 e^{-a^2 k_p t}$, one can take n large enough for the average error in the temperature at time t,

$$S(t) := \int_D \left[V(t,x) - \sum_{p=1}^{n} J_p(t) U_p(x) \right]^2 dx,$$

to be as small as one likes. We see in those considerations the first use of the *Hilbert norm* in place of a uniform norm in the study of a parabolic problem.

Finally, to conclude, we point to the interesting note [**28**] in the *Comptes Rendus* of December 26, 1893, in which Poincaré gives the first general solution to the initial value problem for the *telegraph equation*

$$\frac{\partial^2 V}{\partial t^2} + 2\frac{\partial V}{\partial t} = \frac{\partial^2 V}{\partial x^2}, \tag{13.51}$$

which models, as its name indicates, the propagation of a current in a conducting wire. This equation was introduced in 1857 by Kirchhoff [**13**], starting from Weber's electromagnetic theory, and derived from Maxwell's theory by Heaviside [**11**] in 1876. These authors had only given particular solutions. Poincaré solved (13.51) by the method of Fourier transformations, and a detailed analysis of the solution allowed him to describe precisely the behavior of an initial perturbation, bounded in space:

> *"We see first of all that the head of the perturbation propagates with a certain speed, in such a way that, in front of this head, the perturbation is zero, contrasting with what happens in Fourier's theory of heat, and conforming to the laws of the propagation of light or sound by planar waves, derived from the equation of vibrating strings. But there is an important difference with this latter case, because the perturbation, while*

> *propagating, leaves behind it a non-zero residue [...] and can then disturb the observations."*

Poincaré adds, in his analysis of his scientific work [**41**]:

> *"The telegraph equation by its form is half-way between the equation of vibrating strings and the heat equation. The results are also intermediate to the ones given by those two equations. In the case of vibrating strings, the wave propagates with constant speed without becoming deformed, without spreading out, without sending a signal ahead or leaving one behind. In the case of heat, the wave spreads out immediately over the entire conducting wire, so that one can assign to it neither a beginning nor an end. In the case of electricity, the head of the wave advances regularly with the speed of light; but the wave leaves behind it a residue, a kind of tail."*

The study of the telegraph equation is taken up in greater detail in [**30, 31**].

Poincaré's contributions to the equations of mathematical physics would have sufficed to place him among the greatest mathematicians of the end of the 19th and the beginning of the 20th century.

Bibliography

[1] T. Archibald, *From attraction theory to existence proofs: the evolution of potential-theoretic methods in the study of boundary value problems, 1860–1890*, Revue Hist. Math. **2** (1996), pp. 67–93.

[2] M. Atten, *La nomination de H. Poincaré à la chaire de physique mathématique et calcul des probabilités de la Sorbonne*, Cahiers du Séminaire d'Histoire des Mathématiques **9** (1988), pp. 221–230.

[3] M. Bebendorf, *A note on the Poincaré inequality for convex domains*, Z. Anal. Anwend. **22** (2003), pp. 751–756.

[4] M. Brelot, *La théorie moderne du potentiel*, Ann. Institut Fourier **4** (1952), pp. 113–140.

[5] H. Brezis and F. E. Browder, *Partial differential equations in the 20th century*, Advances in Math. **135** (1998), pp. 76–144.

[6] J. Dieudonné, *History of functional analysis*, North-Holland Math. Studies No. 49, 1981.

[7] E. Fischer, *Ueber quadratische Formen mit reellen Koeffizienten*, Monatshetfe für Math. Phys. **16** (1905), pp. 234–249.

[8] I. Fredholm, *Sur une nouvelle méthode pour la résolution du problème de Dirichlet*, Öfversigt Kongl. Vetenskaps-Akad. Förhandlingar **57** (1900), p. 39-46. Œuvres complètes, pp. 61–68.

[9] S. H. Gould, *Variational methods for eigenvalue problems*, University of Toronto Press, Toronto, 1957.

[10] J. Gray, *Poincaré and electromagnetism theory*: pp. 193–208 of *Henri Poincaré: science et philosophie* (Nancy, 1994), Akademie Verlag (Berlin) and Libr. Albert Blanchard (Paris), 1996.

[11] O. Heaviside, *On the extra current*, Phil. Mag. (5) **2** (1876), p. 135. Electrical papers, I, pp. 53–61.

[12] S. Hildebrandt, *On Dirichlet principle and Poincaré's méthode de balayage*, Math. Nachr. **278** (2005), pp. 141–144.

[13] G. Kirchhoff, *Ueber die Bewegung der Elektrizität in Drähten*, Pogg. Ann. der Phys. **100** (1857), p. 193 and p. 251. Ges. Abh. p. 131.

[14] J. Mawhin, *Henri Poincaré. A life in the service of science*, Notices Amer. Math. Soc. **52** (2005), pp. 1036–1044.

[15] C. Neumann, *Untersuchungen über das Logarithmische und Newton'sche Potential*, Math. Ann. **13** (1878), pp. 255–300.

[16] L. Nirenberg, *Partial differential equations in the first half of the century*: pp. 479–515 of *Development of Mathematics 1900–1950*, under the direction of J. P. Pier, Birkhäuser, Bâle, 1994.

[17] L. E. Payne and H. F. Weinberger, *An optimal Poincaré inequality for convex domains*, Arch. Rat. Mech. Anal. **5** (1960), pp. 286–292.

[18] J. B. Pécot, *Les théories spectrales de Poincaré*, Sciences et Techniques en Perspective **26** (1993), pp. 173–205.

[19] É. Picard, *Sur l'équation aux dérivées partielles qui se présente dans la théorie de la vibration des membranes*, C.R.A.S. Paris **117** (1893), pp. 502–507. Œuvres, tome 2, pp. 545–550.

[20] H. Poincaré, *Sur le problème de la distribution électrique*, C.R.A.S. Paris **104** (1887), pp. 44–46. Œuvres, tome IX, pp. 15–17.

[21] H. Poincaré, *Sur la théorie analytique de la chaleur*, C.R.A.S. Paris **104** (1887), p. 1753-1759. Œuvres, tome IX, pp. 18–23.

[22] H. Poincaré, *Sur la théorie analytique de la chaleur*, C.R.A.S. Paris **107** (1888), pp. 967–971. Œuvres, tome IX, pp. 24–27.

[23] H. Poincaré, *Sur les équations aux dérivées partielles de la physique mathématique*, American J. Math. **12** (1890), pp. 211–294. Œuvres, tome IX, pp. 28–113.

[24] H. Poincaré, *Sur l'équation des vibrations d'une membrane*, C.R.A.S. Paris **118** (1894), pp. 447–451. Œuvres, tome IX, pp. 119–122.

[25] H. Poincaré, *Sur les équations de la physique mathématique*, Rend. Circolo Mat. Palermo **8** (1894), pp. 57–155. Œuvres, tome IX, pp. 123–196.

[26] H. Poincaré, *Sur la méthode de Neumann et le problème de Dirichlet*, C.R.A.S. **120** (1895), pp. 347–352. Œuvres, tome IX, pp. 197–201.

[27] H. Poincaré, *La méthode de Neumann et le problème de Dirichlet*, Acta Math. **20** (1896-97), pp. 59–142. Œuvres, tome IX, pp. 202–272.

[28] H. Poincaré, *Sur la propagation de l'électricité*, C.R.A.S. Paris **117** (1893), pp. 1027–1032. Œuvres, tome IX, pp. 278–283.

[29] H. Poincaré, *Théorie du potentiel newtonien*, Carré, Paris, 1899.

[30] H. Poincaré, *Théorie analytique de la propagation de la chaleur*, Carré, Paris, 1895.

[31] H. Poincaré, *Étude de la propagation d'un courant en période variable sur une ligne munie de récepteur*, L'Éclairage électrique **40** (1904), pp. 121–228; pp. 161–167; pp. 201–212; pp. 241–250. Œuvres, tome X, pp. 445–486.

[32] H. Poincaré, *Remarques sur l'équation de Fredholm*, C.R.A.S. Paris **147** (1908), pp. 1357–1371. Œuvres, tome III, pp. 540–544.

[33] H. Poincaré, *Sur quelques applications de la méthode de Fredholm*, C.R.A.S. Paris **148** (1909), pp. 125–126. Œuvres, tome III, pp. 545–546.

[34] H. Poincaré, *Les ondes hertziennes et l'équation de Fredholm*, C.R.A.S. Paris **148** (1909), pp. 449–453. Œuvres, tome X, pp. 89–92.

[35] H. Poincaré, *Ueber die Fredholmschen Gleichungen*: pp. 1–10 of *Sechs Vorträge über angewählte Gegenstände aus der reine Mathematik und mathematische Physik*, Teubner, Leipzig-Berlin, 1910. Œuvres, tome III, pp. 547–554.

[36] H. Poincaré, *Anwendung der Theorie des Integralgleichungen auf die Flutbewegung des Meeres*: pp. 12–19 of *Sechs Vorträge über angewählte Gegenstände aus der reine Mathematik und mathematische Physik*, Teubner, Leipzig-Berlin, 1910. Œuvres, tome VIII, pp. 289–296.

[37] H. Poincaré, *Anwendung der Integralgleichungen auf Hertzsche Wellen*: pp. 23–31 of *Sechs Vorträge über angewählte gegensände aus der reine Mathematik und mathematische Physik*, Teubner, Leipzig-Berlin, 1910. Œuvres, tome X, pp. 78–88.

[38] H. Poincaré, *Leçons de mécanique céleste, tome III. Théorie des marées*, Gauthier-Villars, Paris, 1910.

[39] H. Poincaré, *Sur la diffraction des ondes hertziennes*, Rend. Circolo Mat. Palermo **29** (1910), pp. 169–259. Œuvres, tome X, pp. 94–203.

[40] H. Poincaré, *Remarques diverses sur l'équation de Fredholm*, Acta Math. **33** (1911), pp. 57–86. Œuvres, tome III, pp. 554–582.

[41] H. Poincaré, *Analyse de ses travaux scientifiques*, Acta Math. **38** (1921), pp. 36–135.

[42] G. Pólya, *Estimates for eigenvalues*: pp. 200–207 of the *Studies in Mathematics and Mechanics presented to R. von Mises*, Academic Press, New York, 1954.

[43] L. Schwartz, *L'œuvre de Poincaré: Équations différentielles de la Physique*: pp. 219–225 of *Le livre du centenaire de la naissance de Henri Poincaré*, Gauthier-Villars, Paris, 1955. Œuvres, tome XI, 2-ième partie, pp. 219–225.

[44] H. A. Schwarz, *Ueber ein die Flächen kleinsten Flächeninhalts betreffended Problem der Variationsrechnung*: pp. 315–262 of the *Festchrift zum Jubelgeburtstage des Herrn Weierstrass*, Acta Societatis Scientiarum Fennicae **15** (1885).

[45] H. Weber, *Ueber die Integration der partiellen Differentialgleichungs* $\frac{\partial^2 u}{\partial x^2} + \frac{\partial^2 u}{\partial y^2} + k^2 u = 0$, Math. Ann. **1** (1869), pp. 1–36.

[46] A. Weinstein and W. Stenger, *Methods of intermediate problems for eigenvalues. Theory and ramifications*, Academic Press, New York, 1972.

CHAPTER 14

Poincaré's *Calculus of Probabilities*

PIERRE CARTIER

14.1. The book's composition

Probability occupies a rather minor place in Poincaré's work. I will here occupy myself with his book *Calcul des Probabilités* [**19**]. As is known, Poincaré for a long time held the chair of Probability and Mathematical physics at the Faculté des Sciences de Paris. There he gave a new course almost every year, and most of them were written up and published. There he considered a great variety of questions (fluid mechanics, celestial mechanics, etc.), and his treatise on probabilities is one of these courses. The first edition of 1896 was written by Albert Quiquet, "former student at the École Normale Supérieure". The *Supplement historique 2005* of the directory of the École Normale Supérieure [**25**] mentions a Jules Théaub (sic!) Quiquet, of the class of 1883, deceased in 1934, an actuary for the insurance company La Nationale. He was therefore neither a student nor a pupil of Poincaré (there were hardly any) but a practitioner of statistics, who came to learn from the great master and lent his pen to Poincaré.

14.2. A physicist's work

This book is a pivotal text. The first edition is plainly a 19th century work, in the wake of a long tradition of textbooks, for example that of Joseph Bertrand [**2**], which was a classic at the time and was cited by Poincaré. A very interesting project for a historian of science would be to make a detailed comparison of Bertrand's textbook and the two successive versions (1896 and 1912) of Poincaré's.[1] My thesis will be that Poincaré must be read as a *physicist*, given the constant references to the kinetic theory of gases (and in accordance with his title as the chair of "probability and mathematical physics"). Obviously his book should not be judged as a "Treatise on Probability" along the lines of the numerous "Treatises on Analysis" from this time period.

Between 1896 and 1912, the story of probability's involvement in physics accelerates: there is the work of Gibbs, completing that of Maxwell and Boltzmann; there is the research of Planck in thermodynamics, and the law of black-body radiation (1900). Above all stand the efforts of Einstein (and those of Smoluchowski) on Brownian motion, theoretical work that was extended by the experiments of Jean Perrin. The computation of the Avogadro–Loschmidt number that resulted gave a solid empirical basis for the still theoretical perspective of statistical physics. Of none of these is there any mention in Poincaré's 1912 book. One may imagine that,

[1] One can find several pieces of comparison in a few of the notes of Bernard Bru's article [**9**].

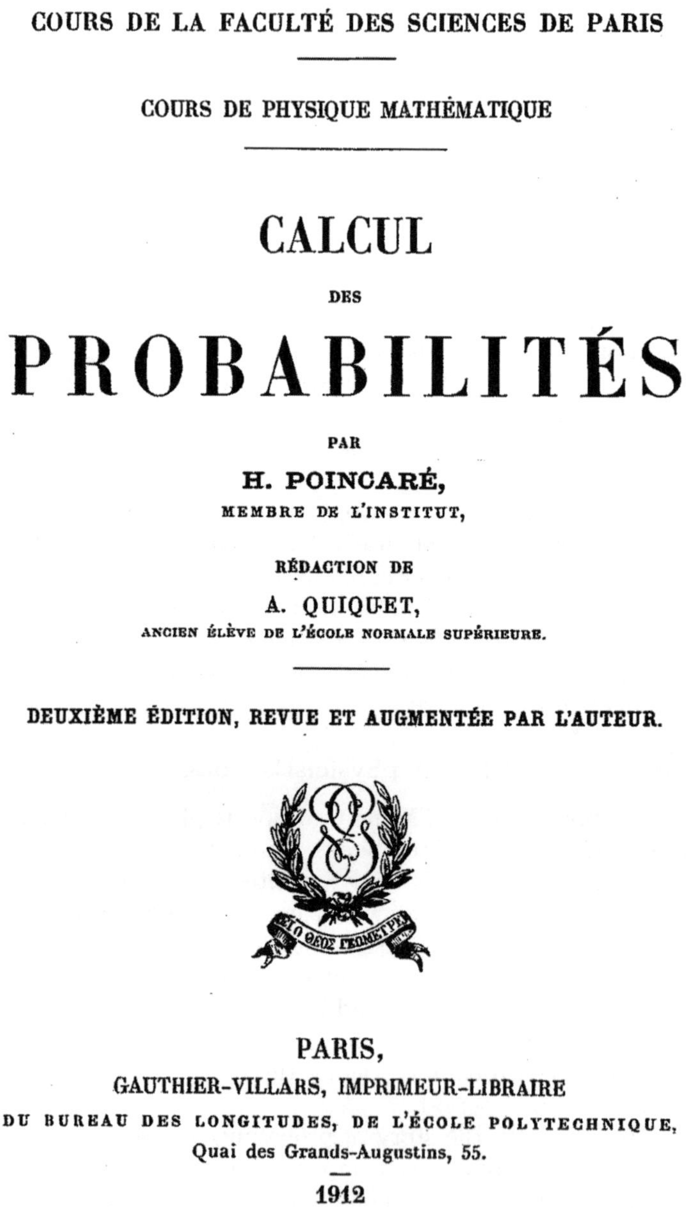

COURS DE LA FACULTÉ DES SCIENCES DE PARIS

COURS DE PHYSIQUE MATHÉMATIQUE

CALCUL

DES

PROBABILITÉS

PAR

H. POINCARÉ,

MEMBRE DE L'INSTITUT,

RÉDACTION DE

A. QUIQUET,

ANCIEN ÉLÈVE DE L'ÉCOLE NORMALE SUPÉRIEURE.

DEUXIÈME ÉDITION, REVUE ET AUGMENTÉE PAR L'AUTEUR.

PARIS,

GAUTHIER-VILLARS, IMPRIMEUR-LIBRAIRE

DU BUREAU DES LONGITUDES, DE L'ÉCOLE POLYTECHNIQUE,

Quai des Grands-Augustins, 55.

1912

FIGURE 14.1. Title page of Poincaré's *Calcul des probabilités*, 1912 edition.

pressed by time (he died in 1912), in the midst of multiple projects, Poincaré did not attempt to present the recent advances in his book.

14.3. Mathematical methods of probability

From the mathematical perspective, as well, the beginning of the 20th century saw a profound upheaval in the calculus of probability. The thesis of Louis Bachelier [**1**] (for which Henri Poincaré was a reporting jury member) inaugurated a long line of works on Brownian motion, which would be pursued by Norbert Wiener [**28**] in 1923 and Paul Lévy [**16**] in 1925. (We observe, however, that Bachelier's work, having passed unnoticed, had no influence on the theory's development and was not rediscovered until much later.) But perhaps the most profound work of this period was that of Émile Borel, first his *Remarques sur certaines questions de probabilités* [*Remarks on some questions of probability*] (1905) [**3**]—where, for the first time, he applied his measure theory *and the Lebesgue integral* to the theory of probabilities[2]—then his proof of the Maxwell–Boltzmann law in 1906 [**4**], and finally his fundamental article [**5**] in 1909 on countable probabilities. After the abstract formulation of measure theory and the Lebesgue integral by Fréchet in 1915, the power of this approach to probability fully revealed itself. It would be pursued by Hugo Steinhaus and the Lvov school in 1923 [**24**], and would lead to Kolmogorov's famous axiomatic system[3] [**15**] in 1933. We also mention the long series of *Monographies sur le Calcul des Probabilités*, published under the direction of Émile Borel and to which most of the French probabilists of the period between the World Wars contributed.

In this area, Poincaré made hardly any innovation. No mention of measure theory, and when he needs to introduce a function that will serve as a probability density, he assumes it is "in general continuous" [**19**, p. 145].[4] Indeed, most often he reasons "like a physicist", and does not burden himself with integration by parts or expansion into a convergent Taylor series. He does not even distinguish between the Fourier transform, which always exists, and the Laplace transform, which assumes strong hypotheses on the moments of a random variable. Nothing in all these reasonings would stand up to the critical eye of the German analysts[5]

---

[2]Borel [**3**] attributed to the Swedish mathematician Anders Wiman the idea, dating from 1900 (therefore before the Lebesgue integral) of applying (Borel's) measure theory to probability (on this subject, see Chapter 2 in [**26**]). The measure theory made it possible to give a rigorous mathematical meaning to the notion of "volume" used (without a precise definition) by Poincaré in 1890 to measure probabilities in the three-body problem (see Chapter 8 in this volume).

[3]Bourbaki's position on probability is rather remarkable. In his endeavor to axiomatize *all* of mathematics, it is surprising that he never dreamed of incorporating probability. And yet, after Kolmogorov, it departed from the branches of applied mathematics and mathematical physics to attain the glorious status of a "pure and axiomatic" mathematical theory. Nevertheless, we have a very interesting text by André Weil [**27**] in 1940 where he refers to probability; according to Weil, Bourbaki needed to include "the first principles of probability" in a booklet on integration. He did not do so, but later he would regret this and would publish a supplementary volume [**7**] to his theory of integration in 1969. In it, he recovers Brownian motion in a tightrope walk of an exercise, but, fortunately, he includes a historical study that is both greatly interesting and greatly pertinent to the analytic problems connected with probability.

[4]The page numbers for the citations in [**19**] are those of the 1912 edition.

[5]These traits would contribute in no small way to the relative discredit Poincaré suffered in the French mathematical community from the 1940s to the 1960s. The criticism of the "German analysts" would be amply transmitted by the Bourbaki faithful, before Dieudonné, in his own thundering way, could restore Poincaré's reputation!

of the period (which, by the way, Poincaré knew quite well). But these remarks strengthen my thesis that *Poincaré's book must be read as the work of a physicist, and not of a mathematician.*

14.4. Refutation of the classical theses on chance

What remains in this book? Much, but one must look for the pearls, which are sometimes hidden. We begin with the *conceptual analysis*, not forgetting that Poincaré is hailed (in France) as one of the great philosophers of science.[6]

The introduction to the 1912 edition is reprinted from a 1907 article [**21**] (with three short passages removed), which was already reprinted in the work *Science and Method* [**22**] in 1908. It analyzes the notion of chance and the possibility of "laws of chance". First of all, Poincaré takes the viewpoint of *objective* probabilities and refuses to reduce chance to simple ignorance. Perhaps this is an incursion, made by a careful scholar, into the ideological realm. I would see this confirmed in this remark:

> *"...many people find it wholly natural to pray to have rain or good weather, while they would find it ridiculous to ask for an eclipse by a prayer ... "* [**19**, p. 5].

Even though he did not categorically refute the thesis, dear to Cournot, of temporal series that overlap, he reduced their scope:

> *"Our weakness does not allow us to embrace the entire universe, and obliges us to decompose it into slices. We seek to do so with as little artificiality as possible, and nonetheless, it happens from time to time that two of these slices act upon each other. The effect of this mutual action* appears[7] *to us to be due to chance. Is this a third way of conceiving of chance? Not always ... "* [**19**, p. 11].

The prayer for rain and the tile fallen from the roof[8] being removed from consideration, let us examine the two main characteristics of chance according to our author.

14.5. First characteristic of chance

I cite Poincaré [**19**, pp. 4 and 5]:

> *"A very small cause, which escapes us, determines a significant effect that we cannot miss, and then we say that this effect is due to chance ... it can happen that small differences in the initial conditions generate*

[6] As an effect of differing didactic traditions, the "basic" German scientist is much more versed in *Naturphilosophie*, found in Spinoza and Kant, than the average student at the *École polytechnique*, whose only philosophical opening is often the *positivism* of Auguste Comte. (Poincaré is, here also, an exception.) And yet there was in Paris as great a philosopher as Henri Bergson, who conversed with Einstein; and the *Revue du mois*, published by Émile Borel and his wife Marguerite Appell (Camille Marbo) was a place for numerous ideological and philosophical debates. For his part, Poincaré instead published in the *Revue de métaphysique et de morale*. Nevertheless, it was in the *Revue du mois* that he published in 1907 the article on chance [**21**] that would serve as the introduction to the 1912 edition of his *Calculus of Probabilities*.

[7] Emphasized by Pierre Cartier.

[8] TRANSLATOR'S NOTE: This is a direct translation of the idiom *"la tuile tombée du toit"*, which refers to unfortunate, apparently random events, such as a passerby being struck by a falling tile.

very large ones in the final phenomena ... Prediction becomes impossible and we have the phenomenon of accident."

The important philosophical point is that determinism is not contradicted. Chance does not at all proceed from ignorance, properly called, but from an insufficient precision in the data. In the classical analysis of determinism by Laplace, it is asserted that an intelligence that could apprehend the complete state of the Universe at a given instant, if it were moreover vast enough to submit its data to analysis, could exactly predict the future.[9] But we should instead reason in economic terms: data collection has a cost, and the complete state of the Universe at a given instant has a prohibitive cost. Good forecasting generates profit, which we can sometimes quantify; a good physical theory should have a satisfactory predictions/data ratio (of the type benefit/investment or quality/price).

Laplace could still think that, for systems such as that of the three bodies in astronomy, approximate knowledge of the initial data would allow, for a reasonable cost, decent forecasting over extremely long times. Yet, Poincaré discovered that this system already has chaotic behavior.[10] Today, we are sensitized to the *butterfly effect*, to sensitive dependence on initial conditions, to strange attractors. After the example of the three-body problem, there appeared, in the 1960s, the Lorenz and Hénon attractors, the Moser system [**17**]... In the purely mathematical realm, only shortly after Poincaré's passing, there was the analysis by Pierre Fatou and (independently) Gaston Julia of the iteration of rational transformations (of the type $z \mapsto z^2 + c$, where z is a complex number).

14.6. Second characteristic of chance

"... the causes are complex and they are many ..." [**19**, p. 9].

This second dimension of chance concerns statistical regularities that come from averaging over large sets. Poincaré leans on the kinetic theory of gases (for which Borel [**4**], in 1906, was doubtless the first to propose a rigorous basis from the mathematical point of view), where one deals with a very large number of molecules. The laws of thermodynamics are interpreted statistically, they depend little or not at all on the molecular mechanism at play, or on the initial distribution in the interior of the gas. The many collisions—or interactions—iron out the starting inequities and lead to homogeneous distributions. The same thing happens for a mixture[11] of two powders after a sufficiently long mixing.

---

[9] Editor's Note: *Recherches sur l'intégration des équations différentielles aux différences finies et sur leur usage dans la théorie des hasards* [*Studies on the integration of finite difference equations and their use in the theory of chance*] (1773) and *Essai philosophique sur les probabilités* [*Philosophical essay on probability*] (1814). Laplace specifies: *"The human spirit offers, in the perfection it was able to give to Astronomy, a feeble sketch of that intelligence* [...] *from which it will forever remain infinitely distant."* And he rejoices that we thus owe *"to the weakness of the human spirit one of the most delicate and most ingenious theories of Mathematics, which is the science of chance or probability."*

[10] Editor's Note: See Chapter 8.

[11] A mathematical theory of mixtures, making it possible to understand for example what distinguishes *grey* from the mixture *black-and-white*, still presents a worthwhile challenge today. I myself have made an attempt [**12**] by using methods of non-standard analysis, which allow the formalization of averaging at two different scales, and this is the heart of the problem.

From a mathematical point of view, we are dealing with the *statistical physics of systems in equilibrium*. It is therefore not a question of fluctuations[12] or of evolution toward an equilibrium. Only the characteristics of equilibrium, and the corresponding averages, are taken into account. Here is the germ of the theory of the limit laws for sums of a large number of random variables, ergodic theorems, and Markov processes: all this would make it possible to bypass Gauss's law of error (the Laplace–Gauss normal law), or to transform it into Brownian motion.

14.7. The mechanism of chance

> "... *so that we can predict, if not what are the effects* in each case, *but at least what these effects will be* on average ... " [**19**, p. 11].[13]

Poincaré illuminated the exponential principle in two ways:

> "... *if the first collision multiplied the deviation by a very large number A, after n collisions it will be multiplied by A^n; it will therefore have become very large, not only because A is large ... but because the exponent n is large* ... " [**19**, p. 8].

This is what would become the Lyapunov stability criterion, and would serve as the basis for the definition of stable and unstable manifolds in dynamical systems. For the first time, we see the obvious importance of eigenvalues, that is, multipliers such as A above. This appears in an even more striking way in a second example, a two-state system symbolized as below.

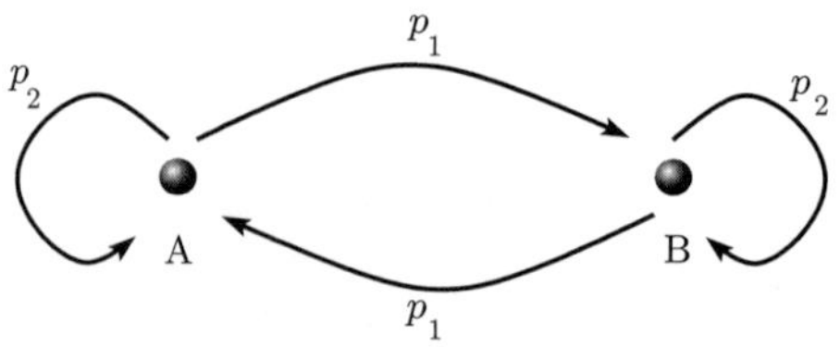

$p_1 + p_2 = 1.$

If we bet one dollar on finding the initial state again after n transitions, the mathematical expectation of the gain will be $G_n = (p_1 - p_2)^n$. For rather large n, and p_1 or p_2 not too close to 1, G_n will be small and *the game will be fair regardless of the values of p_1 and p_2* (Poincaré's conclusion). This is one of the first manifestations of ergodic theorems for Markov chains. We will return to this subject in Section 14.13.

14.8. Definition of probabilities

According to Poincaré, there is none. He begins (following his predecessors, among whom was Joseph Bertrand) by refuting, with simple examples, Laplace's

[12]The theory of fluctuations would not be mathematically understood until the 1950s, thanks in particular to the work of Feller and Spitzer [**13, 23**]. In statistical physics, large scale fluctuations remain a current subject. Note, however, that Bachelier's thesis [**1**] (in 1900) is the first example of analysis of a temporal series.

[13]The phrases "in each case" and "on average" are emphasized by Poincaré.

"definition":

$$\text{probability} = \frac{\text{number of favorable cases}}{\text{number of possible cases, favorable or not}}.$$

This principle appeals to the assumption that the various cases are *equiprobable*, which presupposes the definition of probability! Today every student at the high school level has encountered the *modern definition*:

- one describes the set Ω of possible cases;
- one associates to each element i of Ω a number $p_i \geq 0$, in such a way that the sum of the p_i equals 1

$$\sum_{i \in \Omega} p_i = 1; \tag{14.1}$$

- an event is associated to the set of cases that realize it, which is a part A of Ω, and the corresponding probability is

$$\mathbf{P}(A) = \sum_{i \in A} p_i. \tag{14.2}$$

When the p_i are all equal, we recover Laplace's formula for probability. When the p_i are *rational numbers*, we can reduce to Laplace's "definition" by simple tricks. For Poincaré, the case where the p_i are irrational only appears when passing to a limit (a game played a very large number of times, for example).

In conformity with his general philosophy,[14] Poincaré considers that we must make *conventions*, which could be made legitimate in a variety of ways, then do the mathematicians' work *"that applies to these conventions the rules of calculating"* [**19**, p. 29]. In the dominant philosophies, this could translated by the *axiomatic view*: all that is important are the logical rules for how to employ the basic notions of calculating probabilities. The alternative is the *modeling viewpoint*: we do the calculations, then adjust the free parameters to best make the data fit with experiment or observation. The axiomatician is little preoccupied with what is appropriate to reality, but in this probabilistic realm, the meaning of what is appropriateness is delicate. The most subtle response is what Émile Borel called the *single law of chance*:

> *Any event whose probability is negligible*[15] *does not occur, or rather it is reasonable to behave as though it should not occur.*

14.9. The case of continuous probabilities

When the number of *a priori* possible cases is finite, the description given in the preceding section is sufficient if we add to it the notion of *independence*: when A and B are independent, we have

$$\mathbf{P}(A \cap B) = \mathbf{P}(A) \cdot \mathbf{P}(B) \tag{14.3}$$

(and conversely). After this, it only(!) remains to do some *combinatorics* to calculate complicated finite sums, then some *asymptotic calculation* to obtain the limit probabilities. The work may be difficult, but the framework is elementary and without pitfalls. A good portion of probability theory and statistical physics can be carried out in this context.

---

[14] EDITOR'S NOTE: See Chapter 19.

[15] The delicate point is to specify what is a negligible probability.

But Poincaré knew the traps of "continuous" probabilities, and the "paradoxes" discovered by his predecessors, such as Joseph Bertrand. Even if he was not completely innovative, at least his analysis of the difficulties made its presence felt and made possible a progressive liberation from a persistent (and sterilizing) distrust of continuous probabilities. The definitive modern version says the following:[16]

- to a given experiment (or observation) corresponds its sample space;[17] the points of this space, denoted Ω, correspond to the various *a priori* possibilities;
- Ω is *a priori* an abstract measure space; but in most natural applications, we can reduce to working with a separable complete metric space (a *Polish* space according to Bourbaki), and we can therefore define the notion of a Borel set;
- to each Borel set B of Ω is associated its probability $\mathbf{P}(B)$, satisfying the rules of countable additivity, positivity $\mathbf{P}(B) \geq 0$, and normalization $\mathbf{P}(\Omega) = 1$.

The mathematical model is thus characterized unambiguously, but the explicit construction of the "probability measure" $\mathbf{P}$ on the space Ω is delicate. Thanks to the work of the Russian school (Kolmogorov, Prokhorov, Minlos, Sinai) in the 1960s, we can now make use of powerful analytic techniques.[18]

14.10. Poincaré's response

The preceding analytic methods came much later than Poincaré, and would require the full arsenal of Lebesgue integration, which he did not have the opportunity to study. But his conceptual analysis is right.

The notion of a *uniformly distributed* random variable X over an interval $I = [a, b]$ is classical. If we subdivide the interval into m subintervals of the same length $I_1, \ldots, I_m$, we obtain events $X \in I_1, \ldots, X \in I_m$ that are equiprobable. The probability that we have $c \leq X \leq d$ (with $a \leq c \leq d \leq b$) equals $(d - c)/(b - a)$. It is easy to pass to the case of two independent, uniformly distributed variables X, Y. By introducing the interval J of the variation of Y, subdivided into n subintervals $J_1, \ldots, J_n$, of equal length, the mn events $\{X \in I_i,\ Y \in J_j\}$ are equiprobable. It amounts to the same to define the probability that a point P with random coordinates (X, Y) is located in the subregion A of its *a priori* domain of variation, Ω, as the quotient $|A|/|\Omega|$, where $|A|$ denotes the area of A.[19]

The definition is analogous in three or more dimensions.

The difficulty is that the notion of area is not invariant under a non-linear change of coordinates: what is written as $\iint dx\,dy$ in Cartesian coordinates is written as $\iint r\,dr\,d\theta$ in polar coordinates! The paradoxical examples come from the fact that one can define the same probability by means of two different coordinate systems. On the other hand, if we introduce a probability density in the plane, that is, a positive function $\varphi(x, y)$, the probability defined above can be written more generally as

$$\mathbf{P}(A/\Omega) = \frac{\iint_A \varphi(x, y)\,dx\,dy}{\iint_\Omega \varphi(x, y)\,dx\,dy}. \tag{14.4}$$

---

[16] A quick presentation of this point of view "for physicists" is given in my book [**11**].

[17] TRANSLATOR'S NOTE: Here Cartier uses the phrase *"espace des échantillons"* and includes a footnote saying that in his writing he avoids the use of the English phrase "sample space" in order not to rouse the ire of some of his Francophone colleagues.

[18] See the synthetic exposition given by Bourbaki [**7**] and in particular the study of Brownian motion by these methods.

[19] I follow here the conventions of Michel Mendès France (see his note on p. 295).

We can normalize by imposing the value 1 on the denominator. When we change coordinates, $\varphi(x, y)$ is multiplied by a Jacobian determinant [**19**, p. 121]. In an invariant way, the probability law corresponds to the choice of an exterior differential form ω, which can be written as[20]

$$\omega = \varphi(x, y)\, dx \wedge dy \tag{14.5}$$

in the coordinates x, y (which are now arbitrary).

14.11. Description of a probabilistic model according to Poincaré

From this whole discussion, which occupies part of Chapter 1 and essentially all of Chapters 7 and 8 of [**19**], we will retain the following (in an obviously modernized perspective):

- the mathematical model is defined by the choice, over an appropriate d-dimensional manifold M (the configuration space), of an exterior differential form ω of degree d;
- it is therefore insufficient to fix only some parameters; the choice of ω in a function space depends *a priori* on infinitely many parameters;
- the choice of ω is *a priori* arbitrary, but once it is made, it must be kept;
- ω must satisfy certain principles of symmetry, which are expressed in the form of the invariance of ω under a group G of transformations of M.

In certain cases, the last condition is enough to completely determine ω: these are the interesting cases of *geometric probability*, for which I direct the reader to the chapter by Michel Mendès France (Chapter 15). But without a doubt what is the most important point for Poincaré is that, in certain situations, the final result is independent of the law ω chosen *a priori*; these are the cases where an ergodic theorem applies (the "method of arbitrary functions"). For example, in the problem of the distribution of "small planets" (asteroids) over the zodiac, while Laplace adopted the uniform law in the absence of a more natural hypothesis, Poincaré showed that this law appears *necessarily* "at the end of a very long time", from the fact of the system's evolution, (practically) regardless of the initial distribution [**19**, ch. viii, no. 93, pp. 150–152].

Limitations. The mixed cases where we blend the discrete model of Section 14.8 and the continuous model of Section 14.10 cannot be treated directly. We have also seen that Poincaré assumed the continuity, and even analyticity, of the probability density $\varphi(x, y)$. The case of piecewise regular functions can be handled just as easily and covers most needs, but singular measures in the sense of Lebesgue are ruled out by Poincaré's requirement: for an infinitely small domain, the probability is proportional to the volume. This eliminates self-similar laws.[21] No Stieltjes integrals either!

---

[20] Recall that the invention of the calculus of exterior differential forms is due to Élie Cartan and Henri Poincaré. (EDITOR'S NOTE: See Chapter 11.)

[21] And yet, they exist. When Paul-André Meyer and I were in the service of the National Marine (accompanied by the brother of Michel Mendès France), we amused ourselves by proving that the recommended procedures for anti-submarine fights (copied from the type of instructions used by the U.S. Navy, made by J. von Neumann, Mark Kac, ...) lead to such laws of probability.

14.12. Gauss's law of error

All the works on probability at the end of the 19th century discussed the impact of Gauss's law. The analytic investigations on this subject, which were pursued throughout the 19th century and part of the 20th, led to various formulations of the "central limit theorem".[22] Poincaré also discusses this law with numerous details in Chapter 11 of [**19**], but he remains far from modern analytic rigor. However, as pointed out by Paul Lévy, he introduces—for the first time, it seems, in a course on probability—the tool of the *characteristic function* [**19**, p. 206]. He writes it in one of the forms

$$f(\alpha) = \sum p e^{\alpha x} \qquad \text{discrete case,} \tag{14.6}$$

$$f(\alpha) = \int \varphi(x) e^{\alpha x}\, dx \qquad \text{continuous case,} \tag{14.7}$$

where today we would write $f(\alpha) = \mathbf{E}[e^{\alpha X}]$. He seems to believe that it always has meaning for complex α (Laplace transform), although it must be restricted to the case where $\alpha = it$ with t real (Fourier transform). We will adopt the Fourier-style definition

$$\phi_X(\alpha) = \mathbf{E}[e^{i\alpha X}]. \tag{14.8}$$

The central limit theorem can be formulated thusly: let X be a random variable having finite second moment, with mean $m = \mathbf{E}[X]$ and standard deviation $\sigma = \mathbf{E}[(X-m)^2]^{1/2}$. We take a sample $(X_1, \ldots, X_N)$ formed of N independent random variables, all of the same law as X; we introduce the random variable Θ_N (the normalized random error) by the formula

$$X_1 + \cdots + X_N = mN + \sigma\sqrt{N}\Theta_N. \tag{14.9}$$

Then, *for large N, Θ_N approximately follows the normalized Gauss law*, that is the probability law with density $\gamma(x) = e^{-x^2/2}/\sqrt{2\pi}$.

To carry out the proof, we may assume that $m = 0$ and $\sigma = 1$. Then the characteristic function ϕ of X is of class C^2, with

$$\phi(0) = 1, \qquad \phi'(0) = 0, \qquad \phi''(0) = -1. \tag{14.10}$$

The characteristic function of Θ_N is then $\phi(\alpha/\sqrt{N})^N$. It is an easy exercise to derive from (14.10) the relation

$$\lim_{N\to\infty} \phi(\alpha/\sqrt{N})^N = e^{-\alpha^2/2} \tag{14.11}$$

[22] TRANSLATOR'S NOTE: Here the original text has *"théorème limite central" (ou "théorème central-limite")*, and the editors include the following note.

EDITOR'S NOTE: A theorem discovered by Laplace (1810). It has long borne his name. In 1920 Pólya called it *Zentraler Grenzwertsatz der Wahrscheinlichkeitsrechnung*, that is, "Central limit theorem of the calculus of probabilities". The meaning is clear: it is a limit theorem (a theorem about limits), and it is central in the calculus of probabilities. The English translation of *Zentraler Grenzwertsatz*, which is *Central limit theorem*, was poorly understood (one could not tell whether *Central* was related to *limit* or to *theorem*) and poorly translated into French: hence the gibberish *"Théorème central-limite"* (unfortunately hallowed by usage) and the serious mistranslation *"Théorème de la limite centrale"*.

and to remark that the characteristic function $e^{-\alpha^2/2}$ corresponds to the probability density $\gamma(x)$,

$$\frac{1}{\sqrt{2\pi}}\int_{-\infty}^{+\infty} e^{i\alpha x}e^{-x^2/2}\,dx = e^{-\alpha^2/2}. \tag{14.12}$$

This reasoning[23] is essentially that of Poincaré[24] [**19**, p. 208].

To arrive at the "modern" exposition, we must specify what we mean by "approximately following a probability law". One way to do this is to introduce a notion of convergence in the set of probability measures on $\mathbf{R}$, the *vague* topology (thus named by Henri Cartan, who defined it in 1941 in a more general framework [**10**]).[25] One must also prove a theorem about the continuity of the Fourier transform:

$$\mu \mapsto \phi(\alpha) = \int_{-\infty}^{+\infty} e^{i\alpha x}\,d\mu(x) \qquad (\alpha \text{ is real}), \tag{14.13}$$

which transforms the vague topology of probability measures into simple (or uniform) convergence of characteristic functions. This is the classical theorem of Paul Lévy, proved in 1922 and explained in his 1925 book on probability [**16**]. This second part of the reasoning, absent in Poincaré's writing, necessitates the arsenal of functional analysis, following Lebesgue.

14.13. A jewel of group theory

The final chapter of the book, modestly titled "Various questions", is indubitably the most imaginative. The last section (from no. 237 to the end) on the mixture of liquids is a study of a physical system and the ergodic principle that remains very readable today and shows to what point Poincaré the mechanical physicist had assimilated the ideas of Hamilton.[26] But I would like to pause for a moment on the problem of card-shuffling (nos. 225–232).

Hadamard, in a 1906 article [**14**] on statistical mechanics (a report in Gibbs's *Elementary Principles in Statistical Mechanics*), set forth the idea of comparing the evolution of a system to shuffling cards, to make it understood how a reversible process could give rise to an irreversible evolution (a homogenization). In his article *Le hasard* [*Chance*] (1907), Poincaré first mentioned a mathematical treatment in the case of two cards; in the passage from his 1912 book dealt with here, he proposed a solution to the general case, by a different method.

Poincaré envisions a finite group G, whose elements he denotes $S_0, S_1, \ldots, S_r$ (with S_0 the identity). This group can be assumed to act by permutations on a set of cards, and we consider a random permutation corresponding to the probabilistic scheme

$$\begin{pmatrix} S_0 & S_1 & \cdots & S_r \\ p_0 & p_1 & \cdots & p_r \end{pmatrix},$$

with $p_i \geq 0$, $p_0 + p_1 + \cdots + p_r = 1$.

---

[23] EDITOR'S NOTE: Whose principle goes back to Laplace, and whose presentation had already been simplified by several authors (notably Poisson and Cauchy).

[24] But Poincaré uses the finiteness of all the moments $\mathbf{E}[X^n]$ and makes expansions into Taylor series, even though a finite expansion suffices to pass from (14.10) to (14.11).

[25] Perhaps a more appealing perspective (which was that of Lévy and his contemporaries) is to say that the distribution functions converge to that of the normal law.

[26] Which was not the case of the professors at the Sorbonne who claimed to teach me Mechanics in the 1950s!

He introduces the group algebra in these terms:

> *"We know that there have been inaugurated complex numbers of the form*
>
> $$\mathbf{X} = x_0e_0 + x_1e_1 + \cdots + x_re_r,$$
>
> *where the x are ordinary quantities and the e complex units."* [**19**, p. 302]

He then explains that the e_i multiply like the corresponding group elements S_i and concludes:

> *"We can then represent symbolically the probability law under consideration by the complex number*
>
> $$\mathbf{P} = p_0e_0 + p_1e_1 + \cdots + p_re_r."$$
>
> [**19**, p. 303]

He explains that the game is to prove a limit theorem of the kind

$$\lim_{n\to\infty} \mathbf{P}^n = \frac{1}{r+1}(e_0 + e_1 + \cdots + e_r).$$

Poincaré has just invented random walks on a group, and formulated and proved an ergodic theorem. The proof uses the spectrum of $\mathbf{P}$ and essentially proves the Perron–Frobenius theorem on matrices with positive entries.

Poincaré himself brought many improvements in the theory of Lie groups;[27] he knew the results of Élie Cartan on the subject, and among French mathematicians, he was one of those familiar with the German works. Nevertheless, for an analyst or a mechanical physicist to cite Frobenius on group algebras was hardly commonplace.

The model of card shuffling would be taken up from a more probabilistic angle by Borel in 1912 (a work that passed unnoticed), by Lévy in 1925, and especially by Hadamard and Hostinský in 1927 and 1928, who discovered at the 1928 International Congress of Mathematicians in Bologna[28] (where Hadamard made a presentation on card shuffling) that their results recovered those obtained 20 years earlier by Markov when he introduced the chain model. Starting from 1929, discrete Markov processes, then continuous ones (notably with Kolmogorov's revolutionary article in 1931), would become one of the major subjects of research in probability.

14.14. Conclusion

Obviously, this book of Poincaré on probability has neither the depth nor the novelty of the *Méthodes Nouvelles de la Mécanique Céleste* [**20**] or of the *Analysis Situs* [**18**]. As for its analytic methods, it is still a 19th century work, but it is one of the first works to have entered into the games of statistical physics, well beyond the usual combinatorial exercises connected with games of chance. In this sense, this book already belongs to the twentieth century, paving the way for Einstein, Ehrenfest, Wiener, Landau, and many others... Finally, the persevering reader will discover within [**19**] lovely pearls, such as the use of the Padé approximant (very recent at that time) in Chapter 15. See also, at the end of Chapter 12, a very clear definition of the *dispersion ellipse*, that is, the fact that Gauss's law in several dimensions depends on the choice of a positive definite quadratic form on the configuration space.

---

[27] EDITOR'S NOTE: See Chapter 16.

[28] Cf. [**8**].

Bibliography

[1] L. Bachelier, *Théorie de la spéculation*, Ann. Sci. École Norm. Sup. Paris (3) **17** (1900), pp. 21–86. Reprinted by Éditions J. Gabay (Paris), 1995.

[2] J. Bertrand, *Calcul des probabilités*, Gauthier-Villars (Paris), 1889.

[3] É. Borel, *Remarques sur certaines questions de probabilités*, Bull. Soc. Math. France **33** (1905), pp. 123–128. Reproduced in: *Œuvres* d'Émile Borel, vol. 2, Éd. du CNRS (Paris), 1972, pp. 985–990.

[4] É. Borel, *Sur les principes de la théorie cinétique des gaz*, Ann. École Norm. Sup. **23** (1906), pp. 9–32. Reproduced in: *Œuvres* d'Émile Borel, vol. 3, Éd. du CNRS (Paris), 1972, pp. 1669–1692. And also in: É. Borel, *Introduction géométrique à quelques théories physiques*, Gauthier-Villars (Paris), 1914.

[5] É. Borel, *Les probabilités dénombrables et leurs applications arithmétiques*, Rend. Circ. Mat. Palermo **27** (1909), pp. 247–271. Reproduced in: *Œuvres* d'Émile Borel, vol. 2, Éd. du CNRS (Paris), 1972, pp. 1055–1079.

[6] N. Bourbaki, *Éléments d'histoire des mathématiques*, Hermann (Paris), 1974. Reprinted: Masson (Paris), 1984.

[7] N. Bourbaki, *Éléments de mathématique. Intégration, chap. 9: Intégration sur les Espaces topologiques séparés. Livre VI. Fasc. XXXV*, Hermann (Paris), 1969. (Actualités scientifiques et industrielles, vol. 1343.)

[8] B. Bru, *Souvenirs de Bologne*, Journal de la Société Française de Statistique **144** (2003), pp. 135–226.

[9] B. Bru, *Les leçons de calcul des probabilités de Joseph Bertrand*, Electronic J. Hist. Proba. Stat. **2**.2 (Dec. 2006), pp. 1–44. `http://www.jehps.net/Decembre2006/Bru.pdf`

[10] H. Cartan, *Sur les fondements de la théorie du potentiel*, Bull. Soc. Math. France **69** (1941), pp. 71–96. Reproduced in: *Œuvres* d'Henri Cartan, volume 3, Springer (Berlin), 1979, pp. 1023–1048.

[11] P. Cartier and C. DeWitt-Morette, *Functional integration. Action and symmetries*, Cambridge University Press, 2006.

[12] P. Cartier and Y. Perrin, *Integration over finite sets*, pp. 185–204 of: *Nonstandard Analysis in Practice*, edited by Francine Diener and Marc Diener, Universitext, Springer-Verlag (Berlin), 1995.

[13] W. Feller, *An introduction to probability theory and its applications*, 2 volumes, Wiley, 1966, 1971.

[14] J. Hadamard, *La mécanique statistique*, Bull. Amer. Math. Soc. **12** (1906), pp. 194–210. `http://www.ams.org/bull/1906-12-04/` Run in: Bull. Sci. Math. **30** (1906), pp. 161–179.

[15] A. N. Kolmogorov, *Grundbegriffe der Wahrscheinlichkeitsrechnung*, Springer (Berlin), 1933. English translation: *Foundations of the theory of probability*, Chelsea, 1950.

[16] P. Lévy, *Calcul des probabilités*, Gauthier-Villars (Paris), 1925.

[17] J. Moser, *Stable and random motions in dynamical systems*, Princeton University Press, 1973.

[18] H. Poincaré, *Analysis Situs et ses compléments*: *Œuvres*, tome 6, Gauthier-Villars (Paris), 1953.

[19] H. Poincaré, *Calcul des probabilités*, Gauthier-Villars (Paris), 1896. Second edition, 1912. Reprint of the 1912 edition by Éditions J. Gabay (Paris), 1987.

[20] H. Poincaré, *Les méthodes nouvelles de la mécanique céleste* (3 volumes), Gauthier-Villars (Paris), 1892–1899. Reprints: Dover Publications Inc. (New York), 1957, and Blanchard (Paris), 1987.

[21] H. Poincaré, *Le hasard*, Revue du mois **3** (1907), pp. 257–276.

[22] H. Poincaré, *Science et méthode*, Flammarion (Paris), 1908. Reprinted by Éditions Kimé (Paris), 1999.

[23] F. Spitzer, *Principles of random walks*, D. Van Nostrand Co. (Princeton N.J.), 1964.

[24] H. Steinhaus, *Les probabilités dénombrables et leur rapport à la théorie de la mesure*, Fund. Math. **4** (1923), pp. 286–310.

[25] Supplément historique de l'annuaire de l'École normale supérieure (2005).

[26] J. Von Plato, *Creating modern probability: Its mathematics, physics and philosophy in historical perspective*. Cambridge University Press, 1994.

[27] A. Weil, *Calcul des probabilités, méthode axiomatique, intégration*, Revue Scient. **78** (1940), pp. 201–208. Republished in the *Œuvres scientifiques* of André Weil, vol. 1, pp. 260–272, Springer (Berlin), 1979.

[28] N. Wiener, *Differential space*, J. Math. and Phys. **2** (1923), pp. 131–174.

CHAPTER 15

Poincaré and Geometric Probability

Michel Mendès France

15.1. Introduction

Henri Poincaré is famous for his many contributions in the areas of mathematics as well as physics, not to mention his works on the philosophy of science. However, his work in probability theory is relatively little known even though it seems [**28**] that his course on this subject was published starting in 1896, completed in 1912 [**27**] (the year of his death), and republished in 1923. Paul Lévy, in the preface to his own course on probability [**23**], which appeared in 1925, praised that of Poincaré while still expressing regret that the latter did not make systematic use of the characteristic function (the Fourier transform of the probability measure), whose capital importance is well known today.[1]

Poincaré did not revolutionize the theory of probability, but his work in it is nonetheless original.[2] The present chapter in fact concerns only one result on geometric probability (that is, probability applied to geometric situations: typically, one looks for a way to calculate the average number of intersections of curves placed randomly in a space ...). The idea is to extend in an important way the problem of Buffon's needle (1733), of which we will give here two applications discovered at the end of the 20th century. The first is due to A. Edelman and E. Kostlan [**11**], who rediscovered a very lovely result of M. Kac [**17, 18**], according to which a random polynomial of degree n admits on average $\left(\frac{2}{\pi} + o(1)\right)\log n$ real zeroes $(n \to \infty)$.[3] The second, which is my own [**24, 25**], illustrates the subtle connection that exists between disorder and "confusion".

I would like to thank Pierre Cartier and Jia-Yan Yao for their careful reading of the French version of this text.

15.2. Buffon and his needle

Born in 1707, died in 1788, Georges Louis Leclerc, count of Buffon, was interested in mathematics from a young age. In 1733, at the age of 26, he composed his "Essai d'arithmétique morale" ["Essay on moral arithmetic"] [**3, 5**], which he sent to the Académie des Sciences. This important work contains in particular the famous chapter "Mémoire sur le jeu du franc-carreau" ["Memoir on the game of

---

[1] Editor's Note: Poincaré, as Lévy says, *"indicates the principle of the method in a few lines"*. We remark that it is apparently to Poincaré that we owe the name *characteristic function* (with this meaning).

[2] See Chapter 14.

[3] Throughout this chapter, log denotes the natural logarithm.

franc-carreau"], in which the so-called Buffon's needle problem is treated.[4] In fact, he solved a more complex problem because it dealt with studying the intersection of a needle and a grid, say $\mathbb{Z}^2$ (more precisely, the union of the lines $x = m$ and $y = n$, where $m, n \in \mathbb{Z}$). His memoir would not be published until 1777.

Buffon's classical theorem is the following.

THEOREM 15.2.1 (Buffon). *Consider a network of parallel straight lines in the plane whose spacing is a. The probability that a needle of length $\ell < a$ thrown at random cuts one of these lines is*

$$p = 2\ell/\pi a. \tag{15.1}$$

PROOF. The remarkably beautiful proof that we present here is due to E. Barbier [**2**], while he was still a student at the École Normale Supérieure. Before proving anything, we obviously must define the probability law, that is, specify the meaning of "throwing at random"! The position of the needle is determined in part by its angle θ (modulo π) with the direction of the parallel lines, and in part by the distance h from its midpoint A to the line of the network situated just below.

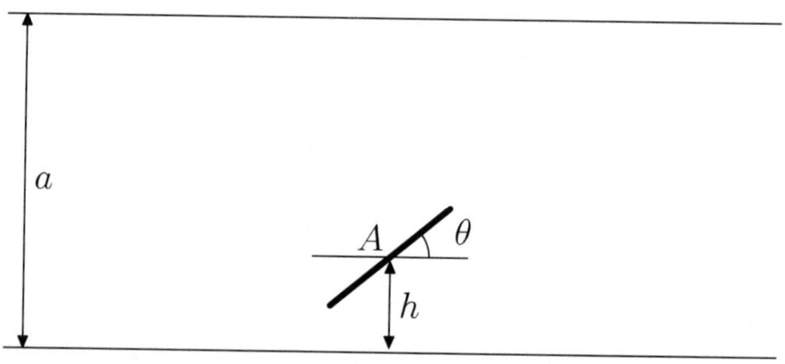

FIGURE 15.1. Definition of the needle parameters h, θ.

The chosen probability law is the uniform measure

$$dp = \frac{dh}{a} \cdot \frac{d\theta}{2\pi}.$$

Barbier then argues as follows. Let p_n be the probability that the needle cuts exactly n lines. We must calculate p_1. The mathematical expectation $\mathcal{E}$ of the number of points of intersection is therefore

$$\mathcal{E} = \sum_{n=0}^{\infty} n p_n.$$

But $\ell < a$; therefore $p_n = 0$ for all $n \geq 2$, and it follows that $\mathcal{E} = p_1$. On the other hand, mathematical expectation enjoys very remarkable linearity properties! Let $A_1, A_2, \ldots, A_k$ be a set of k needles, which we subject to the same random displacement. The expectation of the number of intersection points with the network

[4]EDITOR'S NOTE: Buffon deplored the fact that combinatorial analysis was *"still the only tool used up to this day in the science of probability"* and proposed *"therefore to place Geometry where it rightfully belongs in the science of chance"*. He thus created from nothing the theory of geometric probability.

of parallel lines is

$$\mathcal{E}(A_1, A_2, \ldots, A_k) = \sum_{i=1}^{k} \mathcal{E}(A_i). \tag{15.2}$$

Thanks to the uniformity of the probability law, $\mathcal{E}(A_i)$ only depends on the length ℓ_i of the needle; by abuse of notation we write $\mathcal{E}(A_i) = \mathcal{E}(\ell_i)$. In particular, let us consider two needles A_1 and A_2 rigidly connected to each other so that one extends the other. We then have

$$\mathcal{E}(\ell_1 + \ell_2) = \mathcal{E}(\ell_1) + \mathcal{E}(\ell_2).$$

Given that the function $\mathcal{E}$ is increasing, this functional equation implies that

$$\mathcal{E}(\ell) = c \cdot \ell, \tag{15.3}$$

where c is a constant to be determined. The equality (15.2) then shows that if the needles $A_1, A_2, \ldots, A_k$ constitute a polygonal line of length ℓ,

$$\mathcal{E}(A_1, A_2, \ldots, A_k) = c \cdot \ell.$$

Now we assume that $k \to \infty$ and that this polygonal line tends to a circle of diameter a, and whose circumference therefore measures πa. By passing to the limit, which we leave to the reader to justify with care,

$$\mathcal{E}(\text{circumference}) = c \cdot \pi a.$$

But obviously such a circumference cuts the network of lines in exactly 2 points, therefore $c\pi a = 2$, from which $c = 2/\pi a$. For the needle of length ℓ:

$$p = p_1 = \frac{2}{\pi a}\ell. \qquad \square$$

Remark: Buffon needle at the Bologna Congress. In 1917, Hostinský ([**15, 16**]; for a historical account see [**14**]) replaced the questionable uniformity hypothesis by Poincaré's method of arbitrary functions (see Chapter 14 of the present volume). This uniformity hypothesis was indeed criticized in 1912 by Carvallo [**6**] who realized that boundary effects were ignored. An elementary ergodic theorem enables Hostinský to show that practically any probability law for the needle problem asymptotically leads to the same result as Buffon's. (It is interesting to note that Poincaré had used the same ergodic theorem to show that the distribution of asteroids is asymptotically uniform for essentially any initial distribution.) It is the very idea that inspired Hostinský's interest in the ergodic theorem and Markovian models. In that his contribution could be thought as instrumental (together with Poincaré's approach to the shuffling of cards: see Chapter 14) for the renewal and consecration of the theory of Markov chains in the International Congress in Bologna in 1928 (cf. [**4**]).

The following classical result (close to the statements of Cauchy [**7**] and Crofton [**9**]; see Santaló [**29**]) is essentially contained in the preceding proof. Let C be a planar curve of finite length $|C|$. Let K be its convex hull, ∂K the boundary of K, and $|\partial K|$ the length of K, called the perimeter of C.[5]

[5]In general, in what follows the notation $|A|$ will represent the cardinality of A if A is a finite set, the length of A if A is a rectifiable curve, the surface area of A if A is a 2-dimensional manifold, etc.

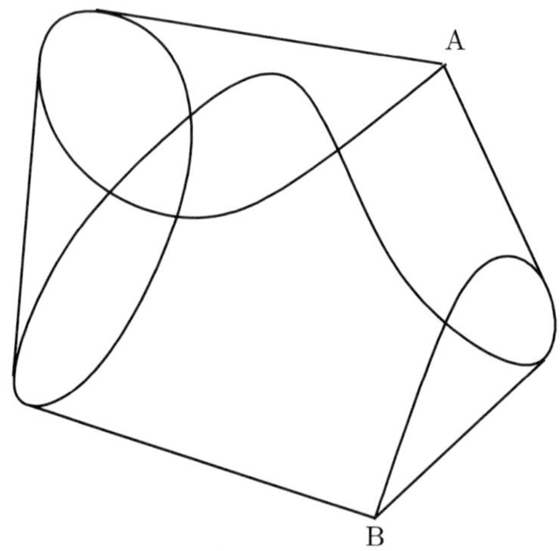

FIGURE 15.2. A curve joining A to B and its convex hull.

THEOREM 15.2.2. *Consider the family of all the lines that intersect C. The average number of points of intersection of these lines with C is*

$$2|C|/|\partial K|. \tag{15.4}$$

PROOF. As in the case of the needle, the idea is first to make precise the probability measure on the family of lines, in order to give a meaning to the "average" number. Let G be the group of motions in the plane, that is the (semidirect) product[6] of the group of rotations (isomorphic to $\mathbb{R}/2\pi\mathbb{Z}$) and the group of translations (isomorphic to $\mathbb{R}^2$). We endow G with Haar measure, that is, the product of the Lebesgue measures on $\mathbb{R}^2$ and $\mathbb{R}/2\pi\mathbb{Z}$. Let L be a line in the plane. When g runs through G, gL runs through the set of all lines. As before, there exists a constant $c > 0$ such that

$$\int_G |C \cap gL|\, dg = c\,|C|$$

(where $|C \cap gL|$ denotes the number of points of intersection of C with the line gL, $g \in G$).[7] The average number of intersection points is obtained by dividing the equality by the measure of the family of lines that meet C, that is, by

$$\int_{gL\cap C\neq\varnothing} dg = \frac{1}{2}\int_G |\partial K \cap gL|\, dg = \frac{1}{2}c\,|\partial K|.$$

The average is therefore indeed $2|C|/|\partial K|$. □

---

[6]EDITOR'S NOTE: An element of G is a pair (t, θ) with $t \in \mathbb{R}^2$ identified with the translation by the vector t and $\theta \in \mathbb{R}/2\pi\mathbb{Z}$ identified with the rotation by angle θ; with the obvious notation, (t, θ) operates on a point $z \in \mathbb{R}^2 \cong \mathbb{C}$ by $(t, \theta) \cdot z = e^{i\theta} z + t$. By applying another element (t', θ') of G one obtains

$$e^{i\theta'}(e^{i\theta} z + t) + t' = e^{i(\theta'+\theta)} z + e^{i\theta'} t + t',$$

from which comes the group law of G: $(t', \theta')(t, \theta) = (e^{i\theta'} t + t', \theta' + \theta)$.

[7]In the preceding case of equation (15.3), the network of lines (the parquet) was fixed and we moved at random a polygonal curve (or, by passing to the limit, a rectifiable curve); it comes to the same thing, of course, to leave the curve fixed and to move the network of lines (the principle of relativity!!!). The situation is then the same as here, except that here we only have one line to move.

These two theorems are thoroughly discussed in all classical texts on integral geometry. See in particular Ambartzumian [**1**], Santaló [**29**], and Langevin [**22**].

It is clear that there exists a general common formulation of the two preceding statements. In the first we consider equidistant parallel lines, and in the second the lines that intersect a given curve. One may imagine more complex situations, but in all cases the average number of intersection points is proportional to $|C|$. In the next section, we shall discuss an extension due to Poincaré.

It seems to me neither necessary, nor interesting, nor pertinent to generalize at the risk of hiding the fundamental ideas. One often discovers the deep essence of a result through one or two particular cases, and this is why I have kept theorems 15.2.1 and 15.2.2 just as they are. I do not at all deny that certain general statements cast new light on such and such particular cases. I think for example of the theorems of Green–Riemann, of Stokes, and of Ostogradski that are all contained in the lapidary formula (notably due to Poincaré[8])

$$\int_{\partial\Omega} \omega = \int_{\Omega} d\omega$$

whose beauty is indisputable!

15.3. Poincaré's theorem in the plane

Again, let G be the group of motions in the plane.

THEOREM 15.3.1 (Poincaré). *Let C and C' be two plane curves with respective lengths $|C|$ and $|C'|$. Then*

$$\int_G |gC \cap C'|\, dg = \int_G |C \cap gC'|\, dg = 4|C| \cdot |C'|.$$

PROOF. Taking again Barbier's idea, we obtain

$$\int_G |C \cap gC'|\, dg = c_1|C|, \qquad \int_G |gC \cap C'|\, dg = c_2|C'|.$$

But these two integrals are evidently equal. Therefore there exists an absolute constant $c > 0$ such that

$$\int_G |C \cap gC'|\, dg = c|C| \cdot |C'|.$$

To determine c, specialize C and C' to be circles with the same radius R. Then $|C \cap gC'| = 2$ if and only if $C \cap gC' \neq \varnothing$ (we exclude the cases where C and gC' are tangent or coincide, which have zero probability and therefore do not contribute to the average). Moreover,

$$dg = d(\text{rotation})\, d(\text{translation}) = d\theta\, da\, db;$$

therefore

$$\int_G |C \cap gC'|\, dg = 2 \int d\theta \iint da\, db = 2 \cdot 2\pi \cdot |B(0, 2R)| = 16\pi^2 R^2,$$

where $B(0, 2R)$ represents the ball centered at 0 and having radius $2R$. But this integral also equals $c(2\pi R)^2 = 4\pi^2 R^2 c$, hence $c = 4$. □

---

[8] EDITOR'S NOTE: See Chapter 11.

A particular case of this result will suffice for what follows. Consider the case where C' is a circumference of radius $\varepsilon > 0$ and where as before C is an arbitrary rectifiable curve. The invariance of C' under rotation around its center allows us as before to factor the integral

$$\int dg = 2\pi \iint da\, db.$$

THEOREM 15.3.2. *Let C be a plane curve of length $|C|$. Let $\varepsilon > 0$ and let*

$$C(\varepsilon) = \{x \in \mathbb{R}^2 \mid \text{distance}(x, C) \leq \varepsilon\}$$

be the ε-neighborhood of C. The average number of intersection points of C with the circles having radius ε and cutting C is

$$4\varepsilon|C|/|C(\varepsilon)|$$

provided that ε is sufficiently small (for example $\varepsilon <$ diam(C), where diam(C) denotes the diameter of C, that is, the largest distance between two points of the curve).

PROOF. According to Poincaré's theorem, if C' denotes a circle of radius ε,

$$\int_G |C \cap gC'|\, dg = 4|C|2\pi\varepsilon.$$

Therefore, if T is the set of translations of the plane, then

$$\int_T |C \cap tC'|\, dt = 4|C|\varepsilon.$$

But we have

$$\int_{\{tC' \cap C \neq \varnothing\}} dt = |C(\varepsilon)|$$

whenever $\varepsilon < \text{diam}(C)$. Therefore the average number of intersection points of C with the circumferences of radius ε is

$$4|C|\varepsilon/|C(\varepsilon)|.$$

□

15.4. The Edelman–Kostlan theorem

In fact, Poincaré did not state Theorem 15.3.1 of the previous section on the *plane* but on the *sphere* S^2, and he did not give the explicit multiplicative constant in the final expression (cf. [**27**, no. 84, p. 141]). His result was later generalized to all *homogeneous spaces*, that is, spaces of arbitrary finite dimension that have a continuous group of displacements allowing motion from any point to any other point (cf. [**8**] and [**21**]). Here we will consider a very particular case of this *generalized Poincaré formula*, which is the case we will need in the next section. Let

$$S^n = \{x \in \mathbb{R}^{n+1} \mid \|x\| = 1\}$$

be the n-dimensional sphere, where $x = (x_0, x_1, \ldots, x_n)$ and $\|x\| = \left(\sum_{i=0}^n x_i^2\right)^{1/2}$. We are given a rectifiable curve C on S^n, and we are interested in the average number of intersections of C with a "random" equator of S^n.

Let us be a bit more precise. An "equator" is an intersection of S^n with a hyperplane $a_0x_0 + a_1x_1 + \cdots + a_nx_n = 0$ (that is, a manifold of dimension $n-1$).

We may assume that the vector $a = (a_0, \dots, a_n) \in \mathbb{R}^{n+1}$ has norm 1: it therefore defines a point of S^n. We pass from the equator

$$S^{n-1} = \{x \in \mathbb{R}^{n+1} \mid \|x\| = 1; x_n = 0\} \qquad (\cong \{x' \in \mathbb{R}^n \mid \|x'\| = 1\})$$

(corresponding to the point $a^0 = (0, \dots, 0, 1)$) to any other by a displacement of S^n (sending a^0 to a). By choosing a at random, according to the uniform probability law (normalized Lebesgue measure), on a hemisphere of S^n (this suffices, because two diametrically opposed points define the same equator, and the boundary of a hemisphere has zero measure), we obtain a probability law on the group G of displacements of this hemisphere. A random equation is the image of S^{n-1} by a displacement chosen at random according to this law.

THEOREM 15.4.1 (Edelman, Kostlan [**11**]). *The average number of intersection points of a rectifiable curve C on S^n with a random equator is*

$$\frac{1}{\pi}|C|.$$

PROOF. The (generalized) Poincaré theorem here gives:

$$\int_G |C \cap gS^{n-1}|\, dg \Big/ \int_G dg = k\,|C| \cdot |S^{n-1}|.$$

To determine the constant k, we choose for C a great circle of S^n. Then $|C| = 2\pi$ and for almost all g, $|C \cap gS^{n-1}| = 2$, from which we conclude that $k|S^{n-1}| = 1/\pi$ and obtain the stated result. □

15.5. The real zeroes of a random real polynomial

A random real polynomial of degree n is of the form

$$P(t) = a_n t^n + a_{n-1}t^{n-1} + \cdots + a_1 t + a_0,$$

where $a = (a_0, a_1, \dots, a_n) \in S^n$ is uniformly distributed on S^n.

THEOREM 15.5.1 (Kac [**17, 18**]). *The average number of real zeroes of a random real polynomial is*

$$E_n = \frac{1}{\pi}\int_{-\infty}^{+\infty} \sqrt{\frac{1}{(t^2-1)^2} - \frac{(n+1)^2 t^{2n}}{(t^{2n+2}-1)^2}}\, dt \quad \underset{n\to\infty}{\sim} \quad \frac{2}{\pi}\log n.$$

PROOF. We reproduce here the remarkably simple and beautiful proof of Edelman and Kostlan, based on the Poincaré theorem. The number of real zeroes of the polynomial

$$P(t) = a_n t^n + a_{n-1}t^{n-1} + \cdots + a_1 t + a_0$$

equals the number of intersection points of the plane

$$a_0 x_0 + a_1 x_1 + \cdots + a_n x_n = 0$$

with the curve γ given by the parametric equations

$$x_0 = 1,\ x_1 = t,\ x_2 = t^2, \dots,\ x_n = t^n \qquad (t \in \mathbb{R}).$$

This is nothing other than the number of intersection points of the projection C of γ onto S^n from the origin, with the "equator"

$$\begin{cases} a_0 x_0 + a_1 x_1 + \cdots + a_n x_n = 0, \\ x_0^2 + x_1^2 + \cdots + x_n^2 = 1. \end{cases}$$

The average value of the number of real zeroes of the random polynomial is therefore $\frac{1}{\pi}|C|$. Note that the parametric equations of C are

$$x_0 = (1 + t^2 + \cdots + t^{2n})^{-\frac{1}{2}}, \ \ldots, \ x_n = t^n(1 + t^2 + \cdots + t^{2n})^{-\frac{1}{2}},$$

which are the same as

$$x_0 = \left((t^{2n+1} - 1)/(t^2 - 1)\right)^{-\frac{1}{2}}, \ \ldots, \ x_n = t^n \left((t^{2n+1} - 1)/(t^2 - 1)\right)^{-\frac{1}{2}}.$$

The integral that appears in the statement of the theorem is none other than

$$|C| = \int_{-\infty}^{+\infty} \left(\sum_{i=0}^{n} x_i'(t)^2\right)^{\frac{1}{2}} dt.$$

It then remains to estimate the integral for infinite n, which, even if it is not exactly trivial, is nonetheless classical. Moreover, Edelman and Kostlan succeeded in pushing the calculation further and were able to show that

$$E_n = \frac{2}{\pi} \log n + 0.625\ldots + \frac{2}{\pi n} + O\left(\frac{1}{n^2}\right). \qquad \square$$

15.6. Three remarks

1. Kac observed that

$$p_n(t) = \frac{1}{\pi}\sqrt{\frac{1}{(t^2 - 1)^2} - \frac{(n + 1)^2 t^{2n}}{(t^{2n+2} - 1)^2}} \Big/ E_n$$

is the probability density of the average number of real zeroes of a random real polynomial of degree n. The graph $t \mapsto p_n(t)$ features peaks at $t = +1$ and $t = -1$, and as $n \to \infty$, $p_n(t)$ tends to the sum of Dirac measures:

$$\frac{1}{2}(\delta_{-1} + \delta_{+1}).$$

The real zeroes therefore accumulate on average towards ± 1. This observation confirms the theorem of Erdös et Turán [**13**] that asserts that almost all zeroes, whether real or complex, of almost all complex polynomials of degree n are distributed in a small neighborhood of the circle $|z| = 1$.

2. If the real coefficients $a_0, a_1, \ldots, a_n$ are independent centered random variables distributed either according to the reduced central normal law on $\mathbb{R}$ or uniformly over $[-1, +1]$ or $\{-1, +1\}$, then the average number of real zeroes is again equivalent to $\frac{2}{\pi} \log n$: see Kac [**17, 18**].

3. Erdös and Offord established the following result (much more difficult and delicate than the preceding ones—we shall not prove it here):

THEOREM 15.6.1 (Erdös, Offord [**12**]). *The 2^{n+1} polynomials*

$$P(t) = \sum_{i=0}^{n} a_i t^i, \qquad a_i = \pm 1,$$

all have

$$\frac{2}{\pi} \log n + o\left((\log n)^{\frac{2}{3}} \log \log n\right)$$

real zeroes, except perhaps for at most $o\left(2^n (\log \log n)^{-\frac{1}{3}}\right)$ of these polynomials.

15.7. Curves and entropy

Let C be a rectifiable plane curve. We are interested, as in Theorem 15.2.2, in the family of lines that cut C. If p_n denotes the probability that one of these lines intersects C in exactly n points ($n \geq 1$), then Theorem 15.2.2 asserts that

$$\sum_{n=1}^{\infty} np_n = \frac{2|C|}{|\partial K|},$$

where $|\partial K|$, we recall, is the perimeter of C (cf. footnote 5 on p. 295). We define the elementary entropy of C by

$$h(C) = \sum_{n=1}^{\infty} p_n \log \frac{1}{p_n}$$

with the usual convention $0 \cdot \infty = 0$ (i.e., the function $p \log \frac{1}{p}$ is extended to $p = 0$ by continuity). A simple variational calculation, consisting in maximizing $h(C) = h(p_1, p_2, \dots)$ under the two constraints

$$\sum_{n=1}^{\infty} p_n = 1 \quad \text{and} \quad \sum_{n=1}^{\infty} np_n = \frac{2|C|}{|\partial K|},$$

shows that

$$h(C) \leq \log 2|C|/|\partial K| + 1. \tag{15.5}$$

Now let C be a half-infinite plane curve such that each finite portion is rectifiable. Let C_s be the starting portion of length s. The (relative) entropy of C is by definition

$$H(C) = \limsup_{s \to \infty} \frac{h(C_s)}{\log s}.$$

(In my previous publications, I defined the entropy by lim inf.) The inequality (15.5) implies

$$0 \leq H(C) \leq 1.$$

Loosely speaking, the curves of zero entropy appear to be simple and organized while the curves of maximum entropy will be more complex. Thus, for example, an infinite branch of an algebraic curve has zero entropy. Indeed, if d is its degree, then

$$\sum_{n=1}^{\infty} np_n \leq d,$$

and therefore for all $s > 0$,

$$h(C_s) \leq \log d + 1,$$

from which $H(C) = 0$.

One can show just as easily that a sinusoid or a logarithmic spiral $\rho = \exp a\theta$ ($a > 0$) has zero entropy.

It is easy to construct curves of a given entropy (one can check that the entropy of the spiral $\rho = (\log \theta)^\alpha$, $\alpha > 0$, is $\alpha/(\alpha + 1)$). In particular a sufficiently zigzagging curve will have entropy 1. But this is also the case for spirals of type $\rho = \exp(\log \log \theta)^2$. Such a curve is very regular, but as θ increases, its coils become closer and closer, so that with the naked eye one can no longer distinguish one from the other, leading to confusion. Those curves for which there exist bounded regions of the plane, say discs of radius 1, which contain arbitrarily long portions of the curve, are called *lumpy*. A precise definition will be given below, and we will show in the end that a curve whose entropy is strictly greater than 1/2 is lumpy: disorder creates confusion!

15.8. Confusion

Before passing to the definitions, perhaps it is worth attempting to justify the term "disorder" applied to the spiral $\rho = \log\theta$. We know that a non-laminar fluid flow, which is thus in some sense disordered, presents turbulence. Would it not be of the type $\rho = \log\theta$? In this, such a spiral could be associated to disorder, while in contrast a spiral $\rho = \exp a\theta$ models biological growth, suggesting order... The reader will excuse me, I hope, for mixing fields: a vague phenomenology on one hand and mathematics on the other.

Let us return to mathematics. Let $\varepsilon > 0$ and let $\nu(s,\varepsilon)$ be the average number of intersection points of the initial portion C_s with the circles of radius ε that meet C_s. According to Theorem 15.3.2,

$$\nu(s,\varepsilon) = 4s\varepsilon/|C_s(\varepsilon)|.$$

Let $A > 0$. I say that the curve C presents an (A,ε)-lump in a neighborhood of the point with arclength coordinate s if $\nu(s,\varepsilon) > A$. (This notion is only interesting if C is unbounded.) By definition, C is lumpy if for every $\varepsilon > 0$,

$$\limsup_{s\to\infty} \nu(s,\varepsilon) = \infty.$$

THEOREM 15.8.1. *Every unbounded plane curve with entropy $H > 1/2$ is lumpy. The lower bound 1/2 is optimal; that is, there exist non-lumpy curves with entropy $H = 1/2$.*

PROOF. The Archimedean spiral $\rho = a\theta$ has entropy 1/2 and is not lumpy; therefore the bound 1/2 is optimal. Now let us establish the implication

$$H > \frac{1}{2} \quad\Longrightarrow\quad C \text{ is lumpy.}$$

Let $h(C_s)$ be the elementary entropy of the initial portion C_s, and let us set

$$H(s) = h(C_s)/\log s.$$

By hypothesis,

$$H(C) = \limsup_{s\to\infty} H(s) > \frac{1}{2}.$$

Therefore there exist $\delta > 0$ and an increasing unbounded sequence s_n such that

$$H(s_n) > \frac{1}{2} + \delta.$$

Moreover, if $\operatorname{diam}(C_s)$ and $|\partial K_s|$ are respectively the diameter and the perimeter of C_s, then

$$2\operatorname{diam}(C_s) \le |\partial K_s|,$$

and for every $\varepsilon > 0$,

$$|C_s(\varepsilon)| \le \frac{\pi}{4}(\operatorname{diam}(C_s) + 2\varepsilon)^2,$$

and hence

$$|C_s(\varepsilon)| \le \frac{\pi}{16}|\partial K_s|^2 + \frac{\pi\varepsilon}{2}|\partial K_s| + \pi\varepsilon^2.$$

Because C is unbounded, $|\partial K_s| \to \infty$ as $s \to \infty$, and therefore

$$\limsup_{s\to\infty} \frac{\log |C_s(\varepsilon)|}{\log |\partial K_s|} \le 2.$$

Consequently, there exists $s(\varepsilon, \delta)$ such that for every $s > s(\varepsilon, \delta)$,

$$\frac{\log |C_s(\varepsilon)|}{\log |\partial K_s|} < 2 + 3\delta.$$

By Theorem 15.3.2,

$$|C_s(\varepsilon)| = \frac{4s\varepsilon}{\nu(s, \varepsilon)},$$

and therefore for every $s > s(\varepsilon, \delta)$,

$$\frac{\log(4s\varepsilon)}{\log |\partial K_s|} - \frac{\log \nu(s, \varepsilon)}{\log |\partial K_s|} < 2 + 3\delta,$$

and hence

$$\frac{\log \nu(s, \varepsilon)}{\log |\partial K_s|} > \frac{\log s}{\log |\partial K_s|} - 2 - 3\delta + o(1). \tag{15.6}$$

We recall (see (15.5)) that

$$H(s) \le \frac{\log 2s - \log |\partial K_s| + 1}{\log s} = 1 - \frac{\log |\partial K_s|}{\log s} + o(1);$$

thus,

$$\frac{\log |\partial K_s|}{\log s} \le 1 - H(s) + o(1),$$

$$\frac{\log s}{\log |\partial K_s|} \ge \frac{1}{1 - H(s)} + o(1). \tag{15.7}$$

According to the inequalities (15.6) and (15.7), we have, for all $s > s(\varepsilon, \delta)$,

$$\frac{\log \nu(s, \varepsilon)}{\log |\partial K_s|} > \frac{1}{1 - H(s)} - 2 - 3\delta + o(1),$$

and consequently for $s_n > s(\varepsilon, \delta)$

$$\frac{\log \nu(s_n, \varepsilon)}{\log |\partial K_{s_n}|} > \frac{1}{1 - \frac{1}{2} - \delta} - 2 - 3\delta + o(1) > \delta + o(1).$$

In other terms,

$$\nu(s_n, \varepsilon) > |\partial K_{s_n}|^{\delta + o(1)},$$

and therefore

$$\limsup_{s\to\infty} \nu(s, \varepsilon) = \infty.$$ □

The interest of Theorem 15.8.1, it seems to me, is this. It relates the rather abstract concept of entropy to the more tangible and descriptive notion of lumpiness. However, we must not confuse the two, as the following analysis shows.

Consider the three spirals

$$\begin{aligned} S_0 &: \rho = \log\log\theta, \\ S_\alpha &: \rho = (\log\theta)^\alpha \quad (\alpha > 0), \\ S_\infty &: \rho = \exp(\log\log\theta)^2, \end{aligned}$$

which have for respective entropies $H(S_0) = 0$, $H(S_\alpha) = \alpha/(\alpha+1)$, and $H(S_\infty) = 1$: it is clear that among these curves the confusion decreases when α increases from 0 to ∞, while on the other hand the entropy increases. Here, therefore, entropy and confusion vary in opposite directions!

The calculation shows that $H(S_0) = 0$, but how can we "understand" this? Here is an attempt at an explanation. A curve represented by its polar equation is a function $\mathbb{R}_+ \to \mathbb{R}$. In particular the circle $\theta \mapsto (R\cos\theta, R\sin\theta)$ is covered infinitely many times as θ increases. At the end of exactly N turns $(0 \leq \theta \leq 2N\pi)$, its elementary entropy

$$h_N = \sum_{k=1}^{\infty} p_k \log \frac{1}{p_k}$$

vanishes because $p_{2N} = 1$ and $p_k = 0$ for all $k \neq 2N$, and for a non-integer number of turns it is easy to see that the elementary entropy remains bounded, thus $H = 0$.[9] But a (geometric) point of the circle corresponds to a polar angle θ that is only defined modulo 2π; in this there is confusion.

Up to visual errors(!), this is indeed what happens for the spiral S_0. The function $\log\log$ varies sufficiently slowly that S_0 is comparable with a circle both from the point of view of entropy and from the point of view of confusion.

In a general way, entropy and confusion are two facets of the same intuition. This is what is established by Theorem 15.8.1, but it would be unwise to demand more!

Without major additional difficulty Theorem 15.8.1 extends to curves C in $\mathbb{R}^n$: *If* $H(C) > 1 - 1/n$, *then* C *is lumpy.*

* * * * *

In 1928 my father defended his thesis [**26**] on the financial politics of a certain Poincaré.[10] It is my pleasure, some 78 years later, to discuss the work of the other Poincaré. A bit of atavism, perhaps?

Bibliography

[1] R. V. Ambartzumian, *Combinatorial integral geometry*, John Wiley & Sons, 1982.
[2] É. Barbier, *Le problème de l'aiguille et le jeu du joint couvert*, J. Math. Pures et Appl. **5** (1860), pp. 273–286.
[3] J.-L. Binet and J. Roger, *Un autre Buffon*, Collection *Savoir: Lettres*, Hermann (Paris), 1977.
[4] B. Bru, *Souvenirs de Bologne*, Journal de la Société Française de Statistique **144** (2003), pp. 135-226.
[5] G. Buffon, *Essai d'arithmétique morale.* Tome IV du supplément à l'Histoire naturelle, 1777. Reproduced in [**3**].
[6] E. Carvallo, *Le calcul des probabilités et ses applications*, Gauthier-Villars (Paris), 1912.
[7] A. Cauchy, *Mémoire sur la rectification des courbes*, Mém. Acad. Sci. Paris **22** (1850), pp. 3–15.
[8] S. S. Chern, *On integral geometry in Klein spaces*, Ann. of Math. **43** (1942), pp. 178–189.

[9] If we make $N-1$ full turns and one incomplete turn, then p_k is zero whenever $k \notin \{2N, 2N-1, 2N-2\}$; therefore $h \leq \log 3$.

[10] TRANSLATOR'S NOTE: Raymond Poincaré (1860–1934), cousin of Henri Poincaré. He served as President of France from 1913 to 1920 and held several other ministerial offices, including Prime Minister and Minister of Finance.

[9] M. W. Crofton, *On the theory of local probability, applied to straight lines drawn at random in a plane; the methods used being also extended to the proof of certain new Theorems in the Integral Calculus*, Phil. Trans. Royal Soc. London **158** (1868), pp. 181–199.

[10] C. Doche and M. Mendès France, *Real roots!*: pp. 42–56 of *Analytic and probabilistic methods in number theory* (Proceedings of the 3rd International Conference held in honour of J. Kubilius in Palanga, September 24–28, 2001), edited by A. Dubickas, A. Laurinčikas and E. Manstavičius. TEV, Vilnius, 2002.

[11] A. Edelman and E. Kostlan, *How many zeros of a random polynomial are real?*, Bull. Amer. Math. Soc. **32** (1995), pp. 1–37. [Erratum: Bull. Amer. Math. Soc. **33** (1996), p. 325.]

[12] P. Erdös and A. C. Offord, *On the number of real roots of a random algebraic equation*, Proc. London Math. Soc. **6** (1956), pp. 139–160.

[13] P. Erdös and P. Turán, *On the distribution of roots of polynomials*, Ann. of Math. **51** (1950), pp. 105–119.

[14] V. Havlovà, L. Mazliak and P. Sisma, *Le début des relations mathématiques franco-tchécoslovaques vu à travers la correspondance Hostinský-Fréchet*, Electronic J. Hist. Proba. Stat. **1** (2005), pp. 1–18. `http://www.jehps.net/Mars2005/HavlovaMazliakSisma.pdf`

[15] B. Hostinský, *Nové řešení Buffonovy úlohy o jehle*, Rozpravy Ceské Akad., Roc. **XXVI** (1917), Tr. II, c. 13.

[16] B. Hostinský, *Sur une nouvelle solution du problème de l'aiguille*, Bull. Sci. Math. 44 (1920), pp. 126–136.

[17] M. Kac, *On the average number of real roots of a random algebraic equation*, Bull. Amer. Math. Soc. **49** (1943), pp. 314–320. [Correction: *ibid.*, p. 938.]

[18] M. Kac, *On the average number of real roots of a random algebraic equation II*, Proc. London Math. Soc. **50** (1949), pp. 390–408.

[19] M. Kac, *Probability and related topics in physical sciences*, Lectures in Appl. Math., Interscience (New York), 1959. Republication: Amer. Math. Soc., 1976.

[20] D. A. Klain and G. C. Rota, *Introduction to geometric probability*, Lezioni Lincei, Academia nazionale dei Lincei. Cambridge Univ. Press, 1997.

[21] M. Kurita, *An extension of Poincaré formula in integral geometry*, Nagoya Math. J. **2** (1951), pp. 55–61.

[22] R. Langevin, *Introduction to integral geometry*, 21st Brazilian Mathematics Colloquium, Instituto de Matemática Pura e Aplicada (IMPA), Rio de Janeiro, 1997.

[23] P. Lévy, *Calcul des probabilités*, Gauthier-Villars (Paris), 1925.

[24] M. Mendès France, *Chaos implies confusion*: pp. 137–152 of *Number theory and dynamical systems* (York, 1987), edited by M. Dodson and J. Vickers, London Math. Soc. Lecture note series, vol. **134**, Cambridge Univ. Press, 1989.

[25] M. Mendès France, *The Planck constant of a curve*: pp. 325–366 of *Fractal geometry and analysis* (Montréal, 1989), edited by J. Bélair and S. Dubuc, NATO Adv. Sci. Inst. Ser. C Math. Phys. Sci. **346**, Kluwer Acad. Publishers, Dordrecht, 1991.

[26] P. Mendès France, *L'œuvre financière du gouvernement Poincaré*, Librairie générale de Droit et de Jurisprudence, Paris, 1928.

[27] H. Poincaré, *Calcul des probabilités*, Gauthier-Villars (Paris), 1912. (New edition: 1923.)

[28] H. Poincaré, *L'analyse et la recherche* (texts chosen and arranged by G. Ramunni), Collection *Savoir: Sciences*, Hermann (Paris), 1991. See in particular the note of G. Ramunni on p. 92.

[29] L. A. Santaló, *Integral geometry and geometric probability*, Addison-Wesley, 1976. Republication: Cambridge Univ. Press, 2004.

[30] H. Steinhaus, *Length, shape and area*, Colloq. Math. **3** (1954), pp. 1–13.

CHAPTER 16

Poincaré, and Lie's Third Theorem

Pierre-Paul Grivel

16.1. Introduction

Unless otherwise specified, K denotes the field of real numbers or the field of complex numbers; when the context is purely algebraic, K is simply a commutative ring with unit.

In the analysis of his work [**36**], H. Poincaré wrote that it was his research on algebra and non-Euclidean geometries that led him to consider continuous groups. But it was not until after the death of Lie (1899) that he entered the general theory of these groups. He then proposed to provide Lie's third theorem with *"a simple and direct proof, applicable in all cases, and this thanks to the use,"* so he said, *"of a very abbreviated symbolic notation."*

Lie theory consists in the study of a set of analytic transformations in n variables,

$$x'_i = \phi_i(x_1; \ldots; x_n; a_1; \ldots; a_r) \qquad (1 \leq i \leq n),$$

effectively depending on r real or complex parameters $a_1, \ldots, a_r$. They form what Lie called a *finite and continuous group of transformations* if the composition of two transformations in the set is again a transformation in the set. Lie sought to obtain a classification of these groups. To this end, he associated to such a group certain *infinitesimal transformations* that generate the group. For this, he expanded the ϕ_i into a Taylor series of order 1 in a neighborhood of the identity element $e = (e_1; \ldots; e_r)$ of the group of parameters,[1] which gives

$$x'_i = x_i + \sum_{1 \leq k \leq r} \xi_{ki}(x_1; \ldots; x_n)\, da_k \qquad (1 \leq i \leq n).$$

Thus, let f be an analytic function in a neighborhood of a point $x = (x_1; \ldots; x_n)$. The first derivatives of the composition $f(x'_1; \ldots; x'_n)$ with respect to the parameters, calculated at the point $(x; e)$, lead us to consider the expressions

$$X_k f = \sum_{1 \leq i \leq n} \xi_{ki}(x_1; \ldots; x_n) \frac{\partial f}{\partial x_1} \qquad (1 \leq k \leq r).$$

We thus define the differential operators $X_1, \ldots, X_r$, which are linearly independent. Lie then remarks that the vector space $\mathfrak{g}$ generated by these operators is endowed with an internal composition law: if X and Y are two elements of $\mathfrak{g}$, then the composition $X \circ Y$ is not, in general, an operator of the previous type, but on the

[1] That is, $\phi_i(x_1; \ldots; x_n; e_1; \ldots; e_r) = x_i$ for $1 \leq i \leq n$ and $x = (x_1; \ldots; x_n)$.

other hand the commutator bracket $[X;Y] = X \circ Y - Y \circ X$ is again an element of $\mathfrak{g}$; more precisely, for every $1 \le h \le r$ and $1 \le k \le r$, we have

$$[X_h; X_k] = \sum_{1 \le s \le r} c_{hks} X_s.$$

The coefficients of these linear combinations are the *structure constants*: they form a family $\{c_{hks}\}_{h,k,s}$ of r^3 constants that are antisymmetric with respect to the first two indices and that satisfy the relations deduced from the Jacobi identity. Indeed, the commutator bracket is antisymmetric, that is, $[X;Y] = -[Y;X]$, and satisfies the relation

$$[X;[Y;Z]] + [Y;[Z;X]] + [Z;[X;Y]] = 0,$$

which Jacobi had established in 1842 in his *Vorlesungen über Dynamik*, while he was studying the Poisson bracket.

Thus the vector space $\mathfrak{g}$, whose elements are what Lie called *infinitesimal transformations*, is endowed with a certain multiplicative structure, and Lie theory essentially amounts to the study of this structure. In particular, Lie's third fundamental theorem asserts that every family $\{c_{hks}\}_{h,k,s}$ of r^3 constants that satisfy the preceding properties determines a finite and continuous group of transformations whose structure is given by the c_{hks}.

In 1930, Élie Cartan [**9**] introduced the name *Lie group* and gave the precise definition:

A K-Lie group (of dimension m) is a topological space G endowed with a group structure and the structure of an analytic K-manifold (of dimension m), these two structures being compatible (i.e., the maps $(g, g') \mapsto gg'$ and $g \mapsto g^{-1}$ are analytic).

In a course given in 1934, H. Weyl [**45**] called the set of infinitesimal transformations associated to a continuous group a *Lie algebra.* The abstract definition appeared for the first time in a 1935 article of N. Jacobson [**22**]:

A K-Lie algebra is a K-module $\mathfrak{g}$ endowed with a linear map $[\ ;\ ] : \mathfrak{g} \otimes \mathfrak{g} \to \mathfrak{g}$ that satisfies the identities $[x;x] = 0$ and $[x;[y;z]] + [y;[z;x]] + [z;[x;y]] = 0$, for each $x, y, z \in \mathfrak{g}$.

To every Lie group G we associate the Lie algebra $L(G)$ obtained by endowing the vector space of vector fields that are left-invariant over G with the commutator bracket. As a vector space $L(G)$ is isomorphic to the tangent space $T_e(G)$ to G at the identity element e of G. If $\varphi : G \to H$ is a morphism of Lie groups, then it induces a morphism of Lie algebras $L(\varphi) : L(G) \to L(H)$ in such a way that if φ is an isomorphism then $L(\varphi)$ is also an isomorphism.

Lie theory consists essentially in studying this correspondence, and in particular understanding under what conditions it is invertible. We may immediately note that the correspondence is not faithful. For example, if we consider the Lie group of the circle $\mathbf{S}^1$ and the Lie group of the real line $\mathbb{R}$, then $L(\mathbf{S}^1) = L(\mathbb{R})$ is the unique real Lie algebra of dimension 1, which is necessarily trivial due to the antisymmetry of the bracket. But $\mathbf{S}^1$ (which is not simply connected) and $\mathbb{R}$ are clearly not isomorphic. However, $\mathbf{S}^1$ and $\mathbb{R}$ are locally isomorphic. This remark leads us to consider a slightly weaker notion than the notion of a Lie group, that of a *local Lie group* (also called a *Lie groupuscule*): this is a pointed analytic manifold $(G;e)$ (i.e., we choose the analytic manifold G and the point e), endowed with a multiplication that is only defined in a neighborhood of the identity element e. (See [**5**] or [**38**] for a precise definition of this notion.) For example, if U is a neighborhood of the identity element e of a Lie group G, then U is a local Lie group; in particular, G itself is a local Lie group. To a local Lie group G we associate as before a Lie algebra $L(G)$.

The local version of Lie's third theorem then says that every finite-dimensional Lie algebra is isomorphic to the Lie algebra of a local Lie group. The construction of this local group uses the exponential formula, which we will discuss a little later. To obtain the global version, we must ask if a local Lie group can be extended to a Lie group. The answer to this question was given by É. Cartan, who showed that every local Lie group is the localization of a simply connected Lie group.

By using the language of categories, the previous facts can be expressed more succinctly in the following way. First of all, Lie groups (resp. Lie algebras) form a category denoted $\mathcal{GL}$ (resp. $\mathcal{AL}$). Then there exist a pair of functors

$$\mathcal{AL} \underset{\Lambda}{\overset{L}{\rightleftarrows}} \mathcal{GL}$$

which establish a dictionary between the Lie groups and the Lie algebras. The work of Lie is the starting point for the construction of this dictionary. As we have seen previously, the functor L is easy to obtain. It is the functor Λ, which is in fact Lie's third theorem, that has sparked numerous works and whose construction, according to J.-P. Serre [**42**], cannot be simple because the third theorem is false for Banach–Lie groups.

Now let us return to Poincaré. Between 1899 and 1908 he published four articles on continuous groups. In a note in the *Comptes Rendus de l'Académie des Sciences* of May 1, 1899, [**32**], after enthusiastically praising Lie, he announced that one can construct an algebra of symbolic polynomials (the premise for the symmetric algebra and the tensorial algebra), in which one has $e^V e^T = e^W$ (a product of exponentials is again an exponential), and that from this *exponential formula* he can easily derive Lie's third theorem. The details of this construction appeared in a memoir published in 1899, on the occasion of a celebration of Sir G. G. Stokes [**33**]. This article considers essentially three themes: the construction of an algebra that is isomorphic to what is now called the universal enveloping algebra of a Lie algebra, a study of the exponential formula $e^V e^T = e^W$ (which extends a similar study begun by J. E. Campbell), and a proof of Lie's third theorem. He remarks that the formulas at which he arrives are different from those of Lie, but their comparison is instructive because it allows one to recover certain theorems of Killing. The following two articles, the *Remarques* of 1901 [**34**] and the *Nouvelles remarques* of 1908 [**35**], are devoted essentially to the study of the adjoint group and to some problems raised by the results of Killing.

It must be noted that Poincaré's results on continuous groups appear to have had relatively little influence. They cannot, in general, be considered either complete or definitive because they are solely of a local nature. In the introduction to his 1930 article on continuous groups and analysis situs [**9**], however, É. Cartan cites *"the three penetrating memoirs* [of Poincaré] *that show the role played by the singular transformations of a group in the structure theory of this group,"* but he denounces *"the insufficiency of the ordinary proofs of the third theorem, which only prove the existence of a piece of the group, perhaps incapable of being extended to form a complete group."*

The introduction of the *calculus of symbolic polynomials* in the third section of the 1899 article [**33**] is probably Poincaré's most interesting contribution to the theory of Lie algebras. In fact, the algebra that he constructs is nothing other than the universal enveloping algebra of a finite-dimensional Lie algebra over a field of characteristic zero, and the construction is such that the Lie algebra is embedded in this algebra: this is the Poincaré–Birkhoff–Witt theorem!

It is therefore the second article, the one from the Cambridge Philosophical Society, which will hold our attention here. We will briefly recall Lie theory in the second section. In the third section we will comment on an article by J. E. Campbell [**6**] about the exponential formula, which indisputably influenced Poincaré. In the fourth section we will explain Poincaré's viewpoint on the universal enveloping algebra, and in the fifth we will show how he proved Lie's third theorem starting from his considerations on the exponential. The following section will be devoted to the developments that appeared during the 20th century in two subjects broached by Poincaré, namely the exponential formula and the enveloping algebra. Finally, in the last section we will show how a notion analogous to the enveloping algebra can be defined in the framework of Leibniz algebras, which are generalizations of Lie algebras.

16.2. Lie's three fundamental theorems

The theory developed by S. Lie in his three-volume work [**26**] that appeared between 1888 and 1893 is devoted to the study of transformations of n variables and r essential parameters of the form

$$x'_i = \phi_i(x_1; \dots; x_n; a_1; \dots; a_r) \qquad (1 \leq i \leq n).$$

The variables $x = (x_1; \dots; x_n)$ are taken in K^n and the parameters $a = (a_1; \dots; a_r)$ are taken in K^r; in addition, the functions ϕ_i are assumed to be analytic in their $n + r$ arguments. Lie called a set of such functions a *finite and continuous group of transformations* if it is closed under composition and if it contains the identity transformation and the inverse of each transformation.

If we define ϕ by setting $\mathrm{pr}_i \circ \phi = \phi_i$ for $1 \leq i \leq n$, where pr_i denotes the projection onto the ith component, we can consider the map

$$\phi : K^n \times K^r \to K^n$$

which satisfies the following condition: for all $a, b \in K^r$ there exists a unique element $c \in K^r$ such that $\phi(\phi(x; a); b) = \phi(x; c)$ for all $x \in K^n$.

Lie's idea was to associate to such a group of transformations a *group of infinitesimal transformations* that generate the given group.

For example, Lie considered the action of the projective group on the Riemann sphere to be the transformation $x' = \frac{x+b}{cx+d}$ $(bc - d \neq 0)$ of $\mathbb{C}$, to which he associated the linearly independent differential operators $X_1 = \partial/\partial x$, $X_2 = x\,\partial/\partial x$, and $X_3 = x^2\,\partial/\partial x$.

But, even though Lie was never explicit on this matter, it is clear that in general the functions ϕ_i are not globally defined; we must therefore restrict ourselves to open subsets of K^n and K^r.

We will briefly explain Lie's fundamental results by using modern concepts and notation. (For all general questions on Lie groups and Lie algebras, the reader may consult the numerous classical works on the subject, whose levels of abstraction vary greatly, going from the most elementary, such as [**15**] or [**38**], to the more abstract, such as [**4**] or [**5**].)

Let G be an r-dimensional K-Lie groupuscule, M an n-dimensional analytic K-manifold, and $\phi : M \times G \to M$ a local operation law of G on M; we will always assume that G acts effectively on M. (See [**5**] for the precise definition of a local operation law; it is enough to know that the action of G on M is defined only in an open subset of $M \times G$ containing $M \times \{e\}$, e being the identity element of G.)

We denote by $\mathfrak{X}(M)$ the Lie algebra of vector fields on M, and by $L(G)$ the Lie algebra of G; recall that $L(G)$ is a Lie sub-algebra of $\mathfrak{X}(G)$ formed by the left-invariant vector fields and that, as a vector space, $L(G)$ is isomorphic to the tangent space $T_e(G)$.

For every $x \in M$ we define an analytic map $\phi_x : G \to M$ by setting $\phi_x(a) = \phi(x; a)$ for every $a \in G$ sufficiently close to e. In addition, we consider the two maps $(d\phi_x)_a : T_a(G) \to T_{\phi(x;a)}(M)$ and $(d\phi_{\phi(x;a)})_e : T_e(G) \to T_{\phi(x;a)}(M)$.

LIE'S FIRST FUNDAMENTAL THEOREM. *Each vector $X \in T_e(G)$ extends uniquely to a vector field $\overline{X} \in \mathfrak{X}(G)$ such that, for every $a \in G$ and every $x \in M$, we have*

$$(d\phi_x)_a(\overline{X}_a) = (d\phi_{\phi(x;a)})_e(X).$$

It follows that the vector field $\overline{X}$ is left-invariant, so that the differential $(d\phi_x)_e : T_e(G) \to T_x(M)$ induces a map

$$\tau_\phi : L(G) \to \mathfrak{X}(M).$$

In addition, the image of τ_ϕ completely determines the local operation law ϕ.

We can make a few remarks on the content of this statement. First of all, we have

$$(d\phi_{\phi(x;a)})_e = (d\phi_x)_a \circ (dL_a)_e,$$

where $L_a : G \to G$ is the left translation defined by the element $a \in G$. Then this formula allows us to define the field $\overline{X}$ by setting $\overline{X}_a = (dL_a)_e(X)$ for all $a \in G$, and this shows that $\overline{X}$ is invariant.

Now we choose a system of local coordinates on G and on M; they induce bases $\frac{\partial}{\partial a_1}, \dots, \frac{\partial}{\partial a_r}$ of $\mathfrak{X}(G)$ and $\frac{\partial}{\partial x_1}, \dots, \frac{\partial}{\partial x_n}$ of $\mathfrak{X}(M)$. The theorem asserts that there exist r linearly independent vector fields $\overline{X}_1, \dots, \overline{X}_r$ on G such that $\overline{X}_j(e) = \frac{\partial}{\partial a_j}(e)$ and $(d\phi_x)_a(\overline{X}_j(a)) = (d\phi_{\phi(x;a)})_e(\overline{X}_j(e))$ (here we denote by $Y(z)$, instead of Y_z, the value of a field Y at a point z). We can therefore write

$$\frac{\partial}{\partial a_j}(a) = \sum_{1 \le k \le r} \psi_{kj}(a)\, \overline{X}_k(a)$$

for all $1 \le j \le r$; we note that the matrix (ψ_{kj}) is non-singular in a neighborhood of the identity, because it equals the identity at this point. We define r linearly independent vector fields $X_1, \dots, X_r$ on M by setting $X_k(x) = (d\phi_x)_e(\overline{X}_k(e))$ for every $x \in M$. In coordinates, we therefore have

$$X_k(x) = \sum_{1 \le i \le n} \xi_{ki}(x) \frac{\partial}{\partial x_i}$$

for every $1 \le k \le r$. A simple calculation then gives, for $1 \le j \le r$,

$$(d\phi_x)_a\left(\frac{\partial}{\partial a_j}(a)\right) = \sum_{1\le k\le r,\, 1\le i\le n} \psi_{kj}(a)\,\xi_{ki}(\phi(x;a))\,\frac{\partial}{\partial x_i}.$$

Let us apply this formula to the ith coordinate function x_i; if we set $x'_i = (\phi_x)_i = x_i \circ \phi_x$, we obtain

$$\frac{\partial x'_i}{\partial a_j} = \sum_{1\le k\le r} \psi_{kj}(a_1;\ldots;a_r)\,\xi_{ki}(x'_1;\ldots;x'_n),$$

which is indeed the formula given by Lie in the statement of his first theorem (see [**18**] or [**26**]).

Let $\mathfrak{g}$ be a finite-dimensional K-Lie algebra; a morphism of Lie algebras

$$\tau : \mathfrak{g} \to \mathfrak{X}(M)$$

is called a *right infinitesimal law of* $\mathfrak{g}$ *in* M.

LIE'S SECOND FUNDAMENTAL THEOREM. *The map*

$$\tau_\phi : L(G) \to \mathfrak{X}(M)$$

is a right infinitesimal law of $L(G)$ *in* M.

Conversely, if

$$\tau : L(G) \to \mathfrak{X}(M)$$

is a right infinitesimal law of $L(G)$ *in* M*, then for every point* $x_0 \in M$ *there exists an open neighborhood* V *of* x_0 *in* M *and a local right operation law* ϕ *of* G *in* V *such that* $\tau_\phi(a) = \tau(a)|_V$ *for every* $a \in L(G)$.

LIE'S THIRD FUNDAMENTAL THEOREM. *Let* $\tau : \mathfrak{g} \to \mathfrak{X}(M)$ *be a right infinitesimal law of a finite-dimensional Lie algebra* $\mathfrak{g}$ *in an analytic manifold* M*. Then there exists a Lie groupuscule* G*, a local right operation law* ϕ *of* G *in* M*, and an isomorphism of Lie algebras* $\psi : L(G) \to \mathfrak{g}$ *such that* $\tau_\phi = \tau \circ \psi$.

Clearly, an important question is to know if these theorems have global versions. This is the case, for example, for the second theorem when G is a simply connected Lie group and M is a compact manifold. On the other hand, it is clear that a Lie group (or groupuscule) acts on itself by multiplication. Thus each of the preceding theorems has a (local or global) version that gives a relation between a Lie group and its Lie algebra. Here we give only the most important result, which is the global version of the third theorem.

THEOREM. *Let* $\mathfrak{g}$ *be a finite-dimensional* K*-Lie algebra. There exists a simply connected* K*-Lie group* G *such that* $L(G)$ *is isomorphic to* $\mathfrak{g}$.

This theorem was proved in 1930 by É. Cartan [**8, 9**]. We note that between 1894 (the year of his thesis) and 1930, É. Cartan published at least twenty articles on the subject of finite and continuous groups, but he only cited Poincaré once on this matter.

16.3. The exponential formula at the end of the 19th century

In Section 12 of Chapter 3 in the first volume of his monumental work **[26]**, S. Lie considers the following situation. Let X be a vector field on K^n, and $x = (x_1;\dots;x_n)$ a point of K^n. If we consider x_i as the ith coordinate function, we may set

$$x_i' = x_i'(t) = x_i + \frac{t}{1!}X(x_i) + \frac{t^2}{2!}X^2(x_i) + \cdots \qquad (1 \le i \le n),$$

or more briefly $x_i' = e^{tX}(x_i)$, which defines a transformation that has the effect of translating x by the one-parameter group associated to the differential equation $\frac{dx_i}{dt} = X(x_i)$.

Now let f be an analytic function in a neighborhood of the point x. Then $f(x_1';\dots;x_n')$ is a function f' of the variable t that can be expanded into a Taylor series; we therefore have

$$f'(t) = f'(0) + \frac{t}{1!}\frac{df'}{dt}\bigg|_{t=0} + \frac{t^2}{2!}\frac{d^2f'}{dt^2}\bigg|_{t=0} + \cdots .$$

Let us assume that in a system of local coordinates we have

$$X = \sum_{1\le i\le n} \xi_i(x_1;\dots;x_n)\,\frac{\partial}{\partial x_i}.$$

Lie (along with Campbell, as we shall see a little later) denotes by X' the expression obtained by replacing x_k by x_k' $(1 \le k \le n)$ everywhere in the previous equality. The formula for the change of coordinates given by $x_i' = e^{tX}(x_i)$ shows that $X(x_i') = X'(x_i')$ for all i such that $1 \le i \le n$. Since $\dfrac{dx_i'}{dt} = X(x_i')$, we therefore have

$$\frac{df'}{dt} = \sum_{1\le i\le n} X(x_i')\,\frac{\partial f}{\partial x_i'} = \sum_{1\le i\le n} X'(x_i')\,\frac{\partial f}{\partial x_i'} = X'(f).$$

By induction we immediately obtain $\frac{d^n f'}{dt^n} = X'^n(f')$. But for $t = 0$, we have $x' = x$, from which $X' = X$; we therefore have finally

$$\text{(a)} \qquad f(x_1';\dots;x_n') = f(x_1;\dots;x_n) + \frac{t}{1!}X(f) + \frac{t^2}{2!}X^2(f) + \cdots$$

for sufficiently small t. If we globalize the notation by setting $x' = e^{tX}(x)$, we therefore see that Lie obtains the formula

$$f(e^{tX}(x)) = \left(e^{tX}f\right)(x).$$

On the other hand, we have

$$\text{(b)} \qquad x_i = e^{-tX'}(x_i') \qquad (1 \le i \le n).$$

Indeed, formula (a) generalizes to $X'^j f(x') = \left(e^{tX}X^j f\right)(x)$ for all $j \ge 0$, and we then have

$$\begin{aligned} e^{-tX'}(f(x')) &= \sum_{j\ge 0}(-1)^j\,\frac{t^j}{j!}\,X'^j f(x') = \sum_{j\ge 0}(-1)^j\,\frac{t^j}{j!}\,e^{tX}X^j f(x) \\ &= e^{tX}\sum_{j\ge 0}(-1)^j\,\frac{t^j}{j!}\,X^j f(x) = e^{tX}e^{-tX}f(x) = f(x). \end{aligned}$$

It therefore suffices to apply this formula to the ith coordinate function x_i.

A little later, in Section 18 of Chapter 4, Lie gives a result that could be called *Lie's exponential map theorem.* We will state it using the concepts and notation of Section 16.2.

Consider first of all an analytic n-dimensional K-manifold M. Let $x \in M$, $X \in \mathfrak{X}(M)$, and let $\pi_{X,x} : I \to M$ be the integral curve of the field X such that $\pi_{X,x}(0) = x$. We assume that 1 belongs to the interval I (this is the case, for example, if X is sufficiently small). We then define a map $\Psi : M \times \mathfrak{X}(M) \to M$, at least locally, by setting $\Psi(x; X) = \pi_{X,x}(1)$. We note that if f is an analytic function in a neighborhood of the point $\Psi(x; X) \in M$, we have $f(\Psi(x; X)) = (e^X f)(x)$ (see for example [**38**]). We can then write Lie's result in the following way.

Let $\phi : M \times G \to M$ be a local right operation law of G in M. There exists a map $\exp : L(G) \to G$, *which is a diffeomorphism from a neighborhood of* 0 *in* $L(G)$ *to a neighborhood of* e *in* G, *such that the following diagram commutes:*

$$\begin{array}{ccc} M \times L(G) & \xrightarrow{1_M \times \tau_\phi} & M \times \mathfrak{X}(M) \\ {\scriptstyle 1_M \times \exp}\Big\downarrow & & \Big\downarrow{\scriptstyle \Psi} \\ M \times G & \xrightarrow[\phi]{} & M \end{array}$$

We define the map exp by setting $\exp Y = \pi_{Y,e}(1)$ for all $Y \in L(G)$, since the field Y is completely determined by its value at the point e.

Now let $x \in M$, and let a be an element of G sufficiently close to e; there exists, therefore, a unique element $Y \in L(G)$, close to 0, such that $a = \exp Y$. Set $X = \tau_\phi Y \in \mathfrak{X}(M)$. The commutativity of the diagram gives $\phi(x; a) = \phi(x; \exp Y) = \Psi(x; X) = \pi_{X,x}(1)$, and if f is an analytic function in a neighborhood of the point $\phi(x; a)$, we therefore have $f(\phi(x; a)) = (e^X f)(x)$. In particular, if f is the ith coordinate function and if we set $x'_i = \phi_i(x; a)$, we obtain

$$x'_i = x_i + X(x_i) + \frac{1}{2!} X^2(x_i) + \cdots ;$$

this is the formula that Lie gave in his theorem.

Thus Lie is certainly the first to have had the idea of exponentiating a differential operator (he did not, however, use either the term or the notation of the exponential). The immediate heirs of Lie, in particular Campbell and Poincaré, remarked that Lie's theory evidently suggests that the product of two exponentials e^X and e^Y must be an exponential e^Z. However, since the product of two operators X and Y does not, in general, commute, it is clear that the expressions

$$e^X e^Y = \sum_{m,n \geq 0} \frac{X^m Y^n}{m! n!} \qquad \text{and} \qquad e^{X+Y} = \sum_{k \geq 0} \frac{(X+Y)^k}{k!}$$

are not, in general, equal. But one will note that the difference between $(X+Y)^2$ and $X^2 + 2XY + Y^2$ is the commutator bracket $[X; Y] = XY - YX$.

The clarification and the proof of the assertion that $e^X e^Y = e^Z$ would become the object of numerous works, and Poincaré would use it to give a new proof of Lie's third theorem, at least locally.

The problem posed by the exponential of an operator is therefore the following:

(1) One must show, by placing oneself in the appropriate algebraic framework, that $e^X e^Y$ can be written as e^Z for a certain Z, given by a formal series in iterated brackets of X and Y.

(2) One must show that the formal series defining Z is convergent.
(3) One must give an explicit formula for Z.

Campbell partially solved the problem, but completely neglected part (2). In contrast, Poincaré gave a method to check the convergence of Z, but he was little interested in giving an explicit expression, even less so because such an expression is not necessary for his proof of Lie's third theorem. In fact, none of these authors desired to make precise the algebraic framework in which they worked, and thus the results they left give little satisfaction.

We will briefly explain Campbell's ideas that first appeared in an article of March 11, 1897 [**6**].

Even though, as we have just remarked, Campbell does not mention it, we will here consider the completion $\widehat{\mathcal{A}}$ of the free associative unitary K-algebra $\mathcal{A}$ over a set of two elements x and y; we can therefore identify $\widehat{\mathcal{A}}$ with the algebra of formal series in two indeterminates x and y. The closure $\widehat{L}$ in $\widehat{\mathcal{A}}$ of the free K-Lie algebra over the set of two elements x and y is a Lie subalgebra of $\widehat{\mathcal{A}}$. Let $\mathfrak{m}$ be the maximal ideal of $\widehat{\mathcal{A}}$. If $z \in \mathfrak{m}$, we have $e^z = \sum_{k\geq 0} \frac{z^k}{k!} \in 1 + \mathfrak{m}$.

Now if u and v are two elements of $\widehat{\mathcal{A}}$ and r is a positive integer, to simplify notation we will denote by $\{vu^r\}$ the polynomial expression $\sum_{0\leq i\leq r} u^i v u^{r-i}$. This being given, Campbell set

$$y_0 = y, \qquad y_1 = [y; x], \qquad \text{and for all } r \geq 2, \quad y_r = [y_{r-1}; x];$$

in other words, $y_r = (-1)^r \operatorname{ad}_x^r(y)$, where ad_x^r consists of iterating r times the *adjoint map of* x, that is, the linear map $\operatorname{ad}_x : \mathfrak{g} \to \mathfrak{g}$ defined by $\operatorname{ad}_x(y) = [x; y]$ for all $y \in \mathfrak{g}$ (see [**4**]).

For all $r \geq 1$ and $m \geq 0$ we have

$$x^r y_m + \{y_m x^{r-1}\} x = \{y_m x^r\} \qquad \text{and} \qquad y_m x^r = x^r y_m + \{y_{m+1} x^{r-1}\}.$$

The first equality is obtained by direct calculation, and the second can be shown by induction on r. By using these two formulas and performing induction on r, we then establish the following result:

For all $r \geq 0$ we have

$$\frac{yx^r}{r!} = \sum_{0\leq j\leq r} a_j \left\{ y_j \frac{x^{r-j}}{(r-j+1)!} \right\},$$

where $a_0 = 1$, $a_1 = \frac{1}{2}$, $a_{2n} = (-1)^{n-1} \frac{B_{2n-1}}{(2n)!}$, and $a_{2n+1} = 0$ for $n \geq 1$ (with B_k denoting the kth Bernoulli number).

Taking into account this equality and the associativity and distributivity of the operations of the algebra $\widehat{\mathcal{A}}$, one proves the following fundamental formula:

Let μ be a constant, and set $z = \sum_{k\geq 0} a_k y_k$; we have

$$(1 + \mu y)e^x \equiv e^{x+\mu z} \pmod{\mu^2}. \tag{16.1}$$

On the other hand, one can easily verify the following result by induction:

Let $t \in \mathbb{R}$; we have

$$e^{tx} y e^{-tx} = \sum_{k\geq 0} (-1)^k \frac{t^k}{k!} y_k. \tag{16.2}$$

Campbell proposed using (16.1) and (16.2) to prove what he called *Lie's theorem that forms the foundation of his theory*, that is to say:

Let

$$X_k = \sum_{1\le i\le n} \xi_{ki}(x_1;\dots;x_n)\,\frac{\partial}{\partial x_i} \qquad (1\le k\le r)$$

be r linear operators such that

$$[X_k;X_h] = \sum_{1\le s\le r} c_{khs}X_s.$$

Let

$$X = \sum_{1\le k\le r} \lambda_k X_k \qquad \text{and} \qquad Y = \sum_{1\le k\le r} \mu_k X_k.$$

Given a point $x=(x_1;\dots;x_n)$, set $x' = e^X(x)$ and $x'' = e^{Y'}(x')$, where

$$Y' = \sum_{1\le k\le r} \mu_k X'_k$$

is the operator formed from Y by replacing x by x'. Then, if Y is sufficiently small, there exists an operator

$$Z = \sum_{1\le k\le r} \nu_k X_k$$

such that $x'' = e^Z(x)$.

We will now consider the Lie algebra of linear differential operators, and we can substitute the operators X and Y for the indeterminates x and y that occur in the preceding formulas. In particular Y_h, being by definition an iterated bracket in X and Y, can be written, by hypothesis, as a linear combination of $X_1,\dots,X_r$. Campbell first remarks that, according to Lie, we have

$$Y'(x_i) = e^{tX}Ye^{-tX}(x_i).$$

(This formula probably does not appear in Lie's text, but it is an immediate consequence of the formulas (a) and (b) above. In fact, one must apply (a) to the function $f(x) = Ye^{-tX}(x_i)$ and recall that X' and Y' are obtained from X and Y by making the change of coordinates given by $x' = e^{tX}(x)$.) He concludes, from formula (16.2), that

$$Y' = \sum_{h\ge 0}(-1)^h\frac{t^h}{h!}Y_h$$

and consequently

$$Y' = \sum_{1\le k\le r} \rho_k X_k.$$

Now the formula to be proved,

$$x''_i = e^Z(x_i),$$

is equivalent, if the μ_i *are sufficiently small that we can neglect their squares*, as Campbell says, to

$$(1+Y')(x'_i) = e^Z(x_i),$$

or again, according to the preceding, to

$$(1+k_1X_1+\dots+k_rX_r)(x'_i) = e^Z(x_i)$$

with the k_i very small. Finally, it therefore suffices to prove a formula of type

$$(1+kY)e^X(x_i) = e^Z(x_i)$$

with k very small. According to (16.1) we can set $Z = X + k\overline{Y}$, where $\overline{Y} = \sum_{h\geq 0} a_h Y_h$, and thus Z is indeed a linear combination of $X_1, \ldots, X_r$.

Exactly eight months later, on November 11, 1897, Campbell submitted a second article, carrying the same title [**7**]. In it, he established, after a very long and fastidious calculation and using a rather laborious notation, the formula $e^y e^x = e^w$ by giving w as a function of x and y.

16.4. The universal enveloping algebra according to Poincaré

Let $\mathfrak{g}$ be a K-Lie algebra. The *universal enveloping algebra* $U(\mathfrak{g})$ of $\mathfrak{g}$ is the quotient of the tensorial algebra $T(\mathfrak{g})$ by the two-sided ideal J generated by the set

$$\{x \otimes y - y \otimes x - [x; y] \mid x, y \in \mathfrak{g}\}.$$

The commutator bracket endows the associative algebra $U(\mathfrak{g})$ with a Lie algebra structure, which is sometimes denoted $U(\mathfrak{g})_L$. The *canonical map* $\sigma : \mathfrak{g} \to U(\mathfrak{g})_L$ is the morphism of Lie algebras defined by setting $\sigma = \psi \circ i$, where $i : \mathfrak{g} \to T(\mathfrak{g})$ is the inclusion and $\psi : T(\mathfrak{g}) \to U(\mathfrak{g})$ is the canonical projection.

The algebra $U(\mathfrak{g})$ is filtered by taking the image under ψ of the canonical filtration of the graded algebra $T(\mathfrak{g})$.[2] We can therefore consider the graded algebra $\operatorname{gr} U(\mathfrak{g})$ associated to $U(\mathfrak{g})$. Let $\varphi : T(\mathfrak{g}) \to \operatorname{gr} U(\mathfrak{g})$ be the morphism of graded algebras that in degree n is the restriction of ψ to $T^n(\mathfrak{g})$ composed with the canonical projection $U_n(\mathfrak{g}) \to U_n(\mathfrak{g})/U_{n-1}(\mathfrak{g})$. In addition, let $S(\mathfrak{g})$ be the symmetric algebra of the module $\mathfrak{g}$, and let $\tau : T(\mathfrak{g}) \to S(\mathfrak{g})$ denote the canonical projection. There exists a unique surjective morphism of graded algebras $\omega : S(\mathfrak{g}) \to \operatorname{gr} U(\mathfrak{g})$ such that $\omega \circ \tau = \varphi$.

Now consider the following three assertions:

(1) The morphism $\omega : S(\mathfrak{g}) \to \operatorname{gr} U(\mathfrak{g})$ is an isomorphism.
(2) If $\{x_\lambda\}_{\lambda \in \Lambda}$ is a totally ordered basis for the K-module $\mathfrak{g}$, then the set

$$\{1\} \cup \{x_{\lambda_1} \cdots x_{\lambda_n} \mid (\lambda_1; \ldots; \lambda_n) \in \Lambda^n,\ \lambda_1 \leq \cdots \leq \lambda_n,\ n \geq 1\}$$

is a basis for the K-module $U(\mathfrak{g})$.
(3) The canonical map $\sigma : \mathfrak{g} \to U(\mathfrak{g})$ is injective.

Assertions (1) and (2) are equivalent, and they imply assertion (3). If K is a field of characteristic zero, then these three assertions are equivalent.

The Poincaré–Birkhoff–Witt theorem states that *if $\mathfrak{g}$ is a free K-module* (in particular if K is a field), *then the morphism $\omega : S(\mathfrak{g}) \to \operatorname{gr} U(\mathfrak{g})$ is an isomorphism of graded algebras, and consequently the Lie algebra $\mathfrak{g}$ injects into the Lie algebra $U(\mathfrak{g})_L$.*

---

[2]A *gradation* of an algebra A is given by a family $\{A^n\}_{n\geq 0}$ of submodules of A such that $A = \bigoplus_{n\geq 0} A^n$ and $A^m A^n \subset A^{m+n}$ for all $m \geq 0$ and $n \geq 0$. In addition, $1 \in A^0$ if the algebra is unitary.

An *increasing filtration* of an algebra A is given by a family $\{A_n\}_{n\geq 0}$ of sub-modules of A such that $A_n \subset A_{n+1}$ for all $n \geq 0$, $A = \bigcup_{n\geq 0} A_n$, and $A_m A_n \subset A_{m+n}$ for all $m \geq 0$ and $n \geq 0$. In addition, $1 \in A_0$ if the algebra is unitary.

Every gradation of A induces a *canonical filtration*, defined by setting $A_n = \bigoplus_{0\leq k\leq n} A^k$ for all $n \geq 0$. Conversely, every filtration of A gives rise to an *associated graded algebra*, denoted $\operatorname{gr}(A)$ and defined by setting $\operatorname{gr}(A)^0 = A_0$ and $\operatorname{gr}(A)^n = A_n/A_{n-1}$ for all $n \geq 1$.

If we assume that A is the tensorial algebra over a finite set, then we can identify A with an algebra of non-commutative polynomials. In this case, A^n is the set of homogeneous polynomials of degree n, and A_n is the set of polynomials of degree less than or equal to n.

The content of Section III of Poincaré's article [**33**] can be considered as the first result in the direction of the preceding theorem. In fact, one can interpret Poincaré's ideas as saying that he constructs the image of a section of the projection ψ. By structure transport he thus obtains an algebra that is isomorphic to $U(\mathfrak{g})$, and his construction is done in such a way that it is immediate that $\mathfrak{g}$ is embedded in this algebra; this follows from the fact that this algebra is also the image of a section of the canonical projection τ. We will analyze Poincaré's construction more closely, modernizing the language and the notation but conserving the spirit of the article.

Let $\mathfrak{g}$ be a finite-dimensional Lie algebra over a field K of characteristic zero, and let $\{X_1; \ldots; X_n\}$ be a basis of $\mathfrak{g}$. (For Poincaré, the X_i were elementary differential operators such that their commutator bracket is a linear combination of these operators, but this particular interpretation does not enter into his reasoning, which is entirely algebraic.) The tensorial algebra $T(\mathfrak{g})$ can then be identified with the algebra $\mathcal{A} = K\{X_1; \ldots; X_n\}$ of non-commutative polynomials in the indeterminates $X_1, \ldots, X_n$ with coefficients in K. We denote by $\mathcal{A}^m$ the subspace of elements of $\mathcal{A}$ that are homogeneous of degree m.

Two monomials M and N of $\mathcal{A}$ are called *equipollent*, and we write $M \equiv N$, if they differ only in the order of their factors. This definition extends in an obvious way to two polynomials P and Q of $\mathcal{A}$.

We say that a polynomial $P \in \mathcal{A}$ is *regular* if we can write

$$P = \sum_i \gamma_i(\alpha_{i,1}X_1 + \cdots + \alpha_{i,n}X_n)^{p_i},$$

where $\alpha_{i,k}, \gamma_i \in K$ and $p_i \in \mathbb{N}$. We denote by $\mathcal{R}$ (resp. $\mathcal{R}^m$) the subspace of regular polynomials in $\mathcal{A}$ (resp. of regular homogeneous polynomials of degree m in $\mathcal{A}^m$).

Poincaré then states without proof the following two properties of regular polynomials:

(a) a necessary and sufficient condition for a polynomial to be regular is that, if it contains among its terms a certain monomial, then all monomials equipollent to this one appear in this polynomial with the same coefficient;

(b) among the polynomials equipollent to a given polynomial there is one and only one regular polynomial.

The proof of these assertions is elementary but not completely trivial. The essential idea is to introduce the *symmetrization* of a monomial. Let Λ_m be the set of multi-indices $(\alpha) = (\alpha_1; \ldots; \alpha_n)$ such that $\alpha_i \in \mathbb{N}$ and $\sum_{1 \le i \le n} \alpha_i = m$. If $(\alpha) \in \Lambda_m$, we denote by $X^{(\alpha)}$ the monomial $X_1^{\alpha_1} \cdots X_n^{\alpha_n}$ of $\mathcal{A}^m$. To this monomial we associate the polynomial $\widetilde{X}^{(\alpha)} \in \mathcal{A}$ determined in the following way. We consider m variables X'_j defined thus: for $1 \le j \le \alpha_1$ we set $X'_j = X_1$, and for $\alpha_1 + \cdots + \alpha_{k-1} \le j \le \alpha_1 + \cdots + \alpha_k$ we set $X'_j = X_k$ (if $\alpha_i = 0$ then the term X_i does not appear). We then set

$$\widetilde{X}^{(\alpha)} = \frac{1}{m!} \sum_{\sigma \in \Sigma_m} X'_{\sigma(1)} \cdots X'_{\sigma(m)}$$

(where Σ_m is the symmetric group on m letters). The terms of the polynomial $\frac{m!}{\alpha_1! \cdots \alpha_n!}\widetilde{X}^{(\alpha)}$ constitute all of the distinct monomials of $\mathcal{A}^m$ that are equipollent to the monomial $X^{(\alpha)}$. In addition, one shows that the set $\{\widetilde{X}^{(\alpha)} \mid (\alpha) \in \Lambda_m\}$ is a K-basis for $\mathcal{R}^m$. From this, one can prove the assertions (a) and (b). (For details, see [**18, 44**].)

Then Poincaré introduces a notion that will play the role of the ideal J that we defined at the beginning of this section. We call a polynomial in $\mathcal{A}$ of the form $P(XY - YX - [X;Y])Q$, where $X, Y \in \mathfrak{g}$ and $P, Q \in \mathcal{A}$, a *trinomial product*. We denote by $\mathcal{I}$ the two-sided ideal of $\mathcal{A}$ generated by the trinomial products. Then let $P, Q \in \mathcal{A}$; we say that P *is equivalent to* Q, and we write $P \sim Q$, if we have $P - Q \in \mathcal{I}$.

For technical reasons, we need one more definition. We say that a polynomial $S \in \mathcal{A}$ is a *regular sum* if S is a linear combination of trinomial products that is a regular polynomial; in other words if $S \in \mathcal{I} \cap \mathcal{R}$.

We now arrive at the main result of Section III of Poincaré's article.

THEOREM. *In the algebra $\mathcal{A}$, each polynomial is equivalent to only one regular polynomial.*

PROOF. The proof of existence is done in three steps.

(i) *Let $P, Q \in \mathcal{A}$; if $P \equiv Q$ then there exists $R \in \mathcal{A}$ such that $P - Q \sim R$ and* $\deg R < \deg(P - Q)$.

It is obviously sufficient to check this assertion in the case where P and Q are monomials.

If $P = AXYB$ and $Q = AYXB$, we take $R = A[X;Y]B$.

Now let P and Q be two arbitrary equipollent monomials. There exists a finite sequence of monomials $M_0, M_1, \ldots, M_p, M_{p+1}$ where $M_0 = P$ and $M_{p+1} = Q$ such that, for $1 \le i \le p+1$, M_i differs from M_{i-1} only in the order of two consecutive factors. We have

$$P - Q = \sum_{1 \le i \le p+1} (M_{i-1} - M_i).$$

But according to the preceding, there exist monomials N_i such that $M_{i-1} - M_i \sim N_i$ and $\deg N_i < \deg(M_{i-1} - M_i)$. We set $R = \sum_{1 \le i \le p+1} N_i$.

(ii) *If $P_m \in \mathcal{A}$ is a polynomial of degree $m \ge 2$, then there exists a unique polynomial $P'_m \in \mathcal{R}$ of degree m and a polynomial $P_{m-1} \in \mathcal{A}$ of degree $m-1$ such that $P_m - P_{m-1} \sim P'_m$.*

In fact, according to property (b) of regular polynomials, there exists a unique polynomial $P'_m \in \mathcal{R}$ such that $P_m \equiv P'_m$. Additionally, according to assertion (i) above, there exists a polynomial $P_{m-1} \in \mathcal{A}$ such that $P_m - P'_m \sim P_{m-1}$. Since the equivalence relation $\sim$ is compatible with addition in $\mathcal{A}$, we conclude that $P_m - P_{m-1} \sim P'_m$.

(iii) So let $P \in \mathcal{A}$ be a polynomial of degree n. If $n \le 1$, we obviously have $P \in \mathcal{R}$. We can therefore assume that $n \ge 2$. According to assertion (ii) above, we can construct by induction a sequence of polynomials $\{P_k\}_{n \ge k \ge 1}$ with $P_n = P$ and $\deg P_k = k$, and a sequence of regular polynomials $\{P'_k\}_{n \ge k \ge 1}$ with $\deg P'_k = k$, such that $P_k - P_{k-1} \sim P'_k$ for all k such that $n \ge k \ge 2$. As before we conclude that

$$P_n - P_1 \sim P'_n + \cdots + P'_2,$$

and it follows that

$$P_n \sim P'_n + \cdots + P'_2 + P_1.$$

We set $Q = P'_n + \cdots + P'_2 + P_1$; we have thus constructed a regular polynomial Q such that $P \sim Q$.

The proof of uniqueness is more delicate. Let $P \in \mathcal{A}$, and assume that there exist two polynomials $Q, Q' \in \mathcal{R}$ such that $P \sim Q$ and $P \sim Q'$. We therefore have $Q - Q' \in \mathcal{R}$ and $Q - Q' \in \mathcal{I}$; in other words $Q - Q' \in \mathcal{I} \cap \mathcal{R}$. Everything therefore reduces to the following result. □

PROPOSITION 1. *Every regular sum is identically zero; that is,* $\mathcal{I} \cap \mathcal{R} = 0$.

Poincaré develops a long seven-page argument to prove this assertion. In a certain sense, his calculations are quite close to those one makes to reduce a word in a free group. We will not give the proof of the proposition here (see [**33**], [**44**] or [**18**] for details), but we will show that the previous theorem is indeed one form of what is today called the Poincaré–Birkhoff–Witt theorem.

Let $\pi : \mathcal{A} \to \mathcal{A}/\mathcal{I}$ be the canonical projection; then the obvious isomorphism $\theta : T(\mathfrak{g}) \to \mathcal{A}$ induces an isomorphism $\bar{\theta} : U(\mathfrak{g}) \to \mathcal{A}/\mathcal{I}$ such that $\bar{\theta} \circ \psi = \pi \circ \theta$. Now according to the theorem we can define a surjective linear map $\varphi : \mathcal{A} \to \mathcal{R}$ whose kernel is $\mathcal{I}$; therefore, there exists an isomorphism $\bar{\varphi} : \mathcal{A}/\mathcal{I} \to \mathcal{R}$ such that $\bar{\varphi} \circ \pi = \varphi$. By structure transport with the help of $\bar{\varphi}$, we endow $\mathcal{R}$ with the structure of a unitary associative algebra, so that $\bar{\varphi}$ becomes an isomorphism of algebras. Thus the algebras $U(\mathfrak{g})$ and $\mathcal{R}$ are isomorphic.

In addition, let $\sigma : \mathfrak{g} \to \mathcal{R}$ be the map obtained by composing the obvious inclusion $i : \mathfrak{g} \to \mathcal{A}$ with the map φ. We endow $\mathcal{R}$ with its Lie algebra structure given by the commutator bracket. We have

$$\sigma([X;Y]) = [\sigma(X);\sigma(Y)]$$

for all $X, Y \in \mathfrak{g}$. Indeed,

$$\begin{aligned}\sigma([X;Y]) &= \varphi(i([X;Y])) = \bar{\varphi}\pi([X;Y] - XY + YX + XY - YX)\\ &= \bar{\varphi}\pi(XY - YX) = \bar{\varphi}\pi(X)\,\bar{\varphi}\pi(Y) - \bar{\varphi}\pi(Y)\bar{\varphi}\pi(X)\\ &= [\bar{\varphi}\pi(X);\bar{\varphi}\pi(Y)] = [\varphi i(X);\varphi i(Y)] = [\sigma(X);\sigma(Y)].\end{aligned}$$

Finally, we have seen previously that the set $\{\widetilde{X}^{(\alpha)} \mid (\alpha) \in \Lambda_1\}$ is a basis for $\mathcal{R}^1$. But this set is nothing other than the set $\{X_1;\ldots;X_n\}$, and we have $\sigma(X_i) = X_i$ for all $1 \le i \le n$. Therefore σ is an injection.

16.5. Poincaré's proof of Lie's third theorem

In his third fundamental theorem, Lie asserts that if one considers a family $\{c_{ijk}\}_{i,j,k}$ of r^3 constants from K that are anti-symmetric with respect to the first two indices and that satisfy the relations imposed by the Jacobi identity, then there exists a finite and continuous group of transformations such that the r infinitesimal transformations associated to it admit the elements of the family $\{c_{ijk}\}_{i,j,k}$ as structure constants.

Lie gave several proofs of this result. For example, let $\mathfrak{g}$ be a finite-dimensional Lie algebra whose center is zero; then the adjoint representation makes it possible to identify $\mathfrak{g}$ with a Lie algebra of matrices, and by taking the exponential of these matrices we obtain the desired group. As Poincaré says [**33**], *"this proof leaves nothing to be desired by its simplicity;"* but in the general case *"Lie gave a completely different proof, much less simple, but which however allows the formation of the operators $X_i(f)$ by integration of ordinary differential equations."*

Poincaré therefore proposed giving a new proof of this theorem, while remarking, however, that his method is close to, but more developed than, results obtained previously by F. Schur [**40**] and by J. E. Campbell [**6**]. Poincaré's method consists in making a study of the formula for the exponential, more detailed than that of his predecessors, and he succeeded thanks to the introduction of a suitable algebraic framework, the algebra of symbolic regular polynomials, in other words, the universal enveloping algebra of a Lie algebra. We will explain Poincaré's argument, while lightly modifying the notation and the logical order of the original text.

Let $\mathfrak{g}$ be a finite-dimensional K-Lie algebra having a basis $X_1; \ldots; X_r$. If $T = \sum_{1\leq i\leq r} t_i X_i \in \mathfrak{g}$, we can consider $e^T = \sum_{n\geq 0} \frac{T^n}{n!}$ as an element of the completion of the algebra of regular polynomials $\mathcal{R}$. Then let $T = \sum_{1\leq i\leq r} t_i X_i$ and $V = \sum_{1\leq i\leq r} v_i X_i$ be two elements of $\mathfrak{g}$. In the completion of the free algebra over two indeterminates V and T we have

$$e^V e^T = 1 + \sum_{k\geq 1} \sum_{0\leq n\leq k} \frac{V^{k-n}T^n}{(k-n)!n!}.$$

If V and T commute, this formula shows that we have $e^V e^T = e^{V+T}$; but this is not at all the case in general, so that the problem is to know if $e^V e^T$ is an exponential, that is, if there exists an element $W = \sum_{1\leq i\leq r} w_i X_i \in \mathfrak{g}$ such that $e^V e^T = e^W$. Of course, W is a function $H(V;T)$ of V and T, or if we prefer, $w = (w_1; \ldots; w_r)$ is a function $H(v;t)$ of $v = (v_1; \ldots; v_r)$ and $t = (t_1; \ldots; t_r)$.

THEOREM. *There exists a neighborhood of* 0 *in* $\mathfrak{g}$ *such that, if* V *and* T *belong to this neighborhood, then we can construct a unique element* W *of* $\mathfrak{g}$ *such that* $e^V e^T = e^W$.

More precisely, there exists a neighborhood G *of* 0 *in* $\mathfrak{g}$ *and a map* H, *defined in a neighborhood of* $\{0\} \times \{0\}$ *in* $G \times G$ *with values in* G, *that makes* G *into a Lie groupuscule for which* 0 *is the identity element and* $-V$ *is the inverse of* V *for all* $V \in G$.

In a certain sense, the uniqueness of W relies on the Poincaré–Birkhoff–Witt theorem, and the existence of W relies on the Hausdorff–Campbell–Baker formula.

First of all, according to Poincaré's theory, we can write

$$\sum_{0\leq n\leq k} \frac{V^{k-n}T^n}{(k-n)!n!} = \sum_{1\leq p\leq k} W(p;k),$$

where $W(p;k)$ is a regular polynomial, uniquely determined, and homogeneous of degree p. Set

$$W_0 = 1 \qquad \text{and, for all } p \geq 1, \qquad W_p = \sum_{k\geq 1} W(p;k).$$

We then have

$$e^V e^T = \sum_{p\geq 0} W_p.$$

As we shall see, the component W_1 plays a particularly important role. It is then an interesting exercise to calculate its first terms, strictly using the reduction to regular polynomials. With a bit of patience, we find

$$W_1 = V + T + \frac{1}{2}[V;T] + \frac{1}{12}([V;[V;T]] + [T;[T;V]) + \cdots.$$

Now if W exists, we have

$$e^V e^T = e^W = \sum_{p\geq 0} \frac{W^p}{p!},$$

therefore we must have, for all $p \geq 0$,

$$W_p = \frac{W^p}{p!},$$

and in particular

$$W = W_1.$$

This shows that if W exists, then it is unique.

Poincaré begins the proof of the existence of W by studying the adjoint representation, which he denotes θ. More precisely, let $V = \sum_{1\leq i\leq r} v_i X_i \in \mathfrak{g}$, and define $\theta : \mathfrak{g} \to \mathfrak{g}$ by setting $\theta(T) = [V;T]$ for all $T \in \mathfrak{g}$. We note that θ is completely determined by the matrix $(b_{ij})_{i,j}$ where $\theta(X_i) = \sum_{1\leq j\leq r} b_{ij} X_j$ $(1 \leq i \leq r)$, and that $b_{ij} = \sum_{1\leq k\leq r} c_{kij} v_k$ $(1 \leq i, j \leq r)$. Let $F(\xi)$ be the determinant of $\xi I - \theta$ and $P(\xi)$ be the matrix of cofactors of $\xi I - \theta$.

We assume first of all that θ has r distinct eigenvalues $\theta_1, \ldots, \theta_r$, and we denote by $Y_1, \ldots, Y_r$ the corresponding eigenvectors. We define θ^m inductively by setting $\theta^0 = I$ and $\theta^m = \theta^{m-1} \circ \theta$ for all $m \geq 1$. Let Φ be a function of one variable admitting a series expansion $\Phi(x) = \sum_{m\geq 0} g_m x^m$. We can then consider the operator defined by the formal series $\Phi(\theta) = \sum_{m\geq 0} g_m \theta^m$. (We recognize here a generalization of the expression $z = \sum_{k\geq 0} a_k y_k$ considered by Campbell.)

PROPOSITION 2. *Let Φ be a holomorphic function on a disk D centered at the origin and containing the roots $\theta_1, \ldots, \theta_r$ of $F(\xi)$. Then Φ determines a well-defined operator $\Phi(\theta) : \mathfrak{g} \to \mathfrak{g}$ given by*

$$\Phi(\theta) = \frac{1}{2i\pi} \int_\gamma \frac{P(\xi)\Phi(\theta)}{F(\xi)}\, d\xi,$$

where γ is the boundary of the disk D.

PROOF. Let $T = \sum_{1\leq k\leq r} t'_k Y_k \in \mathfrak{g}$. We have $\Phi(\theta)(T) = \sum_{1\leq k\leq r} t'_k \Phi(\theta_k) Y_k$. A simple linear algebra calculation shows that $(\xi I - \theta)^{-1}(T) = \sum_{1\leq k\leq r} \frac{t'_k}{\xi - \theta_k} Y_k$. Then a residue calculation gives

$$\frac{1}{2i\pi} \int_\gamma (\xi I - \theta)^{-1}(T)\Phi(\xi)\, d\xi = \Phi(\theta)(T),$$

and we finish by invoking the formula for inverting a matrix. □

COROLLARY. *Let $\Phi(\theta)(T) = \sum_{1\leq i\leq r} h_i X_i$. If Φ is an entire function, then the coefficients h_i are entire functions of $v = (v_1; \ldots; v_r)$.*

Finally, again taking the expansion that is found in Campbell's article [**6**], we obtain the following formula, which is well-defined according to the previous proposition.

PROPOSITION 3. *Let $V, W \in \mathfrak{g}$, $\beta \in K$, and consider the entire function $\Phi(x) = \frac{1-e^{-x}}{x}$. Set $Y = \Phi(\theta)(W)$. We then have*

$$e^{V+\beta W} \equiv e^V e^{\beta Y} \pmod{\beta^2}.$$

Note that, since Φ does not vanish in the disk D, we have $\Phi(\theta)^{-1} = \Phi^{-1}(\theta)$, this latter expression being well-defined.

We can now finish the proof of the theorem. Let $V, T \in \mathfrak{g}$; set

$$H(\beta) = e^V e^{\beta T}.$$

We know that $H(\beta)$ is a formal series whose terms are regular polynomials whose coefficients are series in β. We have

$$H(0) = e^V \qquad \text{and} \qquad H(\beta + d\beta) = H(\beta)\, e^{d\beta T}$$

because

$$H(\beta + d\beta) = e^V e^{(\beta+d\beta)T} = e^V e^{\beta T} e^{d\beta T} = H(\beta)\, e^{d\beta T}.$$

To solve this equation, we look for a vector $W = W(\beta) = \sum_{1\leq i\leq r} w_i(\beta) X_i$ such that $H(\beta) = e^W$ and $H(\beta + d\beta) = e^{W+dW}$; in addition, we must have $W(0) = V$. By substituting into the equation, we must have

$$e^{W+dW} = e^W e^{d\beta T}.$$

Since $d\beta$ is infinitely small, Proposition 3 says that this condition is satisfied if

$$d\beta T = \Phi(\eta)(dW),$$

where $\eta = \mathrm{ad}_W$ and $\Phi(x) = \frac{1-e^{-x}}{x}$. We therefore deduce the differential equation

$$\frac{dW(\beta)}{d\beta} = \frac{\mathrm{ad}_{W(\beta)}}{1 - e^{-\mathrm{ad}_{W(\beta)}}}(T).$$

If β is in a small neighborhood of 0, this equation admits a solution with the initial condition $W(0) = V$. By taking V and T sufficiently small, we can make it so that 1 belongs to this neighborhood. Thus the vector $W = W(1)$ satisfies the relation

$$e^V e^T = e^W.$$

Poincaré then remarks that if the equation $F(\xi) = 0$ has multiple roots, then the preceding results remain true as long as we introduce a few derivatives of Φ.

The precise version of the theorem is not explicit in Poincaré's article, but there is no doubt that he had in his mind the idea of canonical coordinates: to sufficiently small elements of $\mathfrak{g}$, which can be written as $U = \sum_{1\leq i\leq r} u_i X_i$, we associate the set G of points of K^r whose coordinates are $(u_1; \ldots; u_r)$.

The theorem therefore associates a Lie groupuscule G to a Lie algebra $\mathfrak{g}$. To obtain Lie's third fundamental theorem, it remains to check that the Lie algebra of G is isomorphic to $\mathfrak{g}$. But we can identify $L(G)$ with $\mathfrak{g}$ as vector spaces (because the differential of the inclusion of G into $\mathfrak{g}$ is the identity). We will denote the bracket of $L(G)$ by $[\,;\,]_{L(G)}$ and that of $\mathfrak{g}$ by $[\,;\,]_{\mathfrak{g}}$.

Let $X, Y \in L(G)$. In the universal enveloping algebra $U(L(G))$ we have, modulo the ideal of the algebra $\widehat{\mathcal{A}}(L(G))$ generated by the terms of degree greater than 2,

$$[X; Y]_{L(G)} = \log(e^X e^Y e^{-X} e^{-Y}),$$

where log denotes the inverse of exp, defined by the series

$$\log Z = \sum_{n\geq 1} (-1)^{n-1} \frac{(Z-1)^{n-1}}{n}.$$

Indeed, by carrying out the series multiplication we obtain

$$e^X e^Y e^{-X} e^{-Y} = 1 + (XY - YX) + \cdots,$$

therefore

$$\log(e^X e^Y e^{-X} e^{-Y}) = XY - YX + \cdots .$$

And in $U(L(G))$ we have $XY - YX = [X;Y]_{L(G)}$.

On the other hand, by the definition of multiplication in the groupuscule G, we have

$$e^X e^Y e^{-X} e^{-Y} = e^{H(X;Y)} e^{H(-X;-Y)} = e^{H(H(X;Y);H(-X;-Y))},$$

and by a previous calculation, we have

$$\begin{aligned} H(H(X;Y);H(-X;-Y)) &= H(X;Y) + H(-X;-Y) + \cdots \\ &= X + Y + \frac{1}{2}[X;Y]_{\mathfrak{g}} - X - Y + \frac{1}{2}[X;Y]_{\mathfrak{g}} + \cdots = [X;Y]_{\mathfrak{g}} + \cdots . \end{aligned}$$

Therefore we also have $\log(e^X e^Y e^{-X} e^{-Y}) = [X;Y]_{\mathfrak{g}}$.

It follows from this that the structure constants of $L(G)$ are the same as those of $\mathfrak{g}$, and thus the converse of Lie's third theorem is established.

16.6. Poincaré's inheritors

There would be several trails to follow in order to study the evolution of certain notions that are found in Poincaré's work on continuous groups. We will only examine two of them: that of the exponential formula and that of the enveloping algebra.

(1) A *group law* is a formula that makes it possible to define a group structure on a module, on which one already knows how to carry out certain calculations.

The prototype of a group law is the *Hausdorff–Campbell–Baker formula*; this is an element H of the Lie algebra $\widehat{L}(\{U;V\})$ that admits a series expansion $H = \sum_{r,s\geq 0} H_{r,s}$, where $H_{r,s}$ is the component of H having bidegree $(r;s)$. It makes possible the definition of an associative multiplication starting from a Lie algebra bracket.

We have already said that J. E. Campbell was the one to begin the study of this element in an article from March 11, 1897, by calculating $\sum_{s\geq 0} H_{1,s}$. Then in the article of November 11 of the same year, he indicated an inductive procedure to pass from $\sum_{s\geq 0} H_{r,s}$ to $\sum_{s\geq 0} H_{r+1,s}$ for all $r \geq 0$.

We have also seen that H. Poincaré proved the existence and uniqueness of H but was not concerned with a way of effectively calculating it; in fact, he only used $H_{1,1}$ in his proof of Lie's theorems.

The study of H was pursued at the beginning of 1905 by H. F. Baker [**2**], who gave, in a somewhat confused manner, a procedure for calculating H_{n+1} starting from H_n, where H_k is the homogeneous degree k component of H, that is, $H_k = \sum_{r+s=k,\, r,s\geq 0} H_{r,s}$. For this, he used a substitution operator that behaves like a derivation.

In 1906, F. Hausdorff [**20**] took up this operator and considerably simplified Baker's arguments to obtain an inductive procedure for calculating H_n.

Finally, in 1947, E. B. Dynkin [**14**] gave an explicit formula for calculating H_n for all $n \geq 0$, and in 1963, M. Lazard [**25**] studied the p-adic behavior of the coefficients of Dynkin's formula.

Returning to group laws, it was this same M. Lazard who, in 1955, made a study of them [**24**]. For this, he introduced the notion of an *analyzer*, which is a particular case of the more general notion of a *compositor*, this latter being nothing other than the first example of an *operad* (see [**19, 31**] for details on this notion).

(2) We say that a K-Lie algebra $\mathfrak{g}$ is *special* [**13**] if the canonical map $\sigma : \mathfrak{g} \to U(\mathfrak{g})$ is injective. Poincaré therefore proved that if $\mathfrak{g}$ is a finite-dimensional Lie algebra over a field of characteristic zero, then $\mathfrak{g}$ is special. This same result was proved in 1936, with different methods, by G. Birkhoff [**3**] and independently by E. Witt [**46**]. Neither of these authors cites the other, nor do they seem to know Poincaré's result.

The first session of the Seminar Sophus Lie was given by P. Cartier on November 9, 1954 [**41**], and had as its subject *the Poincaré–Birkhoff–Witt theorem.* This was the first time that these three names were united and that Poincaré was given due credit for his discovery of the enveloping algebra. The theorem's name was definitively adopted with the appearance in 1960 of Chapter I of N. Bourbaki's book on Lie groups and Lie algebras [**4**]. His method of proof is that of Cartan–Eilenberg [**11**], which claims to have come from an idea of K. Iwasawa [**21**].

An example of a Lie algebra that is not special was given by A. I. Širšov in 1953 [**43**]; then a simpler example was proposed by Cartier in 1958 [**12**]. Throughout the second half of the 20th century, several efforts were made to look for sufficient conditions, on either K or the structure of $\mathfrak{g}$, for the Lie algebra $\mathfrak{g}$ to be special (see [**18**] for references on this subject).

16.7. From Lie to Leibniz

In the course of his research on problems of periodicity in algebraic K-theory, J. L. Loday discovered in 1989 the *Leibniz algebras* [**27**], and a little later, the *dialgebras* [**28**]. Numerous applications have been made of these algebras, for example in differential geometry [**23**], and they have great interest in operad theory [**29**].

Leibniz algebras generalize Lie algebras in the sense that their bracket is not necessarily anti-symmetric and it satisfies only a particular form of the Jacobi identity. As for dialgebras, they generalize associative algebras because they are modules endowed with two associative operations that satisfy the compatibility conditions. On the other hand, dialgebras play the role for Leibniz algebras that associative algebras play for Lie algebras.

More precisely, we have the following definitions.

A *Leibniz algebra* is a K-module $\mathfrak{g}$ endowed with an operation $[\ ;\ ] : \mathfrak{g} \otimes \mathfrak{g} \to \mathfrak{g}$, not necessarily anti-symmetric, that satisfies the *Leibniz identity*, which is a variant of the Jacobi identity; namely for all $x, y, z \in \mathfrak{g}$ we have

$$[x; [y; z]] = [[x; y]; z] - [[x; z]; y].$$

To give a somewhat general example, consider a non-commutative associative K-algebra A, an A-bimodule M, and a morphism of A-bimodules $f : M \to A$. (We could take $A = T(V)$, where V is a K-module, $M = T(V) \otimes V \otimes T(V)$, and f is concatenation.) Then the K-module M, equipped with the bracket defined by setting $[x; y] = xf(y) - f(y)x$ for all $x, y \in M$, is a Leibniz algebra that is not, in general, a Lie algebra.

The Leibniz algebras form a category denoted $\mathcal{L}eib$, and there is a pair of adjoint functors

$$\mathcal{AL} \overset{\longleftarrow}{\longrightarrow} \mathcal{L}eib$$

because a Lie algebra is obviously a Leibniz algebra, and conversely to every Leibniz algebra $\mathfrak{g}$ we can associate the Lie algebra $\mathfrak{g}_{\mathrm{Lie}} = \mathfrak{g}/\mathfrak{g}^{\mathrm{ann}}$, where $\mathfrak{g}^{\mathrm{ann}}$ is the two-sided ideal of $\mathfrak{g}$, in the sense of Leibniz algebras, generated by the set $\{[x;x] \mid x \in \mathfrak{g}\}$.

A *dialgebra* is a K-module D endowed with two associative operations

$$\dashv : D \otimes D \to D \qquad \text{and} \qquad \vdash : D \otimes D \to D$$

that satisfy the following compatibility conditions: for all $x, y, z \in D$, we have

$$(x \dashv y) \dashv z = x \dashv (y \vdash z), \qquad (x \vdash y) \dashv z = x \vdash (y \dashv z)$$
$$\text{and} \qquad (x \dashv y) \vdash z = x \vdash (y \vdash z).$$

The dialgebras form a category denoted $\mathcal{D}ias$, and there is a new pair of adjoint functors

$$\mathcal{A}ssoc \leftrightarrows \mathcal{D}ias$$

because an associative algebra is trivially a dialgebra, and conversely to every dialgebra D we can associate the associative algebra $D_A = D/D^{\mathrm{ann}}$, where D^{ann} is the ideal of D, in the sense of dialgebras, generated by the set $\{x \dashv y - x \vdash y \mid x, y \in D\}$. In addition, we have a functor

$$\mathcal{D}ias \longrightarrow \mathcal{L}eib$$

that associates to the dialgebra $(D; \dashv; \vdash)$ the Leibniz algebra $(D; [\ ;\ ])$, where the bracket is defined by setting $[x;y] = x \dashv y - y \vdash x$ for all $x, y \in D$.

But we can go further: the preceding functor admits a left adjoint [**28**]

$$\mathrm{Ud} : \mathcal{L}eib \longrightarrow \mathcal{D}ias.$$

We thus obtain a diagram of categories and pairs of adjoint functors, which commutes up to isomorphism [**29**]

$$\begin{array}{ccc} \mathcal{D}ias & \leftrightarrows & \mathcal{A}ssoc \\ \downarrow\uparrow & & \downarrow\uparrow \\ \mathcal{L}eib & \leftrightarrows & \mathcal{AL} \end{array}$$

and which thus shows a lovely correspondence between the theory of Lie algebras and the theory of Leibniz algebras (see [**17**] for more details on these notions).

For example, if $\mathfrak{g}$ is a Leibniz algebra, then $\mathrm{Ud}(\mathfrak{g})$ is the *enveloping dialgebra of* $\mathfrak{g}$, and we have a Poincaré–Birkhoff–Witt-type theorem for Leibniz algebras.

THEOREM ([**1**]). *Let $\mathfrak{g}$ be a Leibniz algebra such that the K-module subjacent to $\mathfrak{g}$ is free. Then there exists an isomorphism of graded dialgebras*

$$\Omega : S(\mathfrak{g}_{\mathrm{Lie}}) \otimes \mathfrak{g} \to \mathrm{gr}\,\mathrm{Ud}(\mathfrak{g}).$$

The study of the pair of adjoint functors that connect the categories $\mathcal{D}ias$ and $\mathcal{A}ssoc$ leads to the following conjecture [**1**]:

CONJECTURE. *Let $\mathfrak{h}$ be a Lie algebra. Assume that there exists a Leibniz algebra $\mathfrak{g}$ such that $\mathfrak{g}$ is a free K-module and that $\mathfrak{g}_{\mathrm{Lie}}$ is isomorphic to $\mathfrak{h}$. Then there exists an isomorphism of graded associative algebras $S(\mathfrak{h}) = \mathrm{gr}\,U(\mathfrak{h})$.*

We would thus obtain a new case of a special Lie algebra, where the condition would deal with the structure of the Lie algebra and not with the ring of coefficients.

Let us point out, without entering into details, that the structure of Leibniz algebras and of dialgebras has been much studied recently, which has led to generalizations that have given a glimpse of a general theory still to be unearthed (see [**16, 29, 30, 39**]).

In conclusion, we can also ask the following question, which currently seems to be completely open. Consider the diagram of categories and functors

$$\begin{array}{ccc} \mathcal{GL} & & \\ \downarrow\uparrow & & \\ \mathcal{AL} & \underset{\longrightarrow}{\longleftarrow} & \mathcal{L}eib \end{array}$$

Will the *coquecigrues* [**27**], those mocking birds of popular legend,[3] come make their nest at the top right of this diagram? In other words, do there exist objects that will play a role for Leibniz algebras similar to that played by Lie groups for Lie algebras? One could then imagine that a pair of vertical arrows on the right of the diagram would be a kind of "Leibnizification" of Lie's theorems, and one could dream of seeing Poincaré's shadow floating there.

* * * * *

The author warmly thanks André Haefliger, with whom he had several conversations that made possible a better understanding of the theories of Poincaré and Campbell, especially on the subject of the exponential function.

Bibliography

[1] M. Aymon and P. P. Grivel, *Un théorème de Poincaré–Birkhoff–Witt pour les algèbres de Leibniz*, Communications in Algebra **31** (2003), no. 2, pp. 527–544.

[2] H. F. Baker, *Alternants and continuous groups*, Proc. London Math. Soc. (2) **III** (1905), pp. 24–47.

[3] G. Birkhoff, *Representability of Lie algebras and Lie groups by matrices*, Ann. of Math. **38** (1937), no. 2, pp. 526–532.

[4] N. Bourbaki, *Groupes et algèbres de Lie*, Chapter I. Hermann, Paris, 1960.

[5] N. Bourbaki, *Groupes et algèbres de Lie*, Chapters II and III. Hermann, Paris, 1972.

[6] J. E. Campbell, *On a law of combination of operators bearing on the theory of continuous transformation groups*, Proc. London Math. Soc. **XXVIII** (1897), pp. 381–390.

[7] J. E. Campbell, *On a law of combination of operators (Second paper)*, Proc. London Math. Soc. **XXIX** (1898), pp. 14–32.

[8] É. Cartan, *Le troisième théorème fondamental de Lie*, Comptes Rendus Acad. Sci. Paris **190** (1930), pp. 914–916 and 1005–1007.

[9] É. Cartan, *La théorie des groupes finis et continus et l'analysis situs*, Mémorial Sci. Math. **XLII**, 1930, Gauthier-Villars.

[10] É. Cartan, *Œuvres complètes*, part I. Éditions du CNRS, 1984.

[11] H. Cartan and S. Eilenberg, *Homological Algebra*, Princeton University Press, 1956.

[12] P. Cartier, *Remarques sur le théorème de Birkhoff–Witt*, Annali della Scuola Norm. Sup. di Pisa (série III) **XII** (1958), pp. 1–4.

[13] P. M. Cohn, *Universal Algebra*, Harper and Row, 1965.

[14] E. B. Dynkin, *Calculation of the coefficients in the Campbell–Hausdorff formula*, Dokl. Akad. Nauk **LVII** (1947), pp. 323–326; translated in: Dynkin et al., *Selected papers of E. B. Dynkin with commentary*, Amer. Math. Soc., 2000.

[3] Translator's Note: The "coquecigrue" is an imaginary bird—perhaps intended to be a mix of a rooster (*"coq"*), a swan (*"cygne"* or *"cigogne"*), and a crane (*"grue"*)—but in any case fabulous and unbelievable. Loday's use of the word in the theory of Leibniz algebras has been adopted in English, as well, to mean precisely the kind of object necessary to complete the diagram.

[15] R. Godement, *Introduction à la théorie des groupes de Lie*, volumes I and II. Publications mathématiques de l'Université Paris VII, 1982.
[16] F. Goichot, *Un théorème de Milnor–Moore pour les algèbres de Leibniz*: pp. 111–133 of *Dialgebras and Related Operads*, Lecture Notes in Mathematics **1763**, Springer, 2001.
[17] P. P. Grivel, *Une introduction aux algèbres de Leibniz*, Preprint, 2001.
[18] P. P. Grivel, *Une histoire du théorème de Poincaré–Birkhoff–Witt*, Expositiones Mathematicae **22** (2004), pp. 145–184.
[19] P. P. Grivel, *L'origine des opérades*, In preparation.
[20] F. Hausdorff, *Die symbolische Exponentialformel in der Gruppentheorie*, Leipziger Ber. **LVIII** (1906), pp. 19–48.
[21] K. Iwasawa, *On the representation of Lie algebras*, Japan. J. Math. **19** (1948), pp. 405–426.
[22] N. Jacobson, *Rational methods in the theory of Lie algebras*, Ann. of Math. **36** (1935), no. 4, pp. 875–881.
[23] Y. Kosmann-Schwarzbach, *From Poisson algebras to Gerstenhaber algebras*, Ann. Ins. Fourier Grenoble **46** (1996), no. 5, pp. 1243–1274.
[24] M. Lazard, *Lois de groupes et analyseurs*, Annales École Norm. Sup. Paris **72** (1955), pp. 299–400.
[25] M. Lazard, *Quelques calculs concernant la formule de Hausdorff*, Bull. Soc. Math. France **XCI** (1963), pp. 435–451.
[26] S. Lie, *Theorie der Transformationsgruppen* (in three volumes), Leipzig, 1888, 1890, 1893.
[27] J.-L. Loday, *Une version non commutative des algèbres de Lie: les algèbres de Leibniz*, L'enseignement Mathématique **39** (1993), pp. 269–293.
[28] J.-L. Loday, *Algèbres ayant deux opérations associatives (digèbres)*, Comptes Rendus Acad. Sci. Paris **321** série I (1995), pp. 141–146.
[29] J.-L. Loday, *Dialgebras*: pp. 7–66 of *Dialgebras and Related Operads*, Lecture Notes in Mathematics **1763**, Springer, 2001.
[30] J.-L. Loday and M. O. Ronco, *Hopf algebra of the planar binary trees*, Adv. in Math. **139** (1998), pp. 293–309.
[31] M. Markl, S. Shnider, and J. Stasheff, *Operads in Algebra, Topology and Physics*, Mathematical Survey and Monographs **96**, Amer. Math. Soc. 2002.
[32] H. Poincaré, *Sur les groupes continus*, Comptes Rendus Acad. Sci. Paris **128** (1899), pp. 1065–1069.
[33] H. Poincaré, *Sur les groupes continus*, Cambridge Philosophical Transaction **18** (1899), pp. 220–255.
[34] H. Poincaré, *Quelques remarques sur les groupes continus*, Rendiconti del Circolo Matematico di Palermo **15** (1901), pp. 213–260.
[35] H. Poincaré, *Nouvelles remarques sur les groupes continus*, Rendiconti del Circolo Matematico di Palermo **125** (1908), pp. 261–321.
[36] H. Poincaré, *Analyse des travaux de H. Poincaré sur les groupes continus faite par lui-même*, Acta Mathematica **39** (1913), pp. 90–92.
[37] H. Poincaré, *Œuvres complètes*, volume III, Gauthier-Villars, 1965.
[38] M. Postnikov, *Leçons de géométrie: groupes et algèbres de Lie*, Mir, Moscow, 1985.
[39] M. O. Ronco, *A Milnor–Moore theorem for dendriform Hopf algebras*, Comptes Rendus Acad. Sci. Paris Série I Math. **332** (2000), pp. 109–114.
[40] F. Schur, *Über den analytischen Character der eine endliche continuierliche Transformationsgruppe darstellende Funktionen*, Math. Ann. **XLI** (1893), pp. 509–538.
[41] Séminaire Sophus Lie, *Théorie des algèbres de Lie, topologie des groupes de Lie*, École Norm. Sup. Paris, 1955.
[42] J.-P. Serre, *Lie Algebras and Lie Groups*, Lecture Notes in Mathematics **1500**, Springer, 1992.
[43] A. I. Širšov, *On the representation of Lie rings as associative rings*, Uspekhi Matem. Nauk (N.S.) **8** (1953), no. 5, pp. 173–175. (Russian)
[44] T. Ton That and T. D. Tran, *Poincaré's proof of the so-called Birkhoff-Witt theorem*, Revue d'histoire des mathématiques **5** (1999), pp. 249–284.
[45] H. Weyl, *The Structure and Representation of Continuous Groups*, Lithographed notes by R. Brauer of Weyl's lectures at the Institute for Advanced Study, Princeton, 1934–1935.
[46] E. Witt, *Treue Darstellung Liescher Ringe*, J. Reine Angew. Math. **177** (1937), pp. 152–160.

CHAPTER 17

The Poincaré Group

Michel Le Bellac

After examining Poincaré's contribution to the genesis of the theory of special relativity, I will explain the importance of the Poincaré group as a fundamental symmetry group of physics. I will describe the possible (or impossible) extensions of the Poincaré group, and finally I will give an argument showing why Galilean electromagnetism was not discovered until after special relativity.

17.1. Introduction

Before getting into the thick of things, I would like to say a word about the origin of the expression "Poincaré group". This group is the set of displacements of space-time, that is, the inhomogeneous Lorentz group: translations, rotations, and pure Lorentz transformations, or "boosts",[1] which is today thought to be the fundamental invariance group of physics, at least in a local region of space-time. In 1939, in a fundamental article, Wigner was the first to classify the irreducible representations of this group, essential for relativistic quantum physics, and it was he who, with Bargman and Wightman, christened it the "Poincaré group" [**28**]. This was in homage to the work that had, in particular, identified the structure of the group of homogeneous transformations of space-time, the rotations and the boosts, which form the homogeneous Lorentz group. As we shall see in Section 17.3, it was Poincaré who gave Lorentz's name to the group that he himself had discovered!

The outline of this article will be the following. In the first part, I will strive to characterize Poincaré's own contribution to the foundations of special relativity, before passing in the second part to the properties of the group of Lorentz transformations. Then I will summarize Wigner's article, and its fundamental consequence: elementary particles are characterized by their mass and their spin. I will next examine the possible extensions of the Poincaré group: such extensions were thought for several years to be excluded, by reason of an impossibility theorem due to Coleman and Mandula [**2**]. It was Haag, Łopuszański, and Sohnius who discovered in 1975 how to get around this theorem by showing that the most general extension of the Poincaré group is the one that includes supersymmetry, introduced one year earlier [**27**]. Finally I will indulge myself in a bit of historical revision, by explaining how the physicists of the end of the 19th century could have

[1] Terminology borrowed from the "booster" of a rocket.

written down a Galileo-invariant electromagnetism, instead of passing directly to Lorentz invariance.

17.2. Genesis of the Poincaré group (1895/1905)

In 1895, motivated by the failure of various attempts to demonstrate the Earth's motion with respect to the "aether", Lorentz endeavored to obtain a formal equivalence between the descriptions of electromagnetic phenomena in the various possible reference frames [**4, 25**]. For Lorentz, the Maxwell equations of electromagnetism should be written in an absolute referential frame, that of the aether. But the Earth moves in this reference frame, and he therefore had to use a Galilean transformation of velocity v, written for simplicity with a single spatial dimension

$$x' = x - vt, \qquad t' = t. \tag{17.1}$$

For Lorentz, this transformation is the only one that has physical meaning. However, it is insufficient, because it is, for example, immediate to check that the equation of electromagnetic wave propagation is not invariant under this transformation. Lorentz therefore added two supplementary transformations, which had for him a non-physical (fictive) character, but which made it possible for him to recover the correct form of the equations at the order v/c, where c is the speed of light in a vacuum.

(1) He carried out a transformation of the electric and magnetic fields, which will be written later in (17.42).
(2) He introduced a "local time" t' by writing, instead of (17.1),

$$x' = x - vt, \qquad t' = t - \frac{vx}{c^2}. \tag{17.2}$$

This double transformation leaves the Maxwell equations invariant to first order in v/c.

Lorentz thus placed the "real" states of the initial physical system in correspondence with the states of a system fictively at rest in the aether, where Maxwell's equations have the correct form. The introduction of local time ensures the invariance at order v/c of these equations, but, as I have already mentioned, Lorentz did not attribute any physical meaning to it; he saw in it a simple calculational trick. In addition—but he did not make this observation, and if he had it would hardly have bothered him—the transformation (17.2) does not define a group law! In order to take into account the negative results of the Michelson–Morley experiment, he moreover had to add the Lorentz–Fitzgerald contraction of lengths, which is an effect of order $(v/c)^2$. For Lorentz, this contraction resulted from the fact that the effect of Earth's motion on the forces of molecular cohesion that determine the dimensions of solids is the same as its effect on the electrostatic forces.

Starting from 1895, Poincaré—in contrast with Lorentz, who believed only in an approximate compensation of the Earth's motion—was firmly convinced of the validity of a principle of relativity and began thinking about the invariance of physics under a change of reference frame. At that time, all physicists were convinced of the validity of a (Galilean) principle of relativity for mechanical phenomena, but it was not the same for electromagnetic phenomena. At the end of the 19th century, Poincaré was the only theoretical physicist to envision the extension of such a principle to electromagnetic phenomena. Several citations show that his reflections became more and more precise. In 1895, he wrote [**18**]:

> *"It is impossible to make manifest the absolute motion of matter with respect to the aether: all that one can demonstrate is the motion of weighty matter with respect to weighty matter..."*

Then in 1901 [**20**]:

> *"...I consider it very likely that optical phenomena depend only on the relative motions of present bodies, light sources, or optical devices..."*

In 1900, he also gave a physical interpretation of Lorentz's local time [**19**]:

> *"I assume that observers placed at different points adjust their watches with the help of light signals; they seek to correct the transmission time of these signals, but being unaware of the translation motion which they are following, and consequently believing that the signals are transmitted at the same speed in the two directions, they are limited to passing their observations back and forth, by sending a signal from A to B, then another from B to A. The local time t' is the time marked by the watches so adjusted."*

He writes in *Science and Hypothesis* [**21**]:

> *"There is no absolute time. To say that two durations are equal is an assertion that has no meaning of itself, and can only acquire meaning by convention."*

In 1904 he participated in the International Congress of Arts and Sciences that was held in St. Louis, and there he gave a much remarked-upon (and remarkable!) presentation, where he defined the principle of relativity thusly:

> *"... according to which the laws of physical phenomena must be the same for a stationary observer as for one carried along in a uniform motion of translation, so that we have no means, and can have none, of determining whether or not we are being carried along in such a motion."* [**22**]

This quotation shows, obviously, that Poincaré took for granted that no physical, mechanical, or optical experiment would make it possible to detect a uniform motion of translation with respect to the aether.

On May 27, 1904, an article of Lorentz appeared in the *Proceedings of the Royal Netherlands Academy of Arts and Sciences*: "Electromagnetic phenomena in a system moving with any velocity smaller than that of light" [**16**]. In this article, Lorentz sought to make the equations of electromagnetism invariant at all orders of v/c, and not only at the first order, and he proposed the following transformation (which I write in modern notation)

$$x' = \ell\gamma(x - vt), \qquad y' = \ell y, \qquad z' = \ell z, \qquad t' = \ell\gamma(t - \beta x/c), \tag{17.3}$$

with

$$\beta = v/c, \qquad \gamma = (1 - \beta^2)^{-1/2},$$

and he gave complicated arguments in favor of the choice $\ell = 1$. The Lorentz transformation was born![2] Over the course of the 1904/1905 winter,[3] Poincaré wrote a letter to Lorentz, commenting on this article. This letter contains an essential observation: in order for the transformations to form a group that respects invariance under rotation, it is necessary that $\ell = 1$.

17.3. The 1905 articles and the Lorentz group

On June 5, 1905, Poincaré published an article, five pages in length, in the *Comptes Rendus de l'Académie des Sciences* [**23**], which summarized a long article (57 pages), "Sur la dynamique de l'électron" [On the dynamics of the electron], which was received by the journal on July 23, 1905, and published in January 1906 [**24**]. In the first article, Poincaré states in the clearest way possible his conception of the principle of relativity: *"It seems that this impossibility to demonstrate experimentally the Earth's motion is a general law of nature; we are naturally carried to admit this law, which we will call the* Postulate of Relativity, *and to admit it without restriction."* He applied this principle to electromagnetism: *"... the equations of an electromagnetic medium are not changed by certain transformations which we will call* Lorentz transformations; *two systems, one immobile, and the other undergoing translation, thus become exact images of each other."* Poincaré had the insight that transformations should form a group, since reciprocity implies the existence of an inverse transformation. On June 30, 1905, Einstein sent to the *Annalen der Physik* his fundamental article "On the electrodynamics of moving bodies" [**6**].

Poincaré's second article is long and dense, and I will restrict myself to the following point: why, according to Poincaré, must we have $\ell = 1$ in (17.3)? In Section 4 of his article, entitled "The Lorentz group", Poincaré writes: *"We are led to envision a continuous group that* we will call the Lorentz group *and that admits as infinitesimal transformations*

(1) *The transformation T_0, permutable to all the rest"* (dilations).
(2) *"The transformations T_x, T_y, T_z"* (boosts along the axes Ox, Oy, Oz).
(3) *"The three rotations $[T_x, T_y], [T_y, T_z], [T_z, T_x]$*

$$[T_x, T_y] = x\frac{\partial}{\partial y} - y\frac{\partial}{\partial x}.\text{"} \tag{17.4}$$

Poincaré, an expert in Lie groups,[4] knew perfectly well that the commutator of the infinitesimal generators of the boosts along the axes Ox and Oy is a rotation around the axis Oz. Now comes the crucial point: *"if we turn the system by* 180 *degrees around Oy, we must again find a transformation. But this rotation amounts to changing the sign of x, x', z, z'"*

$$x' = \ell\gamma(x + vt), \qquad y' = \ell y, \qquad z' = \ell z, \qquad t' = \ell\gamma(t + \beta x/c). \tag{17.5}$$

On the other hand, the inverse substitution of the original transformation

$$x' = \gamma(x + vt)/\ell, \qquad y' = y/\ell, \qquad z' = z/\ell, \qquad t' = \gamma(t + \beta x/c)/\ell \tag{17.6}$$

---

[2]See, however, Voigt [**26**], who understood that Maxwell's equations are not Galilean invariants. Since before 1900, Lorentz and Larmor had both written pieces of the Lorentz transformation, but it would have to wait for Poincaré and Einstein to correctly express the transformation law of the electromagnetic current and that of the sum of velocities.

[3]Poincaré had the (bad) habit of not dating his letters, and therefore we do not have a more precise date.

[4]EDITOR'S NOTE: See Chapter 16.

"must also be a transformation. We must therefore have $\ell = 1$." This is the act that gives birth to the Lorentz group and its Lie algebra.

This article contains a number of other important results. Even though most of them had already been written by other authors, Lorentz in particular, Poincaré was the first to give a coherent and exhaustive list of them. His article contains:

- the composition law for velocities (correct, although the one written by Lorentz in 1904 was not);
- the transformation laws of the electric and magnetic fields $\{\vec{E}, \vec{B}\}$ and the potentials $\{\varphi, \vec{A}\}$, as well as that of the currents $\{\rho, \vec{\jmath}\}$ (same remark as above);
- four-dimensional notation and the Minkowski metric $x^2 + y^2 + z^2 - t^2$;
- the relativistic invariants $\{\vec{E}^2 - \vec{B}^2, \vec{E} \cdot \vec{B}\}$;
- a variational principle for electromagnetism;
- the necessity to take into account the Lorentz symmetry in all physical phenomena, including gravitation.

For today's physicist, the most remarkable result is probably the discovery of the group structure of Lorentz transformations and the explicit writing of its Lie algebra. It must nevertheless be recognized that this point did not present an enormous amount of interest to the physicists of the time, and that very few understood the importance of this result. In addition, the article followed in line with those of Lorentz and other theoreticians of electromagnetism. Poincaré deliberately remained within the framework defined by his predecessors; he studied in tedious detail electromagnetic models of the electron that hardly hold any more interest today; and he did not call into question the existence of aether, as Einstein did three weeks later. For Poincaré, an observer in motion acts *as though* the speed of light were constant. The coordinate transforms give *apparent* spatio-temporal relations for an observer in motion with respect to the aether. The aether defines a true space-time, and the Lorentz transformation an *apparent* space-time for an observer in motion [**3**, pp. 32–34], [**5**]. From a strictly practical point of view,[5] the results of Lorentz and Poincaré are equivalent to those of Einstein for dynamics and electromagnetism. But they did not benefit from the conceptual revolution brought by Einstein, that of the equivalence of all inertial reference frames. Concerning Poincaré, these works marked a full stop in his understanding of relativistic electromagnetism,[6] while for Einstein they represented a starting point to move toward other fundamental results: the mass-energy "equivalence", the equivalence principle, and general relativity.

Editors' note. Poincaré's rather pragmatic attitude toward theories[7] has often been criticized. Louis de Broglie wrote: *"He had a somewhat skeptical*

[5]FAPP, "for all practical purposes", as John Bell wrote in another context [**1**].

[6]Although he opened paths, for example in gravitation, Poincaré did not pursue the way traced by his 1905 articles. We could discuss forever what status Poincaré accorded to the aether, a simple convenience used to pedagogical ends, or on the contrary a fundamental concept in physics. Regardless, Poincaré never sought to situate his articles in a simpler and more general framework, eliminating the dross of the electromagnetic models.

[7]Poincaré's philosophy of science is explained in Chapter 19.

attitude vis-à-vis *physical theories, considering that there exist, in general, infinitely many different viewpoints, with varied imagery, that are logically equivalent, and between which the scholar can only choose for reasons of convenience"* (*Livre du centenaire, Œuvres* of Henri Poincaré, vol. 11, p. 66). It seemed to us fair to give Poincaré here a kind of chance to respond. We quote a few passages from his book *Science and Hypothesis* (1902), which summarize well his conception of what is (and what is not) a theory.

> "No theory seemed more solid than that of Fresnel, who attributed light to the motion of the aether. However, now we prefer over it that of Maxwell. Does this mean that Fresnel's work was in vain? No, because Fresnel's goal was not to know if there is really an aether, if it is or is not formed of atoms, if these atoms really move about in such and such a way; it was to predict optical phenomena. Yet Fresnel's theory still makes that possible, today as well as before Maxwell. The differential equations are still true; one can still integrate them by the same procedures, and the results of this integration still hold the same value. And let no one say that we are thus reducing the physical theories to the role of simple practical recipes; these equations express relations, and, if the equations remain true, it is because the relations maintain their reality. They teach us, after as before, that there is such a relation between one thing and another thing; only, this thing that we formerly called *motion*, we now call *electric current.* But these names were nothing but images substituted for the real objects that nature will forever hide from us. The true relations between these real objects are the only reality we can achieve, and the only condition is that there be the same relations between these objects as between the images we are forced to put in their place. If these relations are known to us, what does it matter if we find it convenient to replace one image with another?"
>
> (*Science and Hypothesis*, beginning of Chapter 10)

> "Thanks to Fresnel, optics form the most advanced part of physics; the so-called theory of undulations forms a set that is truly satisfying to the spirit, but we should not ask of it what it cannot give. The mathematical theories do not have as their goal to reveal to us the true nature of things; that would be an unreasonable pretension. Their unique goal is to coordinate the physical laws that experiment makes known to us, but that, without the help of mathematics, we would be unable even to state. It matters little to us whether aether really exists; that is the metaphysicians' affair. For us, the essential is that everything happens as though it existed, and that this hypothesis is useful for explaining phenomena. After all, do we have any other reason to believe in the existence of material objects? That, too, is nothing but a useful hypothesis; only it will never cease to be so, whereas a day will surely come where the aether will be rejected as useless. But even that day, the laws of optics and

> the equations that translate them analytically will remain true, at least as a first approximation. It will therefore always be useful to study a doctrine that ties all these equations together. The theory of undulations therefore rests on a molecular hypothesis. For those that believe they thus find the law's underlying cause, it is an advantage; for the others, it is a reason to be suspicious, but this suspicion seems to be as little justified as the former's illusion. These hypotheses play only a secondary role. One may sacrifice them; we do not ordinarily do so because the exposition would lose clarity in so doing, but this is the only reason."
>
> (*Science and Hypothesis*, beginning of Chapter 12)

Another example:

> "Carnot's theory, in its primitive form, expressed, next to true relationships, other, inexact, relations—the debris of old ideas; but the presence of the latter did not change the reality of the others. Clausius had only to cut them out as one prunes dead branches. The result was the second fundamental law of thermodynamics. It still had the same relations, even though these relations no longer held, at least in appearance, between the same objects. That was enough for the principle to retain its value. And even Carnot's reasonings did not perish for this; they applied to a material stained by error, but their forms (that is to say, the essential) remained correct."
>
> (*Science and Hypothesis*, Chapter 10)

17.4. A deduction from the Lorentz group

In this section, I will give a deduction from the Lorentz transformation that rests on somewhat general hypotheses [**11, 14**], which will allow me to make a connection with the reasoning used by Poincaré to show that $\ell = 1$. I will not assume, as is usual, that the speed of light is the same in all inertial reference frames. The precise hypotheses are the following.[8]

(i) Space is assumed to be homogeneous: it has the same properties at every point, or, in other words, translation invariance is valid. In addition, space is assumed to be isotropic: it has the same properties in all directions, or, in other words, rotation invariance is valid.

(ii) The transformations that allow one to pass from one inertial reference frame to another form a group. The physical origin of this condition is the following: the relation between inertial reference frames is an equivalence relation, which implies the existence of an identity element (reflexivity of the equivalence relation), an inverse (reciprocity), and an internal composition law (transitivity).

(iii) We can define a notion of causality: there exist events that are consequences of other events that occurred earlier in some inertial reference frame. For these

[8]For more details, see [**11**], which gives examples of other transformations than that of Lorentz where all of the hypotheses are satisfied, except one.

events, the temporal order must be identical in all reference frames: the cause must precede the effect in all referential frames.[9]

Let us recall the definition of an event: an event is produced at a given point in space at a determined moment. In a given inertial reference frame, it is characterized by a time coordinate t and spatial coordinates $\vec{r} = (x, y, z)$. One could imagine that a reference frame is divided into cubes by a lattice of lines, and that at each node of the lattice a watch is placed: the time of an event that is produced at a point $\vec{r}$ of the lattice is measured by the watch located at that point.[10] To synchronize the watches, one must of course take into account the finite speed c of the signals' propagation. It suffices to send out a reference signal at time $t = 0$ from a node of the lattice chosen to be the origin, and the watch located at a distance r from the reference node is set to time $t = r/c$ at the moment the signal is received.

Let us therefore consider two inertial reference frames R and R', with relative velocity $\vec{v}$ along the axis Ox. We choose the origin of time so that the origins O and O' coincide at time $t = t' = 0$. Let a ruler of length ℓ' be placed along the axis $O'y'$ in R', one endpoint being situated at the origin O'. Once the ruler crosses the axis Oy, it leaves a mark on this axis at the point $y = \ell$, and let us assume that $\ell' \neq \ell$, for example $\ell < \ell'$, so that an observer in R concludes that the rulers in motion, arranged perpendicularly to the relative velocity, are shortened (FIG. 17.1).

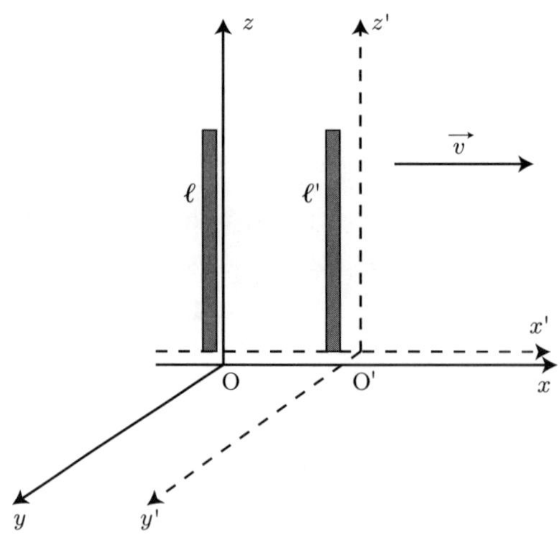

FIGURE 17.1. Reference frame transformation along the axis Ox. The axis $O'x'$ has been slightly displaced with respect to Ox for the sake of clarity in the figure.

[9]Zeeman [**30**] deduces the Lorentz transformation from the requirement of causality, provided that space-time is at least 3-dimensional, but assuming the existence of a light cone.

[10]Everything here assumes that "empty space" has meaning. The concept of an "empty space" is thoroughly undermined by general relativity, for which the structure of space-time is woven from the energy that it contains, and in which such a lattice can only exist, and then only approximately, in limited regions of space-time.

But the *same* experiment[11] allows an observer in R' to conclude that the rulers in motion are lengthened. If that were the case, the two reference frames R and R' would not be equivalent, in contradiction with hypothesis (i), and we must conclude that $\ell = \ell'$: distances perpendicular to the velocity are unchanged.

This argument allows us to restrict ourselves to coordinate transformations along the axis Ox, the transformation law of the other coordinates being $y' = y$ and $z' = z$. The most general transformation law for the coordinates (x, t) is then

$$x' = F(x; t; v), \qquad t' = G(x; t; v), \tag{17.7}$$

where v is the velocity of R' relative to R. According to hypothesis (i), the transformation law for an interval of space-time $\Delta x = x_1 - x$, $\Delta t = t_1 - t$ must be independent of its endpoints x and t. If Δx and Δt are infinitesimal intervals, $\Delta x \to \mathrm{d}x$, $\Delta t \to \mathrm{d}t$, then from (17.7)

$$\mathrm{d}x' = \frac{\partial F}{\partial x}\,\mathrm{d}x + \frac{\partial F}{\partial t}\,\mathrm{d}t, \qquad \mathrm{d}t' = \frac{\partial G}{\partial x}\,\mathrm{d}x + \frac{\partial G}{\partial t}\,\mathrm{d}t. \tag{17.8}$$

We conclude that the quantities $\partial F/\partial x, \ldots, \partial G/\partial t$ are independent of x and t, so that F and G are linear functions of x and t. It follows that the transformation law (17.8) is valid for finite intervals Δx and Δt. Let us rewrite (17.7), taking into account the linearity property, in the form

$$\begin{aligned} x' &= \gamma(v)[x - \nu(v)t], \\ t' &= \gamma(v)[-\mu(v)x + \lambda(v)t]. \end{aligned} \tag{17.9}$$

This transformation law depends *a priori* on four functions of the velocity—$\gamma(v)$, $\nu(v)$, $\mu(v)$, and $\lambda(v)$—and the objective is now to use (i)–(iii) to determine these functions. Let us remark first of all that $x' = 0$ is nothing other than the equation of the motion of O' in R'. This implies that $\nu(v) = v$ because the equation of the motion of O' in R is $x = vt$.

Now let us use the rotation invariance, which translates in this one-dimensional case to the equivalence of the directions x and $-x$, as observed by Poincaré. If (x', t') is obtained from (x, t) by a transformation of velocity v, then $(-x', t')$ must be obtained from $(-x, t)$ by a transformation of velocity $-v$,

$$-x' = \gamma(-v)[-x + vt], \qquad t' = \gamma(-v)[\mu(-v)x + \lambda(-v)t],$$

so that $\gamma(v)$ and $\lambda(v)$ are even functions of v, while $\mu(v)$ is an odd function of v,

$$\gamma(-v) = \gamma(v), \qquad \lambda(-v) = \lambda(v), \qquad \mu(-v) = -\mu(v). \tag{17.10}$$

Let us exploit group property (ii). It will be useful to write the transformation law (17.8) in the matrix form

$$\begin{pmatrix} x' \\ t' \end{pmatrix} = L \begin{pmatrix} x \\ t \end{pmatrix} = \gamma(v) \begin{pmatrix} 1 & -v \\ -\mu(v) & \lambda(v) \end{pmatrix} \begin{pmatrix} x \\ t \end{pmatrix}. \tag{17.11}$$

The inverse matrix L^{-1} is given by

$$L^{-1} = \frac{1}{\gamma(v)[\lambda(v) - v\mu(v)]} \begin{pmatrix} \lambda(v) & v \\ \mu(v) & 1 \end{pmatrix}. \tag{17.12}$$

---

[11] In contrast with the experiment that demonstrates the contraction of lengths parallel to the velocity, where the experiments are *different* when realized in each reference frame, which is what allows the two observers to conclude that the lengths of rulers in motion contract.

Using the relation $L^{-1}(v) = L(-v)$, we obtain

$$(17.13)\qquad \gamma(-v) = \frac{\lambda(v)}{\gamma(v)[\lambda(v) - v\mu(v)]} = \gamma(v), \qquad v\gamma(v) = \frac{v}{\gamma(v)[\lambda(v) - v\mu(v)]},$$

which shows that $\lambda(v) = 1$. Finally we carry out the product of two transformations $R \xrightarrow{v} R'$ of velocity v and $R' \xrightarrow{v'} R''$ of velocity v'. Using the notational abbreviations $\gamma' = \gamma(v')$ and $\mu' = \mu(v')$, we obtain for the product $L'L$

$$(17.14)\qquad L'L = \gamma'\gamma\begin{pmatrix} 1 & -v' \\ -\mu' & 1 \end{pmatrix}\begin{pmatrix} 1 & -v \\ -\mu & 1 \end{pmatrix} = \gamma'\gamma\begin{pmatrix} 1+\mu v' & -(v+v') \\ -(\mu+\mu') & 1+\mu' v \end{pmatrix}.$$

The transformation $L'' = L'L$ must have the same form as L if the transformations form a group. The two diagonal elements of L'' must then be identical, which implies that

$$1 + \mu(v)v' = 1 + \mu(v')v, \qquad \text{from which} \qquad \frac{\mu(v')}{\mu(v)} = \frac{v'}{v}.$$

This relation shows that the function $\mu(v)$ is proportional to v,

$$(17.15)\qquad \mu(v) = \alpha v,$$

where, by dimensional analysis, α has the dimension of the inverse of the velocity squared. At the end of the day, from the expression for L^{-1} we draw

$$(17.16)\qquad \gamma(v) = \frac{1}{\gamma(v)(1-\alpha v^2)}, \qquad \text{whence} \qquad \gamma(v) = \frac{1}{\sqrt{1-\alpha v^2}},$$

where we must choose the positive value for the root because $\gamma(v=0) = 1$. As a bonus, we obtain from (17.13) the addition law for velocities,

$$(17.17)\qquad V = \frac{v+v'}{1+\mu(v)v'} = \frac{v+v'}{1+\alpha v v'}.$$

It is an elementary exercise to check in detail this group law, making sure that the function $\gamma(V)$ of the transformation $R \xrightarrow{V} R''$ has the correct form

$$\gamma(V) = \gamma\gamma'(1+\alpha v v') = \frac{1}{\sqrt{1-\alpha V^2}}.$$

We have now arrived at a crucial point, because three possibilities are open, according to the value of α.

(a) $\alpha = 0$. This case corresponds to the Galilean transformation

$$x' = x - vt, \qquad t' = t.$$

As we have seen, this transformation is in contradiction with experiment since the end of the 19th century.

(b) $\alpha = -\beta < 0$. The transformation law for an interval of space-time $(\Delta t, \Delta x)$ leads to

$$\Delta t' = \frac{\Delta t + \beta v \Delta x}{\sqrt{1+\beta v^2}}.$$

Under these conditions, if Δx is non-zero, it is always possible to find a speed v sufficiently large for $\Delta t'$ to have the same sign as Δx, even if Δt has the opposite sign; there is therefore no causality, and the case $\alpha < 0$ is excluded by hypothesis (iii).

(c) $\alpha > 0$. Since the dimension of α is the inverse of the velocity squared, we can write $\alpha = c^{-2}$, where c has the dimensions of a speed. This gives the Lorentz transformation

$$x' = \frac{x - vt}{\sqrt{1 - v^2/c^2}} = \gamma(v)(x - vt),$$
$$t' = \frac{t - vx/c^2}{\sqrt{1 - v^2/c^2}} = \gamma(v)\left(t - \frac{vx}{c^2}\right). \tag{17.18}$$

It is obvious from the formulas above that we must have $v < c$, and c in this derivation appears as a limiting speed, connected in a fundamental way with the structure of space-time, and not with the speed of propagation of a particular physical phenomenon.[12] In fact, if the photon has zero mass, c is the speed of the photon, or the speed of light, but the notion of limiting speed remains valid even if it turned out that the photon had a very small, but non-zero, mass. Our derivation is therefore not at all tied to a particular physical phenomenon such as electromagnetic radiation. It is of course compatible with the existence of particles with zero mass, propagating at the limiting speed, but it does not require at all the existence of such particles. Special relativity gives us the arena of space-time where physical interactions take place: the electromagnetic, strong, and weak interactions,[13] but it tells us nothing about these interactions.

In conclusion, the only transformations compatible with hypotheses (i)–(iii) are the Lorentz and Galilean transformations: in particular, there is no "low speed limit" compatible with the group structure other than the Galilean transformation. If we admit that the Galilean transformation is in contradiction with experiment, the only possible solution is the Lorentz transformation, which implies a limiting speed.

17.5. The Poincaré group in quantum physics: Wigner's analysis (1939)

Because of the linearity property of quantum theory and a theorem due to Wigner[14] which I will state below, group theory plays a much more important role in quantum physics than in classical physics. This section is devoted to the consequences of the invariance of physics under the Poincaré group, or more exactly the subgroup denoted $\mathcal{P}_+^\uparrow$, which contains neither inversion of space nor of time. This is therefore the group of displacements of Minkowski space-time, that is, the inhomogeneous Lorentz group. It is composed of translations of space-time (a^μ) and rotations and pure Lorentz transformations (the homogeneous Lorentz group), collectively denoted $\Lambda_+^\uparrow$. The notation is standard: a four-vector x^μ is an object with four components that transforms according to the law

$$x'^\mu = a^\mu + \Lambda^\mu{}_\nu x^\nu, \qquad \Lambda^\mu{}_\nu \in \Lambda_+^\uparrow, \tag{17.19}$$

and the group law is

$$\{a, \Lambda\}\{a', \Lambda'\} = \{a + \Lambda a', \Lambda\Lambda'\}. \tag{17.20}$$

---

[12]It is easy to obtain arbitrarily large speeds; for example the speed of displacement of the Moon's dark spot from a laser beam in rotation can be as large as one likes (insofar as the laser can be rotated, but 1 rad/s suffices to attain c on the Moon). But this spot cannot carry information.

[13]But not gravitational interactions! See however [**12**] for an opposing viewpoint.

[14]One can find a proof of Wigner's theorem and some applications for example in [**9**], Chapter 8 and Appendix A.

The index μ of x^μ goes from 0 to 3. In a system of units where $c = 1$, which we adopt in this section and the following, $x^0 = t$ is the time component of the four-vector x^μ, and each x^i, $i = 1, 2, 3$, is a space component. The metric tensor of the Minkowski metric is denoted $\eta_{\mu\nu} = \mathrm{diag}(1, -1, -1, -1)$, and the scalar product of two four-vectors a^μ and b^μ is

$$a \cdot b = \eta_{\mu\nu} a^\mu b^\nu = a_\nu b^\nu = a^0 b^0 - \vec{a} \cdot \vec{b}.$$

In quantum theory, the state of a physical system is represented mathematically by a vector of unit norm in a Hilbert space $\mathcal{H}$, called the *Hilbert space of states.* More exactly, there exists a bijective correspondence between the physical states and the *rays* of $\mathcal{H}$, that is, vectors up to phase: the vectors Φ and $\mathrm{e}^{\mathrm{i}\alpha}\,\Phi$, which differ by a phase factor, represent the same physical state. A fundamental role is played by the probability amplitudes: if the vectors Φ and Ψ in $\mathcal{H}$ represent two states of a physical system, we define the *probability amplitude of finding the state* Φ *in* Ψ as the scalar product (Ψ, Φ). The probability of finding Φ in Ψ is given by the square modulus of this scalar product,[15]

$$\mathrm{Prob}(\Phi \to \Psi) = |(\Psi, \Phi)|^2. \tag{17.21}$$

Wigner's theorem states a fundamental result when a physical problem is invariant under an operation g, which will then be called a *symmetry* of the problem. The invariance under this symmetry implies the invariance of the probabilities (but not *a priori* that of the probability amplitudes, because only the probabilities are observable). If $\tilde{\Phi}_g$ and $\tilde{\Psi}_g$ are the rays obtained by transforming $\tilde{\Phi}$ and $\tilde{\Psi}$ by the transformation g, then the symmetry implies

$$|(\tilde{\Psi}_g, \tilde{\Phi}_g)|^2 = |(\tilde{\Psi}, \tilde{\Phi})|^2. \tag{17.22}$$

Wigner's remarkable result [**29**] is that the symmetry can be realized on the *vectors* of $\mathcal{H}$. There exists a unitary linear transformation, or a unitary antilinear (anti-unitary)[16] one $U(g)$, unique up to phase, such that

$$\Phi_g = U(g)\,\Phi. \tag{17.23}$$

The truly interesting case is the one where the transformations g form a group $\mathcal{G}$. If this group is continuous and connected, the transformations $U(g)$ are necessarily unitary (and not anti-unitary), because every element g of the group can be continuously connected to the identity element e of the group, and we must have $U(e) = \mathrm{Id}$, where Id is the identity element of the Hilbert state space. The transformations $U(g)$ of the Hilbert space of states must obey the composition law[17]

$$U(g_1)U(g_2) = \omega(g_1, g_2)\,U(g_1 g_2),$$

where $\omega(g_1, g_2)$ is a complex number of unit modulus, because $U(g_1)U(g_2)$ and $U(g_1 g_2)$ can differ by a phase factor. In other words, the unitary operators $U(g)$ must form a projective representation of the group $\mathcal{G}$. When $\omega = 1$, we are dealing with vector representations of the group $\mathcal{G}$.

[15] Even though this definition is far from intuitive, the reader can convince himself that it is at least coherent. The probability of finding a state inside itself is 1, and it is less than 1 for all other states by the Cauchy–Schwarz inequality.

[16] A unitary transformation U obeys $(U\Psi, U\Phi) = (\Psi, \Phi)$, while an anti-unitary transformation $\hat{U}$ obeys $(\hat{U}\Psi, \hat{U}\Phi) = (\Phi, \Psi)$. Anti-unitary transformations only become involved if inverting the direction of time is included in g.

[17] It is easy to verify that the phase factor ω cannot depend on the states Φ and Ψ.

In the case of the Poincaré group, Wigner showed that the phase factor ω can only take the values ± 1: the corresponding representations are called spinorial representations. In fact, the spinorial representations of the Lorentz group $\Lambda_+^\uparrow$ are the vector representations of its universal covering group, the group $\mathrm{SL}(2,\mathbb{C})$ of linear transformations of determinant 1 in the two-dimensional complex space $\mathbb{C}^2$.[18] To a transformation $\Lambda \in \Lambda_+^\uparrow$ correspond two matrices $A \in \mathrm{SL}(2,\mathbb{C})$. It is not very difficult to find the Lie algebra: we write the transformation $U(g)$ corresponding to a translation a^μ and to a transformation $A \in \mathrm{SL}(2,\mathbb{C})$ as a function of the ten infinitesimal generators P_μ (translations) and $M_{\mu\nu} = -M_{\nu\mu}$ (M_{ij} = rotations and M_{0i} = boosts)

$$U(a, A) = \exp(\mathrm{i}P\cdot a)\exp\left(\frac{\mathrm{i}}{2}\,\theta_{\mu\nu}M^{\mu\nu}\right). \tag{17.24}$$

These infinitesimal generators are self-adjoint operators acting on $\mathcal{H}$, and they obey the commutation relations

$$[P_\mu, P_\nu] = 0, \tag{17.25}$$

$$[P_\mu, M_{\nu\lambda}] = \mathrm{i}\,(\eta_{\lambda\mu}P_\nu - \eta_{\nu\mu}P_\lambda), \tag{17.26}$$

$$[M_{\mu\nu}, M_{\rho\sigma}] = \mathrm{i}\,(\eta_{\mu\rho}M_{\nu\sigma} + \eta_{\nu\sigma}M_{\mu\rho} - \eta_{\mu\sigma}M_{\nu\rho} - \eta_{\nu\rho}M_{\mu\sigma}). \tag{17.27}$$

We remark that (17.27) was written by Poincaré: the commutation relation (17.4) of the boosts T_x and T_y in the notation of (17.27) is nothing other than

$$[M_{01}, M_{02}] = \mathrm{i}\,M_{12}.$$

Wigner's objective was to classify all irreducible representations of the Poincaré group, and he began by determining the invariants one can form from the infinitesimal generators. A first invariant is

$$P^2 = (P^0)^2 - \vec{P}^2. \tag{17.28}$$

Let $|p\rangle$ be an "eigenvector"[19] of $\mathcal{H}$ common to the P_μ,

$$P_\mu|p\rangle = p_\mu|p\rangle.$$

The key point of Wigner's analysis consists in finding the transformations $\Lambda(p)$ that leave p_μ invariant (the little group of p). They are of the form (s is a four-vector that is not parallel to p)

$$U[\Lambda(p)] = \exp(-\mathrm{i}\,W_\mu s^\mu), \tag{17.29}$$

$$W_\mu = -\frac{1}{2}\varepsilon_{\mu\nu\rho\sigma}\,M^{\nu\rho}P^\sigma, \tag{17.30}$$

where $\varepsilon_{\mu\nu\rho\sigma}$ is the completely anti-symmetric space-time tensor such that $\varepsilon_{0123} = 1$.

The second invariant is $W^2 = W_\mu W^\mu$. Now, P^2 and W^2 are the only combinations that commute with all the generators, and their eigenvalues label the irreducible representations. In addition,

$$[W_\mu, P_\nu] = 0, \qquad W\cdot P = 0.$$

We can therefore find an orthonormal basis in $\mathcal{H}$ formed of eigenvectors common to P_μ and to one of the components of W_μ. We distinguish two possible cases.

---

[18]The homogeneous Lorentz group is identical to the group denoted $\mathrm{O}(3,1)$.

[19]The quotation marks come from the fact that these vectors are not really vectors of the Hilbert space, but this does not bother physicists. It is not difficult to make the equations mathematically correct. I have used Dirac's notation: $\Phi \to |\Phi\rangle$.

Case 1: $p^2 > 0$, $p^0 > 0$. Here $p^2 = m^2$ *is the square of the particle's mass*, and the little group is then the group of unimodular unitary transformations of $\mathbb{C}^2$, or SU(2), which is the universal covering group of the group of three-dimensional rotations SO(3). To each four-vector p^μ we attach a Lorentz transformation (a tetrad) $[p] \in \mathrm{SL}(2,\mathbb{C})$ which transforms the basis $\{[p^0 = m, \vec{p} = 0], \hat{x}, \hat{y}, \hat{z}\}$ of space-time into the basis $\{p = [p^0, \vec{p}], n_i(p)\}$, where the $n_i(p)$ are the three space-like vectors ($n_i^2(p) < 0$) orthogonal to p and satisfying

$$n_i(p) \cdot n_j(p) = -\delta_{ij}, \qquad p \cdot n_i(p) = 0. \tag{17.31}$$

We decompose $W(p)$ onto the $n_i(p)$,

$$W(p) = m \sum_i S_i(p) \cdot n_i(p). \tag{17.32}$$

The $S_i(p)$ satisfy the commutation rules of the Lie algebra of SU(2), which are those of the angular momentum[20]

$$[S_i(p), S_j(p)] = \mathrm{i}\, \varepsilon_{ijk} S_k(p), \tag{17.33}$$

and these are the infinitesimal generators of the little group of p. In addition, the sum of the $S_i^2(p)$

$$\sum_{i=1}^{3} S_i^2(p) = -\frac{W^2}{m^2}$$

takes the values $s(s+1)$, with s a half-integer ($s = 1/2, 3/2, \dots$) or an integer ($s = 0, 1, 2, \dots$): *s is the spin of the particle.*

The eigenstates common to P^μ, $-W^2/m^2$, and S_3 are denoted $|[p], s, s_3\rangle$, the values of s_3 being integers or half-integers, and their transformation law is

$$U(a, A)\, |[p], s, s_3\rangle = \exp[\mathrm{i}(Ap) \cdot a]\, |[Ap], s, s_3'\rangle\, \mathcal{D}^s_{s_3, s_3'}([Ap]^{-1} A[p]), \tag{17.34}$$

where $[Ap]^{-1}A[p]$ is a rotation, called the Wigner rotation, and where the $\mathcal{D}^s_{i,j}$ are matrices of the s-dimensional representation of the group SU(2), that is, the matrices of a representation of the group of three-dimensional rotations SO(3) that is spinorial when s is a half-integer ($s = 1/2, 3/2, \dots$), and vectorial when s is an integer ($s = 0, 1, 2, \dots$). In summary, the mass is given by the Lorentz invariant $P^2 = m^2$ and the spin by the eigenvector $s(s+1)$ of $-W^2$.

Case 2: $p^2 = 0$, $p^0 > 0$. This is therefore the case of zero mass. The little group is then the Euclidean group $E(2)$ of two-dimensional displacements. The representations of physical interest correspond to $W^2(p) = 0$, $W(p) = \lambda(p)p$, and the basis vectors are written $|p, \lambda\rangle$, where λ is called the helicity. Since $P^2 = 0$, the mass is zero, and since s_3 can only take the single value λ, a particle with zero mass and spin λ has only a single helicity state.[21] The transformation law for states $|p, \lambda\rangle$ is

$$U(a, A)\, |[p], \lambda\rangle = \exp[\mathrm{i}(Ap) \cdot a]\, |[Ap], \lambda'\rangle\, \mathcal{D}^{|\lambda|}_{\lambda\lambda'}([Ap]^{-1} A[p]). \tag{17.35}$$

The case $p^2 < 0$ does not seem to present any physical interest.

---

[20] ε_{ijk} is the completely antisymmetric tensor of three-dimensional space such that $\varepsilon_{123} = 1$.

[21] When the interactions of the particle are invariant under parity, the two helicity states $+|\lambda|$ and $-|\lambda|$ are possible. This is the case for the photon, which has two helicity states: $\lambda = +1$, right-circular polarization, and $\lambda = -1$, left-circular polarization. If the neutrino had zero mass, which for a long time seemed possible, but in fact is not the case, the neutrino, with helicity $\lambda = -1/2$, would be a *different* particle from the antineutrino, a particle with helicity $\lambda = 1/2$.

We must emphasize the fundamental character of Wigner's analysis, because it shows that properties introduced empirically, such as the spin of the electron, in fact find their profound origin in the structure of the symmetry group of space-time. From a more practical viewpoint, Wigner's results are at the heart of the relativistic analysis of reactions between particles, and in particular of the standard helicity formalism. The collision matrix S describing the collisions between particles is invariant under the transformations of the Poincaré group

$$U(a, A)\, S\, U^{-1}(a, A) = S, \tag{17.36}$$

and this invariance is a fundamental symmetry of physics.

17.6. The theorems of Coleman and Mandula (1967) and of Haag, Łopuszański, and Sohnius (1975)

In addition to invariance under the Poincaré group, the physics of elementary particles makes use of internal symmetries, which change the type of particle. These symmetries are often approximate. An example is isospin, which exchanges the neutron and the proton, $n \leftrightarrow p$. This isospin symmetry is obviously not respected by electromagnetism, since the electric charge differentiates the (uncharged) neutron and the (charged) proton. On the other hand, it is thought that nuclear interactions, or strong interactions, do not make any difference between the proton and the neutron, and to the extent that we forget about electromagnetism, we can perfectly well envision states that are linear superpositions of a neutron and a proton

$$\lambda\,|n\rangle + \mu\,|p\rangle.$$

The space of states is a two-dimensional complex space, where we can choose for orthonormal basis vectors the states of a neutron $|n\rangle$ and of a proton $|p\rangle$. The internal symmetry group corresponding to isospin is of type SU(2), which is represented in $\mathcal{H}$ by a unitary operator U.

An internal symmetry group $\mathcal{G}_{\text{int}}$ only acts on the internal degrees of freedom of the particles (α), not on their momentum p or their spin s,

$$\langle \alpha, p, s, s_3 |\, U(g_{\text{int}})\, | \alpha', p', s', s_3' \rangle \propto \delta_{pp'} \delta_{ss'} \delta_{s_3 s_3'}, \qquad g_{\text{int}} \in \mathcal{G}_{\text{int}} \tag{17.37}$$

(recall that $\propto$ means "proportional to"). In the 1960s, the particle physicists wondered if it was possible to combine in a non-trivial way an internal symmetry group $\mathcal{G}_{\text{int}}$ and the Poincaré group. The response to this question is negative: this is the content of the theorem of Coleman and Mandula, whose hypotheses and statement I give below. In a general way, the action of a symmetry $\mathcal{G}$ of the matrix S is represented in $\mathcal{H}$ by a unitary operator U that satisfies $USU^{-1} = S$ and that

(a) contains $\mathcal{P}_+^\uparrow$ as a subgroup;
(b) transforms a one-particle state into a one-particle state;
(c) acts additively on multiparticle states.

In addition, Coleman and Mandula assume that

(i) the matrix S is non-trivial;
(ii) there exists a state with minimal mass;
(iii) there exists a finite number of particles with a given mass.

Given these conditions, they show that $\mathcal{G}$ is necessarily a direct product of the Poincaré group and an internal symmetry group:

$$\mathcal{G} = \mathcal{P}_+^\uparrow \otimes \mathcal{G}_{\text{int}}.$$

If P_μ and $M_{\mu\nu}$ are the infinitesimal generators of $\mathcal{P}_+^\uparrow$, and B_l are those of $\mathcal{G}_{\text{int}}$, then the B_l, which are Lorentz scalars, follow the Lie algebra of the internal symmetry group $\mathcal{G}_{\text{int}}$

$$[B_l, B_m] = \mathrm{i} \sum_k c_{lm}^k B_k. \tag{17.38}$$

In its original definition, an internal symmetry cannot change the spin of a particle (see (17.37)). Quantum physics only admits two broad categories of particles, fermions and bosons: two fermions cannot be in the same quantum state (this is the famous Pauli exclusion principle), while two bosons prefer to be in the same state.[22] A famous theorem, the spin-statistics theorem, states that fermions are particles of half-integer spin ($s = 1/2,\ 3/2,\ \dots$), and bosons are particles of integer spin ($s = 0, 1, 2, \dots$). To circumvent the Coleman–Mandula theorem, one must envision a generalization of the notion of internal symmetry where it is possible to change spin by half-units, by transforming fermions into bosons and vice versa. This generalization is supersymmetry. An infinitesimal generator Q of supersymmetry therefore transforms fermions into bosons and vice versa, and it must have a half-integer spin. Quantum field theory forces the generators with half-integer spin to obey anticommutation relations, and one is led to search for a superalgebra (or graded Lie algebra), which in addition to commutators $[\ ,\ ]$ involves anticommutators $\{\ ,\ \}$,

$$\{A, B\} = AB + BA.$$

The supersymmetry algebra will *a priori* have the schematic form

$$\{Q, Q'\} = X, \qquad [X, X'] = X'', \qquad [Q, X] = Q'',$$

where X represents P_μ, $M_{\mu\nu}$, or B_l. We can always decompose Q into a sum of terms belonging to irreducible representations of $\mathrm{SL}(2, \mathbb{C})$,

$$Q = \sum Q_{\alpha_1,\dots,\alpha_a,\dot\alpha_1,\dots,\dot\alpha_b}, \tag{17.39}$$

where the α are indices of the fundamental representation of $\mathrm{SL}(2, \mathbb{C})$ and the $\dot\alpha$ are those of the complex conjugate representation: $\alpha, \dot\alpha = 1, 2$.[23] Each term must be symmetrized with respect to $\alpha_1, \dots, \alpha_a$ and $\dot\alpha_1, \dots, \dot\alpha_b$. The spin is $(a + b)/2$, and $a + b$ must be odd if we want the Q to be fermionic generators obeying anticommutation relations. In addition to conditions (i)–(iii) imposed by Coleman and Mandula, Haag, Łopuszański, and Sohnius require that

(iv) Q acts on $\mathcal{H}$ endowed with a positive definite scalar product;
(v) Q and $\overline{Q}$ (its Hermitian conjugate) belong to the algebra.

Let the anticommutator be

$$\{Q_{\alpha_1,\dots,\alpha_a,\dot\alpha_1,\dots,\dot\alpha_b}, Q_{\beta_1,\dots,\beta_a,\dot\beta_1,\dots,\dot\beta_b}\}.$$

If we make the choice

$$\alpha_1 = \cdots = \dot\beta_b = 1,$$

the theorem of Coleman and Mandula implies that this anticommutator must be a combination of the P^μ, and the only possibility is $a + b = 1$, so that $a = 1/2$ and

[22] This phenomenon is at the origin of the "Bose–Einstein condensate", whose experimental discovery earned a Nobel prize for Cornell, Ketterle, and Wieman in 2001.

[23] These two representations are equivalent in the case of $\mathrm{SU}(2)$, but not in the case of $\mathrm{SL}(2, \mathbb{C})$.

$b = 1/2$: the Q are necessarily generators with spin $s = 1/2$. The final form of the algebra, written schematically, is

$$\begin{aligned} \{Q^A_\alpha, \overline{Q}_{\dot\beta B}\} &= 2\sigma^{\ \mu}_{\alpha\dot\beta} P_\mu \delta^A_B, \\ \{Q^A_\alpha, Q^B_\beta\} &= \{Q_{A\dot\alpha}, Q_{B\dot\beta}\} = 0, \\ [P_\mu, Q^A_\alpha] &= [P^\mu, Q_{A\dot\alpha}] = 0. \end{aligned} \tag{17.40}$$

The indices A and B are the indices of an internal symmetry and go from 1 to N. Even though there is no experimental indication at all in its favor, supersymmetry is very popular among particle physicists. One of the reasons is that it is impossible to conceive of string theory without supersymmetry.

17.7. Galilean electromagnetism

One might wonder what were the reasons for which the theoreticians of electromagnetism at the end of the 19th century did not attempt to write transformation laws of the fields $\vec{E}$ and $\vec{B}$ that are compatible with the Galilean transformation, and almost directly wrote the relativistic law

$$\begin{aligned} \vec{E}'_\parallel &= \vec{E}_\parallel, \qquad \vec{B}'_\parallel = \vec{B}_\parallel, \\ \vec{E}'_\perp &= \gamma\left(\vec{E}_\perp + \vec{v}\times\vec{B}\right), \\ \vec{B}'_\perp &= \gamma\left(\vec{B}_\perp - \frac{\vec{v}\times\vec{E}}{c^2}\right). \end{aligned} \tag{17.41}$$

In equations (17.41), $\vec{E}_\parallel$ and $\vec{B}_\parallel$ represent the components of the electric field $\vec{E}$ and the magnetic field $\vec{B}$ parallel to the velocity of the Lorentz transformation, and $\vec{E}_\perp$ and $\vec{B}_\perp$ represent their components perpendicular to this velocity. In fact the limit $v/c \ll 1$ of (17.41) is not completely obvious, and it is an interesting and relatively little-known fact that one can obtain a limit of the relativistic transformation law of the electromagnetic field compatible with Galilean relativity. However, (17.41) has not one, but *two* possible limits, both of which turn out to be physically pertinent in well-defined situations, where they represent good approximations of the relativistic transformation law [**10**], while the "low velocity" limit $v \ll c$, which is found in certain books and which was used by Lorentz in 1895,

$$\vec{E}' = \vec{E} + \vec{v}\times\vec{B}, \qquad \vec{B}' = \vec{B} - \frac{\vec{v}\times\vec{E}}{c^2}, \tag{17.42}$$

just as in (17.2), does not define a group law! As we will see below, these two limits can even have practical interest, because Galilean relativity is much simpler than Einsteinian relativity. The set $(\rho, \vec{\jmath})$, or electromagnetic current, forms a four-vector that transforms like $(t, \vec{x})$,

$$\rho' = \gamma\left(\rho - \frac{\vec{v}\cdot\vec{\jmath}}{c^2}\right), \qquad \vec{\jmath} = \gamma(\vec{\jmath} - \rho\vec{v}). \tag{17.43}$$

We can see that there exist two possible limits for this transformation: if $c|\rho| \gg |\vec{\jmath}|$, then $\vec{v}\cdot\vec{\jmath}/c^2 \ll (v/c)|\rho| \ll |\rho|$ is negligible with respect to ρ, while if $c|\rho| \ll |\vec{\jmath}|$, then $|\rho|v \ll (v/c)|\vec{\jmath}| \ll |\vec{\jmath}|$ and $\rho\vec{v}$ is negligible with respect to $\vec{\jmath}$; in addition γ can be replaced by 1. The first limit, or *electrical limit*, corresponds to the case where $c|\rho| \gg |\vec{\jmath}|$, which implies $|\vec{E}| \gg c|\vec{B}|$. The currents and the fields of this approximation will be denoted by an index e. The transformation law of the current four-vector, realized for $v \ll c$ when $c|\rho| \gg |\vec{\jmath}|$, is

$$\rho_e' = \rho_e, \qquad \vec{\jmath}_e{}' = \vec{\jmath}_e - \vec{v}\rho_e, \tag{17.44}$$

while the transformed fields are given by

$$\vec{E}_e' = \vec{E}_e, \qquad \vec{B}_e' = \vec{B}_e - \frac{\vec{v} \times \vec{E}_e}{c^2}. \tag{17.45}$$

The "Galilean Maxwell equations" of the electric limit are

$$\vec{\nabla} \times \vec{E}_e = 0, \qquad \vec{\nabla} \cdot \vec{B}_e = 0, \tag{17.46}$$

$$\vec{\nabla} \cdot \vec{E}_e = \frac{\rho_e}{\varepsilon_0}, \qquad \vec{\nabla} \times \vec{B}_e = \frac{1}{c^2}\frac{\partial \vec{E}_e}{\partial t} + \mu_0 \vec{\jmath}_e. \tag{17.47}$$

It is easy to check that these equations are covariant (formally invariant) under the transformations (17.1).[24] Equation (17.47) leads to the conservation of the current in its usual form

$$\frac{\partial \rho_e}{\partial t} + \nabla \cdot \vec{\jmath}_e = 0. \tag{17.48}$$

The Galilean Maxwell equations of the electric limit contain the displacement current, while Faraday's law of induction is absent; thus as we see in (17.45), the motion of an electric field, or more generally its temporal variation, generates a magnetic field, while a variable magnetic field does not generate an electric field.

In a world where there existed only positive charges, the electric limit would be the only physically pertinent one. However, in everyday life, positive and negative charges offset each other almost exactly, so that $|\vec{\jmath}| \gg c|\rho|$. The *magnetic limit*, labelled by the index m, corresponds to the case $c|\rho| \ll |\vec{\jmath}|$, which implies that $|\vec{E}| \ll c|\vec{B}|$. This is the usual case, because the magnetic effects overcome the electric effects due to the neutrality of ordinary matter. In this limit, the transformation law of currents is written as

$$\rho_m' = \rho_m - \frac{\vec{v} \cdot \vec{\jmath}_m}{c^2}, \qquad \vec{\jmath}_m{}' = \vec{\jmath}_m, \tag{17.49}$$

and that of the fields, as

$$\vec{E}_m' = \vec{E}_m + \vec{v} \times \vec{B}_m, \qquad \vec{B}_m' = \vec{B}_m. \tag{17.50}$$

The Galilean Maxwell equations of the magnetic limit are

$$\vec{\nabla} \times \vec{E}_m = -\frac{\partial \vec{B}_m}{\partial t}, \qquad \vec{\nabla} \cdot \vec{B}_m = 0, \tag{17.51}$$

$$\vec{\nabla} \cdot \vec{E}_m = 0, \qquad \vec{\nabla} \times \vec{B}_m = \mu_0 \vec{\jmath}_m. \tag{17.52}$$

The covariance of these equations is easily checked. They contain Faraday's law of induction, but not the displacement current. We see that in the magnetic limit, the current is not associated to a transport of charges, because according to (17.52)

---

[24] One uses the transformation law

$$\vec{\nabla}' = \vec{\nabla}, \qquad \frac{\partial}{\partial t'} = \frac{\partial}{\partial t} + \vec{v} \cdot \vec{\nabla}.$$

$\nabla \cdot \vec{\jmath}_m = 0$, so that $\partial \rho_m / \partial t = 0$: there can be no accumulation of charges of type m in a given volume, and the conservation of the charge has a global character[25]

$$Q_m(t) = \int \mathrm{d}^3 r \rho_m(\vec{r}, t) = \int \mathrm{d}^3 r' \rho'_m(\vec{r}', t), \qquad \frac{\mathrm{d}Q_m}{\mathrm{d}t} = 0.$$

The existence of a Galilean magnetism might seem to contradict the relativistic origin of magnetism, but the Galilean theory is perfectly coherent, *provided that we do not associate the current to a transport of charge.* The fact that we are obligated to resort to two Galilean limits might seem bizarre, but it is the best we can do without violating the group law.

In order to describe the most general Galilean invariant force law, we assume that there exist two kinds of charge and current density, an electric kind that obeys (17.44) and a magnetic kind that obeys (17.49). The following form of the Lorentz force is Galilean covariant

$$\vec{F} = \int \mathrm{d}^3 r \left(\rho_e \vec{E}_e + \vec{\jmath}_m \times \vec{B}_m + \rho_e \vec{E}_m + \vec{\jmath}_e \times \vec{B}_m + \rho_m \vec{E}_e + \vec{\jmath}_m \times \vec{B}_e \right). \tag{17.53}$$

The missing terms are $\rho_m \vec{E}_m$ and $\vec{\jmath}_e \times \vec{B}_e$, which would break the Galilean covariance. We observe that the electric current in a wire must be of magnetic type(!); otherwise there would not be any force between two wires traversed by a current. The Galilean limit defined above is able to take into account a surprising variety of electromagnetic phenomena, and I leave it to the reader's care to make a list of them. Two effects are obviously absent, because

(i) there is no propagation of electromagnetic waves;
(ii) capacitors do not function in an alternating regime.

Property (i) is not surprising, because no limiting speed exists in the Galilean framework, but (ii) persists even in the so-called quasi-stationary regime: thus as we have already mentioned, there is no relation between the current in the wire and the capacitor's charge.

Can we find physical situations where Galilean electromagnetism shows itself to be pertinent? As neither wave propagation nor retardation effects exist, we must certainly restrict ourselves to the quasi-stationary approximation: if l is a typical size of the system and $\tau = l/c$ is the typical transit time of light, it is necessary that $\tau_c \gg \tau$, where τ_c is a characteristic time of the system's evolution. Let us first consider a fluid to which we apply an electric field, and let us assume that Ohm's law $\vec{\jmath} = \sigma \vec{E}$, where σ is the electric conductivity, is valid. We call ε and μ the electric and magnetic permeabilities of the liquid. One of Maxwell's equations can be written as

$$\nabla \times \vec{B} = \sigma \mu \vec{E} + \varepsilon \mu \frac{\partial \vec{E}}{\partial t}. \tag{17.54}$$

---

[25]In Galilean physics, it is perfectly possible to conserve the charge without conserving it locally: one can imagine that a charge disappears at some point at a given instant and that an equivalent charge appears at *another* point at the same instant. This is impossible in an Einsteinian scheme, where two events that are simultaneous in one reference frame are, in general, not so in another reference frame. The conservation of the charge must then be local

$$\frac{\partial \rho}{\partial t} + \vec{\nabla} \cdot \vec{\jmath} = 0$$

because global conservation will not be satisfied in all inertial reference frames.

This equation implicitly contains two characteristic times τ_e and τ_m. If we can neglect the magnetic field (electric limit), then $\tau_e = \varepsilon/\sigma$ is the characteristic time that controls the time for the electric field, or the charge, to relax to equilibrium: τ_e is called the charge relaxation time. On the other hand, if the displacement current, that is, the last term in (17.54), can be neglected (magnetic limit), then (17.54) combined with Faraday's law leads to a diffusion equation for the magnetic field, with a diffusion constant $D = 1/(\sigma\mu)$,

$$\frac{\partial \vec{B}}{\partial t} = \frac{1}{\sigma\mu}\nabla^2\vec{B} = D\nabla^2\vec{B}. \tag{17.55}$$

The characteristic time is then $\tau_m = \sigma\mu l^2$. If the two terms of the right-hand side of (17.54) have the same order of magnitude, the magnetic field is determined for example by the temporal variation of $\vec{E}$ (that is, by the displacement current), i.e.,

$$\frac{E}{cB} \sim \frac{\tau_e}{c\varepsilon\mu l} \sim \frac{\tau_e}{\tau}.$$

The condition for the quasi-stationary approximation to be valid is $\tau_e/\tau \gg 1$, which implies that $E/(cB) \gg 1$ and $(\varepsilon c)/(\sigma l) \gg 1$. This condition will be valid if the conductivity is sufficiently small. In dielectric liquids characterized by a weak value of conductivity σ, the electric limit will be an excellent approximation, valid for example in the important case of electrohydrodynamics (EHD). If, on the other hand, the magnetic field is determined by $\vec{\jmath}$, $B \sim \sigma\mu El$, then

$$\frac{E}{cB} \sim \frac{E}{c\sigma\mu El} = \frac{1}{c\sigma\mu l} = \frac{\tau}{\tau_m}.$$

The condition $\tau_m/\tau \gg 1$ for the quasi-stationary approximation to be valid is then equivalent to $E/(cB) \ll 1$. In the case of fluids with a high conductivity, such as liquid metals, σ is important, $E/(cB) \ll 1$, and the magnetic limit will be an excellent approximation for magnetohydrodynamics (MHD). The constitutive relation $\vec{\jmath} = \sigma\vec{E}$ must be written in a reference frame where the fluid is instantaneously at rest. Passing to another reference frame, say that of the laboratory, is easy. For example, if the liquid moves with a velocity $\vec{v}$ with respect to the laboratory, the relation $\vec{\jmath}' = \sigma\vec{E}'$ can be written in the laboratory reference frame as

$$\vec{\jmath} = \sigma(\vec{E} + \vec{v} \times \vec{B}).$$

It is worth noting that there is no possible Galilean approximation in the case where E and cB are of the same order of magnitude, $E \sim cB$.

17.8. Conclusion

Einstein's name is rightly attached to the discovery of special relativity, as much in the broad public as among physicists. The expression "Lorentz transformation" gives justice to Lorentz for his contribution to the genesis of this theory. Poincaré's contribution is, to my mind, a bit unjustly passed over in silence,[26] except in the expression "Poincaré group", which is only familiar to particle physicists.[27] In the same vein as Jean-Marc Lévy-Leblond—who, in a well-documented article, asks

[26]For a balanced view of the respective roles of Lorentz, Poincaré, and Einstein in the genesis of relativity, see [**5**] and [**7**].

[27]When they have an idea of the origin of this expression, which is far from being the most frequent case.

the question: "What would relativity have become without Einstein?" [**13**]—I will make a less radical hypothesis, by simply supposing that Einstein had been three years younger. The fundamental article would therefore have appeared in June 1908, and not in June 1905. During these three fictitious years, most physicists would have had the time to convince themselves of the validity of Lorentz transformations and the principle of relativity (which Lorentz, "returning the favor", see Section 17.3, christened meanwhile *Poincaré's principle of relativity*) which must be understood as: there does not exist a physical, mechanical, or electromagnetic experiment that can demonstrate a uniform motion with respect to the aether. As Olivier Darrigol notes [**4**], Einstein's theory at that time only took selective advantage of the Lorentz/Poincaré theory, and we are faced with a typical case in which "the chronological order of the construction of a theory does not coincide with its logical structure" [**11**]. The idea is therefore to set an established theory back on its feet, and the absence of these three fictitious years deprived us of a passionate debate between the proponents of the principle of relativity following Poincaré and those of the principle of relativity following Einstein!

Returning to real history, I will leave the final word to Lorentz, who, after having made up for the "imperfections" of his 1904 work, wrote,

> *"Poincaré, on the other hand, obtained a perfect invariance for the equations of electrodynamics, and he formulated the 'postulate of relativity', terms which he was the first to employ* [...] *Let us add that even while he thus corrected the imperfections of my work, he has never reproached me for them."* [**17**]

Acknowledgements. I thank Olivier Darrigol and Jean-Marc Lévy-Leblond for their critical reading of a preliminary version of this article, and for their suggestions. I am equally grateful to Germain Rousseaux for numerous discussions on Galilean electromagnetism. The final text, of course, falls entirely under my responsibility.

Bibliography

[1] J. Bell, *Against measurement*, Physics World, August 1990, p. 33.
[2] S. Coleman and J. Mandula, *All possible symmetries of the S-matrix*, Phys. Rev. **159** (1967), pp. 1251–1256.
[3] T. Damour, *Si Einstein m'était conté*, Éditions du Cherche Midi, Paris, 2005.
[4] O. Darrigol, *Electrodynamics from Ampère to Einstein*, Oxford University Press, 2000.
[5] O. Darrigol, *The genesis of the theory of relativity*, Séminaire de l'IHP, April 2005, pp. 57–68.
[6] A. Einstein, *Zur elektrodynamic bewegter Körper*, Ann. Ph. **17** (1905), pp. 289–306
[7] P. Galison, *L'empire du temps*, R. Laffont, Paris, 2005.
[8] R. Haag, J. Łopuszański, and M. Sohnius, *All possible generators of supersymmetries of the S-matrix*, Nucl. Phys. **B88** (1975), p. 257.
[9] M. Le Bellac, *Quantum Physics*, Cambridge University Press, 2006.
[10] M. Le Bellac and J.-M. Lévy-Leblond, *Galilean electromagnetism*, Nuovo Cimento **14B** (1973), pp. 217–234.
[11] J.-M. Lévy-Leblond, *One more derivation of the Lorentz group*, American Journal of Physics **44** (1976), p. 271.
[12] J.-M. Lévy-Leblond, *On the conceptual nature of the physical constants*, Riv. Nuovo Cimento **7** (1977), p. 187.
[13] J.-M. Lévy-Leblond, *What if Einstein had never been born? A gedanken experiment in physics history*, Proc. of the 24th Int. Coll. on group-theoretical methods in physics, J.-P. Gazeau *et al.* eds, Institute of Physics, London, 2003.

[14] J.-M. Lévy-Leblond and J.-P. Provost, *Additivity, Rapidity, Relativity*, American Journal of Physics **47** (1979), p. 1045.

[15] H. Lorentz, *Versuch einer Theorie des elektrischen und optischen Erscheinungen in bewegten Körpern* (1895), Collected papers **5**, pp. 1–137.

[16] H. Lorentz, *Elecromagnetic phenomena in a medium moving with any velocity smaller than that of light* (1904), Collected papers **5**, pp. 172–197.

[17] H. Lorentz, *Deux mémoires de Henri Poincaré sur la physique mathématique*, Acta Mathematica **38** (1921), pp. 293–308.

[18] H. Poincaré, *À propos de la théorie de M. Larmor*, in L'éclairage électrique (1895), Œuvres complètes, vol. IX, pp. 369–426

[19] H. Poincaré, *La théorie de Lorentz et le principe de la réaction* (1900), Œuvres complètes, Tome IX, pp. 464–488.

[20] H. Poincaré, *Analyse de ses travaux scientifiques* (1901), Œuvres complètes vol. IX, pp. 1–14

[21] H. Poincaré, *La science et l'hypothèse*, Flammarion, Paris, 1902.

[22] H. Poincaré, *The present and future of mathematical physics*, Conférence de Saint Louis, Am. Inst. Phys. (1904), pp. 281–299.

[23] H. Poincaré, *Sur la dynamique de l'électron*, C. R. Ac. Sci. Paris **140** (1905), pp. 1504–1508.

[24] H. Poincaré, *Sur la dynamique de l'électron*, Rend. Circ. Mat. Palermo **21** (1906), pp. 129–176, Œuvres complètes vol. IX, pp. 494–550.

[25] J. Reignier, *La synchronisation de Poincaré, du temps local au groupe de Lorentz*, contribution to the colloquium of l'École des Mines, October 7, 2004, pp. 41–51.

[26] W. Voigt, *Über das Dopplersche Prinzip*, Gött. Nachr. **14** (1887).

[27] J. Wess and B. Zumino, *Supergauge transformations in four dimensions*, Nucl. Phys. **B70** (1974), p. 39.

[28] A. Wightman, *L'invariance dans la mécanique relativiste*, cours de l'École des Houches : *Relations de dispersion et particules élémentaires*, C. de Witt et R. Omnès, Hermann, Paris, 1960.

[29] E. Wigner, *On unitary representations of the inhomogeneous Lorentz group*, Ann. Math. **40** (1939), p. 149.

[30] E. Zeeman, *Causality implies the Lorentz group*, J. Math. Phys. **5** (1964), p. 490.

CHAPTER 18

Henri Poincaré as an Applied Mathematician

YVES POMEAU

Poincaré's eminent status did not at all prevent him from dedicating a great deal of effort to a problem of applied mathematics, where his deep knowledge of classical analysis allowed him to open, just a short time before his passing, a new area of research, that of small effects of transcendental order (exponentially small effects) in the solutions to differential equations, still an active and fertile topic of research.

18.1. Introduction: The physical problem and its history

In 1901, Marconi [**5**] succeeded in receiving in Newfoundland, Canada, a radio signal sent from Poldhu, in the west of England. The signal was transported by Hertzian waves, whose wavelength, rather poorly defined, must be on the order of a kilometer. Because the transmitter was located at a moderate altitude, it seemed that there could only be reception in the zone of diffraction, propagation in a straight line being impossible given the curvature of the Earth's surface. This remark elicited a great interest among researchers, because one could hope to calculate the amplitude of the diffracted wave by using methods developed over the course of the 19th century to solve the wave equation, beginning of course with the contribution of Fresnel. In fact, history shows that the perfection of mathematical analysis was more of an obstacle than a real help in obtaining the solution to the problem. Indeed, Macdonald rather quickly succeeded in finding an explicit solution in the form of a series (equation (18.7) below) that correctly solved the wave equation in spherical geometry. This solution is very difficult to handle, particularly in the limit of short waves (which is also called the WKB limit[1]), which is very singular and does not lead to easy approximations starting from the expression for the exact solution. Macdonald himself deduced from his result a grossly incorrect value of the transmission coefficient. The origin of this error was quickly discovered by Poincaré when he started to be interested in the problem around 1908–1909. He therefore took up the challenge posed by the mathematicians of the English school, turned toward mathematical questions often formulated in a very concrete version. Poincaré's genius allowed him to find the behavior of the transmitted amplitude in the short wave limit before anyone else. The summary that he wrote on the matter, published in Palermo's *Rendiconti* [**12**], is a monument of classical analysis.

[1] EDITOR'S NOTE: From the names of the physicists Gregor Wentzel, Hendrick Anton Kramers, and Léon Brillouin, who (simultaneously and independently) studied this limit for the Schrödinger equation in 1926. But, in fact, the study of this type of limit is much older.

Before returning to the mathematical question, I will mention what followed historically, from the viewpoint of the signal's physical transmission. Poincaré's essential result is the existence of an exponentially weak amplitude in the shadow zone. From a physical viewpoint, one quickly perceives that the amplitudes thus calculated would be much too weak to be detected with the techniques of the time. The observed variability of the quality of detection over a period of 24 hours made it possible to understand that the trans-Atlantic transmission took place in a sort of "wave guide" bounded on one side by the ocean's conductive surface and on the other by an ionosphere made conductive at high altitudes by the fact of its ionization by the Sun's rays. This explanation seems to have been proposed for the first time by Russian scholars working for the imperial Navy, even though, according to Yeang [**5**], "in the air" from the beginning, it was put in writing for the first time by Heaviside. Without going into the priority disputes, of little importance in hindsight, I will note only that at the very end of the memoir in the *Rendiconti* (p. 258), Poincaré mentions—following his conclusions, probably an afterthought to the proofs—this hypothesis of skywave propagation bounded by the ionosphere and the ocean. Without giving further details, he indicates that this hypothesis is one that is "often proposed". We remark, to finish this digression, that a motive of interest in these questions was that naval forces saw (quite rightly) the possibility for communicating by Hertzian waves with ships abroad, an important observation in those times preceding the first World War.

In Section 18.2, I explain how Poincaré found the explicit series expansion for the charge density induced by a wave on the surface of a conductor, the wave being emitted by a vertical dipole outside of the conducting sphere. Poincaré's calculation is very detailed, and I will only reproduce the principal steps. The final result is presented in the form of a series that, even if it is completely explicit, remains particularly inconvenient to manipulate in the physical limit of short waves (of wavelength much smaller than the radius of the Earth). This limit necessitates a double application of the method of steepest descent, which I will describe in broad strokes.

The diffusion of an EM (electromagnetic) wave by a conducting sphere constitutes what is today called the Mie problem (Gustav Mie being the one who posed it in complete generality for the first time). We can pass without too much difficulty from Mie's formulas to those of Poincaré by writing the field emitted by the dipole as a superposition of planar waves intersecting on the conducting sphere. For Mie's formulas, as for those of Macdonald and Poincaré, the approximation in the short wave limit remains difficult. In fact, this limit can be more easily studied by abandoning the completely explicit solution and returning to the wave equation (the Helmholtz equation in this case; cf. *infra*). As shown by Fock [**7**], Fresnel's approximation simplifies the wave equation sufficiently for one to be able to solve it completely in the region where the diffraction takes place. I will describe this calculation in Section 18.3.

Poincaré seems to have indeed been the first to find the transcendental order terms in the solutions to differential equations by extension into the complex plane, as claimed in the book by Born and Wolf [**4**]. They also attribute a certain paternity of this idea to Debye, whose contribution in this area seems rather to have consisted of summing up existing ideas. Poincaré's virtuosity, like that of Fock later on, in the analysis of oscillating integrals by no means marks the ultimate point in the development of a theory. On the contrary, it needed to be understood what happens to solutions of differential equations (or even of partial differential equations; see [**2**]) when a small parameter is placed in front of the highest-order derivative, what is in general called a "singular perturbation". In the problem treated by Poincaré, this aspect of things is hidden in the explicit solutions that poorly bring out the existence of very different spatial scales, the Earth's radius and the wavelength of an EM wave, a typical situation in singular perturbations. These ideas on short waves (or more generally of singular perturbations) experienced considerable progress with the 1985 work of Kruskal and Segur [**10**], which eliminated the need to have solutions in the form of explicit integrals, integrals that are at the very start of the analyses of Poincaré and Fock. Kruskal and Segur analyzed the solution of the original equation by an extension into the complex plane. This extension made it possible to understand the crucial role played by the singularities of the non-perturbed solution. This opened a new route of attack on what are now called perturbations "beyond all orders" and allowed the mastery of the divergent series that appear.

These problems on singular perturbations are in fact rather contemporary. They are found, for example, in fluid mechanics, where the viscosity is often formally small, although it appears in front of the highest-order derivatives in the Navier–Stokes equations. It must be acknowledged that, in the case of flows with high Reynolds number (that is, weak viscosity), we still do not understand well the consequences of this remark. Another striking consequence of the ideas of Kruskal and Segur is that certain phenomena that might seem insensitive to singular perturbation turn out instead to be completely changed by it. A good example is that of the non-robustness of certain physical solutions with respect to a singular perturbation. We mention, for example, the problem of the growth of a needle crystal in an undercooled melt. If we consider the equations of solidification without including the capillary effect (via the Gibbs–Thomson condition at the interface), a continuum of solutions exists for a given undercooling of the liquid, with an arbitrary rate of growth. Adding the effect of surface tension, which is a singular perturbation, we find by expanding the solution in increasing powers of the small parameter that this solution continues to exist order by order. Yet Martine Ben Amar and I have shown [**3**] that the result of this calculation order by order does not really have meaning, because it gives a series with zero radius of convergence (to leading order the general term for large n is on the order of $n!\varepsilon^n$, where ε is the small perturbing parameter). Once this divergence is taken into account, we find that no solution exists that includes the "small" capillary effects! The solution is re-established when we take into account the crystalline anisotropic effects in the Gibbs–Thomson condition. A relatively nearby mathematical situation is that of Saffman–Taylor fingers. I direct the reader interested in this question to the original articles [**6**].

I have chosen in Section 18.4 to describe with a few details a slightly different case (and one simpler than that of [**3**]) of singular perturbation, the effects of derivatives having order greater than those of the Korteweg–de Vries equation on localized solitons. This case is interesting, in part because it has been extended to a partial differential equation [**2**] and also because it allows the use of a Borel resummation to obtain the numerical value of a coefficient which is, in general, rather stubborn to analyze!

18.2. Poincaré's calculation: The solution in the form of a series and its estimation in the short wave limit

To summarize the problem formally solved by Macdonald and then by Poincaré, the idea is that an electric dipole oscillating sinusoidally in time is placed vertically at a distance D from the center of the Earth, which is assumed to be perfectly conducting. The exterior, which is what goes from the surface of the Earth to infinity, is assumed to be formed of a dielectric medium perfectly assimilated into the void. Apart from this last hypothesis, the model is relatively realistic because at the frequencies being considered, ocean water is a good conductor. So as not to return to it again, I recall once more that the defective hypothesis (from the physical point of view) is the one that assimilates the atmosphere into a dielectric, while a layer of conducting plasma at a high altitude in fact ensures the propagation of the waves beyond the horizon by sending them back toward the Earth's surface.

We adopt spherical coordinates, denoting by φ the "latitude" angle, measured from the position of the emitting dipole (therefore the emitting dipole is at $\varphi = 0$); the angle is contained between 0 and π. The dipole is located, as we have said, at a distance D from the center of the conducting sphere, a sphere of radius ρ, and we have the condition $D > \rho$. The field is emitted with the frequency $\omega/2\pi$, in such a way that all the quantities we will consider include a temporal phase factor $e^{-i\omega t}$, which we will omit in general.

Poincaré derived everything from Maxwell's equations, written without vector notation. I will content myself with justifying in a somewhat heuristic manner his result from page 192 of the memoir of the *Rendiconti*, equation (9). This equation gives the distribution of the charge created on the conducting sphere by the emitting dipole. Let, therefore, $\mu(\varphi)$ be the charge distribution, which only depends on the angle of latitude, measured starting from zero at the emitter. This distribution does not depend on the other angle (of "longitude") on the sphere because of the rotational symmetry around the axis passing through the center of the Earth and through the emitting dipole. This circumstance makes the intermediate formulas less heavy than Mie's formulas for the diffraction of an incident planar wave by a sphere, where this symmetry is absent. Poincaré, aware of the interest of this symmetry, devoted the entire beginning of his memoir (Section 3) to a detailed derivation of the integral equation to solve for the case of a conducting surface that is symmetric about an axis passing through the emitting dipole. I do not have the impression that such a case (for example, that of an ellipsoid of revolution) elicited any interest afterwards. It could be useful to look at what happens for example in the case of a very elongated conducting ellipsoid, representing the perturbation due to a conducting wire near the emitter.

The problem therefore consists in solving Maxwell's equations, which I will write in modern notation. The electric field is the sum of the field emitted by the dipole, say $\mathbf{E}_{\text{dip}}$, and of the field emitted by the current and charge distribution on the surface of the conductor. As shown for example in Jackson's book [**8**], equation (9.18), the electric field emitted by a dipole $\mathbf{P}e^{i\omega t}$ can be written as

$$\mathbf{E}_{\text{dip}} = k^2 \frac{e^{ikr}}{r}\Big(\mathbf{P} - \mathbf{n}(\mathbf{n}\cdot\mathbf{P})\Big) + \Big(3\mathbf{n}(\mathbf{n}\cdot\mathbf{P}) - \mathbf{P}\Big)\left(\frac{1}{r^3} - \frac{ik}{r^2}\right)e^{ikr}, \tag{18.1}$$

where $k = \omega/c$ (I will choose, as Poincaré did, a system of units such that $c = 1$, c being the speed of light), $\mathbf{n} = \mathbf{r}/r$, and where $\mathbf{r}$ is the vector expressing the difference between the positions of the dipole and the point where the electric field is measured (this is not the same vector as the one introduced a little later, but I am avoiding the proliferation of notation). This formula contains in particular that of the static dipole, for which $k = 0$ and for which the field decreases as the cube of the distance, which is represented by the term $(3\mathbf{n}(\mathbf{n} \cdot \mathbf{P}) - \mathbf{P})/r^3$ in the expression (18.1). For Poincaré, this dipolar field is the field denoted with stars, formula (1) on page 191 of the *Rendiconti* being that of one of the components of the field $\mathbf{E}_{\text{dip}}$ that I have just defined.

To this field must be added a field created by the current and charge distributions over the sphere, assumed to be infinitely conducting. The sum of the fields thus created must be normal to the sphere, assumed to be perfectly conducting.

The so-called *diffused* field (that is, the one emitted by the charges and the currents on the sphere) is written as a function of the currents alone by using the conservation of charge to eliminate the contribution of the charges. This gives:

$$\mathbf{E}_{\text{diff}} = \frac{ik}{c} \int d^2\sigma' \, \frac{e^{ik|\mathbf{r}-\mathbf{r}'|}}{|\mathbf{r}-\mathbf{r}'|} \, \mathbf{J}(\mathbf{r}').$$

This integral is taken over the surface of the conducting sphere, with the area element $d^2\sigma'$, the position being measured by $\mathbf{r}'$. This formula is obtained by applying the method of Green's functions for the Helmholtz equation. Therefore the superficial current $\mathbf{J}(\mathbf{r}')$ figures into this expression. This current, which, by symmetry, only has a component along the latitude, is determined by the condition that the total field $\mathbf{E}_{\text{dip}} + \mathbf{E}_{\text{diff}}$ is normal to the surface of the conducting sphere. In general, for diffraction problems (for example by a planar boundary, by a circular opening, etc.), this condition is translated by an integral equation for $\mathbf{J}(\mathbf{r}')$, which is more or less easy to solve. For our problem, this solution is rather easy to obtain after projecting the equation onto the angular part of the solutions to the Helmholtz equation. This is what Poincaré did, and from this he deduced the expression given later in (18.3) for the charge density over the sphere, the coefficients A_n being determined by projection (in the functional sense) of the components of $\mathbf{E}_{\text{dip}}$ onto the angular parts of the solutions to the Helmholtz equation. This is found in Section 13, p. 216 ff. of Poincaré's memoir, the general formula being found in equation (3) on page 217.

Section 9 of Poincaré's memoir recovers the laws of the propagation of light rays starting from the solution to Maxwell's equations. This result is certainly not original, because the connection between geometrical optics and the wave equation had already been established by Fresnel. Nonetheless, Poincaré's calculation involves a number of innovations, among which is the application of the shortwave method to the case where there is reflection of waves. Poincaré's starting point is equation (7) of Section 3, rewritten at the beginning of Section 9. This equation is the one that must be satisfied by the charge distribution μ over the surface of the conducting sphere, a sphere illuminated by a dipole radiating an EM wave. It is written as

$$-2\pi\mu + \int d^2\sigma'\mu' \left[\omega^2 L + \frac{d}{dn}\left(\frac{e^{-i\omega r}}{r}\right)\right] = N. \tag{18.2}$$

In this equation the integral extends to the entire surface of the sphere, with surface element $d^2\sigma'$, and N is the component of the electric field normal to the sphere due to the dipole. It is equivalent to

$$N = e^{i\omega(t-r)} \frac{i\omega}{r} \sin\theta \sin\xi,$$

where r is the distance MM' between the point M on the sphere's surface that determines the argument of N in equation (18.2) and the point M' where the surface element of the integral is centered. The angle θ is that of the tangent to the meridian passing through M' with the normal to the sphere at M, and ξ is the angle of the radius vector of $\mathbf{r}$ with the normal to the sphere at M. The integral of equation (18.2) represents the contribution of the diffracted field to the normal electric field (this exterior normal field is connected simply to the surface charge density for a conductor). Poincaré, who did not often associate any name to different functions, operators, etc., systematically called this equation a "Fredholm equation". Poincaré shows first that one can recover the laws of geometrical optics in the limit of short wavelengths. In this limit, the integral of equation (18.2) is calculated by the method of steepest descent. Indeed, it is of the type

$$J = \int N' K \, d^2\sigma',$$

where K is the integral kernel of the Fredholm equation (18.2), that is

$$K = i\omega \, \frac{e^{-i\omega r}}{r} \left(\frac{\cos\theta}{\cos\psi} - \cos\xi \right),$$

and where N' is the value of N at the point M'. The use of the method of steepest descent is justified because, in the short wave limit (that is, for very large ω), the most rapidly varying factor of this integral is the phase, which he calls $\omega\zeta$. The explicit solution to the Fredholm equation (18.2), which I will take as a starting point, therefore gives:

$$\mu(\varphi) = e^{i\omega t} \sum_n \frac{A_n}{\rho_0^2 I_n'(\omega\rho)} P_n(\cos\varphi). \tag{18.3}$$

In this equation, two functions appear: one of the angle φ, which is the Legendre polynomial $P_n(\cos\varphi)$, and the function $I_n'(\omega\rho)$. Naturally, these two functions come from the solution to the Helmholtz equation in spherical coordinates. The Legendre polynomial is a solution to

$$(1-\nu^2)\frac{d^2 P_n(\nu)}{d\nu^2} - 2\nu\frac{dP_n(\nu)}{d\nu} + n(n+1)P_n(\nu) = 0. \tag{18.4}$$

The index n is a positive integer, and we have set $\nu = \cos\varphi$ (Poincaré denotes it μ, which could create some slight confusion, since μ has already been used with a completely different meaning; the usage of the same symbol with multiple meanings sometimes makes a detailed reading of Poincaré difficult). The other function that appears in equation (18.3) is the derivative $I_n'(\omega\rho)$ that comes from the radial part of the Helmholtz equation. The argument of this function is the product $\omega\rho$, that is, 2π times the radius of the Earth, divided by the wavelength of the emitted EM wave. One can guess where the difficulties will be located, since this ratio is very large. The function I_n is one of the two solutions to the radial equation (equation (10) on page 189 of the *Rendiconti*):

$$\frac{d^2F(\zeta)}{d\zeta^2} + \left[1 - \frac{n(n+1)}{\zeta^2}\right] F(\zeta) = 0. \tag{18.5}$$

As Poincaré remarks, this equation is very close to the Bessel equation. In reality, it is included in a rather simple extension of the Bessel equations, and we can express its solution in terms of Hankel functions (which Poincaré does not do). The solution to equation (18.5) that must be retained is the one that, at infinity, is *nearly equal* [*sensiblement égale*][2] to $e^{-i\varsigma}$, which defines the outgoing (also called radiating) wave condition.

The coefficients A_n are given by the following general formula (derived from a projection onto the solutions of the Helmholtz equation in spherical coordinates—the *spherical harmonics*—which I have avoided introducing, since Poincaré never spoke of them):

$$\frac{2e^{i\omega t}}{2n+1} A_n J_n(k\rho)\frac{1}{\rho^2} = \int_0^\pi f'^* P_n(\cos\varphi)\sin\varphi\, d\varphi. \tag{18.6}$$

In this expression, f'^* is the component of the electric field of the dipole tangent to the sphere. Given the extreme complexity of how this field depends on the various geometric quantities, it is completely impossible to find a closed expression for (18.6)—that is, an expression with a finite number of terms and only involving elementary functions. On the other hand, in the short wave limit, it is possible to calculate the amplitudes A_n by the method of steepest descent, which Poincaré does. The result (equation (11), p. 240) is

$$A_n = \frac{-i}{4\pi\omega^2 D^2} I_n(\omega D).$$

From this general expression for A_n, Poincaré derives the solution obtained in the limit of geometrical optics. This solution, naturally, has strictly zero amplitude in the shadow zone, which is where the field amplitude resulting from diffraction must be sought.

This concludes the estimate of the exact formal solution to the problem that was posed. The analysis could continue, and that is where Poincaré's essential innovation is found. In fact, within this problem there exists "a small parameter", which is the ratio of the wavelength to the Earth's radius. This small parameter (or rather, its inverse) has already appeared in the argument concerning Hankel functions, in the form of the combination $\omega\rho$, where $k = \omega$ is the wavenumber of the EM wave (recall that $c = 1$) and ρ is the radius of the Earth. This short wave limit would allow Poincaré, using a profoundly original method, to calculate the amplitude of the diffracted wave.

Once the coefficients A_n were known, Poincaré obtained a completely explicit expression for the charge density induced on the sphere by the emitting dipole:

$$\mu(\varphi) = -\frac{i}{4\pi\omega^2 D^2\rho^2}\sum_n n(n+1)(2n+1)\frac{I_n(\omega D)}{I_n'(\omega\rho)} P_n(\cos\varphi), \tag{18.7}$$

where, we recall, $I_n(\cdot)$ is a Hankel function that solves equation (18.5) with the radiating condition at infinity, and P_n is the Legendre polynomial that solves (18.4). The idea is therefore to determine the behavior of $\mu(\varphi)$ in the shadow zone, for a finite value (i.e., not infinitely small) of the angle φ, while the arguments ωD and $\omega\rho$ of the Hankel functions are large and close to each other. To explain how Poincaré

[2]An elegant expression used by Poincaré, a much better turn of phrase, I think, than our endless "tends to" or "is equivalent to".

carried out this calculation, I will paraphrase his conclusion and then fill in the gaps.

This conclusion forms Section 20, page 256 ff. Poincaré first remarks that, in the limit of the physical problem where D is very close to ρ, we can replace the expression for $\mu(\varphi)$ by

$$\mu(\varphi) = -\frac{i}{4\pi\omega^2\rho^4}\sum_n R(n)P_n(\cos\varphi),$$

where the function $R(n)$ is such that

$$R(n) = n(n+1)(2n+1)\frac{I_n(\omega\rho)}{I'_n(\omega\rho)}.$$

In fact, this series for $\mu(\varphi)$ is divergent, but we can consider it to be equal to the limiting value (which exists) when the arguments $\omega\rho$ and ωD of the two Hankel functions are very close; in (18.7) we write

$$\frac{I_n(\omega D)}{I'_n(\omega\rho)} = \frac{I_n(\omega D)}{I'_n(\omega D)}\cdot\frac{I'_n(\omega D)}{I'_n(\omega\rho)}$$

(with $\rho < D$), and Poincaré shows that $I'_n(\omega D)/I'_n(\omega\rho)$ *"for very large n is nearly equal to $(\rho/D)^{n+1}$"*; then he proves that as ρ tends to D, the sum

$$\sum_n n(n+1)(2n+1)\frac{I_n(\omega D)}{I'_n(\omega D)}\left(\frac{\rho}{D}\right)^{n+1}P_n(\cos\varphi)$$

tends to a limit, and that

$$\sum_n n(n+1)(2n+1)\frac{I_n(\omega D)}{I'_n(\omega D)}P_n(\cos\varphi)$$

then tends to the same limit.

The series calculation is done by first observing that $R(n)$ is a meromorphic function of n that satisfies the condition $R(n) = -R(-n-1)$ (because I_n/I'_n and $n(n+1)$ do not change when we change n into $-n-1$, while $2n+1$ changes sign). This function has a discrete set of simple poles; using the residue theorem and the antisymmetry of $R(n)$ to exchange n and $-n-1$, Poincaré obtains the polar decomposition of $R(n)$ in the form

$$R(n) = \Pi(n) + \sum_k\left[\varpi_k(n) + \frac{B_k}{n-\varepsilon_k} - \frac{B_k}{-n-1-\varepsilon_k}\right],$$

where $\Pi(N)$, $\varpi_k(n)$ are second-degree polynomials (the ϖ_k making it possible to render the sum over k convergent), and the ε_k are the poles of the positive imaginary part. To each ε_k corresponds a pole of the negative imaginary part, which is $-1-\varepsilon_k$, with the same residue B_k.

Recall that we are seeking to calculate the sum $\sum_n R(n)P_n(\cos\varphi)$. Poincaré begins by showing that the polynomial $\Pi(n)$ does not contribute to this sum. For this, he remarks that the relation $\Pi(-n-1) = -\Pi(n)$ means that Π is an odd polynomial in the variable $2n+1$, and hence a linear combination of monomials of the form $(2n+1)^{2k+1}$. However, by using the fact that

$$\sum_n P_n(\cos\varphi)z^n = \frac{1}{\sqrt{1-2z\cos\varphi+z^2}} \qquad \text{for } |z|<1$$

(this is the generating function of the Legendre polynomials), Poincaré observed that

$$\sum_n (2n+1)^{2k+1} P_n(\cos\varphi)$$

(defined as the limit of

$$\sum_n (2n+1)^{2k+1} P_n(\cos\varphi)\left(\frac{\rho}{D}\right)^n \qquad \text{as} \quad \rho \to D$$

with $\rho < D$) is zero; therefore $\sum_n \Pi(n) P_n(\cos\varphi) = 0$. In the same way, for all k, $\sum_n \varpi_k(n) P_n(\cos\varphi) = 0$. Therefore all that remains is the contribution

$$\sum_n R(n) P_n(\cos\varphi) = \sum_k B_k \sum_n \left[\frac{1}{n-\varepsilon_k} + \frac{1}{n+1+\varepsilon_k}\right] P_n(\cos\varphi). \tag{18.8}$$

In Section 19, which precedes the conclusion, Poincaré calculates this sum by a very cunning method, of which I will indicate the principal steps. He first uses a sequence of identities that allow him to take advantage, once again, of the fact that the Legendre polynomials are the coefficients of the Taylor series of $1/\sqrt{1-2z\cos\varphi+z^2}$ with respect to z:

$$f(z) = \sum_n P_n(\cos\varphi) z^n = \frac{1}{\sqrt{1-2z\cos\varphi+z^2}}. \tag{18.9}$$

The series involving the coefficients $P_n(\cos\varphi)$ in equation (18.8) is not a geometric series as that of the Legendre formula is, but a series with simple rational coefficients, which may be obtained by an integral transformation and by using the following identity, valid for every sequence A_n provided that the series involved are convergent: If $f(z)$ is such that

$$f(z) = \sum_n A_n z^n,$$

then the function $U(x) = \sum_{n=0}^{\infty} \frac{A_n}{x-n-1}$ is such that

$$U(x) = \lambda(x) \int_C f(z) z^{-x}\, dz,$$

where $\lambda(x) = e^{2i\pi x}/\big(e^{2i\pi x}-1\big)$. The path of integration is found by assuming that the function $f(z)$ is holomorphic in the unit disk: it is a curve that starts from $z=1$ with argument 0, turns around 0 in the counterclockwise direction while staying in the interior of the unit disk, and returns to $z=1$ (with argument 2π). From this, Poincaré concludes that

$$\sum_n \left(\frac{1}{x-n-1} - \frac{1}{x+n}\right) P_n(\cos\varphi) = \lambda(x) \int_{C'} f(z) z^{-x}\, dz,$$

where now the path of integration starts from $z=1$ with argument 0, turns around 0 in the counterclockwise direction in the interior of the unit disk, returns to $z=1$ (with argument 2π), then makes another turn in the other direction in the exterior of the disk and returns to $z=1$ (with argument 0). We study the limit of large x, which is also that of short waves. In this limit, where he uses the method of steepest descent, Poincaré shows that at the dominant order,

$$\sum_n \left(\frac{1}{x-n-1} - \frac{1}{x+n}\right) P_n(\cos\varphi) \approx K' \frac{e^{-ix\varphi}}{\sqrt{(1-e^{-2i\varphi})x}},$$

where K' is a numerical constant. This limiting value is applicable when x has a very negative imaginary part. This is exactly the case if $-x$ is one of the poles ε_k. Poincaré finally obtains the asymptotic behavior of μ (equation (10), p. 257):

$$\mu \approx \frac{K''}{\omega^2\rho^4}\frac{B_1 e^{i\varepsilon_1\varphi}}{\sqrt{(1-e^{-2i\varphi})\varepsilon_1}}. \tag{18.10}$$

We recall that B_1 is the residue of $R(n)$ (viewed as a holomorphic function n) at ε_k. Therefore it remains to find this pole and its residue.

This pole corresponds to values of n of the same order as the large value of the argument of the Hankel function I_n, which is therefore $n \sim \omega\rho$. In this limit Poincaré uses the integral representation

$$y(\zeta) = \int e^{i\zeta x}(x^2-1)\zeta^{n+1}\,dx$$

of the solution to equation (18.5) which defines the Hankel functions, the path of integration going from -1 to infinity in such a way that $e^{i\zeta x}$ tends to zero. It is at this crucial turning point that the estimation by the method of steepest descent is introduced. Indeed, by setting $n = z\zeta$, where z is now of order 1 (so that ζ is assumed to be large), Poincaré transforms the integral giving the Hankel function into

$$\int \eta e^{i\zeta\Theta}\,dx,$$

where η is a function of x that is not important in what follows, while $\Theta(x) = x - iz\ln(x^2-1)$. Of course, we recognize an oscillating integral in the limit of large ζ. The dominant contribution comes in general from the saddle points, that is, the roots of Θ':

$$x = iz \pm \sqrt{1-z^2}.$$

But in the present case, since we are in a neighborhood of $z=1$, we get a double saddle (the second derivative of Θ also vanishes), and we must therefore push the expansion of the argument of the exponential a notch further than in the usual case, that is, to the third order:

$$\Theta \approx \alpha + \beta x^3,$$

α and β being numeric coefficients. The integral becomes $\int e^{i\zeta\beta x^3}\,dx$ (up to a factor), where we recognize the solution of the Airy equation by quadrature, which Poincaré writes, without calling it such (equation (1), p. 211). The various asymptotic estimates of the Airy function give, in the domain of the values of n that are concerned, the behavior $I_n \approx K_1'\zeta^{1/6}F(te^{4i\pi/3})$, where t is an avatar of the variables in the problem. It is defined as $\alpha^2\zeta^{2/3}$ (in contrast with Poincaré, I have kept the initial variable ζ, while Poincaré changed this ζ into ω along the way). The function $I_n'(\omega\rho)$ being found in the denominator in the definition of the $R(n)$, we conclude rather easily that the poles ε_n are at the zeroes of I_n. On the other hand, the final result (in the shadow zone) is dominated by the smallest of these poles (as was assumed in order to write (18.10)). Therefore we have

$$\varepsilon_1 \approx \omega D - t_0(\omega D/2)^{1/3},$$

where t_0 is the smallest root of the equation $F'(te^{4i\pi/3}) = 0$, F' being the derivative of the Airy function defined by the integral over the real axis:

$$F(t) = \int_{-\infty}^{+\infty} e^{i\left(tZ - \frac{Z^3}{3}\right)}\, dZ.$$

As claimed, the amplitude of the signal decreases exponentially in the shadow zone, with a factor $e^{-it_0(\omega D/2)^{1/3}\varphi}$ that is exponentially small since $\operatorname{Im} t_0 < 0$ (because $\operatorname{Im} \varepsilon_1 > 0$).

This is the final result of Poincaré's efforts, a great work of art by a great analyst.

18.3. Fock's analysis

Fock [**7**], without citing Poincaré on this matter, took up the problem of the diffraction of an EM wave by a conducting sphere in the limit where the wavelength is much less than the sphere's radius, the same limit as was studied by Poincaré. Fock used a method inspired by Fresnel, which consisted in looking for the solution in the form of a planar wave with a slowly varying complex amplitude. Like Fock, I will restrict myself to examining what happens close to the boundary between the illuminated and non-illuminated regions on the sphere (in the limit of geometrical optics). The idea is to consider an incident planar wave that arrives along the x-direction from $-\infty$. The conductive sphere is locally a surface whose normal is close to the z-axis. It has the local Cartesian equation

$$z(x, y) + \frac{1}{2}(ax^2 + 2bxy + cy^2) = 0.$$

The sphere is assumed to be located in the negative z half-space. This implies the inequalities $a > 0$, $b > 0$, and $ac - b^2 > 0$. For a sphere of radius ρ,

$$a = \frac{1}{2\rho} = c, \qquad b = 0.$$

In fact, the variable y will only play a marginal role in this problem, and from now on I will forget it, locally replacing the sphere by the cylinder having equation

$$z(x, y) + \frac{1}{2}ax^2 = 0. \tag{18.11}$$

The reader interested in the more general situation can turn to [**7**].

The idea is therefore to solve the Helmholtz equation (the connection with the complete problem of solving Maxwell's equation can be found in Fock's book [**7**])

$$\nabla^2\Psi + k^2\Psi = 0 \tag{18.12}$$

with appropriate boundary conditions on the surface of Cartesian equation (18.11). Fresnel's approximation reduces to assuming that the solution to this problem is of the type

$$\Psi(x, z) = e^{ikx}\Psi^*(x, z),$$

where the function $\Psi^*(x, z)$ varies much more slowly as a function of x and of z than the phase factor e^{ikx}. This hypothesis (whose consistency is to be verified *a posteriori*) leads to neglecting the second derivative $\partial^2\Psi^*(x, z)/\partial x^2$ in front of $2ik\partial\Psi^*(x, z)/\partial x$ in the equation for $\Psi^*(x, z)$ deduced from (18.12). Starting with equation (18.12) and the definition of $\Psi^*(x,z)$, we therefore find Fresnel's equation for $\Psi^*(x,z)$:

$$\frac{\partial^2 \Psi^*}{\partial z^2} + 2ik\frac{\partial \Psi^*}{\partial x} = 0. \tag{18.13}$$

The boundary conditions are that, for x tending to $-\infty$, $\Psi^*(x,z)$ tends to 1 (the incident plane wave), and the Neumann boundary conditions on the surface of the sphere. This equation is written for the original function $\Psi(\mathbf{r})$ as

$$\mathbf{n} \cdot \nabla\Psi(\mathbf{r}) = 0.$$

The vector $\mathbf{n}$ is normal to the surface of the sphere, with components (ax, by, z), up to an unimportant scalar factor. Thanks to this choice, the Neumann condition on the sphere can be written as

$$\frac{\partial \Psi^*}{\partial z} + ikax\Psi^* = 0.$$

This condition is imposed on the surface $z + \frac{1}{2}ax^2 = 0$. The problem's difficulty comes precisely from this condition on a locally parabolic surface.

To get around this, Fock devises a first change of function,

$$\Psi^*(x,z) = \widehat{\Psi}(x,z)e^{-ikaz}.$$

This change of variables transforms equation (18.13) into

$$\frac{\partial^2 \widehat{\Psi}}{\partial z^2} + 2ikax\frac{\partial \widehat{\Psi}}{\partial z} + 2ik\frac{\partial \widehat{\Psi}}{\partial x} - 2k^2 a\left(z + \frac{1}{2}ax^2\right)\widehat{\Psi} = 0.$$

Thanks to this change of variables, the boundary condition simplifies to $\partial\widehat{\Psi}/\partial z = 0$ for $z + \frac{1}{2}ax^2 = 0$. The next step consists in a new change of variables, which will simplify the support of the boundary condition. We take $\zeta = z + \frac{1}{2}ax^2$ as a new variable, which makes it possible to have the boundary condition at $\zeta = 0$, an essential point. The other variable is simply $\xi = x$. We define a new function $W(\xi,\zeta) = \widehat{\Psi}(x,z)$, where the relations between (x,z) and (ξ,ζ) were just given. We find that $W(\xi,\zeta)$ must satisfy at $\zeta = 0$ the boundary condition

$$\frac{\partial W}{\partial \zeta} = 0.$$

The Fresnel equation becomes:

$$\frac{\partial^2 W}{\partial \zeta^2} + 2ika\xi\frac{\partial W}{\partial \zeta} + 2ik\frac{\partial W}{\partial \zeta} - 2k^2 a\zeta W = 0. \tag{18.14}$$

Taking $2^{-4/3}k^{-1/3}a^{-2/3}$ as the unit of length for ξ and $2^{-1/3}k^{-2/3}a^{-1/3}$ for ζ, we eliminate every parameter from equation (18.14), which becomes the numerical equation

$$\frac{\partial^2 W}{\partial \zeta^2} + i\xi\frac{\partial W}{\partial \xi} - \zeta W = 0.$$

In these changes of variables, we find again the 1/3 powers so characteristic of the Fresnel approximation, the Airy equation, etc. Moreover, we note that the length scale in the direction x is, in physical terms, the wavelength λ to the 1/3 power times the radius of the sphere to the 2/3 power, while the length scale in the normal direction to the sphere is as the radius to the 1/3 power times the wavelength to the 2/3 power, therefore much less than in the direction x.

The (last) change of function $V(\xi,\zeta) = e^{i\xi\zeta - \frac{i}{3}\xi^3} W(\xi,\zeta)$ makes it possible to simplify this last equation, giving

$$\frac{\partial^2 V}{\partial \zeta^2} + i\frac{\partial V}{\partial \xi} + \zeta V = 0. \tag{18.15}$$

The boundary condition is $\partial V/\partial\zeta = 0$ for $\zeta = 0$, and there is also a radiating condition for large negative values of ξ, which is that $V = e^{i\xi\zeta - \frac{i}{3}\xi^3} - V^*$, the constant V^* representing the possible contribution of reflected waves in the region of large negative values of ξ, where they can exist for a reflector of a given form.

The equation we have obtained above is fundamentally solvable by Laplace's method, since its coefficients are at most linear in the variable. Fock remarks that equation (18.15) has as a formal solution a function $V_0 = e^{i\xi t} w(t+\zeta)$, where t is a parameter without significance for the moment. Inserting the form of this function into equation (18.15), we find that $w(\cdot)$ must be a solution to the Airy equation:

$$\frac{d^2 w}{dt^2} = t w(t).$$

In Poincaré's calculation, this Airy equation appears, but only after a large number of steps. The formal solution to the Airy equation is well known, and it is written as

$$w(t) = \frac{1}{\sqrt{\pi}} \int_\Gamma d\tilde{z}\, e^{\tilde{z}t - \frac{1}{3}\tilde{z}^3}.$$

This integral is calculated over a contour Γ which, at infinity, must join two sectors where the integrand tends to zero. For these large values of $|\tilde{z}|$, the integrand is dominated by the cubic term. It is therefore necessary that $\tilde{z}^3$ have a real part that is positive or zero. This occurs in the three sectors $\left[-\frac{\pi}{6}, \frac{\pi}{6}\right]$, $\left[\frac{\pi}{2}, \frac{2\pi}{3}\right]$, and $\left[\frac{4\pi}{3}, \frac{3\pi}{2}\right]$. We obtain the two independent solutions to the Airy equation by choosing a branch of Γ, coming for example from the sector $[\pi/2, 2\pi/3]$ and going toward the sector $[-\pi/6, \pi/6]$, and the other leaving from the sector $[4\pi/3, 3\pi/2]$ and going toward the sector $[-\pi/6, \pi/6]$. We call the solution to the Airy equation defined by the first choice $w_1(t)$ and that defined by the second $w_2(t)$. Formally, we obtain a solution to the complete equation for V by arbitrarily superposing solutions of the type $e^{i\xi t} w(t+\zeta)$. This allows us to write the solution in the form

$$V(\xi,\zeta) = \frac{i}{2\sqrt{\pi}} \int_C dt\, e^{i\xi t} [w_2(t+\zeta) - f(t) w_1(t+\zeta)].$$

The contour of integration (now in terms of the variable t) will be determined a little later, once we have found the function $f(t)$. In this latter representation of the solution, we have used the fact that, once endowed with the proper multiplicative coefficient, the part of the integral that is proportional to w_2 represents well the boundary condition of the incident wave (we see this by analyzing its phase at a great distance) while the one proportional to w_1 represents the reflected wave.

It remains to satisfy the condition on the conducting surface. This condition is written as $\partial V/\partial\zeta = 0$ for $\zeta = 0$. It is satisfied for $f(t) = w_2'(t)/w_1'(t)$, where the primes denote the derivatives. We are close to our goal, since we know the explicit form of $V(\cdot,\cdot)$! In fact we are looking for the asymptotic behavior of $V(\xi, \zeta = 0)$ for large ξ, with

$$V(\xi, 0) = \frac{1}{\sqrt{\pi}} \int_C dt\, \frac{e^{i\xi t}}{w_1'(t)}.$$

When we make the substitutions to obtain this formula, we must find the (constant) value of the Wronskian $w_1'(t)w_2(t)-w_1(t)w_2'(t)$, which equals precisely $-2i$ with our choice of normalization in the definitions of the w. The integral that gives $V(\xi,0)$ converges if we take a path C that goes from the direction with orientation $2\pi/3$ to the one with orientation $-\pi/3$ to ensure convergence at infinity. The behavior of this integral for large ξ is determined by the zero of the denominator that is closest to the real axis, which conforms with Poincaré's result. By returning to the physical quantities, we indeed find the same law of decrease that Poincaré did, with the same coefficient in the exponent.

The radio-technical literature of the 1940s–1960s [**9**] defines Fock functions that equal

$$f(x) = \frac{1}{\sqrt{\pi}}\int_\Gamma \frac{e^{ixt}}{w_1(t)}\,dt \qquad \text{and} \qquad g(x) = \frac{1}{\sqrt{\pi}}\int_\Gamma \frac{e^{ixt}}{w_1'(t)}\,dt,$$

where the function $w_1(t)$ is the one we have introduced and where the circuit of integration comes from infinity along the direction of the half-axis with angle $-2\pi/3$ and goes toward infinity in the direction of the positive real half-axis. These functions were tabulated during the same period.

18.4. Dispersion of solitons in perturbed KdV

An important advance in applied mathematics was the discovery, made by Kruskal and his collaborators, of the connection between certain non-linear partial differential equations and associated linear problems, a beautiful story already narrated by those more knowledgeable than I [**1**]. One of the interesting properties of these systems (called integrable) is that at least some of them possess localized solutions that propagate at a finite speed without being deformed (this deals especially with functions of one variable of time and another of space, these properties tending to disappear when one adds dimensions to space). These structures, christened "solitons", more or less correspond to various observations of waves in shallow channels, although I am not certain of the truly non-linear character of the waves observed. To take an example, the waves of tsunamis are clearly not solitons since, during their offshore propagation, they are so weakly non-linear that this non-linearity has no effect at all. When the tsunamis approach the shore, the variation of depth and the non-linear effects conjugate their effects to give tidal waves, which are also nothing like the soliton.

As the world is imperfect, fortunately, we can legitimately inquire about the persistence of solitary waves in a real world where one would certainly find differences, formally small ones, from the ideal mathematical model. The rather abstract property of integrability is, without a doubt, not one whose existence or absence could influence the manifestations of the real physical world. From this point of view, it seems more important to examine if solitary waves continue to exist or not, once we have taken into account these small differences between the real and the integrable model. These reflections, in brief somewhat banal, have led us to consider the following problem: the archetype of "soliton" integrable partial differential equations is the so-called Korteveg–de Vries (KdV) equation which is written, after changing variables and function (to have a numeric equation, without any parameter in the boundary conditions or in the form of the equation itself) as

$$\frac{\partial u}{\partial t} + u\frac{\partial u}{\partial x} + \frac{\partial^3 u}{\partial x^3} = 0. \tag{18.16}$$

The function $u(x,t)$ can be seen as the height of water in a shallow channel, measured in the coordinate system that moves at the velocity of the gravity waves in this channel, waves that are not dispersive in the first approximation (for a wavelength much larger than the depth of the channel). As shown long ago by Bernoulli, the speed of these waves is $\sqrt{gh_0}$, where g is the acceleration of gravity and h_0 is the depth of the channel at rest. If the wave in question has a height greater than h_0, the maxima of the height will propagate at a faster rate than the Bernoulli value for the non-perturbed height h_0. It will then follow that the wave breaks, according to the Riemann mechanism: the tallest, most rapid perturbations will catch up with the shorter waves in front. This breaking mechanism is well represented by the non-linear term in $u\,\partial u/\partial x$ of equation (18.16).

For equation (18.16) without the third derivative, we know the formal solution, due to Laplace:

$$u(x,t) = u_0(x - ut),$$

u_0 being the initial condition $u(x, t = 0)$. This formal solution presents a breaking wave that disappears if we include the third derivative $\partial^3 u/\partial x^3$ in (18.16). The solution to the complete equation in general evolves towards a family of solitons that are particular solutions of KdV such that $u(x,t)$ depends on x and t through a combination $x - Ct$, where C is an *a priori* arbitrary parameter. The soliton solution, of type $U(x - Ct)$, tends toward zero when its argument $\chi = x - Ct$ tends toward infinity. An elementary calculation shows that $U(\cdot)$ is given by

$$U(\chi) = \frac{C}{2\cosh^2(C^{1/2}\chi/2)}. \tag{18.17}$$

This family with one parameter (C, which must be positive) represents finite-energy solutions, the energy associated to a solution being in general

$$\mathcal{E} = \int_{-\infty}^{+\infty} \frac{u^2}{2}\,dx. \tag{18.18}$$

This magnitude, which is *a priori* a function of time, is in fact a constant of the motion if the solution decreases to zero sufficiently fast at infinity. It is easily checked that the soliton solution indeed has finite energy.

We then ask: do there always exist such localized solutions propagating as a unit at finite speed when we modify the KdV equation "a little"? For such solutions to continue to have a physical meaning, we require at least that their energy be finite. From the nearly infinite number of perturbations one could imagine adding to the KdV equation, we have considered the additions of a fifth spatial derivative with a small coefficient. In principle this perturbation could be found in by systematically developing the dispersion equation of linear waves in the case of large wavelength, the small parameter being the ratio of the depth to the wavelength. In order to be able to change the coefficient of the fifth derivative, we must also introduce the effects of the surface tension, which (practically) is the ratio of the wavelength to the capillary length. For a given ratio of *capillary length over fluid depth*, we can cancel the fifth derivative term. In a neighborhood of this situation, we can actually have a small fifth derivative term with either sign.

I propose to examine, following [**13**], the generalized KdV equation:

$$\frac{\partial u}{\partial t} + u\frac{\partial u}{\partial x} + \frac{\partial^3 u}{\partial x^3} + \delta^2\frac{\partial^5 u}{\partial x^5} = 0. \tag{18.19}$$

In this equation, the coefficient of the fifth derivative has a well-defined sign, except if $\delta = 0$, which I will exclude. This fifth derivative term is written in such a way that it is the only one with an explicit coefficient, since to eliminate it (to reduce it to 1, for example) we would have to make dilations of the other magnitudes in the problem which would introduce coefficients elsewhere in the problem. It is therefore legitimate to consider δ to be small. By transformations we could also make a small coefficient appear in front of the third derivative in the original KdV equation. But this small coefficient could be reduced to unity by dilations, which would prevent us from considering it to be truly small.

In general, a progressive wave of speed C is a solution to the equation (18.19) of the type $U(x - Ct = \chi)$, $U(\cdot)$ being a solution to the ordinary differential equation

$$U\frac{\partial U}{\partial \chi} + \frac{\partial^3 U}{\partial \chi^3} + \delta^2\frac{\partial^5 U}{\partial \chi^5} = C\frac{\partial U}{\partial \chi}, \tag{18.20}$$

or, after an obvious first integration (the constant equals zero for solutions tending to zero at infinity, along with their derivatives), to the equation

$$\frac{U^2}{2} + \frac{\partial^2 U}{\partial \chi^2} + \delta^2\frac{\partial^4 U}{\partial \chi^4} = CU.$$

The question at hand is that of the robustness of the soliton solution (18.17) with respect to the addition of a small perturbation of a higher order of derivation. We are looking for a solution that is *a priori* localized, that is, tending to zero at infinity. To see that this is possible, we reason as follows. If the solution tends to zero at infinity, it must be the solution of the linearized modified KdV equation. For the present case, this is the equation

$$\frac{\partial^3 V}{\partial \chi^3} + \delta^2\frac{\partial^5 V}{\partial \chi^5} = C\frac{\partial V}{\partial \chi}.$$

I have written this equation for an unknown function $V(\chi)$ to emphasize the fact that this concerns only the linearized version of (18.20). This equation has the obvious solutions $e^{\pm iq\chi}$ and $e^{\pm\kappa\chi}$, q^2 and $-\kappa^2$ being the positive (for q) and negative (for κ) roots of the polynomial in X:

$$\delta^2 X^2 + X = C.$$

For small δ, these roots are such that

$$q^2 = \delta^{-2} + \cdots \qquad \text{and} \qquad \kappa^2 = C - \delta^2 C^2 + \cdots .$$

To have a solution that decreases at infinity, we must eliminate, by a choice of initial conditions for the solution to equation (18.20), three prefactors of functions that exponentially increase or oscillate at $+\infty$ and also three factors at $-\infty$. These six conditions cannot therefore be imposed on a fifth-order equation. However, using the same type of argument, we find that by changing the sign of the coefficient of the fifth derivative, the linearized equation has two increasing modes in each direction, which authorizes the choice of initial conditions ensuring a decreasing solution on both sides. As a consequence, in the case examined here, with the positive sign for the coefficient of the fifth derivative, at best we can hope for a solution that does not decrease at infinity but oscillates with a constant non-zero amplitude, the kind of situation we will examine now. We will in particular find the amplitude of these oscillations at infinity.

Following the philosophy of Kruskal and Segur's approach [**10**], we first look at the solution for small δ by a formal expansion in integer powers of δ^2. At the lowest order, the solution to the equation (18.20) can be written as

$$U_0(\chi) = \frac{C}{2\cosh^2\left(C^{1/2}\chi/2\right)};$$

this expression was already given in (18.17), but as an exact solution to the equation without the fifth derivative. If we add this term, by assigning to it a small coefficient δ^2, the solution $U_0(x - Ct)$ is an exact solution only if this coefficient is zero. Otherwise, we must add to it a formal perturbation series obtained by substituting term-by-term this infinite expansion in the complete equation and by solving the equations in χ obtained by eliminating the order δ^{2n} term of equation (18.20) for all integers n. The formal series thus obtained is well defined term-by-term if we impose the condition that it be an even function of χ, and to begin as $U(\chi) = U_0(\chi) + \delta^2 U_1(\chi) + \cdots$, where U_0 is already known, and U_1 is the even solution to

$$U_0\frac{\partial U_1}{\partial \chi} + U_1\frac{\partial U_0}{\partial \chi}\frac{\partial^3 U_1}{\partial \chi^3} + \delta^2\frac{\partial^5 U_0}{\partial \chi^5} = C\frac{\partial U_1}{\partial \chi}.$$

This equation can be solved by quadrature because of the algebraic properties of the functions $\cosh^{-2}\chi$, the sought-after even solution being a superposition of such functions. Naturally, this formal expansion nowhere involves the fact that the solution thus found neither grows nor oscillates at infinity. More exactly, we never encounter an integration condition that imposes the elimination of the solutions that grow or oscillate at infinity. The convergence of the expansion is not at all assured, either.

The problem treated by Poincaré has a mathematical structure close to the present problem but for the fact that the solution is given explicitly and it only remains to calculate the integrals that it gives in the short wave limit. This strategy is impossible in the "soliton" case of the modified KdV since we do not know the solution in a closed form. The method of Kruskal and Segur consists in reconsidering the formal series in powers of δ^2, looking for where this series could become ill conditioned, and understanding how these deviations could affect the perturbation series in the region where it is formally well conditioned (by well or ill conditioned, I mean that successive terms decrease or not, respectively).

The series becomes ill conditioned near points of the complex χ-plane where the order 0 solution (which is U_0) has a singularity, here a double pole. Near such a pole the perturbation, which is a derivative of higher order than those of the order 0 equation, becomes more singular than the other terms of the equation, which prohibits considering it as a small perturbation. More precisely, the poles of U_0 are at the zeroes of $\cosh(\chi C^{1/2}/2)$, that is, at $\chi_n = i\pi(2n+1)C^{-1/2}$, n being an integer. "Near" these poles the perturbation series becomes ill conditioned because the successive terms have poles of increasing order as we consider higher and higher order terms in this series. If we define the local variable as $\widetilde{\chi} = \chi - i\pi(2n+1)C^{-1/2}$, the function U_0 diverges like $(\widetilde{\chi})^{-2}$ close to $\widetilde{\chi} = 0$, while U_1 diverges like $(\widetilde{\chi})^{-4}$, U_2 like $(\widetilde{\chi})^{-6}$, etc. The successive divergent contributions of this expansion are themselves affected by coefficients with increasing powers of δ. The natural idea is therefore to regroup these two effects—increasing order in the small parameter and increasing order of divergence—by defining a "stretched" variable that stays finite when $(\widetilde{\chi})^{-2}$ diverges and δ tends to zero, in such a way that the successive terms of the expansion of the solution $U(\chi)$ of the complete solution all remain finite order. This happens when the ratio $\widetilde{\chi}/\delta$ remains finite. We therefore take $\zeta = \widetilde{\chi}/\delta$ as our stretched variable. The order zero solution diverges like $(\widetilde{\chi})^{-2}$ near double poles, and so the function itself is of order δ^{-2} when $\widetilde{\chi}$ tends to zero while being of order δ. In order for the transformed function to have formal order zero with respect to the small parameter once the change of scale has been carried out, we define the function $u = U/\delta^2$ as a new unknown function. Writing equation (18.20) with the variable ζ for the function $u(\zeta)$, we obtain a new equation where the small coefficient is now in front of the lowest-order derivative, the other terms no longer making a coefficient depending on δ appear. We thus obtain the so-called "inner" equation that formally regroups all the most divergent terms of the perturbation series in powers of δ for the formal solution. This equation is written as

$$\frac{\partial^3 u}{\partial \zeta^3} + \frac{\partial^5 u}{\partial \zeta^5} + u\frac{\partial u}{\partial \zeta} = 0. \tag{18.21}$$

This equation is the same for a neighborhood of all the double poles of U_0. Nonetheless, the double pole that will dominate the possible singular contributions on the real axis is the one that is closest to the real axis, which is the double pole with index $n = 1$, located at $\chi = i\pi C^{-1/2}$.

This "inner" equation (18.21) cannot be solved analytically, but we can obtain a "quasi-solution" by formal summation of its Laurent expansion at large arguments. This expansion is written as

$$u = -12\zeta^{-2} + \sum_{n=2}^{\infty} a_n \zeta^{-2n}. \tag{18.22}$$

In this formal series the coefficients can be computed at every finite order by iteration. We thus obtain $a_2 = -180$, etc. The recurrence relation for the a_n is non-linear and not local in n. On the other hand, in the limit of large n, it reduces to the contributions of only the linear terms, since the non-linear contribution, coming from the term $u\,\partial u/\partial\zeta$, implicates orders of derivation much smaller than the linear terms and therefore becomes negligible at large n.

Therefore, by only retaining the linear terms in the relations between the a_n, we find that $a_n = (-2n+2)(-2n+1)a_{n-1}$, which is solved by $a_n \approx K(-1)^n(2n-1)!$, K being an arbitrary constant. This constant can be approximated by comparison with terms whose order is finite, but as large as possible, from the sequence of the a_n, with the asymptotic estimation. This gives $K \approx 19.969\ldots$. This estimate (crucial for what follows) is much better than that which would result from a numerical calculation by direct solution of the "inner" differential equation.

The (temporary) conclusion is that the formal perturbation series for the solution of the inner equation, that is the series (18.22), does not have meaning, since it diverges for all non-zero values of $1/\zeta$. This only reflects the property of the general solution of the inner equation, which is to also have $e^{\pm i\zeta}$ as possible asymptotic behavior. Symbolically, we can therefore write the general solution at large ζ as a function whose Laurent expansion begins with $-12\zeta^{-2}$ added to the sum of the functions $A_\pm e^{\pm i\zeta}$, where $A_\pm$ are two constants, arbitrary for the moment, which are (each) defined in the Stokes sectors of the complex ζ-plane where the exponential decreases at a large distance. The Stokes lines are lines such that $\operatorname{Re}\zeta = 0$, and hence the real axis encompasses exponentials of two types, since neither one increases nor decreases at infinity in the real direction. The choice of these constants gives a certain freedom in the asymptotic behavior of the solution for which we are looking. Nonetheless this behavior will still include an oscillating solution, either at $+\infty$ or at $-\infty$. The details of this calculation can be found in [**13**].

It is, all the same, interesting to come back to this phase of calculation that is connected at least in part to that of Poincaré. The latter used in an essential way the solution to the Airy equation by quadrature, derived from the Laplace method (which applies to any linear equation with affine coefficients in the variable). In the present problem, what takes the place of this Laplace solution is the expression of the (divergent) sum of the Laurent series for $u(\zeta)$ as a Borel transform of an analytic function in the unit disk. This poses, as it seems to me, a rather deep question, namely, what is the meaning of this method: Can we ensure that the Borel transform (or perhaps transforms) is indeed a solution to the initial equation? This does not appear at all certain to me. I will assume that this summation does indeed give the solution, which therefore is written as

$$u(\zeta) = \int_0^\infty d\tau\, e^{-\tau} v(\tau/\zeta).$$

The function $v(z)$ is defined by its Taylor series:

$$v(z) = \sum_{n=1}^{\infty} b_n z^{2n}.$$

The coefficients b_n are derived term by term from the coefficients of the series (18.22) for $u(\zeta)$. They are therefore such that

$$b_n = \frac{a_n}{(2n)!}.$$

The function $v(z)$ is analytic in a neighborhood of the origin, and it has a logarithmic branch point at $z = \pm i$. At this point it behaves like $v(z) \approx K \ln(1 + i\tau/\zeta)$ (where K is the constant that appears in the asymptotic formula of a_n for large n: $a_n \approx K(-1)^n(2n-1)!$). We are interested in the behavior of $u(\zeta)$ for infinite ζ, which gives access to the behavior of the original function $U(\chi)$ at the infinities of the real axis, for which we are looking. As in Poincaré's problem, it is the choice of an integration path for a "formal" solution (the solution by quadrature via the Laplace method, or the Borel transform of the divergent series for the perturbed soliton problem) that will make it possible to specify the boundary conditions satisfied by the "formal" solution, the solution obtained by quadrature still giving global access, in some sense, to the solution.

In the present problem, we determine the solution's behavior at infinity by choosing the path of integration of the Borel transform, that is, the way in which this path passes from one sheet to another on the Riemann surface of the logarithm of $v(z)$. If we wish for a function without oscillations at large negative χ, we keep the logarithm's contribution $i\pi$ for large positive t. This gives a solution that oscillates with a constant amplitude as χ tends to $+\infty$, the amplitude of the oscillations being

$$U \approx \frac{4\pi}{\delta^2} \sin\left(\frac{\chi}{\delta}\right) e^{-\pi\delta/C}.$$

This amplitude is of transcendent order with respect to the small parameter δ, and we recognize the "rapid" phase factor $\sin(\chi/\delta)$, which is one of the possible asymptotic behaviors for solutions to the linearized asymptotic equation. These oscillations do not die off, as announced, and so no finite-energy soliton solution exists for the perturbed KdV equation.

18.5. Final remarks and conclusion

I have attempted to show Henri Poincaré's role in the formation of what is sometimes called English applied mathematics, that is with a so-called "practical" spirit, seeking to solve completely explicit problems, without any confessed pretension to generality. As his calculation unfolded, he was—it seems to me—one of the first, if not the first, to use the powerful means of classical analysis with finesse and efficiency, showing, to paraphrase (I believe) Hadamard, that the solution to a real problem is often in the complex domain.[3] This combines with one of the first developments using, at least implicitly, the idea of a boundary layer, which had just been systematized by Prandtl for fluid mechanics in the high Reynolds number limit. This deep idea of non-uniform approximation in a small parameter seems to me to have still attracted too little attention.

Another lesson that we can draw from this work of Poincaré, one of his last (he passed away in 1912), is a magnificent example of a researcher's spirit: surrounded by all the honors that went with his status as the greatest mathematician of his time, Poincaré was not content to reflect upon great problems such as that of space and of time (his decisive contribution to relativity is narrated elsewhere in this volume[4]); he did not disdain also passing time, much time if we can judge by his memoir in the *Rendiconti*, on an "applied" problem of estimating complicated oscillating integrals.

Bibliography

[1] M. J. Ablowitz and H. Segur, *Solitons and the inverse scattering transform*, SIAM Studies in Appl. Math., Philadelphia (1981).

[2] T. Akylas and R. Grimshaw, *Solitary internal waves with oscillatory tails*, J. of Fluid Mech. **242** (1992), p. 279.

[3] M. Ben Amar and Y. Pomeau, *Theory of dendritic growth in a weakly undercooled melt*, Europhys. Lett. **2** (1986), p. 307.

[4] M. Born and E. Wolf, *Principles of optics*, Pergamon, London, 1969.

[5] Chen-Pang Yeang, *The study of long distance radio-wave propagation*, Hist. Studies in the Phys. and Bio. Sciences **33** (2003), p. 369.

[6] R. Combescot, V. Hakim, T. Dombre, A. Pumir, and Y. Pomeau, *Analytic theory of the Saffman-Taylor fingers*, Phys. Rev. **A 37** (1988), p. 1270.

[7] V. A. Fock, *Electromagnetic diffraction and propagation problems*, J. of Phys. USSR **10** (1946), p. 399. Pergamon, Oxford (1965).

[8] J. D. Jackson, *Classical electrodynamics*, 2nd ed., John Wiley & Sons, New York, 1975.

[9] G. L. James, *Geometrical theory of diffraction for electromagnetic waves*, IEEE Electromagnetic series 1 (1970), Peter Peregrinus, Southgate House.

[3]EDITOR'S NOTE: One finds this famous adage in a notice by Paul Painlevé, which received limited distribution in 1900: *"It appeared that, between two truths of the real domain, the easiest and shortest path very often passes by the complex domain."* (*Œuvres* of Paul Painlevé, vol. 1, p. 78) It was Hadamard who made it known: he wrote at the end of Chapter VIII of *An essay on the psychology of invention in the mathematical field* (1949, in English; 1959, in French): *"It has been written that the shortest and best way between two truths of the real domain often passes through the imaginary one."*

[4]EDITOR'S NOTE: Chapter 17.

[10] M. Kruskal and H. Segur, *Asymptotics beyond all orders in a model of crystal growth*, Aero. Res. Ass. Princeton Tech. Memo (1985); Stud. Appl. Math. **85** (1991), p. 129.
[11] H. M. Macdonald, *The bending of electromagnetic waves round a conducting obstacle*, Proc. Roy. Soc. **71** (1903), p. 251.
[12] H. Poincaré, *Sur la diffraction des ondes hertziennes*, Rend. Circ. Matem. Palermo **29** (1910), p. 169. (Reproduced in *Œuvres*, vol. 10, p. 94.)
[13] Y. Pomeau, A. Ramani, and B. Grammaticos, *Structural stability of the Korteweg–de Vries solitons under a singular perturbation*, Physica **D 31** (1988), p. 127.

CHAPTER 19

Henri Poincaré and His Thoughts on the Philosophy of Science

GERHARD HEINZMANN

19.1. Philosophy and Science

Henri Poincaré was obviously a mathematician and a physicist, but also a philosopher—a very great philosopher, the greatest philosopher of science according to Karl Popper [**40**, p. 2]. Is this not an excessive judgment? Where is his philosophical work? Is it not exhausted in four or five collections of articles? How can he be compared to other scholars whose names evoke, beyond any doubt, the quality of "very great philosopher" such as Locke, Comte, Peirce, Bergson, Russell, or Quine? And finally, there are plenty of judgments that contradict Sir Popper's opinion: according to Philip Jourdain [**16**, p. 482], Poincaré is without a doubt one of the greatest mathematicians, but his case rather confirms the fact that a great mathematician is not necessarily either a great philosopher or a great logician! To see what is going on, we should perhaps first determine what distinguishes a great philosopher, or what questions he can solve. Within the realm of the theory of knowledge—which interested Poincaré—I propose the following response to the latter question: the philosopher can solve the decision problem[1] concerning the conceptual framework of the knowledge of scientific objects. Thus, the philosopher should

- devise the conceptual assumptions that are at the root of our scientific knowledge, and
- clarify this activity in its symbolic form.

My object is to show, firstly, that Poincaré's conception of the philosophy of science corresponds to the two requirements mentioned and, secondly, that Poincaré devised a sturdy framework.

To emphasize this last point, I will give some indications concerning the current situation in the philosophy of science. According to the naturalist position,[2] the normative approach[3] to questions concerning the foundations of science must be corrected. For the three classical positions—idealism, realism, and formalism (or

[1]To find criteria making it possible to decide if the adoption of conceptual means is sufficient. Example: what is the object of a perceptual experience? Is it a physical object, sensory data, or only a mental disposition? How can we decide among these different possibilities?

[2]A position according to which the scientific method is the paradigm of rationality, so that the autonomous status of philosophy is called into question.

[3]The normative approach consists in fixing conditions for what must be a science. In contrast, a purely descriptive approach limits itself to describing science in action.

structuralism)[4]—have failed to establish truth and ontology in a convincing way, each time starting from metaphysical or empirical criteria, but totally separated from the practice of science. Hence it seems reasonable to suspect with Quine not only the well-foundedness of the difference in principle between *analytic*[5] and *synthetic*,[6] but equally to suspect the *normative/descriptive* dichotomy.[7] What consequence should we draw from this second suspicion? Excluding a *"philosophy-first"* normative position, we have the choice between the identification of epistemology and cognitive psychology, and a *"philosophy-in-between"* viewpoint where, according to Stewart Shapiro's expression, science does not dominate philosophy nor philosophy science [**37**, p. 15]. Adopting this latter position of "moderated naturalism", we indicate that philosophy does not have an access to reality independent of the conceptual schemes used in the sciences. In other words, there do not exist foundations of science that are based on an indubitable philosophical intuition. In this sense, the epistemological justification of beliefs and propositions has its origin in scientific practice. To accept this practice as a fact does not, however, at all imply a utilitarian realism. The difference finds expression by the distinction between the acceptance of common meaning, which can be of an elevated technical level, and the *philosophical* question of the justification of a scientific practice.

19.2. Poincaré's pragmatic occasionalism

Poincaré, at the beginning of the 20th century, gave an original and useful contribution to the current reflection. In a schematic way, it consists in the invention of a process that connects the ontological question and the epistemological question to moderated naturalism through 4 steps:

- In accordance with the empirical maxim, the construction of reality is effected starting from sensations; the construction must be guided by experience.
- In accordance with the idealist viewpoint, experience is not sufficient, but is only the occasion to become aware of certain categories[8] of the mind with which we must accord our experience by decision (or convention).
- In accordance with the realist viewpoint, relations are the only knowable objects. The *relata*[9] remain, on the other hand, inaccessible to human knowledge: cf. [**40**, p. 37] (this is the opposite of Quine's slogan: *"to be is to be the value of a first-order*[10] *variable"*).
- In accordance with moderated naturalism, scientific development and *prediction* serve as supplementary arbitrators of the suitability of the chosen conventions.

[4]Schematically: according to idealism, science only gives knowledge of things grasped *within* the mind; according to realism, science gives approximate knowledge of phenomena that exist independently of any theory; and according to formalism, science does not implicate an ontology of individualizable objects.

[5]Analytic: that which can be verified by examining only the meanings of terms, without recourse to experience.

[6]Synthetic: that which can only be verified by experience.

[7]See footnote 3.

[8]Fundamental forms of thought.

[9]*Relata:* the objects between which these relations take place.

[10]In the sense of *first-order* logic, where the quantifiers (for all, there exists) can only apply to a single type of object (for example, numbers but not functions of these numbers).

In reference to Charles Sanders Peirce, we will call such a process *pragmatic*[11] because the construction of the scientific object starting from sensations is done simultaneously with the constitution of the descriptive language appropriate to it—in other words, because the experience of common sense is the *occasion* for a process of linguistic apprenticeship. To call Poincaré's philosophy "conventionalism" is a poor choice leading to the dogmas of logical empiricism.[12] Thus, I propose to call it "pragmatic occasionalism". Poincaré is not sometimes Kantian, sometimes conventionalist, sometimes empiricist, sometimes predicativist. He always defends the same philosophy: in a process of going-and-coming, the intuition relates at the same time to that which one can do and, at a higher level, to a propositional knowledge. The construction and the description of objects are not independent. In this way, the apparent incoherencies in his philosophy will disappear, although they have prevented so many philosophers from appreciating Poincaré's philosophical views for their true value: they criticize Poincaré (apparently rightly) of being at the same time an intuitionist and a conventionalist, for example, or of refuting actual infinity while still admitting real numbers as infinite classes of Cauchy sequences [**40**, p. 3].

To this contextualization of Poincaré's philosophy of science, we must still add two remarks:

1. It is obvious that Poincaré's philosophical background belongs to the Kantian tradition: "To look for whether [the accord of a theory with experience] takes place or whether it falls short, this is therefore our goal. [...] To neglect one for the other would be nonsense. Isolated, theory would be empty, experience would be myopic; both would therefore be useless and uninteresting." [**23**, p. 298] However, Poincaré does not at all solve the problem of the unity of spontaneity and receptivity[13] by the introduction of a pure sensibility. Rather, he changes the terms of the Kantian opposition: what is important for him is the balance between exactitude and objectivity, the latter being an attribute of reality. He expresses the lost balance in a formula well-known by its popularization under Einstein's pen: "what they [mathematics] have gained in rigor, they have lost in objectivity. It is by distancing themselves from reality that they acquired this perfect purity." [**23**, p. 131][14]

---

[11]By "pragmatism", I understand a philosophical position that is characterized by at least five factors:

- Pragmatism is a method (it is not a system).
- Abandoning every empirical criterion or absolute rationale, it takes observations and statements of common sense for its basis.
- It admits a convergence between ontology and the theory of knowledge and considers fruitfulness as an important criterion of the acceptability of propositions.
- It refuses a separation of principle between the context of justification and the context of invention.
- It admits a correlation between knowledge and value.

[12]These dogmas are the analytic/synthetic dichotomy and reductionism (according to which every empirical proposition is reducible to an observational fact).

[13]According to Kant, knowledge requires at the same time understanding—that is, the spontaneous capacity to create representations (concepts)—and sensibility—that is, the receptive capacity to "intuit" (singular) objects. One of the central questions of his theory of knowledge concerns the problem of the *synthesis* of sensibility (intuition) and understanding (concept).

[14]See [**6**, p. 414]: "As far as the propositions of mathematics refer to reality, they are not certain; and as far as they are certain, they do not refer to reality."

2. Henri Poincaré is often considered to be the father of mathematical conventionalism. In his famous article "Les hypothèses fondamentales de la géométrie" ["The fundamental hypotheses of geometry"] (1887), he compares for the first time the choice between Euclidean geometry and that of Lobachevsky[15] to the choice between coordinate systems and observes that the hypotheses in question are neither "experimental facts nor analytic or synthetic judgments *a priori*" but rather the result of a choice guided by observation of certain physical phenomena [**25**, pp. 90–91]. The choice is therefore not at all arbitrary, and he equally specifies that the adoption of this conception is starting "to become commonplace". Indeed, we observe conventionalist tendencies in the works, more or less contemporary, of Ernst Mach and Emile Boutroux. However, it is no longer possible to say that conventionalism is nothing but a philosophical consequence of the discovery of non-Euclidean geometries since, in his course on analytic mechanics (1847/48), the German Carl Gustav Jacob Jacobi supported theses in his field that strongly resemble Poincaré's conventionalism in physics (see [**34**]).

It is now necessary to understand exactly how Poincaré distinguishes *a priori* principles (for example, complete induction in arithmetic) from conventions as apparent hypotheses (for example, the parallel axiom in geometry) and from true hypotheses (for example, the hypothesis of central forces in physics). Placed side by side, all of these notions express local forms, specific to the different fields considered, of Poincaré's pragmatic occasionalism, whose principal subject is (and I emphasize) the epistemological question of the relational form of scientific knowledge. In order to make his thought process understood, Poincaré proposes in *Science and Hypothesis* to examine the hierarchy of the sciences, from arithmetic to physics, passing by geometry. I will give my own opinion of this path, proposed by Poincaré, adding his logic at the end.

19.3. Arithmetic

According to Poincaré, a reflection on the logical genesis of the principle of *recurrence* (we also say *complete induction*) shows that it possesses the status of an *a priori* synthetic principle:

(1) "We have the faculty to conceive that a unit can be added to a collection of units; *it is thanks to experience that we have the occasion to exercise this faculty and that we become aware of it.*" [**22**, p. 53] (emphasis by G. Heinzmann)
(2) "[Complete induction] is nothing but *the affirmation* of the power of the mind which knows itself capable of conceiving the indefinite repetition of the same act when this act is possible once. *The mind has of this power a direct intuition and experience can be for it nothing but an occasion to make use of it and by this to become aware of it.*" [**22**, p. 41] (emphasis by G. Heinzmann)
(3) "A construction becomes [...] interesting only when we can place it next to other, analogous constructions, forming species of the same genus. [...] The analytic procedure "by construction" does not oblige us to descend to it, but it leaves us at the same level. We can only raise ourselves by mathematical induction." [**22**, pp. 44–45]

[15]EDITOR'S NOTE: Hyperbolic geometry: see Chapters 1 and 2.

In his review of *Science and Hypothesis*, Russell (see [**36**, p. 413]) remarks that the necessary power of the mind should consist of the capacity to add I to an *arbitrary* integer, a fact that is expressed by the pragmatic *schema* **R**:

R: a) $\longrightarrow I$ [we can construct I];
b) $n \longrightarrow nI$ [if we have n, we can construct nI];

which itself is suggested by the experience of being able to add I to I: II; I to II: III; I to III: IIII, and so on, as experience confirms (without use of the schematic parameter n). We can doubtless consider the *schema* **R** as synthetic, but the synthetic character of complete induction is not exhausted in this argument. In fact, the essential element is missing, that is, the final clause:

c) We can obtain *all* the natural integers by application of the schema **R**.

Yet, this clause c) does not follow analytically from the clauses a) and b) of **R**, but, according to the mind of Poincaré, it is the result of a non-logical—that is, synthetic—*reflection* on the construction **R**.

We have thus returned to what is called a recursive definition of the natural integer numbers:

R′: a) $N(I)$ [I is a number];
b) $N(n) \longrightarrow N(nI)$ [if n is a number, its successor is equally a number];
c) We can obtain all numbers by application of a) and b).

Yet, as Poincaré correctly observes:

> "[...] the principle of induction does not mean that every integer number can be obtained by successive additions; it means that for all numbers that one can obtain by successive additions, one can prove any property by way of recurrence." [**29**, p. 835]

Therefore complete induction means, for any property E, that

CI: $$[E(I) \wedge \forall n(N(n) \wedge E(n) \longrightarrow E(nI))] \longrightarrow \forall n(N(n) \longrightarrow E(n)).$$

Yet, the justification of **CI** is correlative to **R′** (see for example [**17**, pp. 21–22]), which means especially that the final clause c) cannot be deduced from the clauses a) and b): an application of **CI** is needed. In other words, **CI** implies **R′** and **R′** implies **CI**. Where is the synthetic element of **CI**, more exactly? Without a doubt it lies in the fact that it is impossible to give a schematic proof for **CI**. For an arbitrary number n, constructed according to **R′**, we can certainly indicate an operation that leads to a proof of $E(n)$, but it is impossible to indicate a uniform form of proof. Because the length of the proof, that is, the number of applications of the *modus ponens*, depends on n in such a way that a singular proof, for a certain argument, cannot be considered, given its internal structure as an *instance* of the scheme of proof:

> "If instead of showing that our theorem is true for all numbers, we wanted only to make it seen that it is true for the number 6 for example, it would suffice for us to establish the first 5 syllogisms of our cascade. [...] No matter how far we thus went along, we would never raise ourselves to the general theorem, applicable to all numbers, which alone can be the object of science. To arrive at that, we would have to have infinitely many syllogisms, we would have to cross a gulf that the patience of the analyst, reduced only to the resources of formal logic, would never succeed in filling in." [**22**, pp. 39–40]

According to Poincaré, in order to cross this gulf, pure intuition is necessary. A "true" intuition is distinguished from simple evidence by the fact that it refers to what one can do in place of what exists. In this sense, pure intuition is not directed to the same object as sensory intuition or imagination (see [**24**, p. 39]). It is the awareness of a capacity of the mind, and experience gives us the occasion to use this capacity. Even though this intuition is *occasioned* by experience, it relates to the scheme of action **R** which is *a priori* because it is the result of our own creativity. The intuition is indispensable, since the closure of the schemes **R** and **R**′ is not created, but only represented by an indefinite repetition relating to different levels: the scheme consists, firstly, in an overview of a potential reiteration at the level of objects and, secondly, in an overview of a potential reiteration of a pragmatic order for **R**, of the *modus ponens* for **R**′. This sort of pure intuition allows Poincaré to say that reasoning by recurrence is the expression of an infinite number of hypothetical syllogisms, "condensed to be thus said in a single formula" [**22**, pp. 38–39].

The negation of the principle is not conceivable for Poincaré, because the principle itself is a form of understanding in the sense where "we cannot conceive of an experience that, interpreted most favorably, could violate it" (see [**7**, p. 32]). Experience is the *"raison de connaître"*[16] of the principle of induction, and this inasmuch as it is a structuring component of different empirical constructions, the theoretical part of our knowledge.

It is well known, according to Poincaré, that the axiom of comprehension is defective because the predicate involved does not obey the principle of bivalence, because of a phenomenon of partial information: the classification is changed by the introduction of new elements. But the "predicative"[17] restrictions invented by Poincaré are too "strong" and even exclude elementary arithmetic:

> "[We must regard] the principle of induction as a synthetic judgment and not as a definition, because this definition would be non-predicative."[18] [**30**, p. 314]

Poincaré therefore postulated, in 1906, complete induction as a synthetic principle *a priori*, because, without induction, arithmetic would be impossible. In other words, an intellectual category, which is the principle of induction, replaces the methodical requirement of predicativity: every reduction of the principle of induction to set theory presupposes non-predicative methods. In the same way the problem of actual infinity, necessary for the definition of the real numbers, is solved: the *definition* of an infinite set in its current meaning in a formula containing a bounded number of words is impossible. Complete induction is only the simplest procedure among other analogous postulates, forming species of the same genus [**23**, pp. 149–150] and "presenting the same essential characteristics" [**29**, p. 818]. One of these postulates is shown in the awareness of our capacity to construct a

[16] "Reason for knowing" (or *ratio cognoscendi*) of a thing: that which reveals it to us, that which allows us to know it. [TRANSLATOR'S NOTE: I have left the phrase untranslated in the main text to emphasize its similarity to the phrase *raison d'être*, with which English speakers are quite familiar.]

[17] This term will be explained in footnote 30, p. 388.

[18] We can prove complete induction relative to finite sets, but we are then obliged to use non-predicative procedures for an "intuitive" definition of finite sets or for the proof that a potential predicative definition is equivalent to an "intuitive" definition of finitude (for example that given by Dedekind: a set is *finite* if it is not in bijection with any of its proper subsets).

continuum in each dimension, called *topological intuition*, another in the awareness of our capacity to conceive of groups, called *algebraic intuition*. On the one hand, the awareness of these two capacities is occasioned us by experience, and, on the other hand, this awareness "recalls" preexisting categories in our mind as forms of understanding [**20**, p. 134 ff., p. 157], [**22**, p. 107]. Even though we therefore use neither hypothesis nor conventions in arithmetic, we shall see in the next section that certain hypotheses in geometry play the same intermediary role as our intuitive capacities between experience and principle.

19.4. Geometry

To introduce the reflections on the foundations of geometry, it is useful to recall three results that are often mentioned in philosophical studies of Poincaré's work.

1. In *Science and Hypothesis*, Poincaré gives [**22**, pp. 24, 166] a classification of hypotheses, in which he distinguishes

H1: verifiable and fruitful hypotheses (*descriptive hypotheses*);
H2: useful hypotheses for fixing our thoughts (*indifferent hypotheses* such as mechanical models);
H3: apparent hypotheses that reduce to definitions or to disguised conventions (*conventions*);
H4: natural hypotheses from which we can hardly escape (*explicative hypotheses*).

According to Poincaré, every generalization is a hypothesis [**22**, p. 165], but not every hypothesis is a generalization: this is shown by **H2** and, if the context is experimental, equally by **H4** because a natural hypothesis can be an indispensable condition for science without being experimentally accessible, such as for example the hypothesis that effect is a continuous function of its cause (see [**22**, p. 166]).

2. A convention that is characterized as an apparent hypothesis does not seem to be a hypothesis: it is only *identified* with a hypothesis. It is clear that not all conventions can be hypotheses or identified with them: a nominal definition is not a hypothesis. It is necessary for conventions to be at least considered in an epistemological interest. And, in fact, certain conventions are not only pertinent as instruments, but they also have an explicative role. On the other hand, hypotheses that are not generalizations do not necessarily include a conventional aspect.

3. This point concerns the relation between definition and convention. The apparent hypotheses can be reduced to definitions or to disguised conventions (**H3**), and a disguised definition is a convention [**22**, pp. 75–76]. Poincaré uses the term *disguised definition* before 1899, the year of Hilbert's famous *Foundations*,[19] to express that the language used descriptively in appearance is not so in truth: certain axioms are used descriptively only in appearance, but *define* in truth the objectivity of a fact (see [**27**, p. 274]). This fact is found to be defined only up to structure, but

[19]Hans Freudenthal has shown that the viewpoints that interpret the geometric axioms as implicit definitions are founded on an anachronistic misunderstanding that wrongly considers Hilbert's views as an adequate instrument for the interpretation of Poincaré's book *Science and Hypothesis* (see [**8**]). Even though this latter was not published until 1902, three years after Hilbert's *The Foundations of Geometry* [**15**], it is in reality a collection that assembles articles written in the 1890s. Yet, in these articles, Poincaré does not anticipate Hilbert's view, according to which the axioms of a formal system are nothing other than the *forms* of statements and, for this reason, devoid of truth-values.

it reflects well the truth of certain relations between objects whose quality remains—as with Helmholtz and others—unknowable [**24**, p. 174].[20] A disguised definition or (disguised) convention is not opposed to an explicit definition, because it *is* explicit and not implicit, but it is opposed to an intuitive description of things. The term "disguised definition" characterizes an explicit definition of a "real" structure, disguised as an intuitive axiom about things.

Thus, in geometry, one easily risks conflating convention and intuitive axioms. In physics, by contrast, we shall see that one rather has a tendency to conflate convention and hypothesis.

It is well known that, according to Poincaré, certain geometric propositions can be neither true nor false because they contain elements of convention. Yet Poincaré emphasizes that one can very well speak of *truth*, after having fixed conventions.[21] What should we retain from this? It is not the place to note simply a trivial inconsistency: we are rather faced with different sorts of conventions. The arguments of David Stump and Michael Friedman, concluding that Poincaré's conventionalism does not at all reduce to Duhem's thesis, are quite correct: *conventions* do not concern only the choice between theories, but they already come into play in classifications with respect to the interpretation of the basic concepts of each theory.

The classifications, insofar as they are mediating elements that we are seeking to fix between sensation and observation, have a character of decision-making. However, these conventions, already necessary in the language describing our sensations, obviously do not determine the geometric language.

Now let us pass to a more technical description. First we must emphasize that, in contrast to Hilbert, Poincaré ties his research into the foundations of geometry[22] to the concept of a group, which, since 1880, has been considered the invariant form of different geometries [**10**, p. 76]. Geometry is nothing other than the study of a group. And "since the existence of a group is not incompatible with that of another group", we cannot say that one geometry is true and that another is false: the choice between different geometries is comparable to that of a coordinate system [**25**, p. 90 ff.]. At the start, our body plays this role of a coordinate system with respect to which we locate an object in space [**23**, p. 104]. Locating an object in the representative space means for Poincaré reflecting on the sequence of actions needed to reach this object, that is, reflecting on sequences of muscular, and not spatial, sensations.

To classify these sensations, Poincaré introduces the essentially vague category of representative space [**21**, p. 7 ff.]: there is neither measure nor the possibility of speaking of constant axes with respect to our body, but thanks to it, we can compare sensations of the same kind and observe the proximity of two objects. In itself, all the sensations are different, since they are accompanied, for example, by

[20] "that which [science] can attain is not the things themselves, as the naive dogmatists think, but only the relations between the things; apart from these relations, there is no knowable reality." [**22**, p. 25]

[21] [**24**, pp. 158–159]: "If you ask me the question: is this fact true? I will begin by asking you, if necessary, to specify the conventions—by asking you, in other words, what language you have spoken; then, once this point has been fixed, I will query my senses and I will answer, yes or no."

[22] See [**22**, Chap. IV] (slightly modified from [**26**]); [**21**] (first publication 1898, in English); [**24**, Chap. II] (extract from [**28**]); [**23**, 2nd book, Chap. I]; and [**20**], article "Pourquoi l'espace a trois dimensions" ["Why space has three dimensions"] (1912).

"various olfactory or auditory sensations" [**20**, p. 142]. Their indistinguishability is a consequence of our abstractive classification. We maintain that the representative space is not formed by a classification starting from motor sensations, but on the contrary, that it is the necessary condition for a classification of motor sensations. It is a form of our understanding and not of our sensibility, since an individual sensation can exist without it [**21**, p. 3].

The construction of geometric space now proceeds from the observable fact that a set of impressions can be modified in two distinct ways: on the one hand without our feeling muscular sensations, and on the other, by a motor action accompanied by muscular sensations. In the first case, Poincaré speaks of an external change, in the second, of an internal change. This observation suggests a conventional classification of external changes: the external changes that can be corrected by an internal change are called changes of position, the others changes of state. By the relation *the external change x can be compensated by the same internal change as the external change y* (see [**21**, p. 15]), we can gather the equivalent external changes into classes called *displacements.* Thus we have the following situation:

1. Two internal changes are considered to be indistinguishable if they provoke the same muscular sensations ([**24**, p. 79]: too close to each other).

2. $$\frac{\alpha, \beta \text{ external changes}}{S \text{ internal change}} \qquad \alpha \sim \beta \quad \text{iff} \quad \exists S \ (S\alpha \doteq S\beta \doteq I),$$

which means: two external changes are equivalent *iff* they have a property in common, i.e. if they are susceptible to being corrected by approximately the same internal change.

3. $$S \approx S' \quad \text{iff} \quad \exists \alpha \ (\alpha S \doteq \alpha S' \doteq I).$$

4. If $\sim$, $\approx$ are equivalence relations, then the equivalence classes of changes of position are called (external and internal) *displacements.* The displacements are therefore identical or disjoint. The definition of displacements as equivalence classes makes it possible to define the identity "and it follows that a displacement can be *repeated* two or several times" [**21**, pp. 16–17].

We note that Poincaré often uses the same word "displacement" for the equivalence classes and for its elements, the changes of position.

5. $\sim$ and $\approx$ are not transitive:

$$\begin{aligned} \alpha \sim \beta \quad &\text{iff} \quad \exists S \ (S\alpha \doteq S\beta \doteq I), \\ \beta \sim \gamma \quad &\text{iff} \quad \exists S' \ (S'\alpha \doteq S'\beta \doteq I), \\ \alpha \sim \gamma \quad &\text{iff} \quad \exists Q \ (Q\alpha \doteq Q\beta \doteq I). \end{aligned}$$

The first two lines do not imply the last.

6. In *The Value of Science* (1903) [**24**, p. 72], Poincaré then formulates the following law:

$$\alpha \neq \beta \qquad \left.\begin{array}{r} S\alpha \doteq I \\ S\beta \doteq I \\ S'\beta \doteq I \end{array}\right\} \implies S'\alpha \doteq I.$$

7. This law guarantees transitivity: $Q = S'$, because by hypothesis we have $S'\gamma = S'\beta = I = S\beta = S\alpha$, and according to the law, $S'\alpha = I$. But what is the

character of the law? Poincaré asserts that it is an empirical fact: "If it were not satisfied, at least approximately, there would not be geometry, there would not be space" [**24**, pp. 72–73]. Torretti claimed that this explanation given by Poincaré was erroneous [**39**, p. 343]. He gave the following counterexample:

external change	internal change
α	S (change of 1 cm)
β	S' (change of 1 cm and k mm)
γ	S'' (change of 1 cm and $2k$ mm)

It could happen that

$$S\alpha \doteq I, \qquad S\beta \doteq I, \qquad S''\beta \doteq I, \qquad \text{but} \quad S''\alpha \not\doteq I.$$

The example is correct, but it does not constitute a refutation of Poincaré. Why? Because Poincaré is not an empiricist:[23] experience teaches us only "that the compensation is approximately produced"; it gives the mind only "the occasion to accomplish this operation", but "the classification is not a raw fact of experience" [**21**, p. 16]. And Poincaré is very precise with respect to the imprecision at this level of reconstruction:

> "When experience teaches us that a certain phenomenon does not correspond *at all* to the indicated laws, we erase the displacements from the list. When it teaches us that a certain change only obeys them *approximately*, we consider this change, *by an artificial convention*, as the result of two other component changes. The first component is regarded as a displacement that *rigorously* satisfies the laws about which I have just spoken, while the second component, which is small, is regarded as a qualitative alteration. Thus we say that natural solids undergo not only large changes of position, but also small fluctuations and small thermal dilations." [**21**, pp. 19–20]

The study of the structure of these displacements suggests other idealizations: we pass to the mathematical concept of a continuous group of transformations G. We then distinguish among the displacements belonging to groups isomorphic to G (some of which may operate on simpler materials than the representative space) those that conserve certain sensations. The most intuitive are the subgroups of rotations. By taking these subgroups into consideration, we obtain a characterization of groups that correspond to geometries of constant curvature. Among these groups, we finally choose the one that allows "the assertion of the existence of an *invariant* subgroup all of whose displacements are interchangeable and which is formed of all the translations" [**21**, p. 34]. In other words, we choose the group that corresponds to Euclidean geometry, because these subgroups are best suggested by experience. In principle, we could have fixed another convention.

Poincaré is well aware that the validity of the group properties in question does not result from an *a priori* reflection, but that it is on the contrary exposed to the danger of an empirical refutation. And yet, "geometry is sheltered from any

[23] "It is interesting to see what has been the role of experience in all of this. It has shown me that a certain law is approximately satisfied. It has not taught me *how* space is and that it satisfies the question at hand. Indeed I knew, before any experience, that space would satisfy this condition or it would not exist. Nor can I say that experience taught me that geometry is possible; I see full well that geometry is possible because it does not imply a contradiction. Experience has taught me only that geometry is useful." [**24**, p. 73]

revision" [**21**, p. 19]. To solve this paradox, Poincaré introduces some conventions, including this one from the first steps of his construction: in effect, in representative space already, changes of position can never be exactly realized since, in general, the starting position does not exactly coincide with the ending position. When the observation of our senses, provoked by an internal change, does not correspond to the expected compensation, it is either struck out or replaced by a new "artificial convention" that considers the change as the result of two components: one rigorously realizing the compensation, the second being a qualitative alteration that is thereafter ignored [**21**, pp. 19–20]. To obtain an exact coincidence, we must therefore read the corrections leading to changes of positions as orders of action in view of an intention. We act as though we could realize this intention as the standard with respect to "ideal bodies [...] taken out of nothing from our mind" [**21**, p. 93]. And yet, what is realized is not this standard, but only the order to realize this standard. Experience here plays a double role: it is at the same time the occasion for introducing the standard, i.e. its *ratio cognoscendi* (see for example [**27**, p. 276]), and the occasion for using the standard in order to conceive of reality [**21**, p. 20]. Like the category of representative space, the general concept of a group is a form of our understanding.

Thus, the usefulness of Euclidean geometry consists first of an epistemological accord founded on previous classifications, carried out according to the principles of occasionalism. It follows that the axiom of Euclidean distance is not exclusively a conventional definition in the linguistic sense. This form of convention where there exists a choice between different possibilities only becomes involved at the last stage. It is for this reason that the claimed axiom of Euclidean distance is not a definition by convention in the proper sense: it is a *disguised* definition or an *apparent* hypothesis (classification **H3**). Holistic conventionalism, discussed by Poincaré in the context of measuring parallax and commented upon at length in studies on Poincaré, is only a particular form of occasionalism and not the expression of a global holism of the Poincareian approach. We are thus very far from the *schemes* of Hilbertian axioms which, because of their non-propositional character, have been interpreted following Louis Rougier as conventions, so that the disguised definitions naturally become implicit definitions.

19.5. Physics

As[24] for the epistemological status of hypotheses in mechanics, Poincaré positions himself—as in his discussion of geometry—with respect to the dichotomy between the so-called empiricism of the English School and the apriorism of the French School [**22**, p. 111]. The principles of mechanics certainly have, according to Poincaré, an empirical origin, but they nonetheless exit the bounds of experimental control, without being *a priori* in the classical sense of the term [**18**, pp. 80, 173]. To understand this medial position, it is important to distinguish "what is experience, what is mathematical reasoning, what is convention, what is hypothesis" [**22**, p. 111]. In fact, the experience that leads us to an experimental result rests on the repetition of a phenomenon, and therefore on physical induction [**22**, p. 26]. Every generalization being a hypothesis, a law or a general hypothesis (**H1** of the classification) is obtained by the fact that each result of empirical physics can be

[24]The next two sections are a modified version of [**13**].

generalized in different ways and that we are indeed obligated to correct the experience [**22**, pp. 159, 161].

However, things are not always so simple. The hypothesis of the law of inertia, for example, cannot be obtained except with respect to the preexisting category of the *differential equation*, which thus replaces the empirical induction that leads to a simple experimental fact. This constitutes the interpretation of the Poincareian *dictum*, which is that the laws are "in sum" differential equations:

> "It is Newton who showed us that a law is only a necessary relation between the present state of the world and its immediately following state. All the other laws, discovered since, are not something else, they are, in sum, differential equations." [**24**, p. 119]

Yet, the analogy and the difference between the geometric axiom as a "disguised definition" and the physical law as a general hypothesis, seem to define Poincaré's occasionalism: in geometry, the convenient conventions (or definitions) are chosen as a function of objects (solid bodies, rays) that are not those of the geometry, and we presuppose the category of group. In mechanics, the conventions are convenient even with respect to mechanical objects [**22**, p. 152] and are for this reason called *hypotheses*. We here presuppose the categories of empirical induction and differential equation. Poincaré's occasionalism therefore integrates specifically Kantian elements while still completely eliminating the transcendental component.

Setting aside the difference of scale, which is that geometry requires a double conventional abstraction, with respect to objects and with respect to laws, in mechanics Poincaré uses a procedure, which we already know from geometry, to pass from empirical laws, understood as verifiable hypotheses, to principles including explicitly conventional elements:[25]

> "When a law has received sufficient confirmation from experience, we can adopt two attitudes, either to leave this law in the fray; it will then remain subject to an incessant revision that will doubtless end by showing that it is only approximate. Or to erect it in *principle*, adopting conventions so that the proposition is certainly true. For this, we always proceed in the same way. The primitive law stated a relation between two raw facts A and B; we introduce between these two facts an abstract intermediary C, more or less fictitious [...]. And then we have a relation between A and C that we can assume to be rigorous and which is the *principle*; and another between C and B that remains a revisable law. The principle, thereafter crystallized so to speak, is no longer subjected to the control of experience. It is not true or false, it is convenient." [**24**, pp. 165–166]

Thus, at first view, the lines for a generalization scheme in physics are presented in the following way: from the phenomenon we pass by physical induction to the experimental result and, thanks to differential equations, to the laws and the general hypotheses which, by decision, can be erected in principle. Certainly, this is only a general scheme and the concrete process of a generalization could include many variations.

However, in physics proper—that is, in optics and in electrodynamics—the conventional elements seem to be weakened. In fact, in his introduction to *Science*

[25] In [**24**, pp. 126–127], six principles are listed: the conservation of energy (Mayer), the degradation of energy (Carnot), the equality of action and reaction (Newton), the principle of relativity, the conservation of mass (Lavoisier), and the principle of least action (Maupertuis).

and Hypothesis [**22**], Poincaré presents the fourth part of his work ("Nature") by referring to a change in method:

> "Up through [mechanics] nominalism triumphs, but we are arriving at the physical sciences proper. Here the scene changes; we encounter another kind of hypothesis, and we see in it all the fruitfulness. Without a doubt, at first appearance, the theories seem fragile to us, and the history of sciences proves to us that they are ephemeral: they do not however die entirely, and from each of them something remains. It is this something that we must seek to untangle, because there, and only there, is the true reality." [**22**, p. 26]

Yet, Poincaré makes explicit in the fourth part that what survives and remains common to the different physical theories is nothing other than what is found to be expressed by differential equations (see [**22**, pp. 173–174]). But, in view of the crisis in physics that was brewing toward the end of the 19th century,[26] this general position would not work without posing problems with respect to concrete principles. It is this fact that Poincaré sensed in his introduction to *Science and Hypothesis*, cited above. We will briefly mention the most interesting case, that of the application of the principle of relativity to electrodynamics. In mechanics, the principle of relative motion says that the form of differential equations remains the same if we change the coordinate axes, whether these axes stay immobile or whether they are moving but "led in a rectilinear and uniform motion" [**22**, p. 130]; see also [**20**, pp. 101–102]. By using in his famous St. Louis conference (1904) the designation "[physical] principle of relativity", which does not apply "to finite equations that are directly observed, but to differential equations" [**20**, p. 103], Poincaré reports that Lorentz introduced the hypotheses of "local time" and of "uniform contraction in the direction of motion" in an attempt to save the principle in its application to the electromagnetic domain (see [**24**, p. 132 ff.]), where it risks losing its universal validity. And yet it was Poincaré who obtained the full compatibility of Lorentz's theory with the principle of relativity. We are thus, Poincaré remarks, *naturally* carried to admit the postulate of relativity in all domains [**31**, p. 495].

According to this argument, the extended form of the principle finds its cause in the Galilean principle and its *reason* in the compatibility with a theory that explains *why* no experiment is able to make known to us the absolute motion of the universe. Like the principle of inertia, it cannot be directly confirmed by experiment, and Kaufmann's negative experiment (1906) even leaves one to believe that it would not have "the rigorous value that one was tempted to attribute to it" [**23**, p. 248]. In the same measure that the complexity of holistic considerations accumulates, the process of equilibrium between theory and experiment becomes more volatile. The change in the status of hypotheses that Poincaré envisioned seems to be the expression of a feeling of resignation to the end of an era, and its appeal to nonetheless discover the invariants of "true reality" sounds as a cry after having been chased from paradise:

> "[...] we may fear that after having made vain efforts to bend nature despite herself to our ideal of unity, tossed about by the waves still rising

[26] On this subject, see [**3**].

> from our new riches, we will have to renounce their classification, abandon our ideal, and reduce science to the recording of innumerable lists." [**22**, p. 183]

Poincaré's model of explanation, founded on a minimum of well-confirmed hypotheses (directly or indirectly by means of fictions) and from which all meaningful propositions can be deduced, is called into question by Maxwell's approach:

> "The English scholar does not look to construct a single, definitive, well-organized edifice. It seems rather that he raises a large number of provisional and independent constructions, between which communication is difficult and sometimes impossible." [**22**, p. 218]

Maxwell's hypotheses were initially "bold" hypotheses, only confirmed "at the end of twenty years" [**22**, p. 239]. Even though he himself was highly involved in this methodological change, he was never satisfied, at least until 1904, either with his own theory nor with that of Lorentz, both containing too many *ad hoc* hypotheses, necessary for the theory's coherence, but unfit for making empirical predictions (see [**40**, p. 17]). According to Poincaré, the reality that science approaches only consists of "true relations" that express the laws and the mathematized principles of physics. History shows us that these relations were never stable. The recent developments of physics teach us the instability of even the most fundamental principles. But "it is impossible for us, at the time that a theory is proposed, to distinguish these true relations from what takes after convention or conjecture" [**4**, p. 22]. Hence the explanation of the fact that Poincaré often refuses, especially in his courses on current subjects, to choose among the multiple proposed theories.

In contrast with Einstein, Poincaré always held to an interdependent equilibrium between theory and experience. For Poincaré, Einstein's new theory was only the expression of a new convention's introduction. He indeed saw that the result that it implied was judged by some to be more convenient, but this did not incite him to abandon the old convention, which was far from arbitrary. Must we criticize him for the fact that he overemphasized the convenience of the system's geometric aspect[27] while neglecting the new convention, since in it space and time are no longer two entirely distinct entities? Indeed, I believe that for Poincaré the decisive element is rather the pragmatic criterion of fruitfulness. With respect to this, the new theory of relativity does not have a clear advantage except in its general version, published well after the death of Poincaré: the geometric concept, in the sense of Helmholtz–Lie–Poincaré (constancy of curvature), is therein definitively abandoned.

19.6. Logic and abstraction

In a famous response to Russell, Poincaré refuses to distinguish psychology from either logic or epistemology [**32**, p. 482]. It would be false, however, to believe that he thus conflates logic and psychology. Indeed, Poincaré does not use the term psychology in the modern sense. He uses it in all situations where he wants to emphasize the dimension of understanding, in particular its historical elements, which he contrasts with the correct logical exposition of the result. It is therefore not surprising that, according to Poincaré, mathematics requires intuition not only

[27] Such an argument is put forth in [**38**, p. 162].

in the context of discovery, but equally in the context of justification.[28] We have already seen that in arithmetic pure intuition is necessary to understand the proofs.

Starting in 1905, Poincaré thus opposed the *logicist* thesis that claims to be able to prove all mathematical truths without recourse to intuition, once the principles of logic are admitted. Yet Poincaré suspected that the logicians were in truth making an equivocal use of the term logic, that they no longer sought the former kind, but a "new logic" containing principles of synthetic proof or formations of non-logical concepts. And he was evidently right. Not only is the modern logic of predicates richer than the traditional (syllogistic) logic, but in order to cope with the idea of reductionism, one is even led to expand it again by certain set-theoretic postulates of existence.

According to Poincaré, the antinomies discovered at the turn of the century—such as Russell's paradox—are the consequence of an improper use of intuition with respect to abstract entities.[29] And this intuitive usage is itself suggested by the erroneous method of conceptual realism (Platonism). Because for an anti-Platonist like Poincaré was from 1906 on, in the given context intuition is not—we have seen this during the discussion on complete induction—an obvious mode for representing an object, but rather concerns our capacity to follow a rule.

The awareness that mastery of a scheme cannot be easily formalized without begging the question (a vicious circle) is indeed taken into consideration by Poincaré in his article "Les mathématiques et la logique" ["Mathematics and logic"] [**29, 30**], where he criticizes logicist reductionism: to produce and understand the definitions in a formal system, he says, we must indeed use the name of a number, numeral adjectives, or at least plurals [**29**, p. 830]. The value of this argument, taken up by Hadamard, Fraenkel, Wittgenstein, and Bernays, naturally depends on its exact form. But Poincaré mentions yet another argument: if we pursue, with the formal system, the intention to clarify or to simplify informal reasoning, we must connect a practical familiarity with its formal characterization. What criterion should be chosen for such a translation? In the second part of his article "Les mathématiques et la logique", Poincaré emphasizes that this translation remains among the *desiderata* even if the logicists succeed in giving a justification of the formalism within their own system, that is, if they find a proof of non-contradiction [**29**, p. 23]. In 1906, therefore, Poincaré began to distance himself from his famous formula, saying that in mathematics, the word "exist" means "exempt from contradiction" [**29**, p. 819].

We know that Poincaré hoped to avoid the known antinomies by limiting himself to predicative definitions (see [**12**]). It was Russell who introduced the terms "predicative" and "non-predicative" to fix the difference between two kinds of propositional functions: those that determine and those that do not determine a

[28]This fact is often ignored. By contrast, the following passage is sometimes cited (see for example [**35**, p. 459]: "Logic, which alone can give certainty, is the instrument of proof: intuition is the instrument of invention" [**24**, p. 37]. But this text only summarizes a discussion on *sensitive intuition* and analytic procedures. Poincaré adds to the lines already cited: "At the moment of formulating this conclusion, I am seized by a scruple", and a few pages later, he emphasizes that pure intuition also gives certainty and makes proof possible [**24**, p. 39].

[29]It should be noted that the term "intuition" is quite ambiguous. This fact is well known and explicitly discussed by Poincaré himself. In *The Value of Science*, he distinguished three kinds of intuition: an appeal to sense and to imagination, generalization by induction, and intuition of pure number—whence comes the axiom of induction. The first two kinds cannot give us certainty, but, he says, "who would seriously doubt the third, who would doubt Arithmetic?" [**24**, p. 33].

class. He called the first "predicative" and the second "non-predicative".[30] Poincaré attributed the mistake of non-predicative definitions to a vicious circle. He formulated a principle to avoid it, which leads directly to the famous Russell's principle. This principle is famous because Russell succeeded in developing a theory that respects it: the ramified type theory. For Poincaré, we find nothing that is comparable: he believed that his position, underlying intuition, kept him safe from definitional mistakes, denounced in his first formulation of the principle. He did not seriously take up the question until he learned from Ernst Zermelo that the proof, proposed by (Argand and) Cauchy, of the fundamental theorem of algebra makes an appeal exactly to the rejected, non-predicative definitions; hence only after the measures taken against the logical antinomies affected "real mathematics"; at this point, he formulated successively different principles. The discussion between Poincaré, Russell, Peano, and Zermelo on the measures to take lasted for six years. The difficulty consisted in the obligation to formulate a principle that is neither too restrictive for the important results in Analysis, nor too liberal with respect to the formations of concept to be excluded by reason of Poincaré's philosophical position. The absence of a truly convincing solution shows that the analysis of the concept of predicativity with the help of informal methods appeared to have reached its limits.

As for mathematical reasoning, Poincaré did not believe that a formal logic could express the essential structure of a proof in view of its understanding (see [**23**, p. 159]). He insists on the non-invariance of mathematical reasoning with respect to its content and advances, so to speak, a *local* conception of reasoning: according to this view, a gap in a proof is no longer a logical gap, but a gap in the mathematical understanding [**5**, pp. 360, 366]. If a universal calculus is, then, no longer the criterion of mathematical rigor, how shall rigor be maintained? Poincaré's response mentions especially an element of condensing. It is obtained by combining Ernst Mach's principle of the economy of thought with the concept of harmony: thanks to "fortunate inventions of language" that introduce a structure of order, the complexity of a domain of objects is made more harmonious (see [**23**, pp. 23–30]). Thus, in mathematical reasoning, the argument is placed in the context of a "mathematical architecture" whose simplest expression is found in complete induction.

In 1912, the year of Poincaré's death, Luitzen Egbertus Jan Brouwer gave his inaugural lecture at Amsterdam with the title "Intuitionism and Formalism". In it, he, an intuitionist, characterized the program of two opposed orientations in mathematics that, at the same time, were represented by the greatest mathematicians of the antagonistic Franco-German pairing: Henri Poincaré and David Hilbert. I

[30]Schematically, we say that a definition is "non-predicative" if the *definiens* (i.e., that which is used to define) contains or relates to a totality of which the *definiendum* (i.e., that which is to be defined) is a part. Let us take as an example the axiom of comprehension:

$$\exists y\ \forall x\ (x \in y \iff R(x)),$$

for an arbitrary $R(x)$. In this axiom, the set y (the set of the x that satisfy the property R) is defined impredicatively because it could be an occurrence of x. The simplest version of a predicative comprehension would be obtained in the following manner: We introduce two kinds of variables: $x, y, \ldots$ for individuals and $X, Y, \ldots$ for predicates. We then pose

$$\exists Y\ \forall x\ (x \in Y \iff R(x)),$$

where in $R(x)$ there is no quantifier that binds predicate variables.

cite Brouwer: "Two points of view can [...] be distinguished, intuitionism (in large part French) and formalism (in large part German). [...] To the question, *where must we locate exactitude in mathematics?*, the two protagonists give a very different response: according to the intuitionist, it is in the human intellect; according to the formalist, it is located on the paper" [**2**, pp. 124–125].

According to the historians, the modernists are those who at the beginning of the 20th century supported an algebraization of the contents up to and including the language of signs, while the conservative tendency finds the justification of mathematical knowledge before language, that is, in the intellect.

The polemics of the period persist today, even though its political evaluation (conservatism/modernism) has rather been reversed: since the death of Bourbaki, it is fashionable to emphasize the intuitive aspects of mathematics. In any case, the historian of mathematics Herbert Mertens places Poincaré on the side of the conservatives (see [**19**, p. 188 ff.]). Yet, it is remarkable that Brouwer himself criticizes a confusion that Poincaré would have made between "the language of mathematics [and] the true mathematical construction" [**1**, p. 176]. And Brouwer is quite right to emphasize that Poincaré does not separate language and mathematics. Poincaré ascribes to language—but not to logic—an essential role in mathematical reasoning. In this sense, we may say that Poincaré is, with respect to Brouwer, a semi-intuitionist[31] who is located exactly between the two fronts of the dispute between conservatives and modernists. For him, mathematical reasonings possess and must conserve an intuitive content that defies the formal contents.

Bibliography

[1] L. E. J. Brouwer, *Over de Grondlagen der Wiskunde* (1907), cited according to the reprinting in: A. Heyting (Ed.), *Brouwer, Collected Works 1*, North-Holland/Elsevier (Amsterdam/Oxford, New York), 1975, pp. 11–101.

[2] L. E. J. Brouwer, *Intuitionism and Formalism (Inaugural address, read 1912)*, reproduced in: A. Heyting (Ed.), *Brouwer, Collected Works 1*, North-Holland/Elsevier (Amsterdam/Oxford, New York), 1975, pp. 123–138.

[3] O. Darrigol, *Henri Poincaré's criticism of fin de siècle electrodynamics*, Stud. Hist. Phil. Mod. Phys. **26**:1 (1995), pp. 1–43.

[4] O. Darrigol, *Diversité et harmonie de la physique mathématique dans les préfaces de Henri Poincaré*, manuscript, 2006.

[31]Poincaré, Borel, Lebesgue, and Baire are equally semi-intuitionist in the sense that they accept, in contrast with Brouwer, the principle of the excluded middle. Here is a list of similarities and differences between Poincaré and Brouwer:

- Poincaré and Brouwer share the conviction that intuition is a guarantee of the certainty inherent in mathematics.
- They share the conviction that complete induction is the "mathematical reasoning *par excellence*".
- In contrast with Poincaré, Brouwer considers intuition as the *unique* basis of mathematical construction.
- Their positions differ with respect to the relation that binds intuition and language.
- The involvement of paradoxes is attributed by Brouwer to the application of logical laws to a linguistic structure that can never be transformed into proper mathematics. According to Brouwer, formalism is useless; according to Poincaré, the Platonistic philosophy that accompanies it must be corrected.
- Poincaré rejects an actual infinity; Brouwer admits it provided that it can be confined to an intuitive construction.

[5] M. Detlefsen, *Poincaré against the Logicicians*, Synthese **90** (1992), pp. 349–378.
[6] A. Einstein, *Geometrie und Erfahrung* (1921), pp. 413–420 of: *Geometrie* (K. Strubecker, ed.), Wissenschaftliche Buchgesellschaft (Darmstadt), 1972.
[7] J. Folina, *Poincaré and the Philosophy of Mathematics*, MacMillan (Houndmills/London), 1992.
[8] H. Freudenthal, *Die Grundlagen der Geometrie um die Wende des 19. Jahrhunderts*, Mathematisch-Physikalische Semesterberichte **7** (1961), pp. 2–25.
[9] M. Friedman, *Poincaré's Conventionalism and the Logical Positivists*, pp. 333–344 de [**11**].
[10] J. J. Gray and S. A. Walter, *Henri Poincaré. Trois suppléments sur la découverte des fonctions fuchsiennes*, Akademie Verlag (Berlin)/Libr. A. Blanchard (Paris), 1997.
[11] J.-L. Greffe, G. Heinzmann and K. Lorenz (eds.), *Henri Poincaré. Wissenschaft und Philosophie*, Akademie Verlag (Berlin)/Libr. A. Blanchard (Paris), 1996.
[12] G. Heinzmann, *Entre intuition et analyse. Poincaré et le concept de prédicativité*, Libr. A. Blanchard (Paris), 1985.
[13] G. Heinzmann, *La Philosophie des sciences de Henri Poincaré*, in: *L'épistémologie française, 1850–1970* (ed. M. Bitbol and J. Gayon), PUF (Paris), 2006, pp. 335–355.
[14] G. Heinzmann (ed.), *Poincaré, Russell, Zermelo et Peano. Textes de la discussion (1906–1912) sur les fondements des mathématiques: des antinomies à la prédicativité*, Libr. A. Blanchard (Paris), 1986.
[15] D. Hilbert, *Grundlagen der Geometrie*, Teubner (Leipzig/Berlin), 1899.
[16] P. Jourdain, *For Logicists*, The Monist **XXII** (1912), pp. 481–483.
[17] S. C. Kleene, *Introduction to Metamathematics* (1952), North-Holland (Amsterdam/Groningen), 1980.
[18] G. Lelièvre, *H. Poincaré: La Science et l'hypothèse*, CNED, Agrégation externe de Philosophie, Institut de Vanves, 2000.
[19] H. Mertens, *Moderne Sprache Mathematik*, Suhrkamp (Frankfort), 1990.
[20] H. Poincaré, *Dernières pensées*, Flammarion (Paris), 1913. Edition cited: 1963.
[21] H. Poincaré, *Des fondements de la géométrie*, The Monist, 1898/99, pp. 1–43. L. Rougier (ed.), Chiron (Paris), 1921.
[22] H. Poincaré, *La science et l'hypothèse*, Flammarion (Paris). Edition cited: 1968.
[23] H. Poincaré, *Science et méthode*, Flammarion (Paris), 1908, edition cited: 1999 (Kimé, Paris).
[24] H. Poincaré, *La valeur de la science*, Flammarion (Paris). Edition cited: 1970.
[25] H. Poincaré, *Sur les hypothèses fondamentales de la géométrie*, Bulletin de la Société mathématique de France **15** (1887), pp. 203–216. Cited according to the reprinting in: (Poincaré 1956 XI: 79–91).
[26] H. Poincaré, *L'espace et la géométrie*, Revue de métaphysique et de morale **3** (1895), pp. 631–646.
[27] H. Poincaré, *Des fondements de la géométrie. À propos d'un livre de M. Russell*, Revue de métaphysique et de morale **7** (1899), pp. 251–279.
[28] H. Poincaré, *L'espace et ses trois dimensions*, Revue de métaphysique et de morale **11** (1903), pp. 281–301.
[29] H. Poincaré, *Les mathématiques et la logique*, Revue de métaphysique et de morale **13** (1905), pp. 815–835 et **14** (1906), pp. 17-34. Reproduced in [**14**], pp. 11–53.
[30] H. Poincaré, *Les mathématiques et la logique*, Revue de métaphysique et de morale **14** (1906), pp. 294–317. Reproduced in [**14**], pp. 79–104.
[31] H. Poincaré, *Sur la dynamique de l'électron*, Rendiconti del Circolo matematico di Palermo **21** (1906), pp. 129–176. Cited according to the reprinting in: Œuvres (1956), vol. IX, pp. 494–550.
[32] H. Poincaré, *La logique de l'infini*, Revue de métaphysique et de morale **17** (1909), pp. 461–482. Reproduced in [**14**], pp. 235–256.
[33] H. Poincaré, *Œuvres*, vol. I–XI, Gauthier-Villars (Paris), 1956.
[34] H. Pulte, *Beyond the edge of certainty: Reflections on the rise of physical conventionalism*, Philosophia Scientiae **4**:1 (2000), pp. 47–68.
[35] M. D. Resnik, *On understanding mathematical proofs* (1996), pp. 459–466 of [**11**].

[36] B. A. W. Russell, *Review of H. Poincaré's Science and Hypothesis*, Mind **14** (1905), pp. 412–418.
[37] S. Shapiro, *Thinking about Mathematics. The Philosophy of Mathematics*, Oxford University Press, 2000.
[38] W. Stegmüller, *Probleme und Resultate der Wissenschaftstheorie und Analytischen Philosophie*, vol. II, Springer (Berlin, Heidelberg, New York), 1970.
[39] R. Torretti, *Philosophy of Geometry from Riemann to Poincaré*, Reidel (Dordrecht), 1978.
[40] E. Zahar, *Poincaré's Philosophy*, Open Court (Chicago/la Salle), 2001.

Henri Poincaré (1854–1912)

Titles in This Series

36 **Éric Charpentier, Étienne Ghys, and Annick Lesne, Editors,** The Scientific Legacy of Poincaré, 2010

35 **William J. Adams,** The life and times of the central limit theorem, Second Edition, 2009

34 **Judy Green and Jeanne LaDuke,** Pioneering women in American mathematics: The pre-1940 PhD's, 2008

33 **Eckart Menzler-Trott,** Logic's lost genius: The life of Gerhard Gentzen, 2007

32 **Karen Hunger Parshall and Jeremy J. Gray, Editors,** Episodes in the history of modern algebra (1800–1950), 2007

31 **Judith R. Goodstein, Editor,** The Volterra chronicles: The life and times of an extraordinary mathematician 1860–1940, 2007

30 **Michael Rosen, Editor,** Exposition by Emil Artin: A selection, 2006

29 **J. L. Berggren and R. S. D. Thomas, Editors,** Euclid's *Phaenomena*: A translation and study of a Hellenistic treatise in spherical astronomy, 2006

28 **Simon Altmann and Eduardo L. Ortiz, Editors,** Mathematics and social utopias in France: Olinde Rodrigues and his times, 2005

27 **Miklós Rédei, Editor,** John von Neumann: Selected letters, 2005

26 **B. N. Delone,** The St. Petersburg school of number theory, 2005

25 **J. M. Plotkin, Editor,** Hausdorff on ordered sets, 2005

24 **Hans Niels Jahnke, Editor,** A history of analysis, 2003

23 **Karen Hunger Parshall and Adrain C. Rice, Editors,** Mathematics unbound: The evolution of an international mathematical research community, 1800–1945, 2002

22 **Bruce C. Berndt and Robert A. Rankin, Editors,** Ramanujan: Essays and surveys, 2001

21 **Armand Borel,** Essays in the history of Lie groups and algebraic groups, 2001

20 Kolmogorov in perspective, 2000

19 **Hermann Grassmann,** Extension theory, 2000

18 **Joe Albree, David C. Arney, and V. Frederick Rickey,** A station favorable to the pursuits of science: Primary materials in the history of mathematics at the United States Military Academy, 2000

17 **Jacques Hadamard (Jeremy J. Gray and Abe Shenitzer, Editors),** Non-Euclidean geometry in the theory of automorphic functions, 1999

16 **P. G. L. Dirichlet (with Supplements by R. Dedekind),** Lectures on number theory, 1999

15 **Charles W. Curtis,** Pioneers of representation theory: Frobenius, Burnside, Schur, and Brauer, 1999

14 **Vladimir Maz′ya and Tatyana Shaposhnikova,** Jacques Hadamard, a universal mathematician, 1998

13 **Lars Gårding,** Mathematics and mathematicians: Mathematics in Sweden before 1950, 1998

12 **Walter Rudin,** The way I remember it, 1997

11 **June Barrow-Green,** Poincaré and the three body problem, 1997

10 **John Stillwell,** Sources of hyperbolic geometry, 1996

9 **Bruce C. Berndt and Robert A. Rankin,** Ramanujan: Letters and commentary, 1995

8 **Karen Hunger Parshall and David E. Rowe,** The emergence of the American mathematical research community, 1876–1900: J. J. Sylvester, Felix Klein, and E. H. Moore, 1994

7 **Henk J. M. Bos,** Lectures in the history of mathematics, 1993

6 **Smilka Zdravkovska and Peter L. Duren, Editors,** Golden years of Moscow mathematics, 1993

5 **George W. Mackey,** The scope and history of commutative and noncommutative harmonic analysis, 1992

TITLES IN THIS SERIES

4 **Charles W. McArthur,** Operations analysis in the U.S. Army Eighth Air Force in World War II, 1990

3 **Peter L. Duren et al., Editors,** A century of mathematics in America, part III, 1989

2 **Peter L. Duren et al., Editors,** A century of mathematics in America, part II, 1989

1 **Peter L. Duren et al., Editors,** A century of mathematics in America, part I, 1988